S0-AKY-144

COOPERATIVE PHENOMENA

Edited by
H. Haken and M. Wagner

With 86 Figures

Springer-Verlag
New York Heidelberg Berlin 1973

PHYSICS

Prof. Dr. HERMANN HAKEN · Prof. Dr. MAX WAGNER

Institut für Theoretische Physik, Universität Stuttgart

ISBN 0-387-06203-3 Springer-Verlag New York Heidelberg Berlin
ISBN 3-540-06203-3 Springer-Verlag Berlin Heidelberg New York

 Library of Congress Catalog Card Number 73-79119. Printed in Germany. Typesetting, printing and binding: Universitätsdruckerei H. Stürtz AG, Würzburg.

This Book is Dedicated to

Herbert Fröhlich, F. R. S.

on the Occasion of his Retirement
from the Chair of Theoretical Physics,
Liverpool University

Herbert Fröhlich

Life as a Collective Phenomenon

"Chance is a cause, but it is inscrutable to human intelligence."
*Democritus**

By F. Fröhlich

If one thinks without preconceptions of collective phenomena in which the discrete constitutive individuals are modified in their behaviour and indeed in their constituting a large collective group, and the whole is more than and different from a simple addition of its parts, living organisms would seem to be the ideal example. ("The universe is an animal", as Plato said.) Most of the contributions to this book are about physics where the collectivity seems a priori less obvious. The billiard-ball model of physics states that if one knows what every independent particle will do, then one can calculate the whole universe past and future; this is the perfect example of a non-collective phenomenon. As Fröhlich has frequently pointed out, it would be practically impossible to calculate all of the paths of all of the particles of a gas. The gas laws would be hidden in this mass of information, but one could not pick out the *relevant* pieces of information to yield their simple relations. At this degree of apparent complication, relevance becomes an essential cognitive property and some sort of overall insight into the whole situation becomes necessary in order to reveal the comprehensible simplicity amid this mind-stunning complexity. So even here, in what at first sight seems the most auspicious field for explaining the whole in terms of its essentially independent parts, there is a practical impossibility in understanding the whole situation in terms of the movements and activities of the parts.

In contrast, let us take the case which is farthest from the random movements and collisions of the essentially homogeneous particles of the ideal gas: the functioning of an organic unit whose parts are essentially differentiated and interact specifically when performing their different functions—the living cell. Here the movements are no longer random but precisely correlated and specific—for example, meiosis which has been poetically called "the dance of the chromosomes", or the meeting of a specific enzyme with its own substrate. In this case there is a whole level of overall organization and specificity which is not to be found in the gas laws. Such specific and precise movements within a non-equilibrium but metastable system would seem to render plausible some form of collective explanation. However, the forms of explanation actually preferred—indeed, considered to be the only "scientific" form of explanation—by today's most popular group of

* Quoted in Aristotle's Physics, Book II, Chapter 4.

biologists, the molecular biologists, are more analogous to those of the homogeneous, independent particles in a gas. This is a fact so implausible that it can perhaps best be understood in a historical context as a reverberation of the old controversy between the mechanists and the vitalists. Vitalism in the nineteenth century and in its modified form at the beginning of the twentieth century asserted that there must be laws other than those of physics to explain the extraordinarily complex phenomena of life.

Such an assumption seemed mystical and unscientific, in direct opposition to the scientific belief or assumption that the laws of physics were adequate to explain everything (especially in the nineteenth century before physics itself became so mysterious that it was forced to take cognizance of its own internal contradictions). While physics provided the model of clarity in explanation, it seemed an insult to logical thought to say that living organisms required a different form of explanation. It seemed almost animistic. The opposing view, that all living processes could be understood in terms of the laws of physics, was called mechanism. Vitalists said there must be laws for organic systems which surpass those of inorganic systems, which perhaps contradict the laws of physics. Mechanists said: "No; everything that happens to living organisms can be explained solely by the laws of physics." This almost temperamental as well as methodological opposition between vitalism and mechanism became historically entangled with the reductionist programme of logical positivism. The aspect seen as particularly antithetical to scientific (positivistic) explanation was any attribution of purpose and significance to living organisms.

However, it seems that this drive to go on looking for explanations of both organic and inorganic phenomena in terms of the fundamental laws of physics had bifurcated into two directions. The one was a purely methodological programme which exhorted those studying living organisms to go on looking for explanations in terms of the laws of physics without having recourse to any special murky or mystical concepts such as entelechy, meaning and purpose. This says nothing about what facts should actually be found. The other fork of the bifurcation concerned the *content* of the possible, or acceptable, laws of biology. It projected both the content and the "Weltanschauung" of current nineteenth-century physics onto any possible future development of biology. It predicted that the laws of biology, when finally understood, would be essentially the same as those of chemistry, making use of local interactions and of transport by diffusion as visualized in nineteenth-century physics. (JOHN LOCKE had already in the seventeenth century theoretically or philosophically planned such a project for explaining the organic phenomena of sensation—perception of sounds, colours, etc., i.e. the secondary qualities—in terms of purely mechanical properties.) This predictive picture of the necessary form of explanation of biological phenomena was covertly made respectable by being entangled with positivistic methodology. It illegitimately borrowed the authority of the scientific method from the other fork of the bifurcation. From this confusion biological research acquired apparent necessary synthetic knowledge about the ultimate forms of biological explanation. It thus asserted as self-evident that, if we find the appropriate independent biological molecules moving by diffusion and interacting by chemical laws, then we have found a complete explanation. This hypothesis, which underlies much of the thinking of molecular biology, has of course been enormously fruitful. But it can limit the type of questions asked and the forms of explanation considered and thus inhibit new directions of thinking. It blocks certain lines of enquiry which might provide facts of a different order, more relevant to treatment by modern physics or to explanations on a higher level of the hierarchy. Thus the paradigm of explanation in molecular biology, the enzyme, might be described in this form: "There are particles, enzymes and other proteins, moving independently and at random within an enclosed area (a cell). All of their movements are to be accounted for by diffusion. On collision, they interact to form chemical bonds, but there are no long-range forces which regulate and direct the overall situation". Whenever there is a problem, a process which is not understood, an enzyme is sought to account for it and very frequently one is found. But how is the finding of a particular enzyme a sufficient explanation of everything that goes on? It is a complete explanation in the sense that it answers the question: What substances are present before and after the interaction? Once a biochemist has found the substance and *identified* it, he feels he has found the sufficient as well as the necessary cause. But is this identification of substances a *sufficient* cause or does it merely create satisfaction (a "There-are-no-more-problems" attitude) within an artificially restricted range of problems? A result of this attitude is that, when one attempts to introduce into biology different forms of explanation, or even to reformulate problems in terms of collective phenomena, the approach is dismissed as superfluous, or even felt to be mysterious and vitalistic. There may be here a certain distant echo of the opposition to NEWTON'S "mystical" forces. Thus paradoxically (by a certain inertia in the change of "Weltanschauung") there has arisen from the programme to apply the scientific method to the explanation of biological phenomena by means of the laws of physics the contrary result that *modern* physics in its role of possible overall coordinating explanation of biological systems is rendered suspect as being vitalistic. This prejudice inhibits such potentially illuminating methods of enquiry of a dynamic nature as measurements of the rate of various processes; furthermore, it blurs the perhaps essential differences between multicellular, differentiated organisms in which movements over long distances must take place relatively rapidly and in a highly coordinated fashion, and very much smaller monocellular organisms where diffusion could plausibly serve as an adequate mechanism of movement. It neglects qualitative differences which might arise from different orders of magnitude in size. For instance, a cell of a higher organism may be one thousand times as large as that of a bacterium so that movement by diffusion across it would take one million times longer, even assuming that the final position is not exactly specified. This is a significant order of magnitude, perhaps requiring the introduction of different kinds of forces. Such physically and dynamically essential differences are disregarded as superfluous. The tacit assumption is that identification is explanation (cherchez l'enzyme). The biological molecule is seen as a mysterious agent whose activities explain everything once its name is known. Thus in present-day molecular biology life is treated, not as a collective phenomenon, but as an aggregate of individual particles travelling over long distances by diffusion and interacting strongly and specifically only on contact. In cases where diffusion quite obviously cannot apply another name, active transport, has introduced (e.g. the "sodium pump" in membranes). Now 'diffusion' is a familiar, but in biology unexamined, physical concept; it requires a certain analysis to reveal the different levels lumped together in it. Diffusion as applied to long-range movements is the physical equivalent of chance. "Chance" itself is a much wider concept which melts together many epistemological distinctions. Deomocritus' statement: "Chance is a cause, but it is inscructable to human intelligence", placed at the head of this paper, applies aptly to the acceptance of diffusion as a means of long-range movement. There are two points of view here: The inside one—the point of view of the moving particle (e.g. enzyme)—and the outside one—the point of view of an observer interested only in the average movement of many such molecules. These might be called the microscopic and the macroscopic points of view. Macroscopically, diffusion is regarded as motion of a particle over a certain distance without reference to its specific final position. Microscopically, however, unless the molecule, say a specific enzyme, meets its specific substrate the required reaction does not take place. Moreover, unless there is a fairly improbable three-cornered meeting between transfer RNA, the appropriate nucleic acid and the enzyme which unites them, and subsequently between this complex and the ribosome and the mRNA, the process of protein formation cannot proceed. It is a wide, complex story with a plot, like history, and it has to have well-defined characters meeting amid the mass of irrelevant molecules, not just random movements of masses of non-differentiated particles. The kings and generals must somehow meet each other at the right time among the milling, nameless armies. To achieve this by diffusion is not impossible in principle, though on investigation it might well turn out to require quite unrealistically long times. In fact, uncritical acceptance of diffusion indicates a lack of interest in the differences between individuals, whereas in biology the differences in types of molecules are vital. Thus it would seem that present-day biology comprises two logically different forms of explanation: 1. the individual molecule as a highly differentiated, active agent (the "leave it to the enzyme" type of explanation) and 2. the diffusion model of overall movement which tends to ignore the differences between individuals. Used simultaneously, they seem to yield more than they actually do—indeed, everything. This engenders the attitude that everything in biology is now in principle explicable—it is only necessary to find more enzymes, proteins, etc., etc. and mix them together in a diffusing soup and everything will be made plain. This attitude discourages the asking of questions answerable only in terms other than those currently in use.

As an example, let us consider enquiring into the speed of reactions. Biologists are now inclined to assume that size makes no essentially qualitative difference — i.e. that collective phenomena are either not present or are only rarely relevant. As mentioned above, however, given that cells of higher organisms can be one thousand times larger than those of the bacteria tacitly assumed to be an adequate model for all living processes, it would take a molecule one million times longer to move across the cell by diffusion, even without a final specific location. This suggests that a range of empirical observations enabling actual speeds of reaction in vivo to be compared with those calculated according to the rate of diffusion alone might be relevant.

If we were to make such comparisons between the actual in-vivo rates of processes and those calculated on the assumption that diffusion is the only moving force and there is no directing force at all, and if we should find a discrepancy of several orders of magnitude, this might provide evidence that some theory of long-range collective forces was required. Then again, let us consider the apparently unnecessarily large size of some biological molecules. To quote FRANCIS CRICK [1]: "It appears to be a general rule that intricate three-dimensional biological structures are always bigger than one might naively expect. The examples of globular proteins, transfer RNA and ribosomes spring to mind." In these cases the economy of biological structures might suggest that this apparently excessive size has some functional role, and we might again be encouraged to look for new forces arising from the very size of biological molecules themselves considered as collective phenomena. Finally, a comparison of the speed of the processes occurring in a metabolizing organism (in terms of physics, a non-equilibrium state being fed with energy) with those taking place among the same materials when the system is not metabolizing might suggest that there is some physical force which is operative only when the system is being fed with energy. An example of this difference is the one-way transport of auxin when the system is respiring, in contrast to its non-directional movement when the system is not respiring.

Such a hypothesis of biological explanation was originally suggested by FRÖHLICH in terms of long-range coherence at the first meeting of L'Institute de la Vie in 1967 [2, 3]. Epistemologically, it has the advantage of unifying the types of explanation from the point of view of the particles and from the point of view of the observer by formulating the microscopic and macroscopic situations in the same field-theoretical terms. Heuristically, it opens up new ranges of dynamic phenomena to be investigated and brings many seemingly unconnected observations within one potential frame of explanation.

A case in which specific long-range forces are obviously required is the pairing of homologous chromosomes in meiosis. This has been interpreted by HOLLAND [4] in terms of these novel concepts to show that coherent oscillations are excited, leading to selective long-range forces [5]. The specific attraction of mRNA to particular sites on the ribosomes might require a similar interpretation.

It would be highly interesting, to attempt to impose the necessary oscillations by external means in the hope of influencing biological developments. Thus, in auxin transport, one might attempt to effect directional transport by induced interactions rather than by respiration. Another possibility would be to try to influence the cambrium, which differentiates differently depending on its position—hence environment—within the plant. Thus, cambrium cells produced by division on the inside of the cambrium become xylem and those produced on the outside become phloem. Let us try to induce such differences in differentiation by growing cambrium in solutions in which appropriate oscillations are imposed from outside. Or let us attempt similarly to influence cell differentiation in small groups of cells of meristem grown in fluids with superimposed vibrational fields. Plant cells grown artificially in a nutrient normally do not differentiate so long as they remain separate or in small groups, but differentiation begins as soon as a mass of cells is formed. This is strongly indicative of a collective phenomenon.

Observations by HELLER [6] could point to influences of imposed fields (microwave region) on biological behaviour. Clearly, extensive experimentation is required to discover the relevant frequencies in various cases. Such experiments might lead to a new approach to the phenomenon of differentiation in terms of collective phenomena.

References

1. Crick, F.: Nature **234**, 25 (1971).
2. Fröhlich, H.: Theoretical physics and biology, p. 13, ed. Marois. Amsterdam: North-Holland 1969.
3. Fröhlich, H.: Intern. J. Quantum Chem. **2**, 641 (1968).
4. Holland, B.: J. Theoret. Biol. **35**, 395 (1972).
5. Fröhlich, H.: Phys. Lett. **39**A, 153 (1972).
6. Heller, J. H., Teixeira, A. A., Pinto: Nature **183**, 905 (1959).

Editors' Foreword

The study of cooperative phenomena is one of the dominant features of contemporary physics. Outside physics it has grown to a huge field of interdisciplinary investigation, involving all the natural sciences from physics via biology to sociology. Yet, during the first few decades following the advent of quantum theory, the pursuit of the single particle or the single atom, as the case may be, has been so fascinating that only a small number of physicists have stressed the importance of collective behaviour.

One outstanding personality among these few is Professor HERBERT FRÖHLICH. He has made an enormous contribution to the modern concept of cooperativity and has stimulated a whole generation of physicists. Therefore, it seemed to the editors very appropriate to dedicate a volume on "cooperative phenomena" to him on the occasion of his official retirement from his university duties. Nevertheless, in the course of carrying out this project, the editors have been somewhat amazed to find that they have covered the essentials of contemporary physics and its impact on other scientific disciplines. It thus becomes clear how much HERBERT FRÖHLICH has inspired research workers and has acted as a stimulating discussion partner for others. FRÖHLICH is one of those exceptional scientists who have worked in quite different fields and given them an enormous impetus. Unfortunately, the number of scientists of such distinctive personality has been decreasing in our century. In recent years FRÖHLICH's energies have been devoted in particular to problems in biology. We are convinced that future generations will be stimulated by his ideas in this field.

The articles in this volume have been written by distinguished scientists. All of them at one time or an other have been taught, stimulated or simply impressed by HERBERT FRÖHLICH. We hope that this book will be especially useful to research workers. We agree with FRÖHLICH that bold ideas are needed and that, if progress is to be achieved, it is much more important to have a qualitative understanding of cooperative effects rather than a fancy mathematical formalism.

The editors greatly appreciate the help of Mrs. U. FUNKE and G. GROSSHANS in preparing the manuscript. They acknowledge the generosity of the Springer-Verlag at all stages of the publication procedure.

Stuttgart, January 1973

H. HAKEN and M. WAGNER

Table of Contents

Chapter I. Quasi-Particles and their Interactions

After the advent of quantum theory, it was applied to problems of solid-state physics and led to the explanation of a wealth of experimental data. At that time the theory was characterized essentially by two approximations: the interaction among electrons is represented by a suitable single electron potential and the interaction between electrons and the lattice vibrations is so small that first-order perturbation theory suffices. This theory describes the scattering of electrons by phonons, a mechanism which eventually leads to the theory of electrical conductivity. In the Soviet Union an important step beyond this scheme was taken by considering the case of strong interaction between an electron and lattice vibrations leading to a polaron: an electron surrounded by a lattice deformation.

A further decisive step was taken by Fröhlich: he cast the problem into a form in which its relevance to problems of elementary particle physics immediately became obvious. In fact, the problem of the Fröhlich polaron inspired a whole generation of quantum-field scientists, including Feynman and his famous approach by means of his path integral formalism. The polaron is the prototype of a quasi-particle: though composed of several more or less strongly interacting particles, it behaves like a single particle with a self-energy, a renormalized mass and a finite lifetime.

Two of the following contributions deal with exciting new aspects of the polaron concept: the polaron glass, and the behaviour of polarons in very high magnetic fields. A further quasi-particle of fundamental importance in semiconductors is the exciton, which is composed of two polarons with opposite charge. At high concentrations, nowadays available by irradiation with laser light, various forms of "exciton matter" can be studied.

The interaction between electrons and phonons not only changes the property of a single electron, it also leads to an effective interaction between them, as was shown by Fröhlich in connection with superconductivity. That this interaction also has an impact on the electronic specific heat of normal metals is perhaps not so well known and is therefore treated in this chapter.

Some Problems about Polarons in Transition Metal Compounds

N. F. Mott

University of Cambridge, Cavendish Laboratory, Cambridge/England

Abstract

It has been suggested by several authors that in Fe_3O_4 and the vanadium bronzes (e.g. $Na_xV_2O_5$) the current carriers are small or intermediate polarons, heavy enough to "crystallize" at low temperatures. This hypothesis is examined in detail, the interaction between polarons and the effect of a high concentration on the polaron's energy and mass being discussed. In Fe_3O_4 it is suggested that there is no contradiction between this model and the "band" model of Cullen and Callen [1]. In $Na_xV_2O_5$ for most values of x which are not a simple fraction of the number of sites, the condensed phase must be disordered (a "polaron glass") and the high linear specific heat at low temperatures is related to the disorder. The blue bronze $K_{0.3}MoO_3$ has electrical properties similar to F_3O_4, though the conductivity is higher and the low magnetic susceptibilities suggest polaron pairing. The relationship of the sharp transition to the Néel point in some "amorphous antiferromagnetics", and to the Wigner transition is discussed.

1. Introduction

The concept of the polaron is particularly associated with the work of Herbert Fröhlich. It is therefore appropriate to write an article about polarons for this "Festschrift", and to suggest ways in which the concept is likely to prove useful in more physical problems than might have been suspected a few years ago. The polarons that I shall discuss are formed round electrons and positive holes in transition metal oxides. I shall discuss particularly materials in which there is a high concentration of current carriers, so that the polarons can interact strongly with each other. Examples are Fe_3O_4, vanadium bronzes (e.g. $Na_xV_2O_5$), more controversially some molybdenum bronzes. We point out that in these materials there is evidence of "polaron band motion" with strong mass enhancement, by a factor of perhaps 10 or more compared with the effective mass in the undistorted crystal. We show that the interaction between polarons, while diminishing the mean polaron energy, does not much affect the hopping energy (W_H) or the mass enhancement. At low temperatures in Fe_3O_4 and the bronzes we envisage a crystallization of polarons, and a disordering of the polarons at the Verwey temperature, to give a random distribution on the available sites. The comments of GOODENOUGH [2] and CULLEN and CALLEN [1, 3] on this model are discussed. Above the Verwey temperature the thermopower obeys the Heikes formula and the polaron gas is, we believe, non-degenerate. In materials in which one constituent is distributed not on a regular lattice (for example the Na^+ ions in $Na_xV_2O_5$), we suppose that at low temperatures the polarons condense into a non-crystalline phase; we note that there appears to be a second-order Verwey transition at a sharp temperature in this case, and discuss the implications of this.

2. The Polaron

Several reviews of polaron behaviour exist; we shall use the notation of AUSTIN and MOTT [4]. The polaron in an ionic material is a carrier (electron or hole) trapped by polarizing the surrounding crystal. In the field surrounding it there is a potential energy $V_p(r)$ given by

$$\left.\begin{aligned} V_p(r) &= -e^2/\varkappa_p r, & r \sim r_p \\ &= -e^2/\varkappa_p r_p & r < r_p \end{aligned}\right\} \tag{1}$$

where

$$1/\varkappa_p = 1/\varkappa_\infty - 1/\varkappa$$

and r_p, the polaron radius, can be obtained by minimising the total energy

$$\frac{\hbar^2 \pi^2}{2 m r_p^2} - \frac{1}{2} \frac{e^2}{\varkappa_p r_p} \tag{2}$$

in a simple approximation; the self-consistent treatment of FRÖHLICH [5] and ALLCOCK [6] somewhat modifies this result. m is the effective mass without polaron formation. This probably does not differ greatly from m_e in transition metal oxides, the narrow band width B (~ 1 eV) arising from the large distance a between ions, using the tight binding formula

$$B = 2z\hbar^2/m a^2. \tag{3}$$

r_p cannot be less than $\frac{1}{2}(\pi/6)^{\frac{1}{3}} a_0$, where a_0^{-3} is the number of metal ions per unit volume. When it reaches this value the polaron is classed as "small".

For large m it is usual to neglect the first term in (2) and to write for the polaron energy $-W_p$, where

$$W_p = \tfrac{1}{2} e^2/\varkappa_p r_p. \tag{4}$$

The polaron can move, either by thermally activated hopping, in which case the mobility is proportional to $\exp(-W_H/kT)$, where

$$W_H = \frac{1}{4} \frac{e^2}{\varkappa_p} \left(\frac{1}{r_p} - \frac{1}{R}\right),$$

R being the hopping distance, or at low temperatures by "polaron band motion". Here the polaron can move with a well-defined wave vector $\boldsymbol{k}$, and large effective mass m_p. The effective mass is given by AUSTIN and MOTT [4] by

$$m_p/m = \{\hbar/2\omega_0 a^2 m\} \exp \gamma' \tag{5}$$

(assuming movement by the adiabatic process), where, if ω is the phonon frequency,

$$\gamma' = W_H/\tfrac{1}{2}\hbar\omega. \tag{6}$$

The factor in the { } is typically about 3. A mass enhancement of about 10 is thus possible with $W_H \sim \frac{1}{2}\hbar\omega$. In this case no thermally activated hopping would be observable at any temperature. We can thus speak of "heavy polarons" which do not show hopping. It may well be that approximations following the work of Feynman [7] and Schultz [8] based on the Fröhlich coupling constant are a better approximation for values of the mass enhancement of up to ~ 20, if r_p is greater than the minimum value. These approximations break down if r_p does have its minimum value, however, and we doubt if they can be pushed to such high masses as in the work of Marshall and Stewart [9].

3. Interaction Between Polarons

The energy due to the polarization of the lattice round an isolated polaron is $-W_p$, where W_p is given by (4). The major part of this energy comes from displacing the ions adjacent to the carrier. It follows that in materials in which there is a high concentration of polarons, together with equivalent positive charges, the interaction between polarons can greatly change this energy. Either the polarons and positive charges will take up an ordered arrangement, or if their distribution is disordered the energy gained by each carrier will be decreased, because two carriers compete for the same ions which would, for an isolated carrier, form the polarisation cloud. Mott [10] put forward the conjecture that in the conduction band of a semiconductor a concentration in which more than about 10% of sites are occupied is unlikely to occur.

An important point not discussed before as far as we know is that the same reduction is *not* to be expected in the hopping energy W_H. The description of hopping given by Austin and Mott, in which two potential wells as illustrated in Fig. 1 have to form such that the energy levels in them are equal still applies, quite independently of the presence of other polarons. Thus at low temperatures the effective mass given by (5) remains large. It may become smaller for very high concentrations of polarons, however, because the appropriate value of $\varkappa_\infty$ in (1) diminishes because of the presence of electrons in the other polarons.

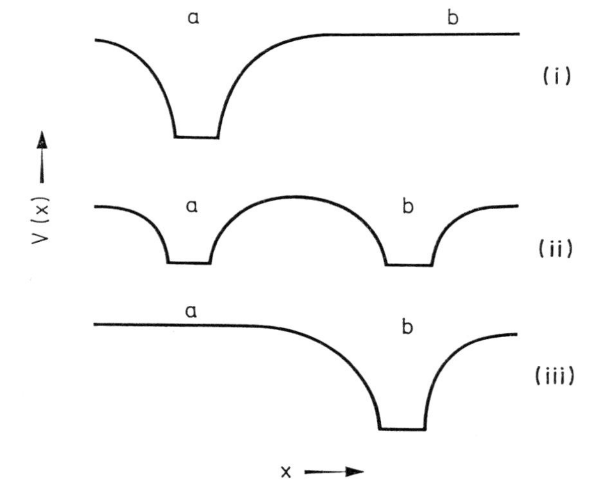

Fig. 1.
Potential energy $V(x)$ of an electron in a polaron well. (i) Before electron starts to move from well a to well b; (ii) Intermediate state; (iii) Final state

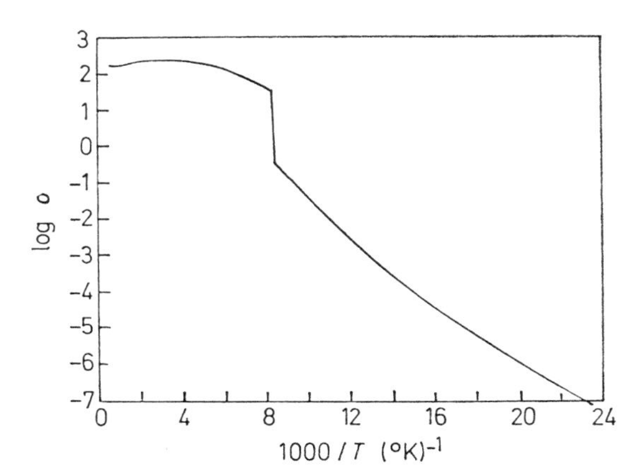

Fig. 2.
Conductivity (in Ω^{-1} cm^{-1}) of Fe_3O_4 as a function of temperature (1 000/T) (CALHOUN [11])

4. Properties of Fe_3O_4

A discussion of the observed properties of this material is given by CALHOUN [11], ADLER [12] and GOODENOUGH [13]; the electrical properties are illustrated in Fig. 2. In the low temperature phase the carriers (Fe^{2+} ions) which occupy half the octahedral sites "crystallize". It was suggested by TANNHAUSER [14], by MOTT [10, 15] and by AUSTIN and MOTT [4] that we should describe the low temperature phase as a kind of Wigner crystallization of polarons, small or intermediate, but in any case heavy. Such a crystallization could occur only if the Madelung energy of the assumed crystal, of order $\sim e^2/\varkappa a$, is considerably greater than the kinetic energy $z\hbar^2/m_p a^2$. This relationship is possible if $\varkappa \sim 20$ and m_p is large (say $\sim 10\, m$). AUSTIN and MOTT queried whether this large polaron mass was compatible with so large a number of carriers: the considerations of the last section show why it can be.

The activation energy in $\ln \sigma$ at low temperatures (0.06 eV) should be half the energy required to put an electron onto a "wrong" site, and should thus be of order $\frac{1}{2} e^2/\varkappa a$. If $\varkappa = 20$, $a = 4$ Å this would be 0.1 eV, in reasonable agreement with the observations.

The sharp change in the conductivity is called a "VERWEY" [16] transition. We think it analogous to the Bragg-Williams order-disorder transition in alloys, which may be first or second order. It is quite different from the "Mott transition", the Hubbard intra-atomic term $U = \langle e^2/r_{12} \rangle$ being always large in our case. Above the Verwey temperature we have a non-degenerate gas of heavy polarons; for such the thermopower should be given by Heikes formula

$$\begin{aligned} \alpha &= \text{const} + (k/e) \ln \{(N - n)/n\} \\ &= \text{const} + (k/e) \ln \{(1/c) - 1\}, \end{aligned} \tag{7}$$

where N is the number of sites and n the number of carriers, $\frac{1}{2}N$ in this case. The observed thermopower is independent of T in the high temperature phase and about 50 μV/degree (TANNHAUSER [14]). This shows that there are more sites for the polaron than octahedral metal sites; GOODENOUGH [13] (p. 308) suggests they may localize on Fe-Fe bonds. We note that the entropy according to this model is $Nc\{\ln(1/c)-1\}$, where N is the number of sites and c the proportion occupied. The model gives an electronic specific heat which tends to zero as kT increases above $\hbar^2/m_p a^2$.

SOKOLOFF [17] has treated a model of this type and given an explanation of the first-order nature of the transition by assuming that the repulsive energy drops with disorder due to some kind of screening. CULLEN and CALLEN [1] take a band model which they claim is different from any model based on polarons. GOODENOUGH [13] stresses that there is no Jahn-Teller distortion round the antiparallel Fe^{2+} ions and that this speaks against a localized polaron model. These authors consider that these effects are not compatible with a polaron model.

We think, however, that, the two points of view can be reconciled. In the ordered phase, the question of whether polarons are present is hardly meaningful; when electrons change places they do not carry a polarization cloud with them. But as soon as an electron moves out of its ordered position onto a "wrong" site, a polarization cloud forms round it with large effective mass. This is why the kinetic energy term $\hbar^2/m_p a^2$ is *not* sufficient to prevent the crystallization.

As regards the low temperature phase, the collective Jahn-Teller distortion found in some Fe^{2+} compounds ($KFeF_3$ or $RbFeF_3$) associated with spin-orbit coupling and collinear spins is absent. The degeneracy of the $e_g(x^2-y^2)$ orbital has to be removed somehow, and as Cullen and Callen point out it is removed in some other way. Possibly there is a superlattice of the two e_g orbitals. This raises a point discussed by MOTT and ZINAMON [18]. Consider a ferromagnetic or antiferromagnetic array of one-electron degenerate ($3d$) atoms, with large Hubbard interaction U. Common sense suggests that this should be an insulator if $U/B \gg 1$, where B is the band width. But according to one-electron band theory it must be a metal, which is what CULLEN and CALLEN say. I believe that, when a situation like this arises, a superlattice of some kind is *always* set up, either by canting of the spins or in some other way. This is because, *at zero temperature*, the descriptions with Bloch or Wannier functions are identical for any insulator, so it should always be possible to set up a band description, with separate wave functions for each spin direction and some superlattice to split the band is essential. A description of how this occurs will be published elsewhere [19].

5. Vanadium Bronze; a "Polaron Glass"?

These materials (e.g. $Na_xV_2O_5$, with x in the range 0.2–0.6) seem to us to fall into the same category as Fe_3O_4 with an important difference; in most (or all) of them the donors (e.g. Na^+) do *not* form a crystalline array.

It is known from the absence of an NMR Knight shift on the Li atom in $Li_xV_2O_5$ (GENDELL, COTTS and SIENKO [20]) that the electrons are in the d band of the V_2O_5 and not located on the Li ions; their wave functions have very little overlap with them. In single crystals of V_2O_5, doped through excess metal, a discussion of the polaron behaviour is given by IOFFE and PATRINA [24]. From their measurements of the thermopower and conductivity it is shown that the mobility has an activation energy of 0.27 eV, decreasing below 200 K. This we take to be W_H (compare the similar behaviour of TiO_2 described by AUSTIN and MOTT). The thermopower α increases below $\sim$300 K, due to recombination be-

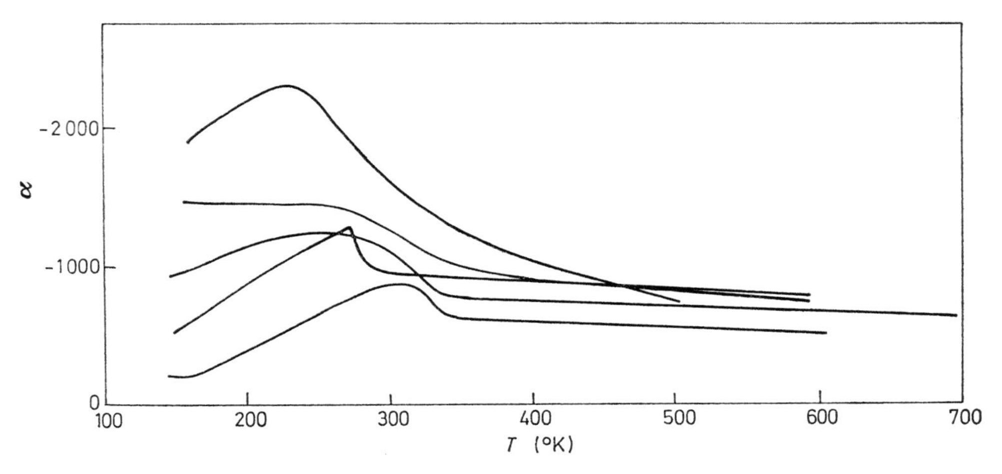

Fig. 3. Thermopower α in μV/degree of n-type V_2O_5 for five specimens (IOFFE and PATRINA [23])

tween the donors and the polarons, and then shows a tendency to change sign, possibly due to impurity conduction (Fig. 3).

Turning now to $Na_xV_2O_5$ (x from 0.2–0.6) in which there is a high concentration of donors, GOODENOUGH [22] shows that at room temperature α obeys the Heikes formula (7) over a wide range of x; this shows, as he points out, that we have a random distribution of heavy polarons on these sites.

It is interesting to estimate the effective mass for band motion. If the hopping energy is the same as for V_2O_5, namely 0.26 eV, and $\hbar\omega \sim 0.1$ eV, the effective mass might be

$$m_p/m \sim 3\, e^{0.26/0.05} = 300.$$

Now in this material, unlike Fe_3O_4, the field is not crystalline because there is no long-range order in the location of the Na^+ ions and the number of carriers need not be a simple fraction of the number of sites. Thus our heavy polarons must condense into a sort of glass. We can if we like think of them as being in Anderson localized states, the random field being due at low temperatures to the Na^+ ions *and* to the other polarons. At high temperatures the metal ions will give a random field, which should be enough to produce Anderson localization for such heavy particles. If the range in energy of these localized states is greater than kT, one does not really have a non-degenerate gas and the applicability of the Heikes formula is surprising. The same difficulty arises in discussing glasses containing V^{4+} and V^{5+} ions, where the Heikes formula is obeyed accurately, although one would expect some spread of levels (AUSTIN and MOTT [4]). It looks as if the range of energies must be not greater than kT, so that

$$e^2/\varkappa a \gtrsim (1/40)\ \text{eV}.$$

The transition in the conductivity is much less marked, the change of slope due to the beginning of ordering being in our view considerably less than the polaron hopping energy W_H, which implies that $\varkappa$ must be 100 at least for V_2O_5.

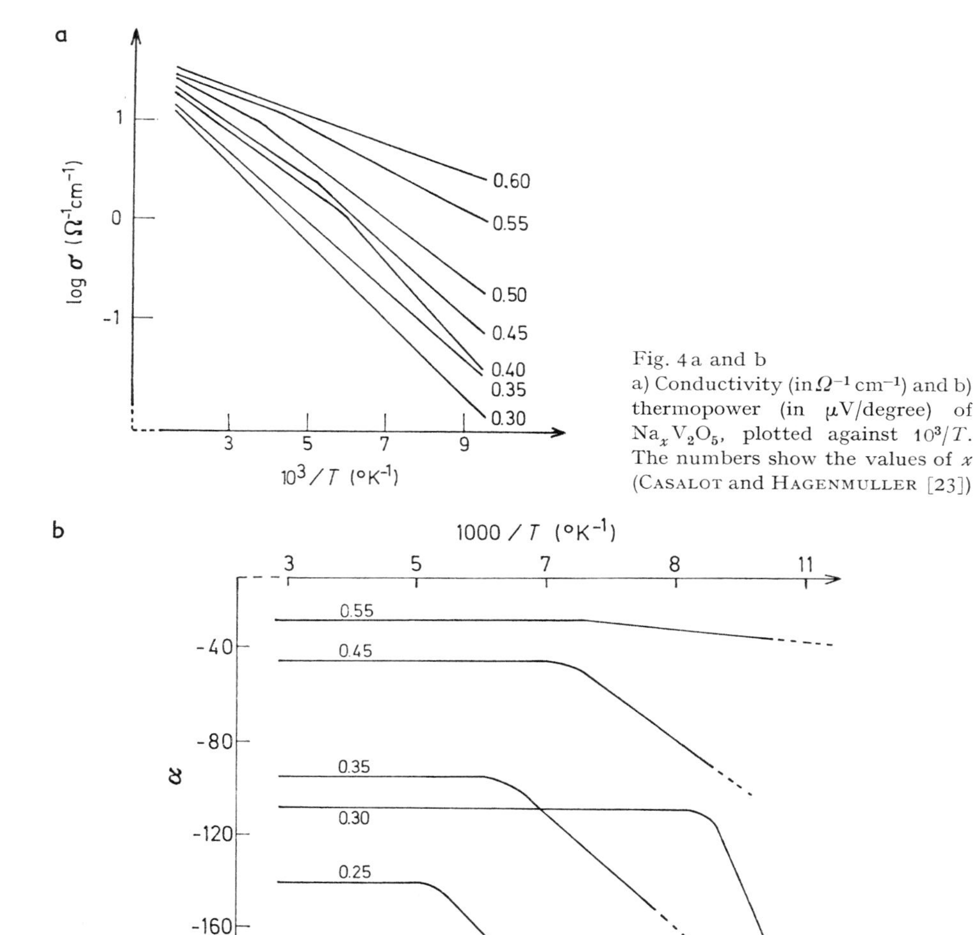

Fig. 4a and b
a) Conductivity (in Ω^{-1} cm^{-1}) and b) thermopower (in μV/degree) of $Na_xV_2O_5$, plotted against $10^3/T$. The numbers show the values of x (CASALOT and HAGENMULLER [23])

Values of this order are necessary to explain the small binding energy of a polaron to a donor in V_2O_5.

On the vanadium bronzes the work of CASALOT and HAGENMULLER [23] establishes the following points.

a) For a wide range of x (0.2–0.5) the high temperature conductivity shows thermally activated hopping; the conductivity and thermopower are shown in Fig. 4.

b) Rather remarkably, the Verwey transition indicated by the drop in the thermopower seems sharp but of second order for all values of x; the "glass" seems to melt at a sharp temperature. As for the vanadium bronzes, there appears

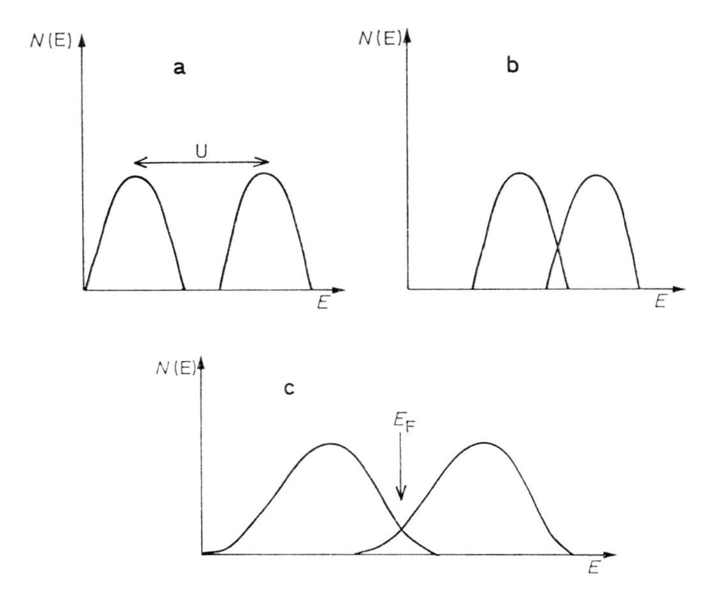

Fig. 5a–c. Suggested energy bands for polaron motion in bronzes. a) No overlap, as for blue $K_{0.3}MoO_3$ at low temperatures; b) Overlap, which may occur as the Verwey temperature is approached; c) Overlap in a "polaron glass" such as $Na_xV_2O_5$

to be a second order transition in the conductivity. We conjecture that here too in the high temperature phase there is a nondegenerate gas of polarons, but that the mass is not high enough to show hopping.

c) At low temperature there is a very large linear specific heat of an order of magnitude that would correspond to a density of states of 1/(0.1 eV per atom).

Now the Na^+ ions, as we have stated, are *not* arrayed on a crystalline lattice so that low temperature phase cannot be crystalline. We suppose that the polarons can be thought of as in localized states of Anderson type, produced by the random field of the Na^+ ions and the other polarons. Due to the disorder, it is likely that the bands for "right" and "wrong" states overlap (Fig. 5c). If so, we expect a linear specific heat of order

$$(m_p/m)\,(k^2T/E)\,g\,.$$

Here E is of the order of the band width and g is the ratio of the polaron density of states to that without condensation. This result does not really depend on the particles being "fermions"; it depends on there being only one particle in each state, due to repulsion.* The large value observed suggests $m_p/m \gg 1$.

If this model is correct the low-temperature conduction is to be thought of as hopping of polarons from one ANDERSON [25] localized state to another; a variation of $\ln \sigma$ with $1/T^{\frac{1}{4}}$ is therefore to be expected at low temperatures.

The fact that the Verwey transition is fairly sharp for a "polaron glass" is remarkable. It looks as if there is a sharp temperature at which the occupation

* Compare the discussion by ANDERSON, HALPERIN and VARMA [24] of the specific heat of glass due to phonons.

number $\langle n/N \rangle$ for a given site reaches the mean value n_0/N; here n_0 is the number of carriers, N the number of sites and n the number of carriers on a given site. The relationship to the Néel temperature of "spin glasses" is discussed in § 8.

6. Blue Molybdenum Bronze

The blue bronze $K_{0.3}MO_3$ has been investigated by BOUCHARD, PERLSTEIN and SIENKO [26], PERLOFF, VLASSO and WOLD [27] and by FOGLE and PERLSTEIN [28]. This is described as "stoichiometric" and the K ions are ordered. The conductio-ity measured by PERLOFF et al. is shown in Fig. 6. As for the vanadium bronzes, there appears to be a second order transition in the conductivity. We conjecture that here too in the high temperature phase there is a non-degenerate gas of polarons, but that the mass is not high enough to show hopping. On the contrary, the conductivity drops rapidly in the range 170–200 K as T rises. We conjecture that we are here in the range just before hopping sets in, where according to HOLSTEIN [29] the band contracts and the mobility drops rapidly. At higher temperatures it ought to increase, if we enter the hopping range, or at least flatten out. The maximum conductivity is very large in the basal plane

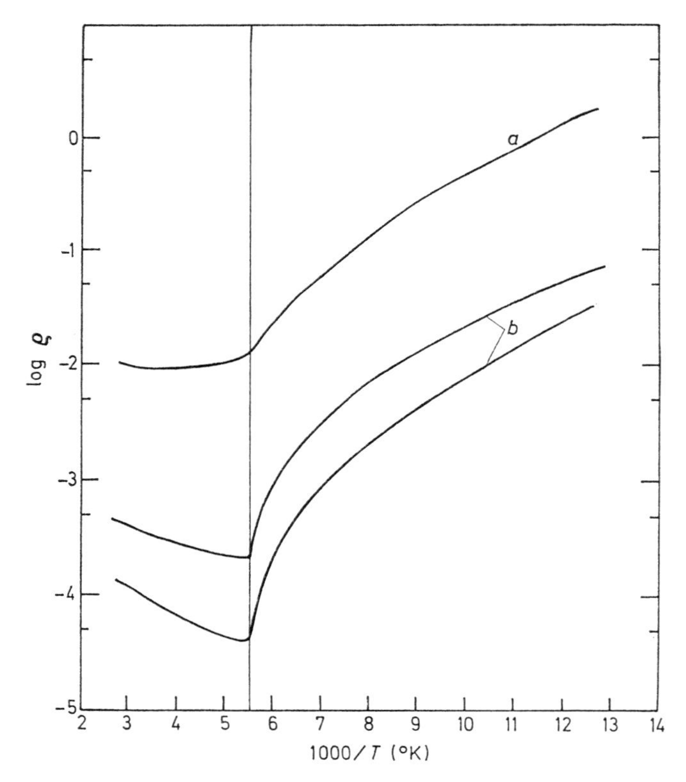

Fig. 6a and b. Log (resistivity) of blue $K_{0.3}MoO_3$ plotted against $1\,000/T$; a) perpendicular and b) two specimens parallel to the b-axis (PERLOFF et al. [27])

($\sim 10^4 \Omega^{-1}$ cm^{-1}), a surprisingly high value if this really is a non-degenerate gas and not a normal metal. On the other hand the thermopower above the transition is very small ($\sim 2\,\mu$V/degree) and seems to change sign as the temperature rises [26]. This seems too small to be explained using the normal metallic formula $\left((\pi^2 k^2 T/3e)\, d \ln \sigma/dE\right)$, unless $d \ln \sigma/dE$ is unexpectedly small for a narrow d-band; we think it means that the Heikes formula (7) must be applied with $n/N = \frac{1}{2}$, the ratio possibly changing slightly with T for free polarons.

A remarkable property of $K_{0.3}MoO_3$ is that it is diamagnetic at low temperatures and only weakly paramagnetic at high. We must conclude from this that the "polarons" which carry the current remain, even at high temperatures, are bound together in diamagnetic pairs. The behaviour is reminiscent of metal-ammonia solutions (MOTT [30]). In this behaviour it is quite different from Fe_3O_4 and the vanadium bronzes. Why they behave differently is not clear; GOODENOUGH (priv. comm.) suggests that it is due to the capture of electrons in molecular orbitals in Mo_{10} clusters.

7. A Degenerate Gas of Small Polarons

Up till now we have considered examples where the polaron gas crystallizes (in the Wigner sense) at low temperatures. In other cases these materials can remain metallic at low T (e.g. Na_xWO_3 for $x > 0.1$). The present author [10] has suggested that in materials showing metallic conduction at low T there are two extreme cases.

a) When interaction with phonons gives the usual mass-enhancement at the Fermi level.

b) When the best approximation is to treat the metal as a degenerate gas of small polarons.

The most obvious group in the second class is non-stoichiometric $SrTiO_3$, where metallic behaviour is observed for concentrations n of electrons above 2×10^{17} cm^{-3} (MOTT [10]); as pointed out by the present author, the semi-empirical relationship

$$n^{\frac{1}{3}} r_0 > 0 \cdot 25 \tag{8}$$

for metallic behaviour thus implies a very large Bohr radius r_0

$$r_0 (= \hbar^2 \varkappa / m_p e^2) \gtrsim 10^{-6} \text{ cm}. \tag{9}$$

Since by hypothesis $m_p > m$, this implies that *very* large values of $\varkappa$ are involved — which exist of course in these nearly ferroelectric materials. But this must mean that the polarons really are "small", in the sense that $r_p < a$; otherwise we could not suppose that the binding energy is given by

$$m_p e^4 / 2\hbar^2 \varkappa^2.$$

If $r_p > a$, the polaron radius would overlap the donor, the force between them would be the much larger value $e^2/\varkappa r^2$, and $\varkappa_\infty$ rather than $\varkappa$ should be inserted in the condition (8), (9) for metallic behaviour.

This in our view is why Na_xWO_3 is metallic for $x > 0.2$, a value which suggests $\varkappa_\infty$ rather than $\varkappa$ in (9). Because of the wide $5d$ band, $r_p > a$; and we do not have here a gas of small polarons.

For equivalent structures tungsten should give a slightly *narrower* d band than molybdenum because of reduced p/d mixing. The assumed narrow d band in the blue *Mo* bronze must be associated with the different crystal structures (DICKENS and NIELD [31]).

8. Comparison with Spin Glasses

We have introduced in § 5 the idea of a "polaron glass"—that is to say a condensation of heavy polarons with a number of sites available greater than and not a simple multiple of the number of polarons. The observations in vanadium and molybdenum bronzes show that the condensed gas has a sharp disordering temperature, like the Verwey temperature in Fe_3O_4, though apparently second order; and in the molybdenum bronzes a linear specific heat is observed. In this section we compare these results with those observed in a rather similar non-crystalline system, the "spin glass" or amorphous antiferromagnet. By this we mean a random array of centres carrying moments, with interaction between them, which will include

a) Solutions such as CuMn, where the mean Ruderman-Kittel-Kasuya-Tosida interaction between the Mn atoms is larger than kT_K, T_K being the Kondo temperature.

b) Heavily doped semiconductors on the non-metal side of the MNM transition.

c) Glasses containing transition metal ions.

We look first at the linear specific heat. A degenerate electron gas gives a linear specific heat whether the states are localized (in the Anderson sense) or not. But we do not think the linear specific heat of vanadium bronze depends on the polarons being fermions—because their mutual repulsion would prevent more than one of them from occupying a given site. It depends on there being a range of energies of these states (determined by the position of the other polarons) which are filled up to a limiting energy, which is analogous to the Fermi energy. As far as we can see, the argument is identical with that given by ANDERSON, HALPERIN and VARMA [24] for the linear term in the vibrational specific heat of glasses, and in our view is a different mechanism to that proposed by MARSHALL [32] for the linear specific heat in spin glasses. We express this as follows. For each spin we suppose that there is a local anisotropy so that only two spin directions need be considered, and a molecular field H due to the Ruderman-Kittel-Kasuya-Yosida field from the others. The internal energy is then

$$\int \frac{N(H)\,\mu H\,dH}{\exp(\mu H/kT)+1},$$

which varies as T^2 if $N(H)$ is finite at the origin. Probably this ought to be generalised, taking Anderson-localized magnons intead of single moments; the Marshall approach corresponds to the "Einstein" specific heat model (see BUYERS et al. [33], LYO [34], ECONOMOU [35]).

Turning now to the bahaviour at the Verwey temperature, it is a striking fact that some spin glasses do show a sharp Néel temperature. The behaviour of Au-Fe observed by CANNELLA et al. [36] and shown in Fig. 7 is typical. Obviously this cannot represent the disappearance of long-range order, since there is

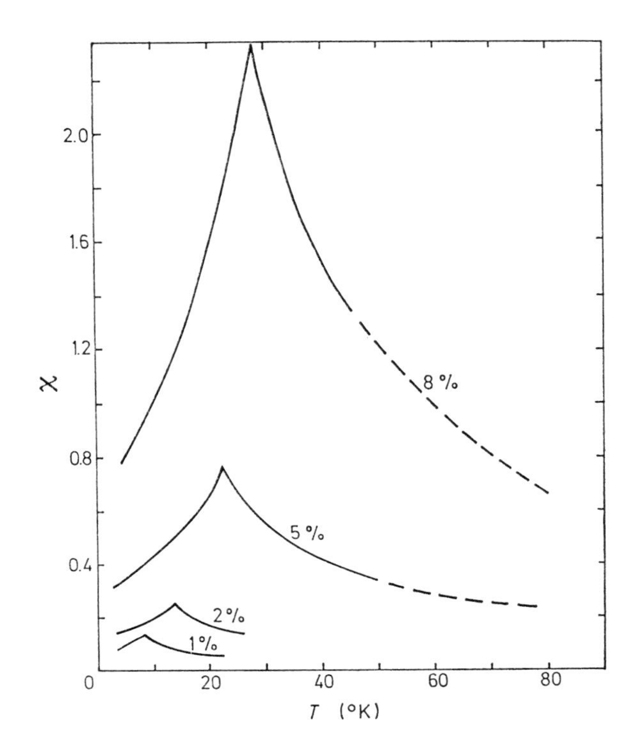

Fig. 7.
Susceptibility ($\chi \cdot 10^{-3}$) of $AuFe$ plotted against T for various concentrations of Fe (CANNELLA et al. [36])

none; it could however represent a temperature at which the time average $\langle\sigma_z\rangle$ of the moment σ_z vanishes for all sites.

Our picture of an amorphous antiferromagnet is thus as follows. There is no long range order; but below the Néel temperature $\langle\sigma_z\rangle$ is not zero—if the time average is over a time short compared with the geological times, in which all spins rotate collectively. But since the molecular fields acting on any one moment are due to all the others, it seems quite plausible that $\langle\sigma_z\rangle$ should vanish at the same temperature for all sites; but it certainly needs proof.

9. Relation to the Wigner Crystallization

WIGNER [37] was the first to point out that an electron gas at low densities could crystallize into a non-conducting state. It was pointed out by the present author [15] that the low temperature phase in Fe_3O_4 is in fact a kind of Wigner crystallization, except that the carriers (Fe^{2+}) form a "crystal" coherent with the lattice. One can of course except a Wigner crystallization *incoherent* with a lattice (as Wigner imagined) and without polaron formation; the present author [15] suggested that this might occur in a highly doped and very strongly compensated semiconductor, where the Bohr radius is big compared with the distance between centres, but where the number of donors can be made as small as one pleases. Recently SOMERFORD [38], using highly compensated n-type InSb, has used strong magnetic fields to induce a Wigner transition, following ideas due to

Durkan, Elliott and March [39], and Care and March [40] give evidence to show that conduction is due to thermally generated vacancies in this lattice. The similarity to the phenomena discussed earlier in this paper is clear.

References

1. Cullen, J. R., Callen, E.: J. Appl. Phys. **41**, 879 (1970).
2. Goodenough, J. B.: J. Solid State Chem. **1**, 349 (1970).
3. Cullen, J. R., Callen, E.: Phys. Rev. Letters **26**, 236 (1971).
4. Austin, I. G., Mott, N. F.: Advan. Phys. **18**, 41 (1969).
5. Fröhlich, H.: Advan. Phys. **3**, 325 (1954).
6. Allcock, G. R.: Advan. Phys. **5**, 412 (1956).
7. Feynman, R. P.: Phys. Rev. **97**, 660 (1955).
8. Schultz, T. D.: Phys. Rev. **116**, 526 (1959).
9. Marshall, J. T., Stewart, B. U.: Phys. Rev. B**2**, 4001 (1970).
10. Mott, N. F.: Phil. Mag. **20**, 1 (1969).
11. Calhoun, B. A.: Phys. Rev. **94**, 1577 (1954).
12. Adler, D.: Solid State Phys. **21**, 1 (1968).
13. Goodenough, J. B.: Metallic Oxides. Progress in Solid State Chemistry **5**, 145 (1971).
14. Tannhauser, D. S.: J. Phys. Chem. Solids **23**, 25 (1962).
15. Mott, N. F.: Advan. in Phys. **16**, 49 (1967).
16. Verwey, E. J. W., Haayman, P. W.: Physica **8**, 979 (1941).
17. Sokoloff, J. B.: Phys. Rev. B**3**, 3162 (1971).
18. Mott, N. F., Zinamon, Z.: Rept. Progr. Phys. **33**, 881 (1970).
19. Mott, N. F.: Metal-Insulator Transitions.: Taylor & Francis 1973. (To be published.)
20. Gendell, J., Cotts, R. M., Sienko, M. J.: J. Chem. Phys. **37**, 220 (1962).
21. Ioffe, V. A., Patrina, I. B.: Phys. Stat. Sol. **40**, 389 (1970).
22. Goodenough, J. B.: Mat. Res. Bull. **6**, 967 (1971).
23. Casalot, A., Hagenmuller, P.: J. Phys. Chem. Solids **30**, 1341 (1969).
24. Anderson, P. W., Halperin, B. I., Varma, C. M.: Phil. Mag. **25**, 1 (1972).
25. Anderson, P. W.: Phys. Rev. **109**, 1492 (1958).
26. Bouchard, G. H., Perlstein, J., Sienko, M. J.: Inorg. Chem. **6**, 1682 (1967).
27. Perloff, D. S., Vlasse, M., Wold, A.: J. Phys. Chem. Solids **30**, 1071 (1969).
28. Fogle, W., Perlstein, J. H.: Phys. Rev. B**6**, 1402 (1972).
29. Holstein, T.: Ann. Phys. (N. Y.) **8**, 343 (1959).
30. Mott, N. F.: Phil. Mag. **24**, 935 (1971).
31. Dickens, P. G., Nield, D. J.: Trans. Faraday. Soc. **64**, 13 (1968).
32. Marshall, W.: Phys. Rev. **118**, 1519 (1960).
33. Buyers, W. J. L., Holden, T. M., Svensson, E. C., Cowley, R. A., Stevenson, R. W. H. Phys. Rev. Letters **27**, 1442 (1971).
34. Lyo, S. K.: Phys. Rev. Letters **28**, 1192 (1972).
35. Economou, E. N.: Phys. Rev. Letters **28**, 1206 (1972).
36. Cannella, V., Mydosh, J. A., Budnick, J. I.: J. Appl. Phys. **42**, 1689 (1971).
37. Wigner, E.: Trans. Faraday. Soc. **34**, 678 (1938).
38. Somerford, D. J.: J. Physics C **4**, 1570 (1971).
39. Durkan, J., Elliott, R. J., March, N. H.: Rev. Mod. Phys. **40**, 812 (1968).
40. Care, C. M., March, N. H.: J. Physics C **4**, L372 (1971).

Light Absorption by Electrons in Polar Crystals in a Strong Magnetic Field

P. G. Harper

Heriot-Watt University, Edinburgh/Scotland

Abstract

An optical phonon structure is proposed for cyclotron absorption in polar crystals at high magnetic fields. This is due to an induced modulation of the optical dipole by the electron phonon interaction. The theory is based on a model which treats the field-transverse lattice motion as adiabatic, resulting in a screened one-dimensional polaron interaction.

The investigation of characteristic polaron properties, both optical and conductive, of electrons (or holes) in polar crystals is greatly assisted by the application of a magnetic field. Its main benefit is to provide a tunable (cyclotron) frequency ω_c which is sensitive to the apparent electron mass. For materials like AgBr or the alkali halides with a high mass (usually >3) the long wave optic mode frequency, ω_0 is at least ten times the frequency ω_c attainable with current magnetic fields, <100 kilogauss, say. For the III–V semiconductors like InSb, InAs, GaAs however, despite their weak polar nature, their low masses have enabled a variety of interesting resonant experiments to be made in which the tuning of ω_c to ω_0 is the dominant feature [1].

Looking forward to the advent of stronger magnetic fields up to 200 kilogauss say, the inequality $\omega_c \gg \omega_0$ becomes practicable, at least for InSb, and it seems worth while to consider the consequent magneto-polaron optics of this ultra-resonant regime. There are two dynamical aspects; on the one hand the electrons are confined to the lowest Landau level so that their free motion is essentially one-dimensional (see Fröhlich's model for a high-field superconductor [2]); on the other hand, the lattice cannot follow the rapid orbital rotation so that it makes a static adjustment to the orbit-averaged charge density. In fact these two aspects are not independent and must be treated simultaneously. Of course for the very high fields produced by adiabatic flux compression, say, non-magnetic forces of any kind become unimportant. This extreme situation is not contemplated.

In Fröhlich's exposition [3] of polaron dynamics, the electron at $\boldsymbol{r}'$ is regarded as a moving point charge source at $\boldsymbol{r}'$. Anticipating now that its subsequent quantum states have a strong axial localisation it is useful to represent the density at $\boldsymbol{r}$ as

$$e\delta(\boldsymbol{r}-\boldsymbol{r}') = e\sum_i \psi_i^*(\boldsymbol{r}')\,\psi_i(\boldsymbol{r})$$

where ψ_i are in principle exact polaron eigenstates. Translational symmetry in the field direction z, say, allows exact total z-momentum denoted by $\hbar k$. An approximate charge density is now proposed of the form

$$\varrho(r, z-z') = e\sum_k \psi_i^*(r, z')\,\psi_k(r, z) \tag{1}$$

where r, z are cylindrical coordinates. The function $\psi_k(r, z)$ is supposed to be a ground state strongly localised near $r=0$. Thus neglecting polaron coupling and radiative perturbation,

$$\psi_k(r, z) = \Phi_0(r)\exp(ikz)$$

and

$$\varrho(r, z-z') = e\,|\Phi_0(r)|^2\,\delta(z-z'). \tag{2}$$

Strictly speaking, the ground state Φ_0 should be self-consistently determined, (see [4]), but it is probably sufficient to simplify even further and replace $|\Phi_0(r)|^2$ by $\delta(r)$. The effect of optic mode coupling is to replace $\delta(z-z')$ by a finite, polaron localisation. This will be discussed later.

Besides the use of the change density (2) in place of $e\delta(\boldsymbol{r}-\boldsymbol{r}')$ it is also supposed that the magnetic Lorentz force is much greater than the reactive lateral polarisation force acting on the electron. Accordingly, the polarisation field components P_x and P_y are neglected and the electrostatic interaction taken as $I=-\int P_z D_z d^3\boldsymbol{r}$ where P_z and D_z are the axial components of lattice polarisation, and electric induction respectively. The latter is due to the source (2) and is conveniently written as

$$D_z(r, z-z') = \Sigma_k \exp[i\,k(z-z')]\,\Delta(k, r).$$

The axial longitudinal polarisation force is similarily expanded in terms of "phonon" amplitudes $b(k)$, $b^\dagger(k)$ as

$$-4\pi P_z = -\Sigma_k\, b(k)\exp(i\,k\,z)\,\lambda^{-1}(k)\,\Delta(k, r) + \text{h.c.}$$

where the effective interaction parameter $\lambda(k)$ has to be found by combining the use of I and the P_z equation of motion, namely

$$\frac{1}{\omega_L^2}\ddot{P}_z + P_z = -\frac{1}{4\pi}\left(\frac{1}{\varepsilon(\infty)} - \frac{1}{\varepsilon(0)}\right)D_z \equiv -\frac{1}{4\pi\varepsilon}D_z. \tag{3}$$

It is then straight forward to obtain the interaction I as

$$I = \Sigma_k\,\lambda(k)\,\{b^\dagger(k)\exp-(ikz') + b(k)\exp(ikz')\} \tag{4}$$

where

$$\lambda^2(k) = \frac{\hbar\omega_0 L}{4\pi\varepsilon}\int |\Delta(k, r)|^2\,d^2r.$$

The effect of the static lattice polarisation in the transverse plane is thus to screen the electron-lattice interaction, the familiar k^{-1} dependence being replaced by $\lambda(k)$ essentially. Taking the $e\,\delta(r)$ approximation to $e\,|\Phi_0(r)|^2$, D_z is given by

$$D(r, z) = -\frac{\partial}{\partial z}\, e\,(r^2+z^2)^{-\frac{1}{2}}$$

so that

$$\Delta(k, r) = 2\pi i e\, k L^{-1} \int_{-\infty}^{+\infty} \exp(-ikz)\,[r^2+z^2]^{-\frac{1}{2}}\, dz.$$

In this special instance λ is seen to be independent of k, but more generally its k-dependence will be simply weaker than k^{-1}.

The interaction I in (4) now describes a screened coupling of the electronic k-states above the lowest magnetic orbital with k-configuration optic mode longitudinal phonons of the statically polarised lattice. The reactive force of these lateral phonon excitations has been disregarded as small compared to the Lorentz force. Consistently with the inequality $\omega_c \gg \omega_0$, coupling to even neighbouring excited magnetic orbits is disregarded.

Polaron corrections to the orbital ground state could now be calculated but it should be recollected that these corrections perturb the charge density in accordance with (1) and therefore themselves contribute to the coupling parameter $\lambda(k)$. It is convenient to consider this induced coupling in the more general problem of optical excitation. Suppose an external radiation field

$$E \exp(-i\omega t) + c \cdot c$$

is applied whose interaction H_1 has dipole matrix element μ between the orbital ground state Φ_0 and next excited state, Φ_1, say. Then for optical (k-conserving) transitions, in the absence of I, the states $\Phi_1(r)\exp(ikz)$ become excited. In consequence, the charge density ϱ acquires oscillatory components proportional to

$$e\,\Phi_0^*\,\Phi_1\,\mu E \exp(-i\omega t)\,\delta(z-z').$$

This oscillating charge density leads to an oscillatory dipole-like D_z which from (3) acts as a forcing term for the P_z modes. Such a forcing term leads to phonon-assisted optical transitions described by the phonon modulated dipole interaction,

$$H_2 = \exp(-i\omega t) \sum_k \gamma(k) \exp(ikz)\, b^\dagger(k) + \text{h.c.} \tag{5}$$

The (0, 1) coupling matrix element $\gamma(k)$ (proportional to E) will be found by using the H_2-perturbed states in (1) to determine self-consistently the oscillatory H_z charge source.

There are thus two optical interactions, H_1, connecting 0 and 1 of same k, with no phonon emission; H_2, connecting 0 and 1 of different k, and with phonon emission. These can act together with the polaron interaction I which connects (0, 0) or (1, 1) of different k, and with phonon emission.

Since there are no degeneracies, it is sufficient to treat the combined optical/polaron interaction by perturbation theory. Thus H_1 acting alone produces the cyclotron absorption at $\omega_c = \omega_c$. Allowing for the perturbation of initial and final states by I, ω_c will be corrected by polaron self-energies. For the weakly polar crystals such as InSb, this correction is small and probably undetectable. A further effect of I would be to produce phonon side band structure, i.e. the combination lines $\omega_c + \omega_0$, $\omega_c + 2\omega_0$ etc. Because the one-phonon line strength is proportional to the Fröhlich coupling constant α, the two-phonon line to α^2 and so on, it is unlikely that they could be observed.

Consider now the phonon modulated interaction H_2. Acting on the initial state $\Phi_0(r) \exp(ikz)$ and followed by I, an oscillatory dipolar contribution to the charge density is produced, which including that due to H_1 may be written, using (1), as

$$\begin{aligned}\Phi_0^* \Phi_1 \exp(-i\omega t) \Big\{ & \frac{\mu E}{\omega_c - \omega} \delta(z - z') \\ & + \sum_{k_1 k'} \frac{\gamma(k')}{\Omega_1(k') + \omega_0 - \Omega_0(k) - \omega} \frac{\lambda(k-k')}{\Omega_1(k) - \Omega_1(k') - \omega_0} \Big\} \exp ik(z - z') \\ & + \text{anti-resonant term in } \exp(i\omega t)\end{aligned} \tag{6}$$

where $\Omega_0(k)$, $\Omega_1(k)$ stand for $\frac{1}{2}\omega_c + \hbar k^2/2m$, $\frac{3}{2}\omega_c + \hbar k^2/2m$ respectively. In contrast to the direct excitation, the virtual electronphonon transitions provide a polaron contribution which is non-local. Actually if the self-energy components were included there would be no truly local charge density.

From (6) may be obtained the oscillatory perturbation of the induction field D_z. Thus the k-Fourier componant of the D_z perturbation (for $\exp(-i\omega t)$) is given as

$$\exp(-ikz') F(k) \Delta_1(r, k) \exp(-i\omega t)$$

where $\exp(-ikz') \Delta_1(r, k)$ analogous to $\exp(-ikz') \Delta_0(r, k)$ is the k-Fourier component of the electric field whose change source is $\Phi_0^*(r) \Phi_1(r) \delta(z - z')$. The abbreviation $F(k)$ stands for the k'-sum in (6).

Finally the optical interaction energy H_2 is given by

$$\begin{aligned} H_2 &= \int (\text{free } P_z)\ (D_z \text{ perturbation})\ d^3 r \\ &= \sum_k A(k) F(k) b(k) \exp(-ikz') + \text{h.c.} \end{aligned} \tag{7}$$

where now $A(k)$ stands for $\lambda^{-1}(k) \int \Delta_0(r, k) \Delta_1(r, k)\, d^2 r$.

Comparing (5) and (7), it follows that the phonon-assisted optical transition matrix element $\gamma(k)$ is determined from the self-consistent condition,

$$\gamma(k) = A(k) \left\{ \frac{\mu E}{\omega_c - \omega} + \sum_{k'} \frac{\lambda(k-k')}{\Omega_1(k) - \Omega_1(k') - \omega_0} \frac{\gamma(k')}{\Omega_1(k') + \omega_0 - \Omega_0(k) - \omega} \right\}. \tag{8}$$

Clearly, $\gamma(k)$ is proportional to E as anticipated, and depends sensitively upon the optical frequency ω.

Even if the functional behaviour of $\lambda(k)$, $A(k)$ were known exactly, the solution of (8) for $\gamma(k)$ could only proceed by approximation. Some general comments however may be made immediately. It is well known that for an inhomogeneous equation like (8), the amplitudes $\gamma(k)$ must become divergent near characteristic solutions $\Gamma_n(k)$ for the corresponding homogeneous equation, namely at those frequencies $\omega = \Omega_n$ which satisfy

$$\Gamma_n(k) = A(k) \sum_{k'} \frac{\lambda(k-k')}{\Omega_1(k) - \Omega_1(k') - \omega_0} \frac{\Gamma_n(k')}{\Omega_1(k') + \omega_0 - \Omega_0(k) - \Omega_n}. \tag{9}$$

That is to say, strong optical absorptions are expected at the frequencies Ω_n. By neglecting the k'-dependence of the energy denominators an approximate eigenequation for Γ_n, Ω_n may be written,

$$(\omega_c + \omega_0 - \Omega_n)\Gamma_n(k) = \frac{A(k)}{\omega_0} \sum_{k'} \lambda(k - k')\,\Gamma_n(k').$$

Thus, the absorbing frequencies are evidently in the vicinity of $\omega_c + \omega_0$. It is doubtful however whether a perturbation treatment would be sufficient to evaluate them since despite the weak nature of $\lambda(k)$, the solutions of (9) are possibly bound states.

References

1. Harper, P. G., Hodby, J. W., Stradling, R. A.: Reports on Progress in Physics **36** (1973).
2. Fröhlich, H.: J. Phys. C **1**, 544 (1968).
3. Fröhlich, H.: Scottisch Universities' Summer School (Ed. C. G. Kuper and G. D. Whitfield), 1962.
4. NATO Advanced Study Institute on Fröhlich Polarons, Antwerp, 1971 (Lecture No. 3, Harper).

Properties of Crystals at High Concentration of Excitons

S. Nikitine

University of Strasbourg, Laboratoire de Spectroscopie et d'Optique du Corps Solide, Groupe de Recherche du CNRS. Strasbourg/France

Introduction

Excitonic Luminescence at Rather Low Excitation Intensities. Before considering the effects at high concentration of excitons, it is useful to give a short review on exciton luminescence at low concentration of excitons.

Excitonic luminescence was first observed by Grillot et al. and simultaneously by the Strasbourg group and by Arkhangelskaia et al. [1].

This kind of emission of light by excitons is of great importance in the subject of this review article.

The spectrum of luminescence is usually excited by irradiating in the continuum. A large number of free, more or less hot electrons and holes are then created. It has been shown by Toyozawa that these charge carriers are thermalized and then combine to excitons. The lifetime of this process can be evaluated to about 10^{-10} to 10^{-11} sec [2].

Excitons can recombine:

1. in a radiationless process,

2. by emission of a resonance line we shall call ν_0, which is however strongly reabsorbed. Another mechanism is the emission of a photon and one or several phonons, generally LO, the emission is then

$$h\nu\,(E_{\text{ex}} - \text{LO}) = E_g - E_{\text{ex}} - E_{\text{LO}} = h(\nu_0 - \nu_{\text{LO}}),$$

where E_{ex} is the binding energy of the exciton in the lowest state, and E_{LO} the energy of the LO phonon [3]. E_g is the energy gap.

3. The exciton may, however, also go down to a metastable state (frequency ν_1, say). Such states have been found in particular in CuCl and are recognized to be paraexcitons ${}^1S_0$ or $E_{\text{ex}}(\uparrow\downarrow)$. These states are below the exciton level (50 cm^{-1}, say) and are stable only at very low temperatures. In CuBr the lowest state has ${}^5S_2$ character and corresponds to a momentum $j=2$ [4].

4. Excitons may be captured by impurities or defects and form Lambert complexes [5]. Three- and four-particle complexes can be formed in this way. (D^+ ex) and (D^0 ex) exciton complexes may recombine and radiate frequencies $\nu(D^+\,\text{ex})$ and $\nu(D^0\,\text{ex})$. Here D^+ and D^0 mean ionized and neutral donor, respec-

tively. Similar considerations can be made for acceptors. These complexes can also be formed by absorption of the same frequencies. According to RASHBA [6], the transition probabilities to these complexes are stronger by a factor 10^3 or 10^4 than transitions to free exciton states (giant oscillator strengths). This estimate may be exaggerated. The luminescence from and absorption to the bound exciton state was first suggested in 1958 [7] by the author. HOPFIELD and THOMAS have carried out fundamental studies on this kind of luminescence [8].

These complexes can also recombine with emission of n LO, where n is usually 1, but could also have a higher value $n=2, 3$, etc. [9].

Some other centers have also been observed; they are less frequent and will not be considered here.

The binding energy of exciton complexes has been calculated taking into account the polarizability of the crystal by EL KOMOSS and by STEBE in the Strasbourg Group [10].

The exciton luminescence can be summarized in a non-exclusive form as follows:

$h\nu_0$	$=E_g - E_{ex}$	resonant emission from the exciton level.
$h\nu_1$	$=E_g - E_{ex} - \Delta E_{jj}$	emission from the forbidden states of exciton resulting from a jj coupling.
$h\nu(D^+ \text{ex})$	$=E_g - E_{ex} - E_{(ex\,3)}$	emission from a 3-particle bound exciton state.
$h\nu(D^0 \text{ex})$	$=E_g - E_{ex} - E_{(ex\,4)}$	emission from a 4-particle bound exciton state.
$h\nu(\text{ex} - n\text{ LO})$	$=E_g - E^i_{ex} - nE_{(LO)}$	emission of a photon and n phonons from exciton states.
$h\nu(D^j_{ex} - n\text{ LO})$	$=E_g - E_{ex} - E_{(ex\,j)} - nE_{(LO)}$	here D^j_{ex} means one of the bound states of the exciton, $E_{(ex\,j)}$ the binding energy of one of these states, $E_{(LO)}$ the energy of a longitudinal phonon.

Fig. 1 gives one of the possible level diagrams with the possible transitions. This figure, too, is not exclusive; some other levels and processes may be involved in some materials.

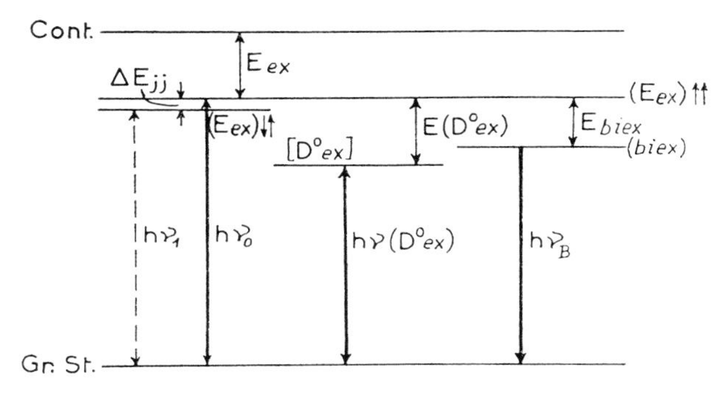

Fig. 1. An exciton luminescence diagram. Other diagrams with different exciton complexes are possible

It can be seen that at low temperatures the luminescence from the lowest states will predominate. The relative values of $E_{(ex\ j)}$ and $E_{(LO)}$ are arbitrary on Fig. 1.

High Concentrations of Excitons

When the concentration of excitons becomes very high, a number of new effects, specific to this situation, can be observed. This situation can be induced by optical excitation with laser light or by electron beams or in *pn* junctions. These effects are of considerable interest and will be described now. The review will concern however only optical excitation.

A. Biexcitons

Biexcitons are molecules formed with two excitons in a similar way to H_2. They were predicted by LAMPERT [5] a long time ago and evidence for the existence of biexcitons has been found experimentally by the Strasbourg Group.

a) Experimental Situation

The experimental situation is as follows. The luminescence of CuCl and CuBr is well known when the excitation is of a conventional type [11]. However the luminescence excited by very intense UV laser radiation has been shown to have various rather remarkable properties [12]. At 77 °K, the spectrum comprises mainly the resonance line ν_0 and a new line ν_B becoming very strong at high intensities of excitation. This line is not observed at low intensities of excitation light i. The difference of $\nu_0 - \nu_B = 350\ \text{cm}^{-1}$ for CuCl and $236\ \text{cm}^{-1}$ for CuBr. The intensity of luminescence light I_B of line ν_B can be represented by the relation

$$\log I_B = m_B \log i + ct. \tag{1}$$

In a wide range of intensities m_B is somewhat lower than 2 but tends progressively to $m_B = 1$ for very high intensities. At 4,2 °K, similar experimental results are obtained. However, at this temperature other lines overlap with the new lines mentioned above, and the variation of their intensity may influence I_B and I_0, the intensity of ν_0. It is assumed that this influence between ν_B and ν_0 is also present at 77 °K and that it is responsible for some perturbations of the variation of both I_B and I_0 observed recently [10] (Figs. 2 and 3).

I_0 is proportional to i for all intensities.

b) Binding Energy of Biexcitons

At very high intensities of excitation in the continuum, a very large number of free carriers are created. As shown by TOYOZAWA, after thermalization, most of them combine at low temperatures to excitons. As the concentration of these excitons is very high, they may associate to form biexcitons.

The recombination process of the biexciton (in CuCl and CuBr) may perhaps be as follows: The biexciton radiates a photon $h\nu_B$ and a free exciton is left.

$$(\text{Biex}) \to (\text{ex}) + h\nu_B. \tag{2}$$

The energy conservation now gives:

$$h\nu_B = E_g - E_{ex} - E_B + E_B^k - E_{ex}^k. \tag{3}$$

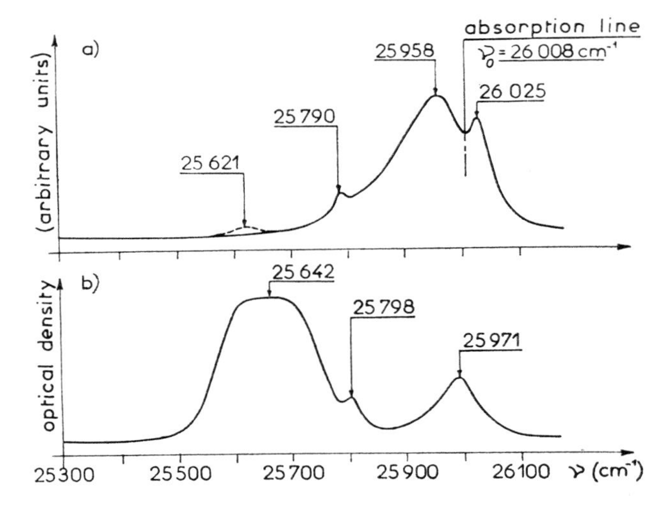

Fig. 2a and b. Luminescence spectrum of CuCl at low temperatures. a) Low-intensity excitation. b) High-intensity excitation; it is suggested that the 25642 band is the ν_B biexciton band

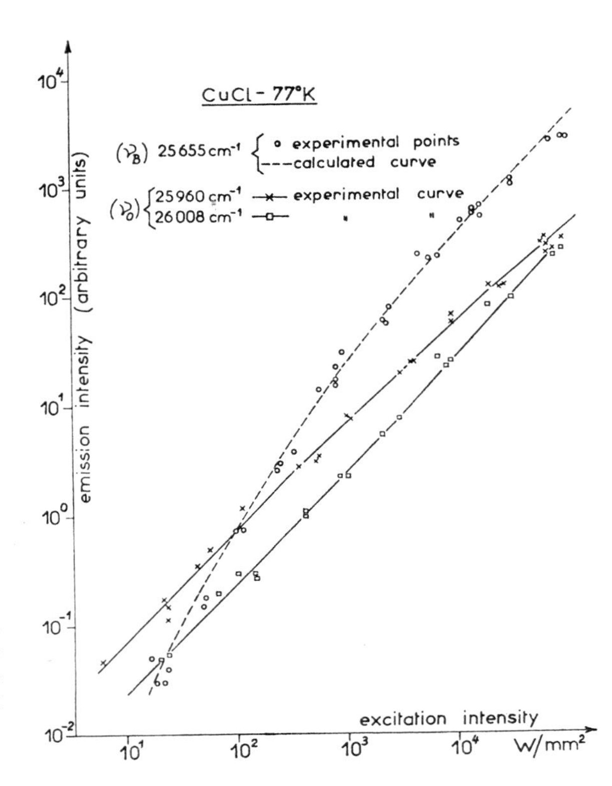

Fig. 3.
The variation of the intensity of the biexciton emission I_B (dotted line) versus the intensity of excitation. Solid lines represent the variation of the intensity I_0 of the resonance line taken for two different frequencies

E_g is the energy gap, E_{ex} the binding energy of the exciton, E_B the binding energy of the biexciton and E_B^k and E_{ex}^k respectively the kinetic energies of the biexciton and of the resulting exciton. E_B^k is probably small at low temperatures, but E_{ex}^k is probably not small. Both account for the breadth of the line ν_B; the maximum of the line is, however, probably obtained in first approximation by neglecting both. This gives:

$$h\nu_B(\max) = E_g - E_{ex} - E_B \tag{4}$$

as $E_g - E_{ex} = h\nu_0$ is well known from resonance emission

$$E_B = h(\nu_0 - \nu_B) \tag{5}$$

E_B has been calculated by Sharma [13] and also by Wehner [14] and recently by Hanamura [15]. The calculations agree approximately for $m_e^*/m_h^* \ll 1$. The calculations are, however, given in the form of curves and no analytical expression could be obtained by Sharma and Hanamura. For $m_e^*/m_h^* \leqq 1$ both theories disagree, the former predicting no bound state at values close to 1, latter predicts bound states for all values of m_e^*/m_h^*.

Wehner gives also calculations for $m_e^*/m_h^* \simeq 1$ and gives approximate expressions for both cases.

b') Approximate Calculation of the Binding Energy of Biexcitons for $m_e^*/m_h^* \ll 1$

A quite good approximate value can be obtained from two simplified independent calculations. Both theories concern small values of m_e^*/m_h^*: the first was given by the author and Grun and Myzyrowicz [16] some time ago; an improved version of this very simple theory is as follows.

Consider the depth of the potential energy well of H_2 and of the biexciton: D_{H_2} and D_B. The problem of the biexcitons is quite similar to that of the hydrogen molecule if $m_e^*/m_h^* \ll 1$, but the energy scale is changed in the ratio of E_{ex}/E_H where E_H is the binding energy of the hydrogen atom. So one can assume in a very good approximation:

$$D_B = D_{H_2} \frac{E_{ex}}{E_H}. \tag{6}$$

In order to obtain from the above relation the binding energy of the biexciton, one has to take into account the zero-point vibration energy E_B^{vo} and the correction of non-adiabatic movement of the holes E_B^{ad}. This supposes that the Born-Oppenheimer approximation is justified. This gives then

$$E_B = D_B - E_B^{vo} - E_B^{ad}. \tag{7}$$

For H_2 the zero-point vibration energy is rather small in comparison to the depth of the potential hole and the non-adiabatic correction is negligible.

For biexcitons, however, the zero-point vibration energy can become quite large in comparison to D_B in some substances. This would give a limitation for the stability of biexcitons which can be written:

$$D_B > (E_B^{vo} + E_B^{ad}). \tag{8}$$

However, if this is not the case, the Born-Oppenheimer approximation collapses and the approximate theory given here is not applicable. Another approximation should then be tried.

Now D_B is given by (6) and E_B^{vo} can be calculated from the relation which is readily found.

$$E_B^{vo} = \frac{E_{ex}}{E_H} \left(\frac{m_e^*}{m_h^*}\right)^{1/2} \left(\frac{M_p}{m_0}\right)^{1/2} E_{H_2}^{vo} \tag{9}$$

where M_p/m_0 is the ratio of the mass of a proton to the mass of the electron at rest and $E_{H_2}^{vo}$ is the zero-point vibration energy of H_2.

The nonadiabatic correction is small. It should, however, not always be neglected. The calculation of this correction has been carried out according to the second theory given by KAZAMANYAN (see below) [17].

The above formula has been applied to some compounds and the following results have been obtained

Table 1. Numerical values used in the calculation of E_B for different substances

	E_{ex} cm^{-1}	m_e^*/m_h^*	D_B cm^{-1}	E_B^{vo} cm^{-1}	ΔE_B^{ad} cm^{-1}	$E_{B\text{calc.}}$ cm^{-1}	$E_{B\text{obs.}}$ cm^{-1}	$E_{B\text{calc.}}$ cm^{-1} evaluated from HANAMURA's curve
Cucl	1523	0.02	560	183	28	359	350	~304
CuBr	873	0.01	320	74	8	238	236	~210
ZnO	338	0.125	124	104	74	Born-Oppenheimer approximation is presumably not applicable		30.4 [a]
CdS	226	0.17	85	81.5	36			17
Ge	34	0.665	12.4	23.4	32			

[a] It has to be remembered that in the experiments on CuCl and CuBr the breadth of the line ν_B is greater than the binding energy for CdS and ZnO as evaluated from HANAMURA's graph. It may be doubted whether, under those conditions and assuming comparable line breadths for CdS and ZnO, bound states are realistic. The situation is to some extent comparable to resonant decaying states in elementary-particle physics.

One can see that the agreement with experiment is quite good for CuCl and CuBr and furthermore that according to the above theory biexcitons cannot be formed in ZnO. The same conclusion is obtained for Ge, CdS, CdSe and GaAs. It seems that the value of about 0.1 for m_e^*/m_h^* should be the limit for the formation of biexcitons in the above theory.

$$m_e^*/m_h^* < 0.1 .$$

But this theory is expected to give reliable results for lower values of m_e^*/m_h^* only. The theory of WEHNER and of HANAMURA predict the stability of biexcitons even in the vicinity of $(m_e^*/m_h^*) \simeq 1$. However, these theories do not consider the complications arising from the corrections due to the polarization of the lattice according to HAKEN. This problem has been considered recently by EL KOMOSS and STÉBÉ [10] for bound excitons and calculations are carried out at present for the biexciton.

It has, however, to be remembered that the above theories would predict very weak binding of biexcitons from about $(m_e^*/m_h^*) \simeq 0.1$ on, and in compounds as ZnO, CdS, CdSe, GaAs, Si and Ge could well not be stable. In fact, in none of these substances does the formation of biexcitons seem to have been detected*. It is possible that other effects compete with the formation of biexcitons. It is also possible that the terms E_B^K and E_{ex}^K which are thought to be responsible for the breadth of the line ν_B give a too high contribution in Eq. (3).

The final result of the second theory due to KAZAMANYAN [17] is equivalent to our version in its present improved form.

In its present form, our theory has an advantage with respect to the theory of KAZAMANYAN as it introduces no ambiguity in the use of the dielectric constant ε, which has to be calculated according to HAKEN's corrections and in the present state will bring many complications.

The introduction of the factor E_{ex}/E_H has the advantage that this quantity is the first information obtained from an exciton spectrum and is unambiguously available from experiment, in an approximate value at least.

c) The Structure of the ν_B Band

A structure in the ν_B band is observed for CuCl and CuBr. This structure can have different origins.

1. Spin-spin Interaction. One origin is as follows: a weak line ν_1 on the low-energy side is observed in some CuCl crystals. It has been assigned to the paraexciton state. So, for this substance a $^1S_0(\Gamma_2)$ state is situated some 50 cm^{-1} below the $^3S_1(\Gamma_4)(\nu_0)$ orthoexciton state**.

Biexcitons can probably be formed from both states. If the binding energy is of the same order in both cases, a structure is expected in the ν_B band. This is really observed and shows a separation of about 50 cm^{-1}.

For CuBr, the lowest state is a $^5S_2$ state splitting into two states, Γ_3 and Γ_5, in the crystal field [3]. The situation is thus much more complicated. No defined structure was observed by GRUN et al., but UETA et al., have reported a structure [19].

We may also ask whether the angular momenta of both holes have a non-negligible interaction, and in how far a hyperfine structure is excluded in this case.

2. Rotational Levels. A second possible interpretation can be suggested. The biexciton could be formed in different rotational energy states***. A rough calculation gives for the rotational energy of a biexciton in CuCl:

$$E_B^r \simeq 30 \text{ cm}^{-1}.$$

d) Kinetics of Formation of Biexcitons

In order to explain the behaviour of the kinetics it was found necessary by KNOX and the author to separate the excitons formed into two parts: the "optical"

* The bibliography taken in account concerns papers known to the author before April 1972. New information has been published since. In particular it has been suggested that in the case m_h^* is anisotropic, its highest value could be taken. This may considerably increase E_B for several compounds [35]. The interpretation of recent experiments on Ge and Si has risen a controvercy which is left out of the scope of this paper.
Note added in proof: Considerably higher values of E_B for $m_e^*/m_h^* \simeq 1$ have been calculated recently by HANAMURA (private communication). This makes the observation of biexcitons more probable when $m_e^*/m_h^* \simeq 1$.

** The notations are those of DRESSELHAUS. See D. CURIE, "Champ cristallin et luminescence", Gauthier Villars, Paris, 1968.

*** Vibrational levels are much higher: for CuCl: 366 cm^{-1} and for CuBr: 148 cm^{-1}.

(density n_0) and "thermal" (density n_1). The biexcitons (density n_2) are assumed to be formed essentially from the very much larger pool of thermal excitons. Both classes of excitons are formed from the recombination of free carriers with rates $A_0 i$ (optical) and $A_1 i$ (thermal); $A_0 i < A_1 i$. The kinetic equations are readily written and the solution is:

$$n_0 = \frac{A_0 \tau i}{1+\tau k}; \qquad n_1 = \frac{k''}{B}\left(\sqrt{\frac{i}{i_0}+1}-1\right); \qquad n_2 = \frac{(k'')^2}{2\,BC}\left(\sqrt{\frac{i}{i_0}+1}-1\right)^2. \quad (10)$$

Here τ is the lifetime of excitons, k the conversion rate of "optical" in "thermal" excitons, k'' rate of decay of thermal excitons (radiationless), B bimolecular collision coefficient and C recombination coefficient of biexcitons. In these formulas the conversion rate k' of "thermal" to "optical" excitons is neglected.

As $I_0 \propto n_0$, this gives the linear dependence of I_0 versus i.

As $I_B \propto n_2$, this can be shown to give for $i \ll i_0$, $I_B \propto i^2$ and for $i \gg i_0$, $I_B \propto i$.

Here:

$$i_0 = \frac{(k'')^2}{2\mathrm{B}}\left(\frac{1+k\,\tau}{A_1+k\,\tau(A_0+A_1)}\right).$$

Though the experimental data give $m_B < 2$, the agreement with the theory seems to be good. As stated above, the discrepancies are very likely to be due to overlapping of the lines ν_B and ν_0 at 77 °K, and to overlapping of these two lines with other lines ν_1 and ν_2 at 4.2 °K.

Therefore it seems that the binding energy of biexcitons and the kinetics of the observed luminescence being in reasonable agreement with the observations, a strong argument in favour of formation of biexcitons is given in this investigation for CuCl and CuBr.

It has also been shown that biexcitons can be formed when the excitation is obtained by the double-photon absorption technique.

The experiments of HAYNES which were originally thought to be a first proof of the formation of biexcitons seem to be ascribed to another process now. This process will be described later. In conclusion, it now seems that, among the substances investigated so far, the formation of biexcitons is plausible in copper halides only (and possibly Cu_2O) *.

B. Exciton-Exciton Interaction without Binding

An exciton-exciton interaction of a special kind has recently been observed by D. MAGDE and H. MAHR [20] (Ithaca), by R. LÉVY, J. B. GRUN and A. BIVAS [21] in the Strasbourg Group and also by BENOIT À LA GUILLAUME et al. [22]. It is not impossible that this is an alternative interaction effect when the biexcitons are not stable.

a) Experimental Situation

The experimental situation is as follows. When CdS, CdSe or ZnO crystals are irradiated with intense laser pulses, a new luminescence appears. This luminescence band M lies below the free exciton level at about the binding energy of the exciton E_{ex} (Fig. 4). The intensity of this ν(ex-ex) band is proportional to n_L^2, where n_L is the number of laser photons reaching the crystal. This suggests a bimolecular

* See however footnote* p. 26.

process. When the intensity of excitation increases the line shifts to lower energies. A saturation takes place for high values of n_L. At these very high concentrations of excitons, some new lines appear. These effects are reported by LÉVY and GRUN and will be dealt with later in this paper.

b) Interpretation

The interpretation can be given according to a mechanism suggested by BENOÎT À LA GUILLAUME [22]: intense laser irradiation of the crystal again produces a large number of free carriers which rapidly combine to give a high concentration of excitons. When two excitons collide, they can form biexcitons if the conditions are suitable for this process. If they are not, the collision of excitons can lead to the following reaction: one exciton is dissociated, taking the energy of the other one. The rest of the energy is radiated.

$$(\mathrm{ex})_1 + (\mathrm{ex})_2 \rightarrow e + h + h\nu\,(\text{ex-ex}).$$

The energy balance is then

$$(E_g - E_{\mathrm{ex}})_1 + (E_g - E_{\mathrm{ex}})_2 = E_g + E_{\mathrm{kin}} + h\nu\,(\text{ex-ex}). \tag{12}$$

This gives then, neglecting E_{kin} in a first approximation

$$E_g - 2E_{\mathrm{ex}} = h\nu\,(\text{ex-ex}) \tag{13}$$

So the radiated frequency should be less than the exciton resonance frequency by another amount of E_{ex}. The kinetic energy gives a small correction of energy of the photon. The description of this process is in good qualitative agreement with experiment. If the intensity of excitation continues to increase, the line ν (ex-ex) is shifted to lower energies. This effect, studied recently in detail by LÉVY and GRUN [23] (Fig. 4) can be explained as follows.

When a large number of excitons are dissociated, the valence band is filled with holes and the conduction band with electrons up to some higher level below and above the extrema of the bands, respectively. Hence the dissociation requires not only the binding energy of the exciton, but a certain amount of kinetic energy. This effect has been calculated [24] and gives for ellipsoïdal bands:

$$\Delta\nu = \left(\frac{1}{m_e^*} + \frac{1}{m_h^*}\right)\frac{h}{8}\left(\frac{3}{8\pi}\right)^{2/3} n^{2/3} \tag{14}$$

here the masses have the following definition:

$$m_e^* = (m_{xe}^* m_{ye}^* m_{ze}^*)^{1/3}$$

where n is the number of electrons (or holes) per unit volume produced by dissociation of excitons in the above process. Now it can be shown that in this case $n \propto i^{1/2}$. Therefore $\Delta\nu \propto i^{1/3}$, where i is the intensity of the excitation light. Fig. 5 shows the results of the experimental measurements. It can be seen that the theory is in good agreement with experiment.

C. Photon and Phonon Emission

Another process of light emission is quite common and is interpreted as follows. An exciton can emit simultaneously a photon and a phonon (in general, a longi-

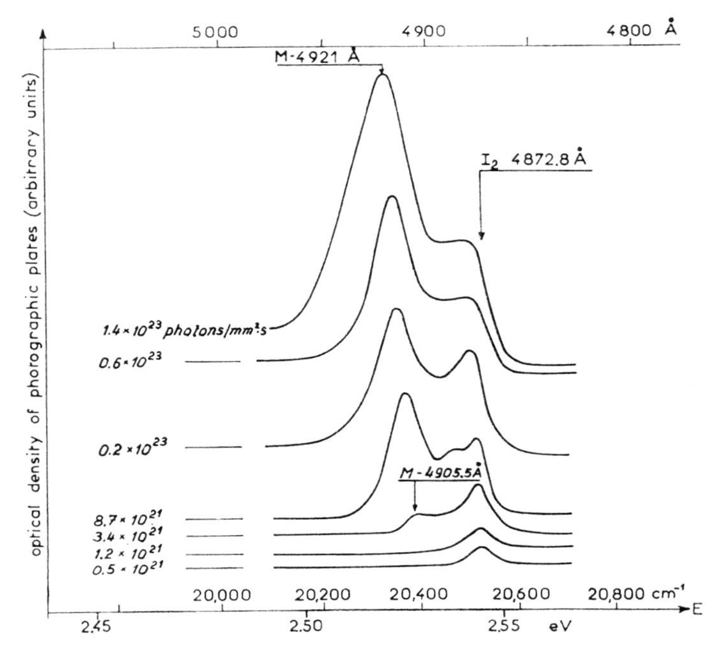

Fig. 4. Variation of the emission spectrum for a rather thick crystal of CdS, showing the shift and intensity of the ν(ex-ex) line (M line) for different excitation intensities (photon $(\mathrm{mm})^2\, s^{-1}$)

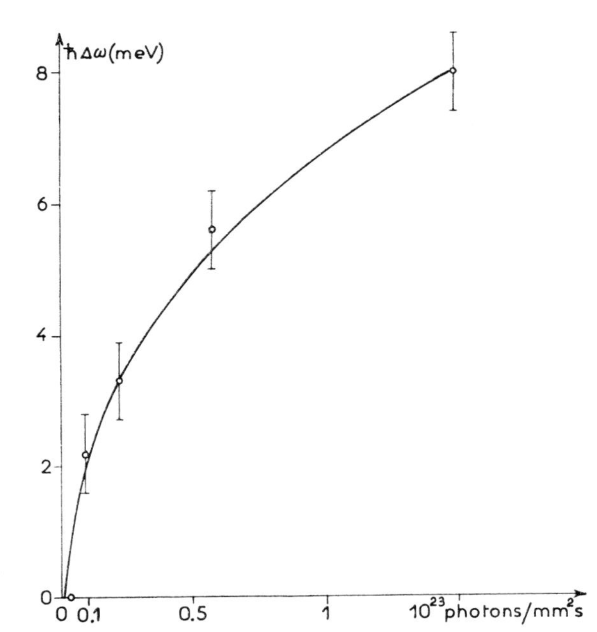

Fig. 5.
"Red" Shift of the ν(ex-ex) line as a function of the excitation intensity (CdS)

tudinal optical phonon). This is quite a common process which has been observed in a number of substances in conventional luminescence (see above). In CuCl, the line is however superimposed upon another very strong line which has been interpreted as a recombination-process emission from a complex (A^0 ex), A^0 being a neutral acceptor. In CdS this line was observed by HOPFIELD and THOMAS [8].

The process can be written as follows:

$$(\text{ex}) \rightarrow \hbar\omega + (\text{LO}) \tag{15}$$

and the energy of the photon is:

$$E_g - E_{\text{ex}} - E(\text{LO}) = \hbar\omega\,(\text{ex-LO}).$$

This usually rather weak emission is, however, of interest, since GRUN, LÉVY and BIVAS [25] have shown that this line becomes very strong when the luminescence is excited with high-intensity laser light. It has been shown that from a certain excitation intensity onwards the emission in this particular process becomes stimulated.

This effect is discussed in the second part of the paper.

The theory of this effect has been given by HAUG [26].

A new development seems to have been observed recently by GRUN, LÉVY and BIVAS. In CdS, this group observed some competition between the (ex-ex) effect and the (ex-LO) effect [27].

D. The Blue Shift and Bleaching of the Excitonic Absorption Line

A fourth effect which occurs at very high concentrations of excitons has recently been observed by the Strasbourg Group [28a].

1. Experimental Situation

The ν_0 resonance absorption line in CuCl is a transition to an $n=1$ orthoexciton state. The emission resonances line should have the same frequency. However, it is reabsorbed and on account of that a minimum of luminescence is observed at ν_0, the luminescence band being broader than the absorption line.

Using a second synchronous light source, it was observed recently that the exciton absorption line ν_0^a is shifted to higher energies while the emission line ν_0^e is not shifted (or only slightly shifted). The shift $\Delta\nu_0^a \propto \sqrt{i}$, where i is the intensity of the laser light. The second source has a much smaller intensity and should not influence the shift. The absorption apparently becomes much weaker at very high intensities.

2. Interpretation

It has first to be remembered that the excitons are strongly polarizable. If their concentration is high, they will change the dielectric constant of the medium and this is bound to give a blue shift of the exciton level, E_{ex} being diminished. This leads to the shift:

$$\frac{\Delta E_{\text{ex}}}{E_{\text{ex}}} = 110\, N_{\text{ex}}\, a_{\text{ex}}^3 \tag{16}$$

where N_{ex} is the concentration of excitons and a_{ex} is the Bohr radius of the exciton.

For high values of the intensities of irradiation $N_{ex} \propto \sqrt{i}$ for thermal excitons [Eq. (10)] and so is the shift of ν_0^a. However, unless the emission has a longer lifetime than the laser pulse, the above theory does not explain why the emission band is not appreciably shifted while the absorption is. Though the above effect does not seem to agree unambiguously with experiment, the effect must exist but is probably smaller than expected.

Another suggestion is, however, more plausible.

The usual selection rules are for the formation of first-class excitons $\boldsymbol{K}=0$, $\Delta \boldsymbol{K}=0$, where $\boldsymbol{K}$ is the total wave vector. When excitons are formed, their wave functions are obtained from a summation over all the values of $\boldsymbol{K}$ of the states of the conduction band. If now the conduction band is full of electrons up to a certain $\boldsymbol{K}$ value, say $\boldsymbol{K}_0$ (and the same with the valence band and holes), the wave functions of the new excitons formed cannot include in the summation all the $\boldsymbol{K}$ states but only those $|K|>|K_0|$. The binding energy of the exciton calculated by a variational principle is now found to be changed by ΔE_{ex}.

$$\frac{\Delta E_{ex}}{E_{ex}}=32\pi\, a_{ex}^3\, n \tag{16'}$$

where n is the density of electrons and is of the same order as N_{ex} per volume. When the exciton recombines on account of the selection rules stated above, $\Delta \boldsymbol{K}=0$ and $\boldsymbol{K}=0$, only the optical excitons will participate in this recombination, and these excitons are made up of all the states. So the binding energy of these excitons is not changed by the above process.

This model probably oversimplifies the situation because the free electrons rapidly form excitons. However, the excitons formed will occupy a part of the states of the conduction band so that any new exciton formed must be built up from the rest of the states. This will probably change the numerical coefficient of the above formula, but not its general character. In this theory it is assumed that the excitons are not bosons*. Furthermore, it has to be emphasized that a non-equilibrium situation is considered.

The shift of the exciton absorption line is obtained from the kinetic equation for the steady state.

$$\alpha i=\beta n^2 \tag{17}$$

where α is the absorption coefficient and β a recombination rate. This gives again a dependence on $\sqrt{i}$. The change of binding energy ΔE_{ex} is of the same order as in the first model, but this second model explains why the luminescence band is not shifted while the absorption is.

Taking realistic values of the parameters, it can be shown that the agreement with experiment is reasonable. The $\sqrt{i}$ variation of $\Delta\nu$ is in good agreement with measurements of GRUN et al. [28a] for not too high values of i. For high values of i a saturation effect is observed, see Fig. 6.

The fundamental characteristic of this effect is that the luminescence band is separated from the absorption line which is shifted to higher energies. So the emission is no longer resonant. This is of importance for what follows. See also footnote p. 34.

---

* This statement means that no Bose condensation is assumed.

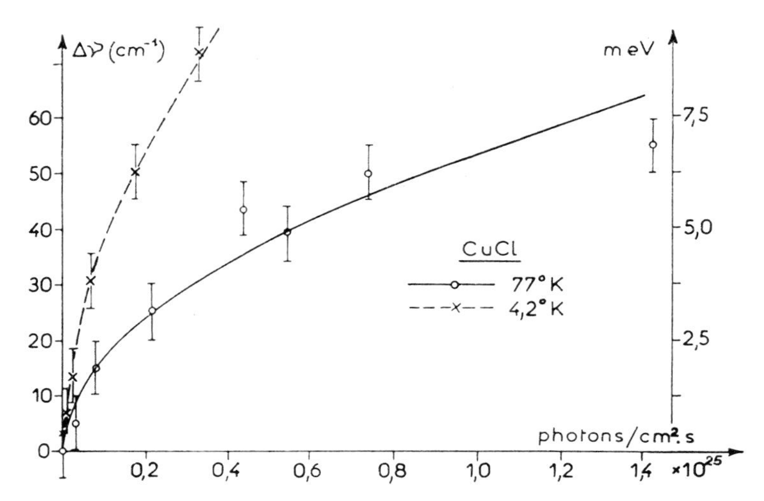

Fig. 6. "Blue" Shift of the $n = 1$ absorption line of CuCl at low temperatures as a function of the excitation intensity

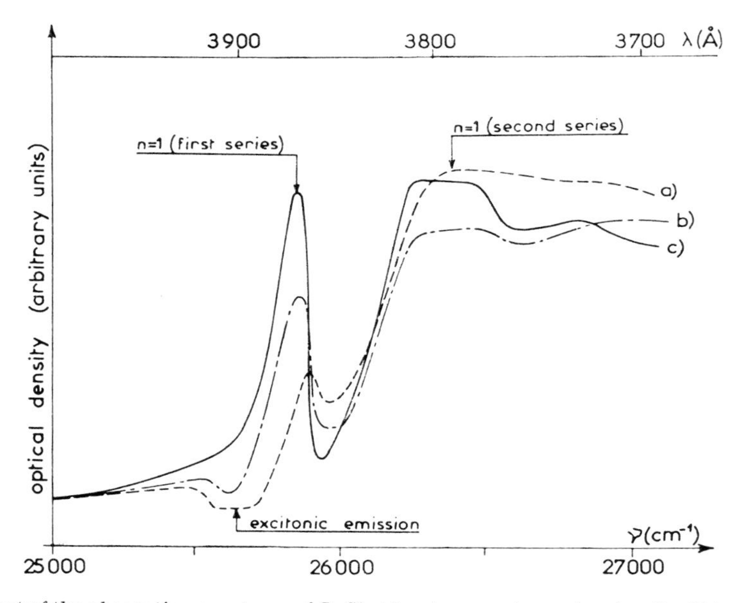

Fig. 7. Part of the absorption spectrum of CuCl at low temperatures showing the "blue" shift of the $n = 1$ "fine" line and the bleaching of this line for different excitation intensities

It has to be emphasized that this effect may turn out to be of considerable interest. Hanamura and Keldysh and Kosloff [28b] have developed a theory assuming Bose condensation and superfluidity of excitons. However, this theory predicts a shift of the exciton level. This means the absorption and emission bands both shift to higher energies in the same way. As this is in opposition to the experimental result, it is tempting to conclude that the experiment is in opposition to Bose condensation of excitons.

3. The Bleaching of the $n=1$ Line in CuCl

It is observed that the exciton $n=1$ absorption line bleaches out considerably when a CuCl sample is irradiated with very intense light (Fig. 7). This effect can be interpreted as follows [29].

As a result of illumination a concentration n of excitons is formed. Let us now assume that n_0 is the highest possible concentration of excitons; a reasonable value for n_0 is say $1/a_{ex}^3$. Then, σ being the absorption cross-section and τ the lifetime of excitons, the following kinetic relation can be written:

$$\frac{dn}{dt} = \sigma i (n_0 - n) - \frac{n}{\tau}.$$

This equation gives a bleaching. The saturation value is given for the steady state:

$$\sigma i (n_0 - n) - \frac{n}{\tau} = 0.$$

If the frequency of the irradiation light coincides with the exciton absorption, the absorption coefficient is $K = \sigma(n_0 - n)$ and $K_0 = \sigma n_0$.

This gives with $\Delta K = K_0 - K$

$$K = \frac{K_0}{1 + \sigma \tau i}$$

and

$$\tau = \frac{\Delta K}{K} \frac{1}{\sigma i}$$

which provides the possibility of measuring the lifetime of the exciton. The bleaching of the $n=1$ line in CuCl has been observed under a somewhat different experimental condition, the irradiation being carried out in the band-to-band transition region of the spectrum. A spectrophotometric study is being carried out at present.

Stimulated Emission from Free and Bound Excitons

The question of stimulated emission from free or bound excitons is under investigation at present and good deal of theoretical and experimental work will be needed before a clear understanding can be obtained. However, a first review on the subject can be attempted now subject to the above restrictions. The aim of this section is to show that in different experiments stimulated emission from free or bound exciton levels has actually been observed and that further significant results can be expected.

1. Stimulated Emission from Biexcitons

The properties of biexcitons are very favourable for stimulated emission. The reverse absorption process of optical formation of biexcitons has not been observed and it has a very low probability. Therefore, the situation is very similar to a three-level laser diagram. If other losses are not too high, it can be predicted that upon high excitation the stimulated emission should be observed. Now, the only substances in which biexcitons have so far been shown to exist with very plausible arguments seem to be CuCl and CuBr crystals*. (See footnote p. 26 however.)

The rate equation for the emission is:

$$\frac{dn_p}{dt} = n_p G\, n_B + G\, n_B - 2K\, n_p. \tag{18}$$

Here n_p is the number of photons in a given mode, G is a gain coefficient, n_B the number of biexcitons per cm^3. The last term describes the losses. This equation shows that from a threshold $G\, n_B > 2K$ the coefficient of n_p is positive, and this indicates stimulated emission.

The observation of stimulated emission is, however, generally rather complicated as the preparation of an optical cavity with CuCl and CuBr is very difficult. Thus the observations by the Strasbourg Group could not be assumed to be a proof of stimulated emission. Neither could intensity measurements.

However, from kinetic equations the intensity radiated can be given in the form:

$$I = \frac{I_{Sp}}{\alpha}\,(e^{\alpha x} - 1) \tag{18'}$$

where α is the net gain coefficient and I_{Sp} is the intensity of the spontaneous emission, $\varkappa$ is the length of the emitting material. A superlinear dependence of I on $\varkappa$ should be a direct indication of stimulated emission.

This way of eliminating the above difficulties was suggested by K. L. SHAKLEE and R. F. LEHENY (Bell. Lab.) and has been applied to CuCl by SHAKLEE, LEHENY and R. E. NAHORY quite recently [33].

These authors prepared a crystal in the form of a cube and irradiated it with a nitrogen laser. A rectangular image of very small width and variable length was formed on an etched surface of the cube close to the perpendicular surface from which the radiation was measured.

The observations give an evident confirmation of stimulated emission in the biexciton (ν_B) emission band. Very considerable gains have been obtained, larger even than in GsAs under similar conditions. The experiments were performed at 2 °K.

2. *Exciton Resonance Emission* ν_0

It was shown recently by H. HAKEN** that the resonance line of exciton spectra cannot be stimulated.

Assuming that the absorption and emission lines coincide and have the frequency ν_0, it can be shown that the kinetic equation of light emission in an exciton resonance line is:

$$\frac{dn_p}{dt} = \frac{2\,|g|^2}{\gamma}\,(n_{ex} - n_p) - 2K\, n_p. \tag{19}$$

* Some results on Cu_2O and Si may also perhaps be interpreted as due to biexcitons [30], [31]. The interpretation of HAYNES is, however, still under discussion.

** Various points in this section were worked out in stimulating discussions with Prof. H. HAKEN and some of the reported theoretical results are due to him. The author thanks him for permission to make use of them here before publication. This will be done only schematically. The full theoretical argument is to be published by H. HAKEN and the author.

Here n_P is the concentration of photons, n_{ex} the concentration of excitons, g is a transition matrix element, γ a damping constant and K a loss coefficient. It can be seen that the terms containing n_P on the right-hand side are negative.

This means that only spontaneous emission takes place.

No stimulated emission is expected therefore in the exciton resonance line, unless other arguments are taken into consideration. No stimulated emission has yet been observed in exciton resonance lines.

3. Emission in Case of Bleaching and Shift of the Absorption Line

We have seen above that when the excitation light intensity becomes very high, the $n=1$ line in CuCl bleaches and is shifted to higher energies whereas the emission line does not shift appreciably. Therefore, it can be shown that the absorption line no longer coincides with the emission line and moreover becomes very much weaker.

Consequently, the validity of the above statement is much reduced and in fact a stimulated emission is not completely improbable. It has, however, not been observed so far, to the author's knowledge.

Another argument, however, has to be advanced against this conclusion.

4. Emission from Forbidden Exciton Levels

It has to be remembered that the exciton $n=1$ level is not the lowest level an exciton can occupy. At low temperatures, as shown above, a considerable number of other lower levels will be occupied and the luminescence will be emitted primarily from these levels. Among these levels, different complexes and biexcitons should be mentioned. They will be considered below.

Let us mention here that the exciton $n=1$ levels usually show a fine structure. So, for CuCl, the $n=1\,(\Gamma_4)$ resonance level is a spectroscopic $^3S_1$ state (ortho-exciton level). Another state $^1S_0$ (paraexciton level Γ_2) lies 50 cm^{-1} below. Disregarding all other possibilities, excitons could accumulate at low temperatures at this paraexcitonic level. Whether this level could radiate stimulated emission is a question that has not yet been considered theoretically. This line is in the wing of the ortho-excitonic absorption line and is very weak except in certain conditions.

In CuBr the lowest state is a $^5S_2$ state. This state is split in two by the crystal field (Γ_3 and Γ_5) but some of these lines are too close to the $^3S_1$ line.

It is possible that the excitation accumulates in still lower states, biexciton levels or complexes of bound excitons, or that the suitable conditions have not been obtained. Anyhow, these states will depopulate at low temperatures the ortho-excitonic level.

5. The (ex-LO) *Luminescence and Stimulated Emission*

This luminescence has been observed in different crystals. In particular, it has been observed in CdS by Hopfield and Thomas [8]. Later on, Lévy et al. [23] have shown that this line can become stimulated when the excitation intensity is high enough.

The theory of this effect has been given by Haug [26] and confirms the stimulated character of the emission.

The rate equation is in this case according to HAKEN in first approximation:

$$\frac{d n_{LO}}{dt} = n_{LO} \beta n_{ex} + \beta n_{ex} - 2K n_{LO}. \tag{20}$$

Here n_{LO} is the number of photons generated in this process. n_{ex} is the number of excitons cm^{-3}, β is the rate coefficient and K is a loss coefficient. It can be seen that the first term represents the stimulated emission. When this term becomes larger than the last term, the coefficient of n_{LO} becomes positive and the laser action starts up from this threshold.

It can be shown experimentally that the intensity of the emission for not too high excitation intensities varies linearly. When a threshold excitation intensity is reached, the variation of the emission intensity becomes superlinear with a slope of about 8 or 10, the value of this slope depending presumably on the quality of the sample. This is an example of stimulated emission from exciton levels first observed by the Strasbourg Group.

For CuCl such an emission has been observed but coincides with the ν_2 line which has been shown to be also an emission from an A^0_{ex} complex. The line probably has a mixed character (ex-LO) and A^0_{ex}. As the latter can be observed also in absorption, no stimulated emission can be observed. Further on, the biexciton level has a still lower energy.

6. Competition between (ex-ex) *and* (ex-LO) *Emission. Stimulated Emission from Different Centers*

In some rather thick CdS crystals it has been shown that the (ex-ex) emission line M increases quadratically with the excitation light intensity and is shifted to lower energies. However, when the threshold for the emission of the (ex-LO) line is passed, a stimulated emission of this line starts and becomes very much stronger than the ν(ex-ex) line which reaches saturation, see Fig. 8.

This effect has been studied recently and the theory of this effect has been developed with HAKEN and EL KOMOSS and will be published later.

The energy balance equation is in this case:

$$\frac{d n_p(\text{ex-ex})}{dt} = \alpha n_{ex}^2 - 2K n_p(\text{ex-ex}). \tag{21}$$

Here n_p(ex-ex) is the number of photons emitted in this process. The first term gives the spontaneous emission. As can be seen, the only term on the right-hand side containing n_p(ex-ex) is negative, and therefore no stimulated emission is expected.

A more elaborate theory of the competition of the two effects by HAKEN and EL KOMOSS is under study and will be published later. Furthermore, it has to be remembered that recent experiments of GRUN and LEVY show some more complicated effects (see below). These effects are going to be included in the more elaborate theory. The above Eq. (21) concerns the spectral condition stipulated in these experiment where the ex-ex and ex-LO effects are in competition.*

In thin platelets the situation is somewhat more complicated. In this case, a number of other lines are stimulated before the line (ex-LO) seets in [23] (Fig. 9).

* See footnote p. 34. Under some conditions, it appears that the (ex-ex) effect can also be stimulated (private communication of Prof. PILKUHN and also Dr. R. LÉVY).

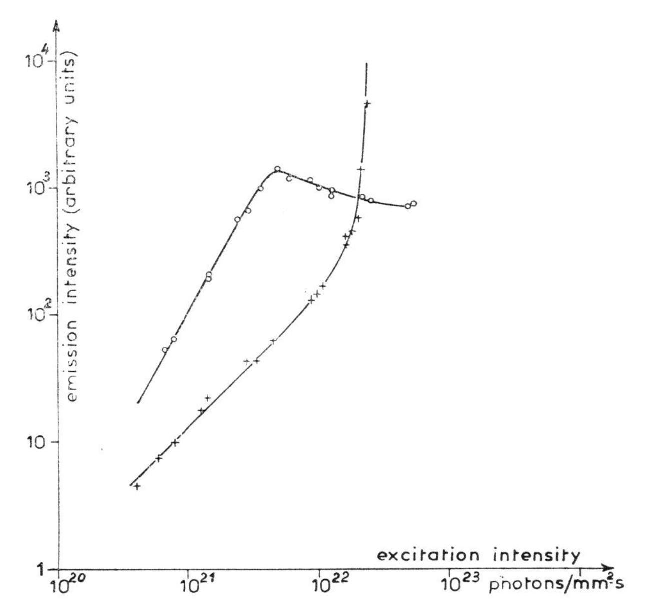

Fig. 8. Competition of the ν(ex-ex) and the ν(ex-LO) emission as a function of excitation intensity in a rather thick sample of CdS

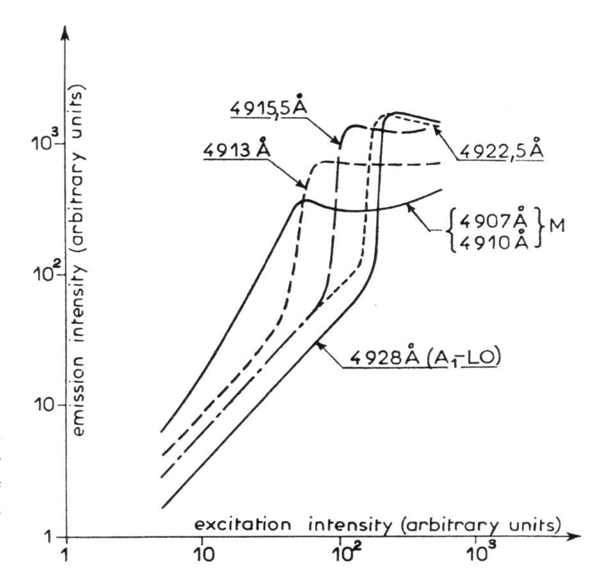

Fig. 9.
Competition of the ν(ex-ex) and other emissions for a thin platelet of CdS at low temperature and as a function of excitation intensities

The origin of these new stimulated lines is not quite clear at present. One process suggested for the two lines I_1 and I_2 (4913 and 4915 Å) by ERA and LANGER [33] is as follows: a bound exciton recombines under the action of a collision with another free exciton but leaves the donor (or acceptor) in an excited state.

Clearly, the emitted photon has an energy which is smaller than the simple emission from a bound exciton complex by the excitation energy of the donor. Let us call these lines ν_{L_i}. The energy emitted is then:

$$h\nu_{L_{ji}} = \hbar\nu_{I_i} - E_{D_i}^{(j)}$$

here ν_{I_i} is the frequency of the radiation from the complex giving rise to the line I_i, $i=1, 2, 3$ and $E_{D_i}^{j}$ is the excitation energy of the j state, say. The theory seems to fit well for two lines 4913 Å and 4915 Å with $i=2$, and the complex responsible for the line ν_{I_i} is the same as the one responsible for I_2. It could be objected that at these concentrations a free donor could still exist. In particular, the I_1, I_2 and I_3 lines are much weaker to the extent that they are not observed.

The three other lines seem to be explained by a mechanism of collision of an exciton with a bound exciton. It is clear that the photon energy of these lines N_1, N_2, N_3 must have an energy lower than the photon of the M line by the amount of the binding energy of the given complex. This gives:

$$\hbar\nu_{N_i} = \hbar\nu_M - E_{D_i}$$

where $i=1, 2, 3$, corresponding to the complexes discussed above.

When on increasing the excitation intensity the last threshold is reached, a sudden explosion of luminescence appears. This effect of sudden ignition of various lines is very striking and it seems likely that all the processes involved are mutually stimulated. We suggest the process be called a luminescence avalanche. Some resemblance to a kind of stimulated Raman effect can be found in this process. The spectrum is shown in Fig. 10. A theory is now being developed on this basis by HAKEN and EL KOMOSS. In one of our first publications on luminescence excited by laser light we reported quite similar ignition of a large number of lines [23]. It can be seen from Fig. 10 that all the emission processes involved in the avalanche concern emission of phonons, and one could suppose that this circumstance is essential for the mutual stimulation of the lines of the luminescence avalanche. This idea has already been suggested some time ago [23].

Anyhow, the successive processes described here are in mutual competition, the higher-energy process being saturated when the low-energy process is stimulated. This is, however, no longer the case when the avalanche threshold is reached. When this threshold is passed, it seems that all the lines emitted are then mutually stimulated.

The progression of the effects below the thresholds may be due to the formation of a broad absorption band overlapping with the emission lines, the losses for each process being gradually enhanced by this absorption. This suggestion was made by LANGE and GOTO [34].

It will be remembered that the behaviour of rather thick crystals and of thin platelets is different. These effects seem therefore to bear some relation to the quality of the crystal.

7. *Condensation and Metallization of Excitons*

Some very important research work concerning the possible condensation of excitons in droplets, transformation of an exciton fluid into an electron-hole plasma, and the radiative effects related to these effects is in full development at present.

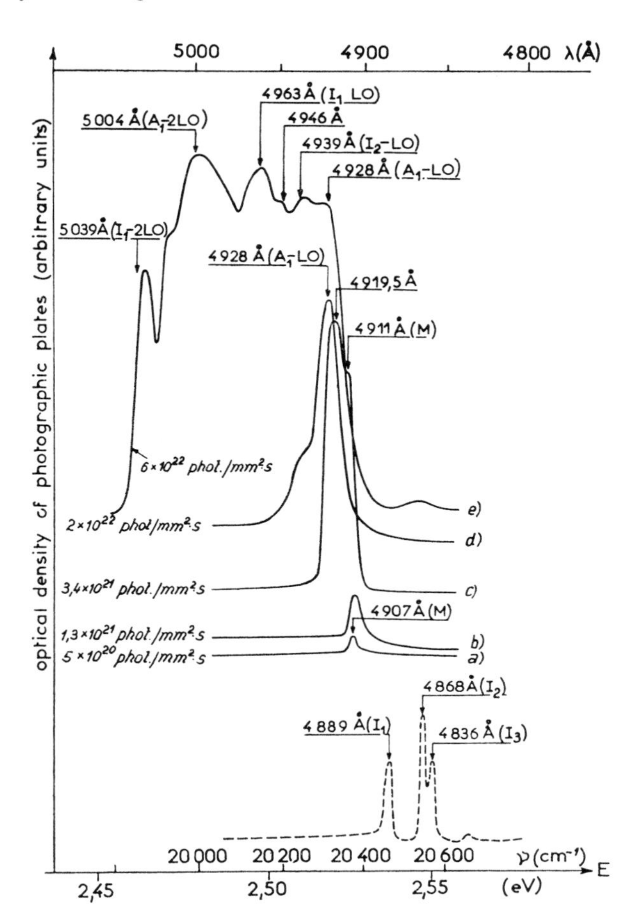

Fig. 10. The variation of the emission spectrum of a thin platelet of CdS at low temperatures for different excitation intensities. Showing the successive ignition of different emission lines and finally the luminescence avalanche

Substantial arguments in favour of condensation of excitons have been advanced by different authors. This very promising subject of research is outside the scope of this paper as it is still in full development. However, is should be emphasized that this important chapter of exciton physics is certainly going to advance and should yield important information.

Conclusion

In this short review, the properties of excitons at very high concentration have been examined, excluding questions related to exciton fluid, condensation or phase changes, as stated above.

The study of excitons at high concentrations is of great importance. Research work performed in the Strasbourg Laboratory has been specially emphasized in this paper.

First, the question of formation of biexcitons was considered. Biexcitons were observed in CuCl and CuBr some time ago, and the interpretation of the related phenomena seems to offer strong arguments in favour of the formation of biexcitons. Both the kinetics of the formation of biexcitons and the calculation of the binding energy are in reasonable agreement with the experiments of Myzyrowicz, Grun, Lévy and Bivas in the Strasbourg Group on the above substances. The existence of biexcitons in other substances has been proposed but it seems that stronger arguments are necessary for proof. The binding energies for substances like CdS or ZnO are very weak and probably smaller than or of the order of the line width of biexcitons, which may well be unstable in some categories of substances. It could be suggested that the conditions for the instability of biexcitons may be $\sigma > 0.1$ or so, and (or) low binding energy of excitons. Another argument, suggested by Benoit à la Guillaume, is that biexcitons should not appear in an indirect gap material in which exciton condensation would be more probable. This interesting idea deserves to be carefully examined. Further, various other radiative processes appearing at high concentrations of excitons have been described.

It is suggested that a first category of effects may correspond to a collision of two excitons without binding. Such an effect has been seen in materials in which biexcitons have not been observed.

Another effect which seems to be quite frequent is the simultaneous emission of a LO phonon and a photon from the lowest exciton level or from an exciton complex. It seems that this emission can also be observed from excited levels of both, or leaving the radiative center in an excited state.

High concentration of excitons can cause a "red" shift of the line ν(ex-ex) or the "blue" shift of the absorption resonance line with respect to the emission line. These "blue" and "red" shifts are of importance and could be explained.

It has been shown that some of the above effects lead to a stimulated emission.

It can be shown that excitons do not emit stimulated emission unless the absorption line is strongly shifted with respect to the emission line.

It can be shown that biexcitons can recombine under stimulated emission; this effect has been observed in CuCl by the Bell Lab. Group.

In the (ex-ex) process, as observed in CdS, no stimulated emission has been observed. This can be justified theoretically. However, in some special conditions and materials, the theoretical conditions may be different and so the conclusions, could be too.

The (ex-LO) effect leads to stimulated emission. In CdS, this effect operates from a threshold excitation intensity in competition with the (ex-ex) effect to give some intermediate effects in between the saturation of the (ex-ex) effect and the threshold of the (ex-LO) effect in some crystals. At very high concentration for CdS a kind of avalanche luminescence sets in, in which a number of phonon-assisted emissions are observed. This was reported some time ago by Myzyrowicz and Grun and has recently been observed again by Lévy and Grun.

Other effects may be observed under different conditions and in different materials.

It can be seen that the subject treated in this paper is of some importance and contributes a number of new effects which can yield important information.

Bibliography

1. Grillot, E. F., Bancie-Grillot, M., Pesteil, P., Zmerli, A.: Comp. Rend. **242**, 1794 (1956).
   Grillot, E. F.: J. Phys. Radium **17**, 822 (1956).
   Nikitine, S., Perny, G., Sieskind, M., Reiss, R.: J. Phys. Radium **17**, 817 (1956).
   Nikitine, S., Perny, G.: J. Phys. Radium **17**, 1017 (1956).
   Nikitine, S., Reiss, R.: Compt. Rend. **244** 2788, (1957); **245**, 52 (1957).
   Arkhangelskaia, V. A., Feofilov, P. P.: J. Phys. Radium **17**, 824 (1956).
2. Toyozawa, Y.: Suppl. Progr. Theoret. Phys. (Kyoto) **12**, 93 (1959).
3. Nikitine, S., Ringeissen, J., Certier, M.: Acta Phys. Polon. **26**, 745 (1964).
   Nikitine, S., Ringeissen, J., Sennet, J.: Conference on Phys. Semiconductors Radiative Recombination, p. 279 (1964).
4. a) Nikitine, S., Reiss, R.: J. Phys. Rad. Letters **20**, 718 (1959).
   b) Myzyrowicz, A., Grun, J. B., Bivas, A., Levy, R., Nikitine, S.: Phys. Letters **25**A 286 (1967).
   c) Grun, J. B., Myzyrowicz, A., Raga, F., Bivas, A., Levy, R., Nikitine, S.: Phys. Stat. Sol. **22**, K 155 (1967).
   d) Levy, R., Bivas, A., Grun, J. B.: J. Phys. **31**, 507 (1970).
   e) See also Hopfield, J. J., and Thomas.
5. Lampert, M. A.: Phys. Rev. Letters **1**, 450 (1958).
6. Rashba, E. I.: Soviet Phys. JETP **4**, 759 (1962).
7. Nikitine, S.: Phil. Mag. **4**, 1 (1959);—J. Chim. Phys. **55**, 43 (1958).
8. Thomas, D. G., Hopfield, J. J.: Phys. Rev. **128**, 2135 (1962).
9. Nikitine, S., Reiss, R.: J. Phys. Rad. Letters **20**, 718 (1959).
   Reiss, R.: Thesis (Strasbourg), Cahiers Phys. **13**, 129 (1959).
10. El Komoss, S. G., Stebe, B.: Nuovo Cimento. J. Phys. (in press).
    El Komoss, S. G.: J. Phys. Chem. Solid. **33**, 750 (1972);—Phys. Rev. **10**, 3411 (1971).
11. Nikitine, S., Ringeissen, J., Certier, M.: Acta Phys. Polon. **26**, 745 (1964).
    Reiss, R.: Thesis (Strasbourg), Cahiers Phys. **13**, 129 (1959).
    Lewonczuk, S., Ringeissen, J., Nikitine, S.: J. Phys. **32**, 941 (1971).
12. a) Myzyrowicz, A., Grun, J. B., Levy, R., Bivas, A., Nikitine, S.: Phys. Letters **26**A 615 (1968).
    b) Nikitine, S., Myzyrowicz, A., Grun, J. B.: Helv. Phys. Acta **41**, 1058 (1968).
    c) Myzyrowicz, A.: Thesis (Strasbourg) (1968).
    d) Bivas, A., Levy, R., Nikitine, S., Grun, J. B.: J. Phys. **31**, 227 (1970).
    e) Bivas, A.: Thesis (3ème Cycle) (Strasbourg) (1969).
    f) Knox, R. S., Nikitine, S., Myzyrowicz, A.: Opt. Comm. p. 19 (1969).
    g) Grun, J. B., Nikitine, S., Bivas, A., Levy, R.: J. Luminescence **1**, **2**, 241 (1970).
13. Sharma, R. R.: Phys. Rev. **170**, 770 (1968).
14. Wehner, R. K.: Solid State Comm. **9**, 457 (1969).
15. Akimoto, O., Hanamura, E.: Solid State Comm. **10**, 253 (1972).
    See also, Adamovski, Bednarek, Suffczynki: Solid State Comm. **9**, 2037 (1971).
16. Nikitine, S., Grun, J. B., Myzyrowicz, A.: Helv. Phys. Acta **41**, 1058 (1968).
    Myzyrowicz, A.: Thesis, Strasbourg (1968).
    Nikitine, S.: Non Linear Optics Colloquium, Titisee (1971).
17. Kazamanyan, Z. A.: Soviet Phys. **1**, 341 (1967).
18. Curie, D.: Champ cristallin et Luminescence, p. 191 and 196. Paris: Gauthier Villars 1968.
19. Souma, H., Koike, H., Kaoru, K., Suzuki, Ueta, M.: J. Phys. Soc. Japan (in press).
    See also, Goto, T., Ueta, M.: J. Phys. Soc. Japan **24**, 656 (1968).
20. Magde, D., Mahr, H.: Phys. Rev. Letters **24**, 890 (1970).
21. Levy, R., Bivas, A., Grun, J. B.: Phys. Letters **36**A, 159 (1971).
22. Benoit à la Guillaume, C., Debever, J., Salvan, F.: Phys. Rev. **177**, 567 (1969).
    See also Benoit à la Guillaume, C., Salvan, F., Voos: Intern. Confer. University Delaware (1969).
23. Levy, R., Grun, J. B.: In press.
24. Levy, R., Grun, J. B., Haken, H., Nikitine, S.: Solid State Comm. **10**, 915 (1972).

25. Levy, R., Bivas, A., Grun, J. B.: J. Phys. **31**, 507 (1970).
26. Haug, H.: J. Appl. Phys. **39**, 4681 (1968).
27. Levy, R., Bivas, A., Grun, J. B.: **36**A, 159 (1971).
28. a) Bivas, A., Levy, R., Grun, J. B., Comte, C., Haken, H., Nikitine, S.: Opt. Comm. **2**, 227 (1970).
b) Keldysh, L. V., Kozlov, A. N.: JETP **27**, 521 (1968).
Hanamura, E.: J. Phys. Soc. Japan **29**, 50 (1970).
29. Nikitine, S., Levy, R., Grun, J. B.: To be published.
30. Haynes, J. R.: Phys. Rev. Letters **17**, 860 (1966).
31. Gross, E. F., Kreingold, F. I.: JETP Letters **12**, 68 (1970).
Lvov, O. I., Pavinskii, P. P.: ZhETF Letters **14**, 253 (1971).
32. Shaklee, K. L., Leheny, R. F., Nahovy, R. E.: Phys. Rev. Letters **26**, 888 (1971).
33. Era, K., Langer, D. W.: J. Appl. Phys. **42** (3), 1021 (1971).
34. Goto, T., Langer, D. W.: Phys. Rev. Letters **27**, 1004 (1971).
35. Shigoe Shionoya, Hiroshi Saito, Eiichi Hanamura and Okikazu Akimoto (to be published).

Some Steady State Effects Associated with Electric Currents in Semiconductors

B. V. PARANJAPE

University of Alberta, Dept. of Physics, Edmonton/Canada

More than 30 years ago in connection with the theory of dielectric breakdown FRÖHLICH [1] investigated the conditions necessary for maintaining a steady state of electrons in a strong electric field. He found that beyond a certain critical value of the field these conditions could not be satisfied. Interest in the behaviour of electrons in fields below the critical field came later with discovery of transistors. Clearly his ideas are applicable in this region also.

FRÖHLICH [2] observed that an electric field acts on the free electrons directly. An external electric field transfers momentum and energy to the free electrons. If a steady state is reached electrons transfer this momentum and energy to the lattice vibrations at the same rate as they receive those from the field. Assuming that the lattice is maintained at a constant temperature through contact with a heat bath the lattice gets rid of this energy and momentum to the bath. Thus in a solid, one has an electron system and a system of lattice vibrations. In equilibrium, interaction between these two systems involving absorption and emission of lattice quanta or phonons, leads to no net exchange of energy or momentum. The electrons have a Maxwell Boltzmann distribution of energy and the phonons have a Planck distribution. In an electric field the electron distribution must deviate from its Maxwellian form if the electrons are to transfer energy and momentum to the lattice. FRÖHLICH [3] proved that a simple model for free electrons interacting with lattice vibrations only, can never lead to a steady state distribution function of electrons if a constant electric field however weak is present. Other processes leading to larger dissipation of energy from the high energy electrons must be introduced if a steady state distribution is to be found. One of these processes is collisions due to the mutual interaction between electrons. If free electron density is sufficiently high the Coulomb interaction between electrons can lead to an effective mechanism for the exchange of energy and momentum between particles of the electron system. FRÖHLICH thus introduced a very simple concept valid at high electron densities, the idea of an electron temperature T which may be higher than T_0 the temperature of the lattice. The system of electrons in an electric field can thus be described through a displaced Maxwell Boltzmann distribution at a temperature $T \neq T_0$. The displacement $\hbar \boldsymbol{k}_0$ in the momentum space is equal to the average momentum of an electron. Description of an electron system through a temperature is a reasonable approximation if the rate of loss of energy of an electron to the other electrons becomes a faster process than the rate of loss of its energy to the lattice vibrations. Using this condition FRÖHLICH showed that the electronic density necessary for the validity of such a model is often found in semiconductors. The parameters introduced in the model can be determined from the condition that the rate of transfer of momentum and energy from the field to the electrons must be equal to the rate of transfer of the same from the electron system to the phonon system. Thus

$$e\boldsymbol{F}n = \sum_{\boldsymbol{k}} \hbar \boldsymbol{k} \left.\frac{\partial f(\boldsymbol{k})}{\partial t}\right|_{\mathrm{ep}} \tag{1}$$

$$eFn\frac{\hbar k_0}{m} = \sum_{\boldsymbol{k}} \frac{\hbar^2 k^2}{2m} \left.\frac{\partial f(\boldsymbol{k})}{\partial t}\right|_{\mathrm{ep}} \tag{2}$$

where $\boldsymbol{F}$ is the external electric field, n is the no. density of electrons, $\boldsymbol{k}$ the wave vector of the electron, $f(\boldsymbol{k})$ the distribution of the electrons (assumed a displaced Maxwell Boltzmann distribution), $\partial f(\boldsymbol{k})/\partial t]_{\mathrm{ep}}$ is the rate of change of the distribution function due to the interaction of the electrons with the phonon system. Eqs. (1) and (2) correspond to the momentum and the energy balance of the electron system with the external field. The left hand side corresponds to the input and the right hand side to the output. Although Eqs. (1) and (2) are exact, detailed calculations for the evaluations of the right hand sides of these equations requires the use of time dependent perturbation theory. It then follows from the calculation of transition probability in which energy and momentum are conserved, that in a scattering process involving an electron and the phonon system the exchange of momentum is of the order of the momentum of the electron, but the exchange of energy is much smaller than the energy of the electron. The process is nearly elastic (more so for interactions with acoustic waves than for optical). Thus Eq. (2) involves smaller quantities compared with those in Eq. (1). Many problems of transport theory can be solved using the Boltzmann transport equation. However this equation cannot be solved exactly. Eqs. (1) and (2) are the conditions that a solution of the Boltzmann equation must satisfy if loss of energy and momentum from the electrons is only to the phonon system. FRÖHLICH has made a good guess of the approximate form of the solution of the Boltzmann's equation and imposed conditions (1) and (2) on the solution. Of the two parameters introduced in the solution the average momentum is more directly measurable than the electron temperature. One therefore uses Eq. (2) to estimate the electron temperature, since the right hand side of Eq. (2) is difficult to estimate accurately because of the small quantities involved. The estimate of temperature thus will not be too accurate. The temperature T will depend on external field and the drift velocity. Having determined T from Eq. (2) one then determines the drift velocity from Eq. (1) in which the temperature of the electron gas appears as a parameter. The dependence of the drift velocity on the applied electric field or the field dependent mobility thus calculated is in very good agreement with measurements of mobility in strong fields. Measurement of electronic temperature is harder, yet theoretical estimates of T are in qualitative agreement with experimental observations. It is surprising that calculations based on the above model give good agreement with experiments performed when the electron density is lower than the critical density.

The theory discussed above assumes that the sample is in a perfect heat bath; thus any excess phonons created by the free electrons are removed from the sample immediately. Such an over-simplified assumption is not always valid. Phonons induced by electrons modify the equilibrium distribution of phonons directly, just as an electric field modifies the distribution function of electrons. Thus in an electric field the phonon distribution function is modified due to a modified electron distribution function with the result that the mobility of electrons is changed. Such an effect was predicted and later confirmed by experiments about ten years ago. Much progress has been made in our understanding through further development of theory and experiments in materials with low thermal conductivity where induced phonon effects are noticeable. To obtain the steady state distribution function of electrons as well as that of phonons one must solve coupled Boltzmann transport equations for electrons and for phonons. Thus in a steady state taking into considerations only collision between electrons and phonons (ep), electrons and electrons (ee) and phonons and phonons (pp) one has

$$\left.\frac{\partial f(\boldsymbol{k})}{\partial t}\right|_{\text{Field}} + \left.\frac{\partial f(\boldsymbol{k})}{\partial t}\right|_{\text{ep}} + \left.\frac{\partial f(\boldsymbol{k})}{\partial t}\right|_{\text{ee}} = 0 \tag{3}$$

and

$$\left.\frac{\partial N_{\boldsymbol{q}}}{\partial t}\right|_{\text{ep}} + \left.\frac{\partial N_{\boldsymbol{q}}}{\partial t}\right|_{\text{pp}} = 0 \tag{4}$$

where $N_{\boldsymbol{q}}$ is the phonon distribution function. The suffixes indicate the cause of the rate of change of the distribution function. Many attempts [4] have been made to solve Eqs. (3) and (4) with partial success. The difficulty arises when the drift current is large and $\partial N_{\boldsymbol{q}}/\partial t|_{\text{pp}}$ is small (electron density of the order of 10^{15} cm^3). One assumes that a phonon-phonon relaxation time exists thus

$$\left.\frac{\partial N_{\boldsymbol{q}}}{\partial t}\right|_{\text{pp}} = -\frac{N_{\boldsymbol{q}} - N_{\boldsymbol{q}}^0}{\bar{\tau}_q} = -\frac{\xi_{\boldsymbol{q}}}{\bar{\tau}_q}\,. \tag{5}$$

Here $N_{\boldsymbol{q}}^0$ is the Planck distribution and $\bar{\tau}_q$ is a relaxation time defined through Eq. (5). $\xi_{\boldsymbol{q}}$ is the excess phonon distribution induced by the electrons, and hence is proportional to the no. density of electrons. ($\xi_q \neq 0$ in a small region of $\boldsymbol{q}$ space.)

If one assumes that phonon-phonon interaction is completely described through Eq. (5) one can always find a drift velocity of electrons above which Eq. (4) breaks down. At $\bar{\tau} \sim 10^{-5}$ sec and $n \sim 10^{15}$ cm^3 the critical drift velocity is of the order of s, the velocity of sound in the medium. At drift velocities $v > s$ the right hand side of Eq. (4) is not zero. Under these conditions one concludes that the induced phonon density increases to such a large value that non-linear process becomes important and approximation made in assuming Eq. (5) becomes invalid. One can thus interpret the observed phonon build up in materials like CdS. There is no systematic calculation that is valid above this critical drift velocity, largely because our knowledge of non-linear phonon-phonon interaction is poor. It is not quite clear how this interaction can be handled. From quite general considerations it can be seen that non-linear interaction is a fourth or a higher order process in calculation of phonon-phonon scattering, and therefore one would expect it to be important only at high phonon density. Because of the lack of detailed knowledge of the non-linear phonon-phonon interaction, conventional perturbation theory calculation cannot be performed.

A simple method of treating the problem of phonon excitation due to high electron drift velocity is proposed. In analogy with the method used by FRÖHLICH one can guess the approximate form of the solution of the phonon Boltzmann equation which will satisfy two momentum equations equivalent to Eqs. (1) and (2). In analogy with Eqs. (1) and (2) one has

$$\sum_{\boldsymbol{q}} \hbar \boldsymbol{q} \left(\frac{\partial N_{\boldsymbol{q}}}{\partial t}\bigg|_{\mathrm{pp}} + \frac{\partial N_{\boldsymbol{q}}}{\partial t}\bigg|_{\mathrm{ep}} \right) = 0 \tag{6}$$

and

$$\sum_{\boldsymbol{q}} \hbar s q \left(\frac{\partial N_{\boldsymbol{q}}}{\partial t}\bigg|_{\mathrm{pp}} + \frac{\partial N_{\boldsymbol{q}}}{\partial t}\bigg|_{\mathrm{ep}} \right) = 0 \tag{7}$$

where it is assumed that the phonons have a momentum $\hbar \boldsymbol{q}$ and energy $\hbar q s$, i.e. acoustic phonons. Since the phonon system receives all its momentum and energy from the electron system the rate of transfer of these by the electrons must be equal to the rate the electron system receives it from the field. Thus assuming that ep scattering is through normal process (not Umklapp),

$$\sum_{\boldsymbol{q}} \hbar \boldsymbol{q} \frac{\partial N_{\boldsymbol{q}}}{\partial t}\bigg|_{\mathrm{pp}} = e \boldsymbol{F} n \tag{8}$$

$$\sum_{\boldsymbol{q}} \hbar q s \frac{\partial N_{\boldsymbol{q}}}{\partial t}\bigg|_{\mathrm{pp}} = e F n v, \qquad [v = (\hbar k_0 | m)]. \tag{9}$$

An arbitrary distribution function of excited phonons can be expanded in terms of Lagendre polynomials. Thus

$$\xi_{\boldsymbol{q}} = \xi_q^0 + \xi_q^1 \cos \theta_{\boldsymbol{q}} + \dots \tag{10}$$

where

$$\cos \theta_{\boldsymbol{q}} = \boldsymbol{q} \cdot \boldsymbol{k}_0 / q k_0 .$$

Assuming (5) and substituting in Eqs. (8) and (9) one obtains

$$\int_0^\infty \left(\frac{1}{3} \xi_q^1 - \frac{s}{v} \xi_q^0 \right) \frac{1}{\tau_q} q^3 \, dq = 0. \tag{11}$$

Since Eq. (11) must hold for arbitrary values of v it is reasonable to assume

$$\xi_q^1 = \frac{3s}{v} \xi_q^0 . \tag{12}$$

For further development one assumes Maxwell Boltzmann distribution for electrons and a distribution of the form (10) for the phonons. Substituting these distributions in Eqs. (1) and (2) using Eq. (12) one obtains two equations containing two unknown parameters v and T, and an unknown function ξ_q^0. These equations can be written in a manageable form if some reasonable approximations are made. Thus

$$e F = \frac{m v}{\tau(T, T_0)} \left(1 + \frac{\hbar s q_0}{k T_0} \xi_{q_0}^0 \left(1 - \frac{3 s^2}{v^2} \right) \right) \tag{13}$$

and

$$e F v = \frac{m v^2}{\tau(T, T_0)} \frac{2 m s^2}{k T} \left(1 + \frac{\hbar s q_0}{4 k T_0} \xi_{q_0}^0 \left[1 + \frac{3s}{v} \sqrt{\frac{8 k T}{m v^2}} \right] \right) + \frac{3 m s^2}{\tau(T, T_0)} \left(\frac{T}{T_0} - 1 \right); \tag{14}$$

where $q_0 = \sqrt{8mkT}/\hbar$ and $\tau(T, T_0)$ is the conductivity relaxation time for electrons of energy kT when interacting with phonons having a Planck distribution at temperature T_0. Thus for pure piezoelectric interaction $\tau = \tau^0 (T/T_0)^{\frac{1}{2}}$ where τ^0 is the conductivity relaxation time due to thermal phonons only (for deformation potential interaction $\tau = \tau_0 (T/T_0)^{-\frac{1}{2}}$). In writing Eqs. (13) and (14) integrals of the form

$$\int_0^\infty \exp\left[-(q^2/q_0^2)\right] f(q)\, q^2\, dq$$

have been replaced by $f(q_0) \cdot q_0^3$. The electron distribution function has been expanded in terms of $(\hbar^2 k_0^2/2mkT)^{\frac{1}{2}}$ and terms up to $(\hbar^2 k_0^2/2mkT)$ have been retained. $\hbar s q_0/kT_0$ is assumed less than unity (high temperature approximation). In $\xi_{\boldsymbol{q}}$ only two terms in the expansion have been retained. One can solve for v or T from Eqs. (13) and (14) if one knew ξ_q^0. It is easy to see that an electron of wave vector $\boldsymbol{k}$ interacts most strongly with phonons of wave vector $2\boldsymbol{k}$. Since the magnitude of the average wave vector of electrons is $\sqrt{2mkT}/\hbar$ one expects that the induced phonons will have a distribution which peaks around $2\sqrt{2mkT}/\hbar = q_0$. As a trial form of the function one may assume a Gaussian distribution of induced phonons

$$\xi_q^0 = A \exp\left[-(q^2/q_0^2)\right], \tag{15}$$

where the constant A is determined from Eqs. (9) and (5). Making further simplification $\bar{\tau}_q$ is independent of q one finds

$$A = \frac{v}{s}\left(\frac{eF\bar{\tau}\, 4\pi^2 n}{\hbar q_0^4}\right). \tag{16}$$

Substituting for ξ_q^0 from Eqs. (15) and (16) in Eqs. (13) and (14) one obtains

$$eF\left(1 - \frac{mv^2}{kT_0}\frac{\bar{\tau}}{\tau} \cdot \frac{4\pi^2 n}{q_0^3}\left(1 - \frac{3s^2}{v^2}\right)\right) = \frac{mv}{\tau} \tag{17}$$

and

$$eFv\left(1 - \frac{mv^2}{kT} \cdot \frac{ms^2}{kT_0}\frac{\bar{\tau}}{\tau}\frac{4\pi^2 n}{q_0^3}\left(1 + \frac{6s}{v}\sqrt{\frac{2kT}{mv^2}}\right)\right) = \frac{ms^2}{\tau}\left(\frac{2mv^2}{kT} + 3\left(\frac{T}{T_0} - 1\right)\right). \tag{18}$$

One can solve for T and v from Eqs. (17) and (18). It is expected that the model discussed above of induced phonons should be good only at large induced phonon densities and therefore at large drift velocities $v > s$. However, it can be shown that one gets good results even at low drift velocities. In the limit of vanishing electric fields one observes from equation (18) that $T \cong T_0$. Substituting $T = T_0$ in Eq. (17) one finds that the low field mobility

$$\mu_0 = \underset{F \to 0}{\mathrm{Lt}} \quad \frac{v}{F} = \frac{e\tau_0}{m}(1 + \alpha) \tag{19}$$

where

$$\alpha = (ms^2\bar{\tau}\, 12\pi^2 n/kT_0\, \tau q_0^3)\big|_{T=T_0}. \tag{20}$$

Eq. (19) agrees well with the results obtained from a more exact calculation. One notes that the effect of induced phonons is noticeable even at low fields if α is greater than or of order 1. In this case one expects that the electric current starts from a low value corresponding to no induced phonons and builds up to a saturation in value. In strong fields explicit dependence of τ on T_0 and T has to be considered to solve Eqs. (17) and (18). One should also consider the possibility that in the case of piezoelectric materials a significant increase in the value of T will change the dominant interaction of the electrons with phonons from piezoelectric to deformation. In the special case when $v^2 = 3s^2$ one can show that for $\alpha \gg 1$ no value of $T > T_0$ exists that will satisfy Eqs. (17) and (18). Thus in this case a steady state solution does not exist. Such a situation arises at a field $F \cong (ms/e\tau_0)$.

References

1. Fröhlich, H.: Proc. Roy. Soc. (London), Ser. A **160**, 280 (1937).
2. Fröhlich, H.: Proc. Roy. Soc. (London), Ser. A **188**, 521 (1947).
3. Fröhlich, H.: Proc. Roy. Soc. (London), Ser. A **188**, 532 (1947).
4. Conwell, E. M.: Solid State Physics, Suppl. 9 (1967).

The Free Energy and Dynamical Structure Factors of a Metallic System

A. B. Bhatia

University of Alberta, Dept. of Physics, Edmonton/Canada

1. Introduction

In 1950, Fröhlich [1] showed that the same Bloch electron-phonon interaction [2] which is responsible for the electrical resistivity of a metal, leads also, if strong enough, to a net attractive interaction between the electrons and is thus responsible for the phenomenon of superconductivity. This fundamental idea, as is, of course, well known, forms the basis of all subsequent work on superconductivity. It introduced a new element in the theory of normal metals also, namely, that the electron-phonon interaction affects the electron density of states of a normal metal and hence its electronic specific heat etc. [1, 3, 4]. The present Chapter is partly concerned with this effect and partly with the change in the electronic specific heat due to alloying.

2. Method and Scope

Our discussion of the aforementioned topics is essentially based on the ideas which underlie the current method [5–8] of claculating the electrical resistivities of simple metals (defined below). This method relates the electrical resistivity of a liquid or solid metal to its respective structure factor and is based on the assumptions: (1) that the wave functions of the conduction (valence) electrons can be described by plane waves, and (2) that the effective interaction between the conduction electrons and the ions can be written as a sum of the effective potentials due to the individual ions in the metal and that it can be considered as weak. On this basis the theory of scattering of the conduction electrons in the metal is the same as the theory of x-ray or neutron scattering and the relevant scattering function may be factored into an atom form factor, which depends only on the properties of the individual ion in the metal, and a structure factor. The latter depends only on the relative positions of the ions and is the same as for x-ray or neutron scattering.

The atom form factors for the valence electrons are now provided with considerable accuracy by the pseudopotential theories initiated by Phillips and Kleinman [9] and Antonick [10] in connection with the band structure calculations. The true potential $V(r)$ in which an electron moves is, of course, not weak near the nucleus of an ion. However, the fact that the wave functions of the valence electrons must be orthogonal to the wave functions of the core electrons is equivalent to a repulsive potential V_R in the core region, which more or less cancels out the true strong attractive interaction near the nucleus. The resulting effective potential $v(r) = V(r) + V_R$ is considerably weaker than $V(r)$ and is usually referred to as the pseudopotential. The pseudopotential atom form factors have been calculated (and tabulated) in the literature under varying sets of approximations and the topic has been excellently reviewed recently by HEINE [11] and COHEN and HEINE [12]. Here we need to make only the following remarks.

In order that the atom form factors represent weak scattering and remain unaffected by any small local deviations in the configuration of the ions from equilibrium (due to the screening effects etc., they depend on the overall density of the metal), it is necessary that the wave functions of the core electrons from the adjacent atoms do not appreciably overlap one another. The metals, namely, the alkalis and the polyvalent metals, for which this "small core" approximation is valid are called the simple metals and have small band gaps. The approximation becomes progressively worse for the noble metals and the transition metals. For the case of alloys of simple metals it has to be remembered in any quantitative calculation, that the atom form factors will depend in general, on the relative concentration of the ions (and also possibly on order parameters), although the actual calculations to date are available only for the dilute alloys [13, 14].

Returning to the problem as to how the electronic density of states (and hence the specific heat etc.) is affected by the thermal motion of the ions and by alloying in the above approach, we have to calculate the contribution to the free energy of the metal (alloy) due to the weak effective (pseudopotential) electron-ion interaction. In the next Section, it is shown that the second order correction to the free energy due to this interaction is related to the scattering function, which determines the inelastic scattering of the conduction electrons, and hence to the dynamical structure factor(s) of a metal (alloy). This is followed by a brief discussion of the structure factors for an alloy (§ 4). Finally, in § 5, we discuss how this formulation compares with and leads to some of the results on electronic density of states and specific heat obtained previously in the literature by other methods.

3. Expression for the Free Energy

Our system consists of the ions and the conduction electrons, whose wave functions are described by plane waves, interacting *via* the weak pseudopotential electron-ion interaction. We denote the Hamiltonian of the ions by $\mathscr{H}_I(\mathbf{R})$, where $\mathbf{R}$ stands for all the ionic coordinates $\mathbf{R}_1, \mathbf{R}_2, \ldots, \mathbf{R}_N$. It is not necessary to specify the form of $\mathscr{H}_I(\mathbf{R})$ at this stage. The interaction energy due to the pseudopotential interaction is of the form

$$H_{\text{int}} = \sum_i \mathscr{H}_{\text{int}}(\mathbf{r}_i, \mathbf{R}); \quad \mathscr{H}_{\text{int}}(\mathbf{r}, \mathbf{R}) = \sum_{j=1}^{N} v_j(\mathbf{r} - \mathbf{R}_j), \tag{1}$$

where $\mathbf{r}_i$ denotes the coordinate of the i-th electron and $v_j(\mathbf{r} - \mathbf{R}_j)$ is the pseudopotential of the j-th ion located at $\mathbf{R}_j$. We write the matrix elements for $v_j(\mathbf{r})$

between the electron plane wave states, normalised to volume V of the system, as

$$w_j(\mathbf{k}, \mathbf{q}) \equiv N V^{-1} \int_V e^{-i\mathbf{k}' \cdot \mathbf{r}} v_j(\mathbf{r})\, e^{i\mathbf{k} \cdot \mathbf{r}}\, d^3 r, \qquad \mathbf{q} = \mathbf{k} - \mathbf{k}', \tag{2}$$

where N is the total number of atoms in volume V. The integral in (2) depends, in general, on both $\mathbf{k}'$ and $\mathbf{k}$ and not just on $\mathbf{q}$ since the pseudopotential $v_j(\mathbf{r})$ is generally non-local. The $w_j(\mathbf{k}, \mathbf{q}=0)$ affect the density of states in the first order of perturbation [13, 15] and we abbreviate

$$W(\mathbf{k}) = N^{-1} \sum_j w_j(\mathbf{k}, \mathbf{q}=0). \tag{3}$$

For $\mathbf{q} \neq 0$, we shall, for simplicity, neglect the effects of non-locality so that $w_j(\mathbf{k}, \mathbf{q}) = w_j(\mathbf{q})$.

It is convenient to work with the total Hamiltonian H of the system using the second quantisation scheme for the electrons. Remembering that the Fermion field operators are

$$\psi(\mathbf{r}) = V^{-\frac{1}{2}} \sum_{\mathbf{k}} C_{\mathbf{k}} \exp[i\mathbf{k} \cdot \mathbf{r}], \qquad \psi^+(\mathbf{r}) = V^{-\frac{1}{2}} \sum_{\mathbf{k}} C^+_{\mathbf{k}} \exp[-i\mathbf{k} \cdot \mathbf{r}],$$

where $C_{\mathbf{k}}$ is the Fermi annihilation operator and $C^+_{\mathbf{k}}$ the creation operator, one may write, using 1–3,

$$H = H_0 + H_1, \tag{4}$$

where

$$H_0 = \mathscr{H}_I(\mathbf{R}) + \sum_{\mathbf{k}} C^+_{\mathbf{k}} C_{\mathbf{k}} E_{\mathbf{k}}, \qquad E_{\mathbf{k}} = \varepsilon_{\mathbf{k}} + W(\mathbf{k}), \tag{5}$$

$$H_1 = N^{-1} \sum_{\mathbf{k}\mathbf{k}'}{}' C^+_{\mathbf{k}'} C_{\mathbf{k}} A(\mathbf{q}, \mathbf{R}), \tag{6}$$

with

$$A(\mathbf{q}, \mathbf{R}) = \sum_j w_j(\mathbf{q}) \exp[i\mathbf{q} \cdot \mathbf{R}_j], \tag{7}$$

and $\varepsilon_{\mathbf{k}} = \hbar^2 k^2 / 2m$. The prime on the summation sign in (6) means that the terms $\mathbf{k} = \mathbf{k}'$ are excluded from the sum. We regard H_1 as a perturbation.

Now quite generally, if ϕ_0 is the free energy of a system with Hamiltonian H_0 and ϕ_1 and ϕ_2 are respectively the first and second order corrections to the free energy due to a perturbation H_1, then ϕ_1 and ϕ_2 are given by (SCHAFROTH [16], BUCKINGHAM and SCHAFROTH [17])

$$\phi_1 = \sum_\alpha \langle\alpha| H_1 |\alpha\rangle \exp[\beta(\phi_0 - E_\alpha)], \tag{8}$$

$$\phi_2 = \frac{1}{2} \beta \phi_1^2 + \sum_{\alpha\alpha'} |\langle\alpha| H_1 |\alpha'\rangle|^2 \left(\frac{1 - \exp[\beta(E_\alpha - E_{\alpha'})]}{2(E_\alpha - E_{\alpha'})}\right) \exp[\beta(\phi_0 - E_\alpha)], \tag{9}$$

where $H_0 |\alpha\rangle = E_\alpha |\alpha\rangle$ and $\beta = (k_B T)^{-1}$.

For H_0 given by (5), the eigenstates $|\alpha\rangle$ are the direct products of the ionic states $|n\rangle$, ($\mathscr{H}_I(\mathbf{R}) |n\rangle = E_n |n\rangle$), and the many body electron states

$$|\sigma_{k_1} \sigma_{k_2} \ldots \sigma_{k_m} \ldots\rangle, \; \sigma_{k_m} = 0 \quad \text{or} \quad 1.$$

Then $\langle\alpha| H_1 |\alpha\rangle = 0$ for all $|\alpha\rangle$ and hence the first order correction to the free energy $\phi_1 = 0$. Next noting that $\exp[\beta(\phi_0 - E_\alpha)]$ is the statistical factor for the unperturbed system and can be factored into the electronic and ionic statistical factors, we obtain (including a factor 2 for spin)

$$\phi_2 = \frac{1}{N^2} \sum_{\mathbf{k}\mathbf{k}'}{}' f_{\mathbf{k}}(1 - f_{\mathbf{k}'}) \sum_{n n'} p_n |\langle n'| A(\mathbf{q}, \mathbf{R}) |n\rangle|^2 \left(\frac{1 - \exp[\beta\hbar(\omega_{\mathbf{k}\mathbf{k}'} - \omega_{n'n})]}{\hbar(\omega_{\mathbf{k}\mathbf{k}'} - \omega_{n'n})}\right), \tag{10}$$

where $f_{\mathbf{k}}$ is the Fermi-Dirac distribution function, $\hbar\omega_{\mathbf{k}\mathbf{k}'} = E_{\mathbf{k}} - E_{\mathbf{k}'}$, $\hbar\omega_{n'n} = E_{n'} - E_n$ and p_n is the probability of finding the ions in state $|n\rangle$.

Now define a function $\Gamma(\mathbf{q}, \omega)$ by

$$\Gamma(\mathbf{q}, \omega) = N^{-1} \sum_{n n'} p_n |\langle n'| A(\mathbf{q}, \mathbf{R}) |n\rangle|^2 \, \delta(\omega - \omega_{n'n}), \tag{11}$$

which, following the arguments given by VAN HOVE [18] in connection with neutron scattering, may be written as

$$\begin{aligned}\Gamma(\mathbf{q}, \omega) &= (2\pi N)^{-1} \int \sum_{n n'} p_n |\langle n'| A(\mathbf{q}, \mathbf{R}) |n\rangle|^2 \{\exp[-i(\omega - \omega_{n'n})\, t\}\, dt \\ &= (2\pi N)^{-1} \int \exp[-i\omega t]\, dt \langle A^+(\mathbf{q}, \mathbf{R}(0))\, A(\mathbf{q}, \mathbf{R}(t))\rangle_T, \end{aligned} \tag{12}$$

where $\langle\ldots\rangle_T$ denotes the ensemble average over the unperturbed ionic system, $A^+(\mathbf{q}, \mathbf{R}(0))$ is the conjugate to the operator $A(\mathbf{q}, \mathbf{R}(0))$, and the ionic coordinates $\mathbf{R}(t)$ are the Heisenberg operators at time t:

$$\mathbf{R}_j(t) = e^{i\mathscr{H}_I t/\hbar}\, \mathbf{R}_j(0)\, e^{-i\mathscr{H}_I t/\hbar}.$$

Using (11), Eq. (10) becomes

$$\phi_2 = \frac{1}{N} \sum_{\mathbf{k}\mathbf{k}'}{}' f_{\mathbf{k}}(1 - f_{\mathbf{k}'}) \int_{-\infty}^{\infty} d\omega \left(\frac{1 - \exp[\beta\hbar(\omega_{\mathbf{k}\mathbf{k}'} - \omega)]}{\hbar(\omega_{\mathbf{k}\mathbf{k}'} - \omega)}\right) \Gamma(\mathbf{q}, \omega), \tag{13}$$

which is the relation we wished to derive here and is due to BHATIA and O'LEARY [19]. (Note that $w_j(\mathbf{q})$ [Eq. (2)] is defined here with a different normalisation factor than in Ref. [19]).

The function $\Gamma(\mathbf{q}, \omega)$, which appears also in the expression for the resistivity of an alloy [20, 21], essentially determines the scattering of conduction electrons or neutrons (with appropriate $w_j(\mathbf{q})$), per unit solid angle, per unit energy range $\hbar\omega$. In fact, for a pure metal with all $w_j(\mathbf{q})$ equal to, say $w(\mathbf{q})$,

$$\Gamma(\mathbf{q}, \omega) = |w(\mathbf{q})|^2\, S(\mathbf{q}, \omega), \tag{14}$$

where

$$S(\mathbf{q}, \omega) = \frac{1}{2\pi N} \int e^{-i\omega t}\, dt \langle \sum_i e^{-i\mathbf{q}\cdot\mathbf{R}_i(0)} \sum_j e^{i\mathbf{q}\cdot\mathbf{R}_j(t)}\rangle_T, \tag{15}$$

is the well known VAN HOVE [18] dynamical structure factor. We express $\Gamma(\mathbf{q}, \omega)$ for an alloy in terms of its dynamical structure factors (DSF) in the next section. Like $S(\mathbf{q}, \omega)$, these DSF are also the property of the unperturbed ionic system alone, and if they are known either theoretically or experimentally from thermal neutron and x-ray scattering data, one may calculate ϕ_2 from (13) and hence the second order corrections to other thermodynamic properties of the system.

It will be apparent that Eq. (13) applies equally to a solid or liquid, but in so far as it is based on perturbative approach, any contribution to $\Gamma(\mathbf{q}, \omega)$ from the elastic Bragg reflections in a solid should be excluded, the effect of these being separately calculated as band structure. In the following this will be understood.

4. Dynamical Structure Factors for an Alloy

A convenient way of formally defining the DSF for an alloy is by considering the local deviations in the number density of the different types of ions in it.

Consider an alloy having N_α, $\alpha=1, 2, \ldots \nu$, atoms of the type α, and let $\bar{n}_\alpha = N_\alpha/V$ be the mean number density of α atoms. The average concentration c_α of α atoms is

$$c_\alpha = N_\alpha/N = \bar{n}_\alpha/\bar{n}, \tag{16}$$

where $N=\sum_\alpha N_\alpha$ and $\bar{n}=\sum_\alpha \bar{n}_\alpha$. If $n_\alpha(\mathbf{r}, t)$ denote the number density operator at time t for the species α, then

$$\begin{aligned} \delta n_\alpha(\mathbf{r}, t) &\equiv n_\alpha(\mathbf{r}, t) - \bar{n}_\alpha \\ &= -\bar{n}_\alpha + \sum_j \delta\left(\mathbf{r} - \mathbf{R}_j^\alpha(t)\right), \end{aligned} \tag{17}$$

where the sum is over all the α atoms. Making the Fourier expansion

$$\delta n_\alpha(\mathbf{r}, t) = (1/V) \sum_{\mathbf{q}} N_\alpha(\mathbf{q}, t) \exp\left[-i\,\mathbf{q}\cdot\mathbf{r}\right], \tag{18}$$

$$\begin{aligned} N_\alpha(\mathbf{q}, t) &= \int \delta n_\alpha(\mathbf{r}, t) \exp\left[i\,\mathbf{q}\cdot\mathbf{r}\right] d^3r \\ &= \sum_j e^{i\,\mathbf{q}\cdot\mathbf{R}_j^\alpha(t)} - N_\alpha \delta_{\mathbf{q},0}. \end{aligned} \tag{19}$$

The reality of $\delta n_\alpha(\mathbf{r}, t)$ requires that $N_\alpha^+(\mathbf{q}, t) = N_\alpha(-\mathbf{q}, t)$.

In terms of $N_\alpha(\mathbf{q}, t)$, the expression (7) for $A(\mathbf{q}, \mathbf{R})$ becomes

$$A(\mathbf{q}, \mathbf{R}(t)) = \sum_\alpha w_\alpha N_\alpha(\mathbf{q}, t), \tag{20}$$

where $w_\alpha \equiv w_\alpha(\mathbf{q})$ and where we have ignored a term containing $\delta_{\mathbf{q},0}$ since its presence or absence does not affect the value of ϕ_2. Then $\Gamma(\mathbf{q}, \omega)$, defined in (12), may be written as ($\alpha, \beta = 1, 2, \ldots, \nu$),

$$\Gamma(\mathbf{q}, \omega) = \sum_{\alpha,\beta=1}^{\nu} w_\alpha^* w_\beta (c_\alpha c_\beta)^{\frac{1}{2}} \left[S_{\alpha\beta}(\mathbf{q}, \omega) + S_{\alpha,\beta}^{(a)}(\mathbf{q}, \omega)\right], \tag{21}$$

where the S's are the dynamical structure factors defined as

$$\begin{aligned} S_{\alpha\beta}(\mathbf{q}, \omega) = (4\pi N)^{-1} (c_\alpha c_\beta)^{-\frac{1}{2}} \int e^{-i\omega t} dt \langle N_\alpha^+(\mathbf{q}, 0)\, N_\beta(\mathbf{q}, t) \\ + N_\beta^+(\mathbf{q}, 0)\, N_\alpha(\mathbf{q}, t)\rangle_T, \end{aligned} \tag{22}$$

and

$$\begin{aligned} S_{\alpha\beta}^{(a)}(\mathbf{q}, \omega) = (4\pi N)^{-1} (c_\alpha c_\beta)^{-\frac{1}{2}} \int e^{-i\omega t} dt \langle N_\alpha^+(\mathbf{q}, 0)\, N_\beta(\mathbf{q}, t) \\ - N_\beta^+(\mathbf{q}, 0)\, N_\alpha(\mathbf{q}, t)\rangle_T. \end{aligned} \tag{23}$$

It may be verified that irrespective of the nature of the ionic system,

$$S_{\alpha\beta}(\mathbf{q}, \omega) \, [= S_{\beta\alpha}(\mathbf{q}, \omega)]$$

are real valued functions and that $S^{(a)}_{\alpha\beta}(\mathbf{q}, \omega) \, [= -S^{(a)}_{\beta\alpha}(\mathbf{q}, \omega)]$ are pure imaginary. The symmetry of the system may, in general, be expected to imply that $S^{(a)}_{\alpha\beta}(\mathbf{q}, \omega) = 0$. Hence (21) becomes

$$\Gamma(\mathbf{q}, \omega) = \sum_{\alpha,\beta=1}^{\nu} w^*_\alpha w_\beta (c_\alpha c_\beta)^{\frac{1}{2}} S_{\alpha\beta}(\mathbf{q}, \omega). \tag{24}$$

If all the w_α have same phase, i.e., $w_\alpha = |w_\alpha| \exp[i\gamma]$, where γ is the same for all α, then (24) is true irrespective of whether $S^{(a)}_{\alpha\beta}$ are zero or not.

The total (elastic + inelastic) scattering in a given direction $\mathbf{q}$ is determined by the function $\Gamma(\mathbf{q}) = \int \Gamma(\mathbf{q}, \omega) \, d\omega$. Since for any smooth function $f(t)$,

$$\int_{-\infty}^{\infty} \exp[-i\omega t] \, f(t) \, d\omega \, dt = 2\pi f(0)$$

one has, from (22) and (24),

$$\Gamma(\mathbf{q}) = \sum_{\alpha,\beta} w^*_\alpha w_\beta (c_\alpha c_\beta)^{\frac{1}{2}} S_{\alpha\beta}(\mathbf{q}), \tag{25}$$

where the static structure factors $S_{\alpha\beta}(\mathbf{q})$ are given by

$$S_{\alpha\beta}(\mathbf{q}) = \int S_{\alpha\beta}(\mathbf{q}, \omega) \, d\omega \tag{26}$$

$$= (2N)^{-1} (c_\alpha c_\beta)^{-\frac{1}{2}} \langle N^*_\alpha(\mathbf{q}) \, N_\beta(\mathbf{q}) + N^*_\beta(\mathbf{q}) \, N_\alpha(\mathbf{q}) \rangle_T. \tag{27}$$

In (27), all the $N(\mathbf{q})$ refer to the same time and hence need not be considered as quantum operators. Since $N^*_\alpha(\mathbf{q}) = N_\alpha(-\mathbf{q})$, we have, by definition, $S_{\alpha\beta}(\mathbf{q}) = S_{\alpha\beta}(-\mathbf{q})$.

For an alloy containing ν types of atoms, the expression (24) for $\Gamma(\mathbf{q}, \omega)$ contains $\frac{1}{2}\nu(\nu+1)$ DSF. The general calculation of these, even for a binary alloy, is quite difficult and we shall consider a few special cases in the next section. For this discussion it will be found useful to introduce a related set of DSF for a binary alloy and consider the high temperature and long wavelength limit ($\mathbf{q} \to 0$) of the corresponding static structure factors [20, 21].

Let $N(\mathbf{q}, t)$ denote the Fourier transform of the local deviation $\delta n(\mathbf{r}, t)$ in the total number density

$$n(\mathbf{r}, t) = n_1(\mathbf{r}, t) + n_2(\mathbf{r}, t),$$

then

$$\begin{aligned} N(\mathbf{q}, t) &= \sum_{j,\alpha=1,2} \exp[i\mathbf{q} \cdot \mathbf{R}^\alpha_j(t)] - N\delta_{\mathbf{q},0} \\ &= N_1(\mathbf{q}, t) + N_2(\mathbf{q}, t). \end{aligned} \tag{28}$$

We next define the local deviation from the mean concentration $c \, (\equiv c_1)$ of species 1 by

$$\delta c(\mathbf{r}, t) = (V/N) \, [(1-c) \, \delta n_1(\mathbf{r}, t) - c \, \delta n_2(\mathbf{r}, t)], \tag{29}$$

so that if δn_1 and δn_2 change in proportion to their respective mean concentration, namely, c and $(1-c)$, then $\delta c(\mathbf{r}, t)=0$. If we make the Fourier expansion

$$\delta c(\mathbf{r}, t)=\sum_{\mathbf{q}} C(\mathbf{q}, t) \exp[-i\,\mathbf{q}\cdot\mathbf{r}], \qquad C^{+}(\mathbf{q}, t)=C(-\mathbf{q}, t), \tag{30}$$

then

$$\begin{aligned} C(\mathbf{q}, t) &= (1/V) \int \delta c(\mathbf{r}, t) \exp[i\,\mathbf{q}\cdot\mathbf{r}]\, d^3r \\ &= (N^{-1})\, [(1-c)\, N_1(\mathbf{q}, t) - c N_2(\mathbf{q}, t)]. \end{aligned} \tag{31}$$

Making use of (28), (31) and (20) in (12), the expression for $\Gamma(\mathbf{q}, \omega)$, corresponding to (24), may be written as

$$\begin{aligned} \Gamma(\mathbf{q}, \omega) &= |\overline{w}|^2\, S_{NN}(\mathbf{q}, \omega) + |w_1 - w_2|^2\, S_{CC}(\mathbf{q}, \omega) \\ &\quad + [\overline{w}\,(w_1^* - w_2^*) + (\overline{w})^*\,(w_1 - w_2)]\, S_{NC}(\mathbf{q}, \omega), \end{aligned} \tag{32}$$

where $\overline{w}=c w_1+(1-c)\, w_2$ and

$$\begin{aligned} S_{NN}(\mathbf{q}, \omega) &= (1/2\,\pi N) \int \exp[-i\omega t]\, dt \langle N^{+}(\mathbf{q}, 0)\, N(\mathbf{q}, t)\rangle_T, \\ S_{CC}(\mathbf{q}, \omega) &= (N/2\,\pi) \int \exp[-i\omega t]\, dt \langle C^{+}(\mathbf{q}, 0)\, C(\mathbf{q}, t)\rangle_T, \\ S_{NC}(\mathbf{q}, \omega) &= (1/4\,\pi) \int \exp[-i\omega t]\, dt \langle N^{+}(\mathbf{q}, 0)\, C(\mathbf{q}, t) + C^{+}(\mathbf{q}, 0)\, N(\mathbf{q}, t)\rangle_T. \end{aligned} \tag{33}$$

Remembering (28), we see that $S_{NN}(\mathbf{q}, \omega)$ is similar to the DSF $S(\mathbf{q}, \omega)$ for a pure metal [Eq. (15)], and it essentially represents correlations between the fluctuations in the number density of the particles. Similarly, $S_{CC}(\mathbf{q}, \omega)$ represents correlations between the fluctuations in the concentration and $S_{NC}(\mathbf{q}, \omega)$ the cross correlation between these two fluctuations. The relations between these number-concentration DSF and $S_{\alpha\beta}(\mathbf{q}, \omega)$, $\alpha, \beta=1, 2$, may be obtained by comparing the coefficients of $w_\alpha^* w_\beta$ in the expressions (24) and (32) for $\Gamma(\mathbf{q}, \omega)$. One obtains $(c_1+c_2=1)$

$$\begin{aligned} c_1 S_{11} &= c_1^2 S_{NN} + S_{CC} + 2 c_1\, S_{NC}, \\ c_2 S_{22} &= c_2^2 S_{NN} + S_{CC} - 2 c_2\, S_{NC}, \\ (c_1 c_2)^{\frac{1}{2}}\, S_{12} &= c_1 c_2 S_{NN} - S_{CC} + (c_2 - c_1)\, S_{NC}. \end{aligned} \tag{34}$$

From the definitions (28) and (31) of $N(\mathbf{q})$ and $C(\mathbf{q})$ respectively it is easy to see that as $\mathbf{q}\to 0$, the static structure factors $S_{NN}(\mathbf{q})=\int S_{NN}(\mathbf{q}, \omega)\, d\omega$ etc. have a simple physical meaning for a fluid, namely,

$$\begin{aligned} &S_{NN}(0)=\langle(\Delta N)^2\rangle/N, \qquad S_{CC}(0)=N\langle(\Delta c)^2\rangle \\ &S_{NC}(0)=\langle \Delta N \Delta c\rangle. \end{aligned} \tag{35}$$

In (35), $\langle(\Delta N)^2\rangle$ is the mean square fluctuation in the number of particles in the volume V of the medium, $\langle(\Delta c)^2\rangle$, the mean square fluctuation in concentration, Δc being defined by $\Delta c=N^{-1}[(1-c)\,\Delta N_1 - c\,\Delta N_2]$, and $\langle \Delta N \Delta c\rangle$ is the correlation between the two fluctuations Δc and ΔN. One can therefore readily calculate $S_{NN}(0)$ etc. from the thermodynamic theory of fluctuations. At high temperatures (above the Debye temperature), $S_{NN}(\mathbf{q}\to 0)$ etc. for a solid also may be calculated from the fluctuation theory, except that one now has to take into account the full complement of the stress and strain variables rather than just pressure and volume as for a fluid. The calculations are lengthy and may be found in BHATIA and THORNTON [21]. We quote the results here.

First, if $\mathbf{y}$ denote the unit vector parallel to $\mathbf{q}$, then as $\mathbf{q}\to 0$, $S_{NN}(\mathbf{y})$ may be written as

$$S_{NN}(\mathbf{y})=(N/V)\,k_B T\varkappa(\mathbf{y})+[\varDelta(\mathbf{y})]^2\,S_{CC}(\mathbf{y}), \tag{36}$$

where

$$\varDelta(\mathbf{y})=S_{NC}(\mathbf{y})/S_{CC}(\mathbf{y}), \tag{37}$$

and

$$\varkappa(\mathbf{y})=\sum_{r=1}^{3}\frac{(\mathbf{e}^{(r)}\cdot\mathbf{y})^2}{\varrho\mathscr{V}_r^2}. \tag{38}$$

In (38), ϱ is the density of the material, $\mathbf{e}^{(r)}\equiv\mathbf{e}^{(r)}(\mathbf{y})$ is the unit polarization (displacement) vector associated with the particular elastic wave propagating along the $\mathbf{y}$-direction whose velocity is $\mathscr{V}_r\equiv\mathscr{V}_r(\mathbf{y})$, $\mathscr{V}_r$ being calculated using the isothermal (rather than adiabatic) elastic constants. We observe that if the second term in (36) is zero, $S_{NN}(\mathbf{y})=(N/V)\,k_B T\varkappa(\mathbf{y})$ which has the form of the well known Faxen-Waller expression for the structure factor (in the limit under consideration) of a pure crystal, usually obtained by expanding the ionic coordinates in terms of the phonon coordinates; for references and for the explicit form of $\varkappa(\mathbf{y})$ in terms of the elastic constants, see [6].

Eqs. (36) and (38) are valid for a crystal of any symmetry. The expressions for $S_{CC}(\mathbf{y})$ and $S_{NC}(\mathbf{y})$ or $\varDelta(\mathbf{y})$ are considerably simpler for the case of a crystal of cubic symmetry with no residual strains and it will suffice here to quote these. One has

$$\varDelta(\mathbf{y})=-\delta\,B_T\,\varkappa(\mathbf{y}), \tag{39}$$

and

$$S_{CC}(\mathbf{y})=\frac{N k_B T}{(\partial^2 G/\partial c^2)_{T,\sigma,N}+V\delta^2\,B_T[1-B_T\varkappa(\mathbf{y})]}\,,\ * \tag{40}$$

where B_T is the isothermal bulk modulus, G is the Gibbs free energy, σ refers to the stress variables, and

$$\delta=(N/V)\,(v_1-v_2), \tag{41}$$

where v_1 and v_2 are the partial molar volumes per atom of the two species at the concentration c.

It will be seen from (36–40) that if $\delta=0$, i.e., if, roughly speakling, the two types of atoms are of the same size and shape, then $\varDelta(\mathbf{y})=S_{NC}(\mathbf{y})=0$, and the fluctuations in the number density and in concentration occur independently of one another, as might be expected intuitively. For this case we may expect quite generally that $S_{NC}(\mathbf{q})\simeq 0$ for all $\mathbf{q}$. Further, if the alloy is completely random, we have, from (40),

$$S_{CC}(\mathbf{y})=N k_B T/(\partial^2 G/\partial c^2)_{T,\sigma,N}=c\,(1-c), \tag{42}$$

using the usual form of G for the ideal random mixtures.** For this case $S_{CC}(\mathbf{q})=c\,(1-c)$ for all $\mathbf{q}$ also.

In general $\varkappa(\mathbf{y})$ and hence $S_{NN}(\mathbf{y})$ etc., all depend on the direction of $\mathbf{y}$. For an elastically isotropic solid, however, $\mathbf{e}^{(r)}\|\mathbf{y}$ for the longitudinal branch of the elastic waves and $\mathbf{e}^{(r)}\perp\mathbf{y}$ for the two transverse branches. Hence $\varkappa(\mathbf{y})=(B_T+\frac{4}{3}\mathscr{G})^{-1}$,

* This expression has also been given by KRIVOGLAZ [22] in another connexion.

** For a discussion of the forms of S_{CC} for different types of mixtures see Ref. [20].

where $\mathscr{G}$ is the shear modulus, is independent of the direction of $\mathbf{y}$. For a fluid $\mathscr{G}=0$ and the expressions for $S_{NN}(0)$ etc. are given by

$$\begin{aligned} S_{NN}(0) &= (N/V)\, k_B T B_T^{-1} + \delta^2 S_{CC}(0) \\ S_{NC}(0) &= -\delta S_{CC}(0) \\ S_{CC}(0) &= N k_B T / (\partial^2 G/\partial c^2)_{T,P,N}. \end{aligned} \tag{43}$$

5. Applications

As mentioned in § 2, the formula (13) for the second order correction to the free energy has application to the problem of how the electronic density of states is affected by the thermal motion of the ions and by alloying. In this Section we examine how this expression for ϕ_2 leads to some of the well known results on these effects and comment on some additional features which it implies [19]. The different cases of (13), of course, arise from specifying the nature of the ionic system or more precisely its Hamiltonian $\mathscr{H}_I(\mathbf{R})$ which determines the various dynamical structure factors.

a) Pure Metals (Crystals)

If we take the Hamiltonian for the ions corresponding to phonons in the harmonic approximation, then in the one phonon approximation $S(\mathbf{q}, \omega)$ is given by (see, for example, MARCH et al. [23])

$$\begin{aligned} S(\mathbf{q}, \omega) = e^{-2D(q)} \sum_{\mathbf{Q}, r, \mathbf{g}} \frac{\hbar\left(\mathbf{q}\cdot\mathbf{e}_{\mathbf{Q}}^{(r)}\right)^2}{2M\omega_{\mathbf{Q}r}} \{(n_{\mathbf{Q}r}+1)\, \delta(\omega-\omega_{\mathbf{Q}r})\, \delta_{\mathbf{q}-\mathbf{Q},\mathbf{g}} \\ + n_{\mathbf{Q}r}\, \delta(\omega+\omega_{\mathbf{Q}r})\, \delta_{\mathbf{Q}+\mathbf{q},\mathbf{g}}\}, \end{aligned} \tag{44}$$

where we have omitted, for reasons discussed earlier, the elastic Bragg reflection terms. In (44), $D(q)$ is the Debye-Waller factor, $\omega_{\mathbf{Q}r}$ is the frequency of the phonon of wave vector $\mathbf{Q}$ and unit polarization vector $\mathbf{e}_{\mathbf{Q}}^{(r)}$, $(r=1, 2, 3)$, M is the ionic mass, $\mathbf{g}$ are the reciprocal lattice vectors and $n_{\mathbf{Q}r}$ is the Planck function:

$$n_{\mathbf{Q}r} = (\exp[\beta\hbar\omega_{\mathbf{Q}r}]-1)^{-1}.$$

The sum over $\mathbf{Q}$ is restricted to the first Brillouin zone. Substituting (44) into (14) and (13), one obtains after some rearrangement of terms

$$\begin{aligned} \phi_2 = 2 \sum_{\mathbf{k}, \mathbf{Q}, r, \mathbf{g}} e^{-2D(q)} |w(\mathbf{q})|^2 \frac{\hbar\left(\mathbf{q}\cdot e_{\mathbf{Q}}^{(r)}\right)^2}{2MN\omega_{\mathbf{Q}r}} \times \\ \times \left(\frac{(n_{\mathbf{Q}r}+1)\, f_{\mathbf{k}+\mathbf{q}}(1-f_{\mathbf{k}}) - n_{\mathbf{Q}r}\, f_{\mathbf{k}}(1-f_{\mathbf{k}+\mathbf{q}})}{E_{\mathbf{k}+\mathbf{q}} - E_{\mathbf{k}} - \hbar\omega_{\mathbf{Q}r}} \right), \end{aligned} \tag{45}$$

where $\mathbf{q}=\mathbf{Q}+\mathbf{g}$.

If we now consider an elastically isotropic crystal in the Debye approximation, neglect the umklapp terms $(\mathbf{g}\neq 0)$ and the Debye-Waller factor*, put

$$E_{\mathbf{k}} = \varepsilon_{\mathbf{k}} = \hbar^2 k^2/2m,$$

* It may be noted that the correction terms to (44) due to the multiphonon processes tend to cancel the effect of the Debye-Waller factor [24].

and replace $w(\mathbf{q})$ by $w(0)=-\frac{2}{3}\xi_0$ (ξ_0, unperturbed Fermi energy at $T=0$ °K), then (45) reduces to the expression (21) of BUCKINGHAM and SCHAFROTH [17], except for a numerical factor of the order of magnitude unity connected with the definition of the constant C in FRÖHLICH's electron-phonon interaction parameter F [1,4]. The discussion of the electronic specific heat on the basis of (45) would follow along the same lines as given by BUCKINGHAM and SCHAFROTH and will not be pursued here; see also ASHCROFT and WILKINS [25].

b) Liquid Metals

For a liquid metal, a useful approximation to make in the expression (13) for ϕ_2 is to assume that $S(q,\omega)\simeq S(q)\,\delta(\omega)$, where $S(q)$, as before, is the static structure factor, i.e. it represents the total (elastic+inelastic) scattering in the direction $\mathbf{q}$. We shall refer to this as the "static approximation".* Then (13) may be written as ($\mathbf{q}=\mathbf{k}-\mathbf{k}'$)

$$\phi_2=(1/N)\sum_{\mathbf{k}\mathbf{k}'}{}' |w(q)|^2 S(q)(f_{\mathbf{k}}-f_{\mathbf{k}'})/(E_{\mathbf{k}}-E_{\mathbf{k}'}). \tag{46}$$

Remembering that $\sum_{\mathbf{k}}=V/(2\pi)^3\int d^3k$, we can write (46) as

$$\phi_2=2\int n_0(E_{\mathbf{k}})\,G(E_{\mathbf{k}})\,f_{\mathbf{k}}\,dE_{\mathbf{k}}, \tag{47}$$

where

$$G(E_{\mathbf{k}})=\frac{1}{4\pi N}\mathscr{P}\int\frac{|w(q)|^2 S(q)}{E_{\mathbf{k}}-E_{\mathbf{k}'}}\,n_0(E_{\mathbf{k}'})\,d\Omega_{\mathbf{k}'}\,dE_{\mathbf{k}'}. \tag{48}$$

In (47) and (48), $n_0(E_{\mathbf{k}})=(V/2\pi^2)\,k^2\,dk/dE_{\mathbf{k}}$ is the electron density of states, per spin, with $E_{\mathbf{k}}$ given by (5). Further, $d\Omega_{\mathbf{k}'}$ is an element of solid angle for $\mathbf{k}'$ and $\mathscr{P}$ denotes the principal value of the integral.

Using the standard expression for an integral involving the Fermi distribution function [28], Eq. (47), to first non-vanishing order in T/ξ_0, becomes

$$\phi_2=2\int_0^{\xi_0} n_0(E_{\mathbf{k}})\,G(E_{\mathbf{k}})\,dE_{\mathbf{k}}+2\alpha T^2 n_0(\xi_0)\,G'(\xi_0), \tag{49}$$

where $\alpha=\frac{1}{6}(\pi k_B)^2$ and $G'(\xi_0)=(dG/dE_{\mathbf{k}})_{E_{\mathbf{k}}=\xi_0}$. The first term in (49) may be interpreted as the second order correction to the energy of the electrons at 0 °K. Hence one may infer from (49) and (5) that the energy $\mathscr{E}(\mathbf{k})$ of an electron in the state $\mathbf{k}$, to second order in perturbation, is

$$\mathscr{E}(\mathbf{k})\simeq E_{\mathbf{k}}+G(E_{\mathbf{k}}),\qquad E_{\mathbf{k}}=\hbar^2k^2/2m+W(\mathbf{k}), \tag{50}$$

which is the standard perturbation theory expression for the energy. For its discussion and references, see MOTT and DAVIS [29].

From (49) we may evaluate the second order correction to the energy U_2 and the specific ΔC_V by the thermodynamic formulae, $U_2=\phi_2-T(\partial\phi_2/\partial T)_V$ and $\Delta C_V=(\partial U_2/\partial T)_V$. Assuming that the temperature dependence of $G(E_{\mathbf{k}})$, via $S(q)$, can be neglected, one has

$$\Delta C_V/C_V^e=-G'(\xi_0), \tag{51}$$

* The effects of the correction terms to this approximation on the resistivity and other transport properties of a liquid metal have been recently considered by RICE [26] and are found to be small; see also GREENE and KOHN [27].

where $C_V^e = 4\alpha T n_0(\xi_0)$ is the electronic specific heat of the unperturbed electronic system ($\mathscr{E}_{\mathbf{k}} = E_{\mathbf{k}}$). For the case under consideration, the whole of ΔC_V may be attributed to the electrons.

c) Binary Alloys

Consider a binary solid alloy and let us first assume that the two types of atoms in it can be considered to be having the same mass, size and force constants, so that the only difference between them lies in their atom form factors w_1 and w_2. Then from the discussion given in § 4, we expect that $S_{NC}(\mathbf{q}, \omega) = 0$ and $S_{NN}(\mathbf{q}, \omega) = S(\mathbf{q}, \omega)$ (apart from the Bragg reflection terms). If we treat $S_{CC}(\mathbf{q}, \omega)$ in the static approximation, $\Gamma(\mathbf{q}, \omega)$, from (32), becomes

$$\Gamma(\mathbf{q}, \omega) \simeq |\overline{w}|^2 S(\mathbf{q}, \omega) + |w_1 - w_2|^2 S_{CC}(\mathbf{q})\, \delta(\omega). \tag{52}$$

The first term is due to the phonons and makes a contribution to ϕ_2 (and hence to the electronic specific heat at low temperatures) identical with (45), except that w is replaced by $\overline{w}$. The second term makes an additional contribution to the electronic specific heat which, following § 5 b) above, may be seen to be

$$(\Delta C_V / C_V^e) = -\left(d G_{CC}(E_{\mathbf{k}}) / d E_{\mathbf{k}}\right)_{E_{\mathbf{k}} = \xi_0}, \tag{53}$$

where

$$G_{CC}(E_{\mathbf{k}}) = \frac{1}{4\pi N} \mathscr{P} \int \frac{|w_1 - w_2|^2 S_{CC}(\mathbf{q})}{E_{\mathbf{k}} - E_{\mathbf{k}'}} n_0(E_{\mathbf{k}'})\, d\Omega_{\mathbf{k}'}\, d E_{\mathbf{k}'}, \tag{54}$$

and where we have again neglected any temperature dependence of G_{CC}. The expressions (53) and (54) are the same as derived by STERN [30] in another manner if we remember that in our notation $S_{CC}(\mathbf{q}) = c(1-c)$ for a random alloy.

An estimate of the correction term to the static approximation

$$S_{CC}(\mathbf{q}, \omega) = S_{CC}(\mathbf{q})\, \delta(\omega)$$

used in (52) may be made as follows. If we expand the expression (12) for $\Gamma(\mathbf{q}, \omega)$ in terms of the displacements of the ions about their equilibrium positions, neglect all the terms higher than the second order in the displacements (and hence phonon coordinates), and assume that the two types of atoms have the same mass etc., as before, and are distributed at random, we may obtain

$$S_{CC}(\mathbf{q}, \omega) = c(1-c)\, \delta(\omega) + c(1-c)\, \eta(\mathbf{q}, \omega), \tag{55}$$

where the correction term $\eta(\mathbf{q}, \omega)$ is

$$\begin{aligned} \eta(\mathbf{q}, \omega) = \sum_{\mathbf{Q}, r} \hbar (2MN\omega_{\mathbf{Q}r})^{-1} (\mathbf{q} \cdot e_{\mathbf{Q}}^{(r)})^2 \{(n_{\mathbf{Q}r} + 1)\, \delta(\omega - \omega_{\mathbf{Q}r}) \\ + n_{\mathbf{Q}r}\, \delta(\omega + \omega_{\mathbf{Q}r}) - (2 n_{\mathbf{Q}r} + 1)\, \delta(\omega)\}, \end{aligned} \tag{56}$$

which has the required property that

$$\int_{-\infty}^{\infty} \eta(\mathbf{q}, \omega)\, d\omega = 0.$$

Although a detailed calculation is lacking at present, one may see that the contribution of (56) to ϕ_2 at low temperatures would be at most μ times the contribution from the first term in (55), where

$$\mu \sim \sum_{\mathbf{Q},r} \frac{\hbar\left(\mathbf{q}_0 \cdot e_{\mathbf{Q}}^{(r)}\right)^2}{2MN\omega_{\mathbf{Q}r}} = \frac{3}{2}\left(\frac{\hbar^2 q_0^2}{2Mk_B\Theta_D}\right). \tag{57}$$

The last equality is obtained by using the Debye approximation. In (57), Θ_D is the Debye temperature and q_0 is a wave number beyond which the pseudopotentials are vanishingly small. Since $q_0 \sim 2k_F$, k_F wave number of the Fermi electrons, one sees that μ itself is a small fraction.

Finally, it may be mentioned that the differences in the mass and (or) force constants and size of the impurity and the host atoms can contribute to the scattering and the resistivity significantly (see Ref. [21, 22, 31, 32] and references given there). In our notation, these contributions to the scattering are described by the DSF $S_{NC}(\mathbf{q}, \omega)$ and the concommitant changes in S_{NN} and S_{CC}. The expression (13) for ϕ_2, in conjunction with (32), thus provides a way of calculating the contributions to the second order corrections to the free energy from these sources also.

d) ν-Component Random Alloys

Let us again assume that the different types of atoms in the alloy can be considered to have the same mass, force constants etc., and that they are distributed at random. Then, in the approximation similar to that made in (52), the DSF $S_{\alpha\beta}(\mathbf{q}, \omega)$ are given by

$$S_{\alpha\beta}(\mathbf{q}, \omega) \simeq (c_\alpha c_\beta)^{\frac{1}{2}}\, S(\mathbf{q}, \omega) + [\delta_{\alpha,\beta} - (c_\alpha c_\beta)^{\frac{1}{2}}]\, \delta(\omega), \tag{58}$$

and hence

$$\Gamma(\mathbf{q}, \omega) \simeq |\overline{w}|^2\, S(\mathbf{q}, \omega) + \delta(\omega)\left(\sum_\alpha c_\alpha |w_\alpha|^2 - |\overline{w}|^2\right), \tag{59}$$

where $\overline{w} = \sum_\alpha c_\alpha w_\alpha$.

The first term in (59) makes a contribution to ϕ_2 similar to the case of a pure metal [Eq. (45)]. The second term makes a contribution ΔC_V to the electronic specific heat given by (53) and (54), except that in the expression for G_{CC}, $|w_1 - w_2|^2\, S_{CC}(\mathbf{q})$ is now replaced by

$$F(\mathbf{q}) \equiv \sum_\alpha c_\alpha |w_\alpha|^2 - (\overline{w})^2. \tag{60}$$

The resulting expression for ΔC_V may, of course, also be obtained by generalising STERN's work [30], quoted earlier, and the special case of ternary random alloys has been recently discussed by DAVIS and RAYNE [33]. If we define the concentrations y and x by writing the ternary alloy in the form $A_y(B_x C_{1-x})_{1-y}$, then

$$F(\mathbf{q}) = y(1-y)\,|w_A - x w_B - (1-x)\, w_C|^2 + (1-y)\, x(1-x)\, |w_B - w_C|^2. \tag{61}$$

The measurements of the electronic specific heat by DAVIS and RAYNE [33] on the alloy system $Ag_y(Au_x Cu_{1-x})_{1-y}$ are found to agree fairly well with the concentration dependence implied by (61).

As regards the numerical magnitude, it should be noticed that if we subtract from the first (electron-phonon) term in (59) the quantity, $\left(\sum_\alpha c_\alpha |w_\alpha|^2\right) S(\mathbf{q}, \omega)$, which represents linear interpolation between the pure metals, then the remainder

[$-F(\mathbf{q})\, S(\mathbf{q}, \omega)$] has the same concentration dependence as the term due to Stern [30] and Davis and Rayne [33]. A recent calculation [34] for binary alloys indicates that both these terms, involving $F(\mathbf{q})$, contribute in general significantly to the low temperature electronic specific heat. It will be remembered that these considerations are necessarily approximate, being based on (58) with $S(\mathbf{q}, \omega)$ assumed to be independent of concentration.

References

1. Fröhlich, H.: Phys. Rev. **79**, 845 (1950).
2. Bloch, F.: Z. Physik **52**, 555 (1928).
3. Buckingham, M. J.: Nature **168**, 281 (1951).
4. Fröhlich, H.: Proc. Roy. Soc. (London), Ser. A **245**, 291 (1952).
5. Krishnan, K. S., Bhatia, A. B.: Nature **156**, 503 (1945).
6. Bhatia, A. B., Krishnan, K. S.: Proc. Roy. Soc. (London), Ser. A **194**, 185 (1948).
7. Gerstenkorn, H.: Ann. Physik **10**, 49 (1952).
8. Ziman, J. M.: Phil. Mag. **6**, 1013 (1961).
9. Phillips, J. C., Kleinman, L.: Phys. Rev. **116**, 287 (1959).
10. Antonick, E.: Phys. Chem. Solids **10**, 314 (1959).
11. Heine, V.: Adv. in Solid State Phys.: Ed. Ehrenreich, H., Seitz, F., Turnbull, D., Vol. 24, p. 1. New York: Academic Press 1971.
12. Cohen, M. H., Heine, V.: Adv. in Solid State Phys.: Ed. Ehrenreich, H., Seitz, F., Turnbull, D., Vol. 24, p. 38. New York: Academic Press 1971.
13. Harrison, W. A.: Pseudopotentials in The Theory of Metals. New York: Benjamin, Inc. 1966.
14. Gupta, O. P.: Phys. Rev. **174**, 668 (1968).
15. Stocks, G. M., Young, W. H., Meyer, A.: Phil. Mag. **18**, 895 (1968).
16. Schafroth, M. R.: Helv. Phys. Acta **24**, 645 (1951).
17. Buckingham, M. J., Schafroth, M. R.: Proc. Phys. Soc. (London) A **67**, 828 (1954).
18. van Hove, L.: Phys. Rev. **95**, 249 (1954).
19. Bhatia, A. B., O'Leary, W. P.: Lettere Nuovo Cimento **3**, 14 (1972).
20. Bhatia, A. B., Thornton, D. E.: Phys. Rev. B **2**, 3004 (1970).
21. Bhatia, A. B., Thornton, D. E.: Phys. Rev. B **4**, 2325 (1971).
22. Krivoglaz, M. A.: Theory of X-ray and Thermal Neutron Scattering by Real Crystals. New York: Plenum Press 1969.
23. March, N. H., Young, W. H., Sampanthar, S.: The Many-Body Problem in Quantum Mechanics. London: Cambridge University Press 1967.
24. Sham, L. J., Ziman, J. M.: Adv. in Solid State Phys. Ed. Seitz, F., Turnbull, D., Vol. 25. New York: Academic Press 1963.
25. Ashcroft, N. W., Wilkins, J. W.: Phys. Letters **14**, 285 (1965).
26. Rice, M. J.: Phys. Rev. B **2**, 4800 (1970).
27. Greene, M. D., Kohn, W.: Phys. Rev. **137**, 513 (1965).
28. Mott, N. F., Jones, H.: The Theory of the Properties of Metals and Alloys, p. 177. London: Oxford University Press 1936.
29. Mott, N. F., Davis, E. A.: Electronic Processes in Non-Crystalline Materials. London: Oxford University Press 1971.
30. Stern, E. A.: Phys. Rev. **144**, 545 (1966).
31. Maradudin, A. A., Montroll, E. W., Weiss, G. H., Ipatova, I. P.: Theory of Lattice Dynamics in the Harmonic Approximation, Second Ed. New York: Academic Press 1971.
32. Kagan, Y., Zhernov, A. P.: Zh. Eksperim. i Teor. Fiz. **50**, 1107 (1966) [English transl: Soviet Phys. JETP **23**, 737 (1966)].
33. Davis, T. H., Rayne, J. A.: Phys. Letters **36** A, 40 (1971).
34. Bhatia, A. B., O'Leary, W. P.: (to be published).

Chapter II. Superconductivity and Superfluidity

Let us introduce this chapter with a personal reminiscence. When one of the editors (H. H.) was a student at Erlangen University in 1948–1951 he participated at two very strange seminars. Professor HILSCH and his coworkers, among them Prof. BUCKEL, who were studying superconductivity in now well-known experiments, had asked the theoreticians to study and digest for them the theories of superconductivity. Trying to understand the theories, most of them stemming from famous experts, drove us poor students nearly crazy. We could not understand any of these papers because in each some inconsistency seemed to appear. So we began to doubt that we would ever become good physicists at all. Now we know that all the microscopic theories were incorrect because they considered the wrong interactions.

At that time FRÖHLICH'S investigations of the phonon-electron interaction as the cause of superconductivity came as a big surprise and might have even seemed somewhat absurd. Why should just a coupling, which causes the electrical resistance, be responsible for the complete elimination of the electrical resistance? The fascinating development of the theory of superconductivity and that of the closely related phenomenon of superfluidity is presented in the following papers.

Electron-Phonon Interactions and Superconductivity*

JOHN BARDEEN

University of Illinois, Department of Physics and Materials Research Laboratory, Urbana Illinois

Introduction

One of FRÖHLICH's major achievements was to point out [1], in 1950, the significance of the electron-phonon interaction in superconductivity. Earlier, interactions between electrons and phonons (the quanta of the lattice vibrations) were taken into account for scattering and resistance; FRÖHLICH showed that the same interaction should modify the energy of the electrons and proposed it as the basis for a theory of superconductivity. He suggested using a Hamiltonian, now called the Fröhlich Hamiltonian, in which Coulomb interactions are disregarded except as they may modify the energies of the individual electrons, but which includes the interaction between electrons and phonons. An effective attractive interaction between the electrons results from exchange of virtual phonons. Strong confirmation of the importance of electron-phonon interactions in superconductivity came from the independent experimental discovery of the isotope effect. Although the theory given at that time was unsuccessful in accounting for superconductivity, FRÖHLICH's formulation of the problem has been the basis for nearly all subsequent theoretical developments.

In this article we shall trace the development of the role of electron-phonon interactions in superconductivity. It is designed for the non-expert and will attempt to give some impression of the striking advances in understanding that have occurred in the past twenty years. Only a brief account will be given of developments from 1950 to 1957, when a microscopic theory was given by COOPER, SCHRIEFFER and the author [3], based on a model of an electron gas with attractive interactions. The magnitude of the required interaction was determined empirically from the critical temperature, T_c.

In subsequent developments [4] by many people, theory and experiment have developed to such an extent that detailed quantitative information on the electron-phonon interaction can be obtained from experiments on superconductors. The most powerful and useful method comes from measurements of the voltage dependence of the tunneling current through a thin insulating barrier separating a normal metal and a superconductor. Analysis of data from such experiments yields the single particle Green's function which can in turn be used to derive in quantitative detail the various thermal and transport properties of super-

* Supported in part by the National Science Foundation NSG GH 33634 (46-22-34-348).

conductors, with excellent agreement with experiment. First principles calculations of electron-phonon interactions which have been carried out for a number of simple metals and alloys are in good agreement with those derived empirically from tunneling data. Not only is there no question about the validity of the theory, but superconductivity provides a valuable method for studying the electron-phonon interaction in metals and alloys.

Very important for these developments has been the application of methods derived from quantum field theory to many-body problems, including thermal Green's functions, Feynman diagrams, Dyson equations and renormalization concepts. FRÖHLICH was one of the first to recognize the value of field theory methods in solid state problems, and made use of the analogies between electrons interacting with phonons in solids and with photons in quantum electrodynamics. In 1950, diagramatic methods were just being introduced to quantum field theory. It was many years before they were developed with full power for applications to quantum statistical mechanics.

FRÖHLICH did his original work on electron-phonon interactions and superconductivity while he was visiting at Purdue University in the spring semester of 1950. Independently, without his knowledge, two groups had been measuring the transition temperatures of separated mercury isotopes and found a positive result that could be interpreted as $T_c M^{\frac{1}{2}}$—const, where M is the isotopic mass. If the mass of the ions is important, their motion and thus the lattice vibrations must be involved. The isotope effect was discovered independently by a group at Rutgers University under the leadership of B. SERIN and by E. MAXWELL, working at the U.S. National Bureau of Standards.

At that time, a great deal of research in low temperature physics in the United States was sponsored by the U.S. Office of Naval Research, and periodic meetings were held to exchange information. The isotope effect was first announced by the two groups at an ONR sponsored meeting held at the Georgia Institute of Technology in March, 1950, but was not publicized outside of the low temperature community. Papers were submitted for publication shortly after the meeting, and appeared in mid-May. It was not until this time, just before he was ready to submit his own paper for publication, that FRÖHLICH learned about this strong experimental confirmation of his ideas.

In early May, not long before the papers on the isotope effect appeared in print, I learned about the exciting results in a telephone call from SERIN. I immediately thought that electron-phonon interactions must be involved and attempted to construct a theory on this basis. The theory was based in large part on an earlier abortive attempt to construct a theory of superconductivity based on energy gaps at the Fermi surface. A short communication was sent in for publication in the latter part of May. About a week later, FRÖHLICH visited the Bell Laboratories in Murray Hill, N. J., where I was working at the time. It was exciting to learn about his ideas on electron-phonon interactions and superconductivity, worked out much more completely than my own. We were both convinced that at last we were on the road to an explanation of superconductivity. Although our approaches were different, mine using a variational method and his perturbation theory, both theories were based on the self-energy of the electrons in the phonon field rather than a true interaction between electrons.

It soon became evident that there were major difficulties in these approaches, not easy to overcome. Aside from mathematical difficulties concerned with the use of perturbation expansions when they are not really valid, it became evident that nearly all of the self-energy occurs in the normal state and changes very little when the metal becomes superconducting. A new idea was needed to understand what happens at the superconducting transition.

The year 1950 was a banner one in other respects as well for superconductivity theory. This was the year that Ginzburg and Landau presented their now famous phenomenological theory of superconductivity, designed to apply to temperatures near T_c. Also in the same year appeared F. LONDON's book [6] on superconductivity. Although mainly concerned with the phenomenological macroscopic theory, he discussed his ideas that superfluids exhibit quantum effects on a macroscopic scale and indicated the general nature of a possible underlying microscopic theory. His ideas were remarkably perceptive. He states that "the long range order of the average momentum is to be considered one of the fundamental properties of the superconducting state" and that superconductivity requires "a solidification or condensation of the average momentum distribution." This is a good description of the many-electron wave functions for a superconductor, but it took a long time to find the correct theory. One is used to dealing with a condensation in real space, but a condensation in momentum space and the reasons for it is much more difficult to visualize.

Work done during the next few years pointed up more sharply the difficulties of any theory based on a perturbation expansion. SCHAFROTH showed that starting with the Fröhlich Hamiltonian, one can not derive the Meissner effect in any order of perturbation. As we shall discuss, Migdal later was able to derive an expression for the self-energy corrections from electron-phonon interactions supposedly correct to order $(m/M)^{\frac{1}{2}}$ and found nothing corresponding to superconductivity.

On the positive side, in 1955 PINES and I carried through a time-dependent self-consistent field calculation of electron-phonon interactions in a model which included Coulomb interactions from the start. One ordinarily thinks of Coulomb interactions as large compared with the phonon-induced interactions between electrons, but this is not true when the energy difference between the electron states involved is small. The Coulomb interaction between normal state quasi-particles near the Fermi surface is via a screened interaction of short range. The field of the moving ions is also strongly screened by the electrons. Such screening must be taken into account in any first principles calculation of the electron-phonon coupling. The calculation helped to confirm that the Fröhlich Hamiltonian should be a reasonable starting point to derive a theory of superconductivity.

Although not a realistic model for superconductivity, FRÖHLICH in 1954 worked out the consequences of a one-dimensional model of an electron gas interacting with lattice vibrations and obtained results of considerable mathematical interest. Thermal properties of the system were derived later by KUPER. FRÖHLICH showed if the electron-phonon-coupling is sufficiently strong, an instability can develop so as to give an energy gap at the Fermi surface. Electrons in states just below the Fermi surface are lowered in energy. The gap varies as $\exp(-a/\lambda)$, where a is a numerical constant and λ is the electron-phonon coupling constant. This function cannot be expanded in powers of λ and so could not be derived in a perturbation expansion of the electron-phonon interaction.*

I was much interested in this calculation because it is closely related to a theory of superconductivity I tried to develop just before the war. Small lattice displacements were to give energy gaps and thus lower the energy of electrons in states near the top of the Fermi distribution. Since the energy gaps would be small, electrons near the Fermi surface would have small effective mass and hopefully a large diamagnetism. Needless to say, the theory ran into difficulties and was never published in full. When I heard about the isotope effect from SERIN, I tried to revive the theory with the energy gap coming from dynamic interactions with the phonons rather than from small static lattice displacements.

Further background to the development of the microscopic theory is the concept of electron pairs. As suggested by GINZBURG and by SCHAFROTH, the idea was that if electrons are associated in pairs, the pairs would obey Einstein-Bose statistics, and that superconductivity, like superfluidity in helium, would be a consequence of Bose condensation. In 1956, COOPER showed that a pair of electrons above the Fermi surface is unstable against the formation of a bound state in the presence of an attractive interaction no matter how weak. The electron-phonon interaction of course provided an effective attraction for electrons near the Fermi surface.

This is a brief account of the background that led to the pairing theory of superconductivity of COOPER, SCHRIEFFER and the author, a preliminary account of which was published in the spring of 1957 and the full paper [3] appeared in the fall of that year. In the ground state, electrons are associated in pairs of opposite spin and momentum, or, when there is current flow, in pairs of exactly the same momentum. Restrictions of Fermi-Dirac statistics are such that in this way maximum advantage can be taken of the phonon-induced attractive interaction between electrons. We also gave the spectrum of quasi-particle excitations in the superconductor and showed that one could derive the various thermal and transport properties without much more difficulty than for normal metals. As envisaged by LONDON [6], there is a "condensation of the average momentum distribution." There is macroscopic occupation of the common momentum of the pairs.

---

* *Added in proof, March* **15, 1973**. Very recently, experimental evidence has been presented which appears to indicate paraconductivity resulting from superconducting fluctuations just above a Peierls soft mode instability in a pseudo one-dimensional organic solid at a temperature of about 60 °K. L. B. COLEMAN, M. J. COHEN, D. J. SANDMAN, F. G. YAMAGISHI, A. F. GARITO and A. J. HEEGER have reported (submitted to Solid State Communications, February 20, 1973) a conductivity maximum of greater than 10^6 (ohm cm)$^{-1}$ (more than that of copper) in crystals of an organic charge transfer salt (TTF) (TCNQ) and related solids. While they attribute the fluctuations giving rise to the high conductivity to pairing (BCS) superconductivity arising from the soft-mode instability, it seems much more likely that it is an example of a Fröhlich one-dimensional superconductor (see J. BARDEEN, submitted to Solid State Communications, March, 1973). In the Fröhlich model [Proc. Roy. Soc. A**223**, 296 (1954)] there is no pairing but a macroscopic occupation of a moving lattice wave that carries the electrons along with it. The paraconductivity arises in a similar way to that in a BCS superconductor and should have the same temperature dependence. Thus it appears that after nearly 20 years, Fröhlichs' theory of superconductivity in one-dimensional systems has found a realization in nature.

The key thing is pairing, not pairs. There is no localized pairing of electrons into "pseudo-molecules" which obey Bose statistics. Although this analogy is often used, particularly by BOGOLIUBOV and coworkers, I think that it is misleading. The reason for the condensation is not Bose-Einstein statistics, but comes from the exclusion principle; pairing allows one to make best use of the available phase space to form a coherent low energy ground state.

After the microscopic theory appeared, alternative and more general formulations were given by a number of people, including ANDERSON, BOGOLIUBOV, GOR'KOV, KADANOFF and MARTIN, NAMBU and others. BOGOLIUBOV [7], and independently VALATIN, introduced a transformation to new fermion variables that represent quasi-particles in the superconducting state. Although equivalent to the excitation spectrum introduced by BCS, it is formally much simpler to use. In inhomogeneous situations, the momentum or wave vector is not a good quantum number for the electrons and one must find the wave functions for the states that are to be paired by a self-consistent field calculation. BOGOLIUBOV generalized the quasi-particle transformation for this situation and gave equations that have since been widely used for deriving quasi-particle wave functions in superconductors.

Perhaps even more important and useful is the method of Green's functions and Feynman diagrams, introduced to superconductivity theory by GOR'KOV [8]. He showed that one could describe pairing in the ground state by giving finite averages to $\langle\psi(\boldsymbol{r}', t')\,\psi(\boldsymbol{r}, t)\rangle$ in addition to the usual single particle Green's function $\langle\psi^*(\boldsymbol{r}', t')\,\psi(\boldsymbol{r}, t)\rangle$. Here $\psi(\boldsymbol{r}, t)$ is the wave field operator for the electrons. An average for the pair field, $\psi\psi$, makes sense only if one uses wave functions with variable numbers of particles in the ground state. We had done this for mathematical convenience in our original paper, but it later turned out to make good physical sense as well, for it enables one to define the phase of the condensate. As first pointed out by ANDERSON, the ground state is degenerate in phase. One can project to states of fixed number of electrons, N, by averaging over the phase, Φ. In a quantum mechanical sense, phase and particle number are conjugate variables, with an uncertainty relation $\Delta\Phi\Delta N \sim 1$. As illustrated by the Josephson effect, the phase is an important variable for superfluids.

The temperature dependent Green's function techniques introduced by MATSUBARA in 1955, were extensively developed in the years 1957–1962, by KUBO, by MARTIN and SCHWINGER, and particularly by the Russian workers, MIGDAL, GALITSKII, GOR'KOV, ABRIKOSOV, DZYALOSHINSKI and others. A great advantage of the method is that it is possible to treat time dependent effects such as retardation and quasi-particle lifetimes in the electron-phonon interaction.

Migdal Theory for Normal Metals

In 1958 MIGDAL [5] gave a solution of the Fröhlich Hamiltonian for normal metals by use of Green's function techniques. His solution involves errors only of order $\hbar\omega_q/E_F$ or of $(m_{\text{electron}}/M_{\text{ion}})^{1/2}$ but is otherwise valid for arbitrarily strong coupling. He found no instabilities in the Fermi surface from self-energy contributions of the type suggested by FRÖHLICH.

The Fröhlich Hamiltonian is

$$H = H_0 + H_1 \tag{1}$$

where H_0 represents bare electrons and phonons

$$H_0 = \Sigma_{\boldsymbol{k}}\, \varepsilon_0(\boldsymbol{k})\, a_k^+ a_k + \Sigma_{\boldsymbol{q}}\, \omega_0(\boldsymbol{q})\, b_q^+ b_q \tag{2}$$

and H_1 is the electron-phonon interaction

$$H_1 = -i \sum_{\boldsymbol{q}<q_m,\, \boldsymbol{k}} \alpha_q a_{k+q}^+ a_k (b_q + b_{-q}^+). \tag{3}$$

Here $\boldsymbol{q}$, the wave vector for the phonons, runs over the Brillouin zone and α_q is the coupling constant. The electron quasi-particle energies, $\varepsilon_0(\boldsymbol{k})$, are relative to the Fermi energy and presumably include band and Coulomb effects.

The electron and phonon Green's functions are defined by averages of time-ordered operators

$$G = -i\langle T\psi(1)\, \psi^+(2)\rangle \quad \text{and} \quad D = -i\langle T\Phi(1)\, \Phi^+(2)\rangle \tag{4}$$

where ψ is the wave field operator for electron quasi-particles and Φ for the phonons

$$\Phi(\boldsymbol{r}, t) = -i\, e^{iHt} \Big\{ \sum_{\boldsymbol{q}<q_m} (b_q + b_{-q}^+)\, e^{i\boldsymbol{q}\cdot\boldsymbol{r}} \Big\}\, e^{-iHt}. \tag{5}$$

The brackets $\langle\ \rangle$ represent thermal averages over an ensemble.

We may express the Fourier transforms of the Green's functions in terms of the Matsubara frequencies.

$$\omega_n = (2n+1)\, \pi i k_B T, \qquad \nu_n = 2n\pi i k_B T \tag{6}$$

and define the four vectors

$$P = \boldsymbol{k}, \omega_n, \qquad Q = \boldsymbol{q}, \nu_n. \tag{7}$$

The thermal Green's functions for H_0 are then

$$G_0(P) = \frac{1}{\omega_n - \varepsilon_0(\boldsymbol{k}) + i\delta_k} \tag{8}$$

$$D_0(Q) = \left\{ \frac{1}{\nu_n - \omega_0(\boldsymbol{q}) + i\delta} - \frac{1}{\nu_n + \omega_0(\boldsymbol{q}) - i\delta} \right\}. \tag{9}$$

The key point of MIGDAL's theory is that vertex corrections are small. In drawing Feynman diagrams, we shall use straight lines to represent electron propagators, curly lines phonon propagators, light lines to represent G_0 and D_0 and heavy lines the fully renormalized propagators. The expansion of the vertex function, Γ, is a series of which the first two terms are illustrated in Fig. 1. MIGDAL

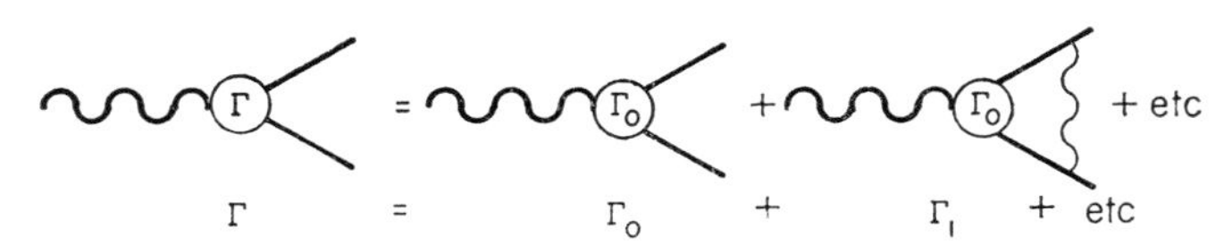

Fig. 1. First two terms in the expansion of the vertex function, Γ. External lines are included only for clarity

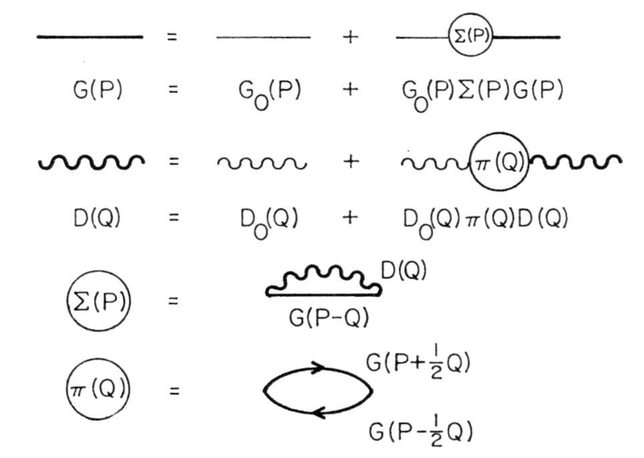

Fig. 2.
Dyson equations for electron and phonon propagators and electron and phonon self-energies

shows that corrections of order Γ_1 and of higher order terms are of order $(m/M)^{1/2}$ compared with Γ_0 and so can be neglected.

The Dyson equations for the renormalized propagators D and G are illustrated in diagramatic form in Fig. 2. Because vertex corrections are small, the self energies may be evaluated from the lowest order diagrams, as illustrated, if we use fully renormalize propagators.

If we set $\Gamma=\Gamma_0$, we have a closed set of equations that can be solved for G and D. The equation for D is trivial and gives only a small renormalization of the phonon frequencies. In regard to the electron self energy, $\Sigma(P)=\Sigma(\boldsymbol{k}, \omega)$ MIGDAL points out that we are interested in states such that $\boldsymbol{k}$ is very close to the Fermi surface, k_F, so that the dependence on $\boldsymbol{k}$ is unimportant, but the frequency dependence is important. To a close approximation, one may take

$$\Sigma(\boldsymbol{k}, \omega)=\Sigma(k_F, \omega)\equiv\Sigma(\omega). \tag{10}$$

The electron quasi-particle energy is then given by a root, ω_k, of

$$\varepsilon(\boldsymbol{k})=\omega_k=\varepsilon_0(\boldsymbol{k})+\Sigma(\omega_k). \tag{11}$$

In the thermal GREEN'S function formalism, one must make an analytic continuation from the imaginary ω_n to the real ω axis to determine $\Sigma(\omega)$.

Although Σ is small near the Fermi surface, it changes rapidly with energy, so that it can have an important effect on the slope $(d\varepsilon/dk)$ and thus on the density of states at the Fermi surface. One may define a quasi-particle renormalization factor

$$Z(k)=m^*/m=(d\varepsilon_0/dk)/(d\varepsilon/dk) \tag{12}$$

where at the Fermi surface

$$Z(k_F)=1+\lambda. \tag{13}$$

The density of states and thus the low temperature electronic specific heat is enhanced by the factor $1+\lambda$.

The expression for λ is in modern notation

$$\lambda = 2 \int_0^\infty d\omega \frac{\alpha^2(\omega) F(\omega)}{\omega} \tag{14}$$

where $F(\omega)$ is the density of states in energy in the phonon spectrum and $\alpha^2(\omega)$ is an average over the polarization directions of the square of the coupling constant. Note that λ is always positive so that the Fermi surface is stable if the lattice is stable. Typical values of λ range from about 0.5 to 1.5. The parameter λ corresponds roughly to the $N(0)\,V$ (density of states times effective interaction) of the BCS theory.

The fact that $\Sigma(\omega)$ depends on ω but not upon $\boldsymbol{k}$ has important experimental consequences. From the form of the single-particle Green's function ($T = 0$ °K)

$$G(\boldsymbol{k}, \omega) = \frac{1}{\omega - \varepsilon_0(\boldsymbol{k}) - \Sigma(\omega) + i\delta} \tag{15}$$

it follows that the spectral weight is

$$A(\boldsymbol{k}, \omega) = 2\pi\,\delta(\omega - \varepsilon_0(\boldsymbol{k}) - \Sigma(\omega)) \tag{16}$$

where the δ-function applies to integration over ε_0, a function of the momentum variable $\boldsymbol{k}$ and not to ω. For integration over ω one must write

$$A(\boldsymbol{k}, \omega) = 2\pi Z_k\, \delta(\omega - \omega_k) \tag{17}$$

where $Z_k = 1 + \lambda_k$ is the wave function renormalization factor. In evaluating transition rates, one often encounters expressions of the form

$$\int \frac{d^3 k\, d\omega}{(2\pi)^4} A(\boldsymbol{k}, \omega) \,. \tag{18}$$

The Z_k factors would enter if one integrates over ω first, and this factor would just cancel the increase in density of states. However, it is often convenient to integrate over ε_0 first, making use of the simple δ-function form of $A(\boldsymbol{k}, \omega)$, and the Z_k factors do not enter at all.

Examples of effects unaffected by the $(1 + \lambda)$ enhancement of the density of states near the Fermi surface are electrical conductivity, nuclear spin relaxation times and tunneling between normal metals. Perhaps the most striking experimental confirmation comes from measurements of the nuclear spin relaxation time T_1. For example, in aluminum, T_1 is inversely proportional to T for temperatures extending from 1 °K to 1000 °K, with no sign of an increased density of states by a factor $1 + \lambda$ expected for $T < \Theta_D$ from a simple quasi-particle model. Similar results have been obtained from other metals.

The quasi-particle velocity at the Fermi surface is decreased by the factor $1/Z(k_F)$. In addition to the electronic specific heat, the factor Z enters the Fermi velocity in experiments such as cyclotron resonance and the Tomasch effect in tunneling.

Nambu-Eliashberg Theory for Superconductors

Although MIGDAL's theory gives an excellent treatment of electron-phonon interactions in normal metals, it gave no clue as to the origin of superconductivity. The difficulty is that with pairing summation of an infinite series of terms add to give a finite result even though each is of order $(m/M)^{1/2}$. Following the variational method used by BCS, BOGOLIUBOV gave a nearly equivalent derivation from the Fröhlich Hamiltonian based on the principle of "compensation of dangerous diagrams." According to the BCS theory, the gap parameter, Δ_0, for $T=0$ °K, is given by

$$\Delta_0=1.75\, k_B T_c=2\omega_c \exp\left[-1/N(0)\, V\right] \tag{19}$$

where ω_c is the cut-off frequency of the phonon-induced interaction. Later, more satisfactory Green's function derivations were given by NAMBU [9] and ELIASHBERG [10]. Although equivalent to the Bogoliubov and BCS versions of the theory when the coupling is weak, they differ in the region of strong coupling. By weak coupling it is meant that the significant phonon energies are very large compared with $k_B T_c$, so that the gap parameter $\Delta(\omega)$ is essentially a constant independent of frequency for energies up to a few times $k_B T_c$. Also to a sufficient approximation, one may take $Z_n=Z_s$. A typical example of a weak coupling superconductor is Al. In strong coupling superconductors, such as Pb, it is important to take the frequency dependence of $\Delta(\omega)$ into account, and both real and imaginary parts of Δ are important. Further, the renormalization factors $Z_n(\omega)$ and $Z_s(\omega)$ may differ in normal and superconducting states.

ELIASHBERG derived a pair of coupled integral equations for the frequency dependent quasi-particle self energy and pair potential accurate to terms of order $(m/M)^{1/2}$ for all coupling. In this section we shall sketch the way the equations are derived without going into detail.

NAMBU used a spinor notation

$$\Psi=\begin{pmatrix}\psi_\uparrow(\boldsymbol{r},t)\\ \psi_\downarrow^+(\boldsymbol{r},t)\end{pmatrix} \tag{20}$$

and a matrix Green's function with components

$$\boldsymbol{G}_{\alpha\beta}=-i\langle T\Psi_\alpha \Psi_\beta^+\rangle. \tag{21}$$

Thus $\boldsymbol{G}_{11}$ and $\boldsymbol{G}_{22}$ are the usual single particle Green's functions for spin up and spin down particles and $\boldsymbol{G}_{12}$ and $\boldsymbol{G}_{21}$ are the anomalous GREEN's functions of GOR'KOV.

$$\boldsymbol{G}_{11}(P)=G(P)=-i\langle T\psi_\uparrow^* \psi_\uparrow\rangle \tag{22}$$

$$\boldsymbol{G}_{12}(P)=F(P)=-i\langle T\psi_\uparrow \psi_\downarrow\rangle=\boldsymbol{G}_{21}^*. \tag{23}$$

The self-energies may be defined by the matrix equation

$$\boldsymbol{G}^{-1}=\boldsymbol{G}_0^{-1}-\boldsymbol{\Sigma}. \tag{24}$$

There are two self-energies, Σ_1 and Σ_2, the first being the diagonal quasi-particle self-energy and the second the off-diagonal term representing the pairing. It is possible to determine these as in MIGDAL's theory by using fully renormalized matrix propagators in the lowest order diagrams shown in Fig. 2, thus omitting vertex corrections. Errors are again of order $(m/M)^{1/2}$. Explicit expressions for the self-energies are

$$\Sigma_{1s}(P)=(1/V\beta)\sum_{P'}\alpha_{p-p'}^2\, G(P')\, D(P-P') \tag{25a}$$

$$\Sigma_{2s}(P) = (1/V\beta) \sum_{P'} \alpha^2_{p-p'} F(P')\, D(P-P') \tag{25b}$$

$$\pi_S(Q) = -(2\alpha_q^2/V\beta) \sum_{P} [G(P+Q)\, G(P) - F(P+Q)\, F(-P)]. \tag{25c}$$

Again the momentum dependence of Σ_1 and Σ_2 is unimportant, and following NAMBU one may define

$$\omega Z_s(\omega) = \omega + \Sigma_1(\omega) \tag{26}$$

$$\Delta(\omega) = \Sigma_2(\omega)/Z(\omega), \tag{27}$$

the latter corresponding to the pair potential of the BCS theory. Note that Δ and Z may be complex and include quasi-particle life-time effects.

ELIASHBERG carried out the integrals over the momenta coordinates and derived a pair of non-linear integral equations for Z_s and $\Delta(\omega)$. We shall give them only for $T=0$ °K. For real ω they are

$$\Delta(\omega)\, Z(\omega) = N(0) \int_{\Delta_0}^{\omega_c} \mathrm{Re}\left\{\frac{\Delta(\omega')}{\omega'^2 - \Delta^2(\omega')}\right\} K_+(\omega, \omega')\, d\omega' \tag{28}$$

$$(1 - Z(\omega))\, \omega = N(0) \int_{\Delta_0}^{\omega_c} \mathrm{Re}\left\{\frac{\omega'}{\omega'^2 - \Delta^2(\omega')}\right\} K_-(\omega, \omega')\, d\omega' \tag{29}$$

where

$$K_\pm(\omega, \omega') = \sum_\lambda \int_0^\infty [\alpha_\lambda^2(\omega_0) F_\lambda(\omega_0)] \left[\frac{1}{\omega' + \omega + \omega_0 + i\delta} \;\; \frac{1}{\omega' - \omega + \omega_0 - i\delta}\right] d\omega_0. \tag{30}$$

Here λ represents the polarization of the phonon, α_λ is the coupling parameter and $F_\lambda(\omega_0)$ is the density of phonon states of energy ω_0. The high frequency cut-off ω_c is an energy large compared to the Debye energy and small compared with E_F. Coulomb interactions may be included by adding a term to K_+.

The Eliashberg equations, while not exact, give a quite accurate quantitative description of superconductors and have been widely used both in the interpretation of experimental data and in basic calculations of superconducting properties from microscopic theory. The first solutions were obtained by MOREL and ANDERSON [11] for an Einstein phonon spectrum. Coulomb interactions were included, following BOGOLIUBOV [7], by introducing a parameter μ^* which is an effective interaction covering the same energy range as the phonon spectrum and depends logarithmically on the phonon frequency cut-off

$$\mu^* = \frac{\langle N(0)\, V_{\text{coul}}\rangle}{1 + \langle N(0)\, V_{\text{coul}}\rangle \ln (E_F/\hbar\omega_c)}\,. \tag{31}$$

The parameter μ^* is essentially independent of frequency. Values of μ^* vary only slowly with electron density and other metallic parameters and are typically of order 0.1–0.2.

The $N(0)\, V$ of the BCS theory is approximately equal to $\lambda - \mu^*$. When the renormalization factor Z is taken into account, this becomes $(\lambda - \mu^*)/(1+\lambda)$. More elaborate expressions have been obtained by MCMILLAN [16] for strongly-coupled superconductors. The fact that μ^* depends on the frequency cut-off means that μ^* depends on isotopic mass, and this can cause departures of the exponent, α, in $T_c M^\alpha = \text{const}$ from the value 1/2. One may use the measured value of α to obtain an empirical estimate of μ^*.

SCHRIEFFER spent the summer of 1961 with a group who were developing methods for computer control using graphical display methods. He decided that the Eliashberg equations would provide a good test of the methods as well as yield valuable information for the physics of superconductivity. He, together with G. J. CULLER, B. D. FRIED and R. W. HUFF [12], calculated the complex $\Delta(\omega)$ as a function of energy for a model with a Debye frequency spectrum.

In the meantime, GIAEVER, HART and MEGERLE [13] had observed anomalies in the tunneling density of states that appeared to be associated with phonons. SCHRIEFFER continued work on the problem with a view to getting more accurate and realistic solutions of the Eliashberg equations to interpret the tunneling data. After going to the University of Pennsylvania in the fall of 1962, he was joined in the work by J. W. WILKINS, then a graduate student, and by D. J. SCALAPINO. They showed [14] that the tunneling density of states is given in terms of a complex $\Delta(\omega)$ by

$$\frac{(dI/dV)_S}{(dI/dV)_N} = \frac{N_S(\omega)}{N(0)} = \mathrm{Re}\left[\frac{\omega}{(\omega^2 - \Delta^2(\omega))^{1/2}}\right]. \tag{32}$$

Here $\hbar\omega = eV$ is the applied voltage on a normal-superconducting tunnel junction and $I(V)$ is the current-voltage characteristic of the junction. By use of Kramers-Kronig relations, both the real and imaginary parts of $\Delta(\omega)$ can be derived from tunneling data covering a complete range of voltages.

SCHRIEFFER et al. were able to get a good fit of experimental data on Pb with a phonon spectrum consisting of two Lorentzian peaks, one for transverse waves and one for longitudinal. By then, more complete and accurate tunneling data had been obtained by J. M. ROWELL. Solutions of the Eliashberg equations were extended to temperatures up to T_c for Pb, Hg and Al by SWIHART, WADA and SCALAPINO [15]. They found good agreement with experiment both for T_c and for the ratio $2\Delta(0)/k_B T_c$. The latter is anomalously large for strong coupling superconductors such as Pb and Hg. Again, a phonon spectrum consisting of two Lorentzian peaks was used, one for longitudinal and another for transverse phonons with positions and widths chosen to fit the tunneling data.

Another important step was made by W. L. MCMILLAN [16], who devised a computer program to derive $\alpha^2(\omega) F(\omega)$ directly from the tunneling data with use of the Eliashberg equations, where

$$\alpha^2(\omega) F(\omega) = \sum_\lambda \alpha_\lambda^2(\omega) F_\lambda(\omega). \tag{33}$$

Fig. 3 gives a comparison of $\alpha^2(\omega) F(\omega)$ derived from tunneling data for Pb by MCMILLAN and ROWELL and the phonon density of states obtained from neutron scattering data.

MCMILLAN'S program since has been used to analyze tunneling data for a number of other metals and alloys, including Al, Hg, Sn, transition elements Ta and Nb [17], a rare earth, La [18], and the compound superconductor, Nb_3Sn [19]. For a time arguments were given that other mechanisms of superconductivity might be important in transition elements and compounds. However, in

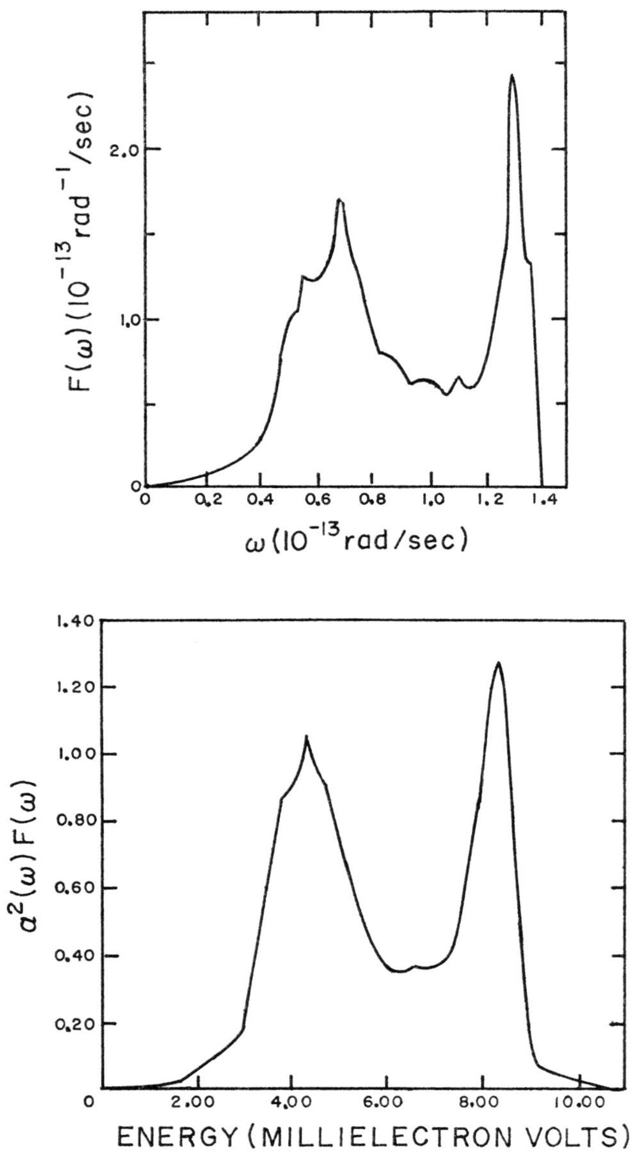

Fig. 3. Comparison between $\alpha^2(\omega)\, F(\omega)$ for Pb derived from tunneling data by McMILLAN and ROWELL and the phonon density of states, $F(\omega)$, derived by STEDMAN et al. from neutron scattering data. (After D. J. SCALAPINO, reference 4, p. 511)

all cases tunneling data indicate that the phonon mechanism is dominant. For example, analysis of the best junction made with tantalum [17] gives $\alpha^2(\omega)\, F(\omega)$ with phonon peaks at energies which agree with neutron scattering data, $\mu^* = 0.11$ and $\lambda = 0.69$, all in excellent accord with the phonon mechanism. Superconductivity in Nb_3Sn is also accounted for by the phonon mechanism with a reasonable μ^*. This is also true from analysis of recent tunneling measurements on lanthanum, for which a mechanism based on virtual excitation of close lying F-levels had once been postulated. If the F-levels are important, they must enter indirectly through the phonon coupling.

From the tunneling data one may derive $Z_s(\omega)$ and $\Delta(\omega)$, and thus $\Sigma_1(\omega)$ and $\Sigma_2(\omega)$ and the matrix Green's function. With knowledge of the Green's functions, one can in turn derive essentially all of the various thermal and transport properties of superconductors, such as specific heats and critical fields, electric and thermal conductivity, nuclear spin relaxation times, ultrasonic absorption, etc. Agreement between theory and experiment has been remarkably good, even in cases where there are large deviations from the law of corresponding states as a result of strong-coupling effects.

In 1962, ELIASHBERG [20] derived an expression for the free energy on the basis of the Fröhlich Hamiltonian; i.e. effects of Coulomb interactions were neglected except as they may occur in the renormalization of the quasi-particle energies. His expression has the useful feature that it is stationary with respect to variations of Σ_1, Σ_2 and π if these quantities satisfy the equations (25). Stephen and the author [21] made use of this property to derive a rapidly convergent expression for the free energy difference between superconducting and normal states, $\Omega_s - \Omega_n = -H_c^2/8\pi$.

$$H_c^2/8\pi = -\operatorname{Re} N(0) \int_0^\infty \Big\{ [Z_s(\omega) + Z_n(\omega)] \Big(-\omega + (\omega^2 - \Delta^2)^{1/2} + \frac{\Delta^2}{2(\omega^2 - \Delta^2)^{1/2}} \Big) + \frac{[Z_n(\omega) - Z_s(\omega)]\Delta^2}{2(\omega^2 - \Delta^2)^{1/2}} \Big\} \tanh \frac{\beta\omega}{2}\, d\omega. \tag{34}$$

This expression converges much more rapidly than one derived independently by WADA [15].

When the coupling is weak, $Z_s(\omega) \simeq Z_n(\omega)$ and $\Delta(\omega) \simeq \Delta(0)$. The expression (34) reduces to the one derived by BCS. At $T = 0$ °K one finds for the condensation energy

$$E_{\mathrm{BCS}} = \tfrac{1}{2} N(0)\, Z_n \Delta_0^2 \tag{35}$$

where $N(0)\, Z_n$ is the phonon enhanced density of states and Δ_0 is the gap parameter at $T = 0$ °K. However, for strong coupling there are significant differences. For example, for Pb the more exact expression (34) gives

$$E_{\mathrm{theor}} = 0.78\, E_{\mathrm{BCS}}. \tag{36}$$

Using experimental values for H_c and Δ_0, we find

$$E_{\mathrm{exp}} = (0.76 \pm 0.02)\, E_{\mathrm{BCS}} \tag{37}$$

in excellent agreement with the theoretical value. The theory also is in very good agreement with the temperature dependence of H_c, which differs significantly from the weak coupling limit.

Derivation of Electron-Phonon Interaction from Microscopic Theory

To be completely convincing and to be sure that nothing has been overlooked one would like to derive the electron-phonon interaction parameter $\alpha^2(\omega) F(\omega)$ from first principles. This is a truly formidable task, because it means that one should be able to derive the complete phonon spectrum $\omega(q)$ from basic theory as well as the electron-phonon interaction. With use of pseudo-potential methods it is possible to do this for simple metals such as Pb and Al. However, in most cases neutron scattering data has been used to obtain the phonon spectrum. Considerable progress has been made in fundamental calculations of the electron-phonon interaction in transition metals. The theory is simplest for "dirty" metals with a short scattering mean free path from impurities. Scattering around the Fermi surface washes out the anisotropy of Δ so that it is a function only of energy, not direction in k-space.

The first calculations of $\alpha^2(\omega) F(\omega)$ by SCALAPINO, WADA and SWIHART made use of pseudo-potential methods to estimate $\alpha_q(\omega)$ but used a simplified two peak phonon spectra derived from neutron scattering data. They found reasonable agreement with values of $\alpha^2(\omega) F(\omega)$ derived from tunneling data. They also calculated the tunneling density of states by solving the Eliashberg equations with an approximation to the theoretical $\alpha^2(\omega) F(\omega)$, and again got good agreement.

The most extensive calculations from fundamental theory have been by J. P. CARBOTTE and co-workers [22]. In most calculations they derived force constants from neutron scattering data measured in symmetry directions and then used these to derive the phonon spectrum for all wave vectors. These were combined with values of α^2 calculated from pseudo-potential theory to obtain $\alpha^2(\omega) F(\omega)$. The Eliashberg equations were then solved for $\Delta(\omega)$ and $Z(\omega)$ for several metals and alloys. For the gap edge, Δ_0, in Al at $T=0$ °K they found 0.19 meV compared with an experimental value of 0.17 meV. For Pb, the theoretical gap edge is 1.49 meV compared with an experimental 1.35 meV. For Na, not observed to be superconducting, the theoretical Δ_0 is less than 10^{-5} meV. They also calculated Δ_0 for Pb-Tl alloys of varying composition, again finding good agreement with experiment. Other calculations have included amorphous Bi and BiTl alloys. A simplified solution of the Eliashberg equations appropriate to weak and moderately coupled superconductors was applied to calculate Δ_0 for Al, Tl, In and Sn, and its variation with pressure, in satisfactory agreement with experimental values. There can thus be little doubt of the basic validity of the theory.

More recent extensions have been to the transition metals. Because of the complexities of the d-band and interactions between s- and d-bands, the theory is, more complicated and less certain, but there can be no doubt that the electron-phonon interaction is the basic mechanism for superconductivity in transition metals and compounds, including those of high T_c, as well as in the simple metals [16, 23].

An interesting application of basic theory is a calculation that indicates that a metallic form of hydrogen should exist at very high pressures and that this form should be superconducting [24]. Estimates of the transition temperature range as high as 200 °K. Several groups are attempting to reach the required pressures of over one million atmospheres.

Conclusions

FRÖHLICH's speculation of twenty odd years ago that electron-phonon interactions are responsible for superconductivity is now well established. Theory and experiment have developed to such an extent that very detailed information about the interaction parameters and their energy dependence can be derived from tunneling data. It is also possible to derive the interaction for simple metals from essentially first principles calculations, with excellent agreement with experiment. The methods are now being extended to more complex metals and metallic compounds, including those with high transition temperatures.

In all materials that have been studied by tunneling, the electron-phonon interaction has been shown to be the dominant mechanism responsible for superconductivity, with Coulomb interactions playing only a minor role. This is not to deny that with appropriate structures, other mechanisms, such a virtual exchange of excitons, may predominate.

In the future we may except to learn much more about the factors that influence the electron phonon interaction in more complex materials. Calculations from first principles are so difficult that theory never will replace experiment in the search for new materials, but with better understanding of the underlying mechanism theory should provide a more useful guide than it has provided in the past.

References

1. FRÖHLICH, H.: Phys. Rev. **79**, 845 (1950); — Proc. Roy. Soc. (London), Ser. A **215**, 291 (1952).
2. For a more complete discussion and references, see J. BARDEEN: Encyclopedia of Physics (S. FLUGGE, Ed.), Vol. XV, p. 274. Berlin: Springer 1956.
3. BARDEEN, J., COOPER, L. N., SCHRIEFFER, J. R.: Phys. Rev. **108**, 1175 (1957).
4. For recent review articles with references, see the chapters by D. J. SCALAPINO and by W. L. MCMILLAN and J. M. ROWELL in: Superconductivity (R. D. PARKS, Ed.), Vol. 1. New York: Marcel Bekker, Inc. 1969. An excellent reference for the theory and earlier experimental work is J. R. SCHRIEFFER: Superconductivity. New York: W. A. Benjamin, Inc. 1964.
5. MIGDAL, A. B.: Zh. Eksperim. i Teor. Fiz. **34**, 1438 (1958). [English transl: Soviet Phys. JETP **7**, 996 (1958).]
6. LONDON, F.: Superfluids. New York: John Wiley and Sons 1950.
7. See BOGOLIUBOV, N. N., TOLMACHEV, V. V., SHIRKOV, D. V.: A New Method in the Theory of Superconductivity. New York: Consultants Bureau, Inc. 1959.
8. GOR'KOV, L. P.: Zh. Eksperim. i Teor. Fiz. **34**, 735 (1958). [English transl. Soviet Phys. JETP **7**, 505 (1958).]
9. NAMBU, Y.: Phys. Rev. **117**, 648 (1960).
10. ELIASHBERG, G. M.: Zh. Eksperim. i Teor. Fiz. **38**, 966 (1960); — Soviet Phys. JETP **11**, 696 (1960).
11. MOREL, P., ANDERSON, P. W.: Phys. Rev. **125**, 1263 (1962).
12. CULLER, G. J., FRIED, B. D., HUFF, R. W., SCHRIEFFER, J. R.: Phys. Rev. Letters **8**, 339 (1962).
13. GIAEVER, I., HART, H. R., MEGERLE, K.: Phys. Rev. **126**, 941 (1962).
14. SCHRIEFFER, J. R., SCALAPINO, D. J., WILKINS, J. W.: Phys. Rev. Letters **10**, 336 (1963). SCALAPINO, D. J., SCHRIEFFER, J. R., WILKINS, J. W.: Phys. Rev. **148**, 263 (1966).

15. For references, see the review article of SCALAPINO, Ref. 4.
16. For references, see the review article of MCMILLAN and ROWELL, Ref. 4. For application to transition metals, see W. L. MCMILLAN, Phys. Rev. **167**, 331 (1968).
17. SHEN, L. Y. L.: Superconductivity in *d*- and *f*-Band Metals (D. H. DOUGLAS, Ed.), p. 31. New York: Amer. Inst. of Physics 1972.
18. LOU, L. F., TOMASCH, W. J.: Phys. Rev. Letters, **29**, 858 (1972).
19. SHEN, L. Y. L.: Phys. Rev. Letters, **29**, 1082 (1972)
20. ELIASHBERG, G. M.: Zh. Eksperim. i Teor. Fiz. **43**, 1005 (1962). [English transl. Soviet Phys. JETP **16**, 780 (1963).]
21. BARDEEN, J., STEPHEN, M.: Phys. Rev. **136**, A1485 (1964).
22. CARBOTTE, J. P.: Superconductivity (P. R. WALLACE, Ed.), Vol. 1, p. 491. New York: Gordon and Breach 1969.
CARBOTTE, J. P., DYNES, R. C.: Phys. Rev. **172**, 476 (1968).
LEAVANS, C. R., CARBOTTE, J. P.: Can. J. Phys. **49**, 724 (1970).
23. See articles by K. W. BRENNEMAN and J. W. GARLAND and by others in: Superconductivity in *d*- and *f*-Band Metals (D. H. DOUGLAS, Ed.). New York: Amer. Inst. of Physics 1972.
24. ASHCROFT, N. W.: Phys. Rev. Letters **21**, 1748 (1968).

The Mechanism for Superconductivity— Twenty-Two Years Later

G. RICKAYZEN

The University of Kent, The Physics Laboratories, Canterbury, Kent/England

1. Introduction

I refer, of course, to the publication in 1950 of the important paper [1] of FRÖHLICH on the possible role of virtual phonons in the mechanism for superconductivity. Up to that time the interaction between the electrons and the ions in a metal was treated in the adiabatic approximation which assumes that the wave function and energy of the electrons depends on the instantaneous positions of the ions. The energy of the electrons then contributes to the potential energy of the ions and determines the lattice vibrations and hence the phonons. The first correction to the approximation leads to the scattering of electrons with the emission and absorption of real phonons. FRÖHLICH pointed out that the then fairly recent developments in field theory showed that there would also be processes in which virtual phonons could be exchanged between electrons and these processes would lead to an interaction between the electrons. In collaboration with others [2], he had already applied these field theoretic ideas to consider the self-energy of an electron due to its interaction with lattice vibrations in a polar crystal. He now applied these ideas to the interaction of electrons with phonons in a metal.

Three important results emerged:

1. part of the resultant interaction between the electrons turned out to be attractive and could lead to a new favoured state of the electrons;

2. because the attraction was greater the stronger the electron-phonon interaction, superconductivity, if this was the correct mechanism, would be favoured in those metals which were poor conductors at room temperature;

3. the critical temperatures and critical magnetic fields of superconductors would depend on the isotopic mass of the ions.

The second result was well-known to be generally true. The third was being confirmed experimentally and independently [3] while FRÖHLICH was writing his paper. This showed clearly that the lattice was involved in the transition and suggested that FRÖHLICH's theory was on the right road. However, it became clear, especially through the work of SCHAFROTH [4], that, although phonons were involved in the transition, the wave function proposed by FRÖHLICH could not describe the electrodynamic properties of superconductors.

In order to separate the problem of finding the wave function for a superconductor from that of dealing with the electron-phonon interaction FRÖHLICH [5] proposed a canonical transformation for eliminating the electron-phonon interaction from the Hamiltonian and replacing it by an electron-electron interaction. The resulting Hamiltonian is

$$H_F = \sum_{\boldsymbol{k},\boldsymbol{\sigma}} \varepsilon_{\boldsymbol{k}} \overset{+}{c}_{\boldsymbol{k}\boldsymbol{\sigma}} c_{\boldsymbol{k}\boldsymbol{\sigma}} + \frac{1}{2} \sum_{\boldsymbol{k},\boldsymbol{k}',\boldsymbol{q}} \frac{\hbar\omega_{\boldsymbol{q}} |g_{\boldsymbol{q}}^2|}{(\varepsilon_{\boldsymbol{k}} - \varepsilon_{\boldsymbol{k}+\boldsymbol{q}})^2 - (\hbar\omega_{\boldsymbol{q}})^2} \overset{+}{c}_{\boldsymbol{k}+\boldsymbol{q}\boldsymbol{\sigma}} \overset{+}{c}_{\boldsymbol{k}'\boldsymbol{\sigma}'} c_{\boldsymbol{k}'+\boldsymbol{q}'\boldsymbol{\sigma}'} c_{\boldsymbol{k}\boldsymbol{\sigma}} \tag{1}$$

where $\overset{+}{c}_{\boldsymbol{k}\boldsymbol{\sigma}}$ is the creation operator for an electron of spin $\boldsymbol{\sigma}$, crystal wave-vector $\boldsymbol{k}$, $\varepsilon_{\boldsymbol{k}}$ is the corresponding Bloch energy, $\omega_{\boldsymbol{q}}$ is the angular frequency of a phonon of wave-vector $\boldsymbol{q}$ and $g_{\boldsymbol{q}}$ is the constant coupling such a phonon to the electrons. The Hamiltonian is widely known as the Fröhlich Hamiltonian. This work was later extended by BARDEEN and PINES [6] to include a microscopic treatment of the Coulomb interaction between the electrons and led to a similar result to FRÖHLICH's with the addition of screened Coulomb interaction:

$$H = H_F + \tfrac{1}{2} \sum_{\substack{\boldsymbol{k},\boldsymbol{k}',\boldsymbol{q} \\ \boldsymbol{\sigma}\boldsymbol{\sigma}'}} V_s(\boldsymbol{q}) \overset{+}{c}_{\boldsymbol{k}+\boldsymbol{q}\boldsymbol{\sigma}} \overset{+}{c}_{\boldsymbol{k}'\boldsymbol{\sigma}'} c_{\boldsymbol{k}'+\boldsymbol{q}'\boldsymbol{\sigma}'} c_{\boldsymbol{k}\boldsymbol{\sigma}} \tag{2}$$

where $V_s(\boldsymbol{q})$ is the Fourier transform of the screened Coulomb interaction. It is now realised that these calculations essentially determine the dynamic frequency, ω, and wave-vector, $\boldsymbol{q}$, dependent dielectric constant of the material $\varepsilon(\boldsymbol{q},\omega)$ and that the Coulomb interaction between the electrons is screened by this. In fact, in the above Hamiltonians the dynamic screening is replaced by a static one and this is achieved, in effect, by the following consideration. The interaction gives rise to the scattering of an electron from a state $\boldsymbol{k}$ to astate $\boldsymbol{k}+\boldsymbol{q}$. Such a real scattering would be accomplished by an external field of angular frequency ω provided

$$\hbar\omega = \varepsilon_{\boldsymbol{k}} - \varepsilon_{\boldsymbol{k}+\boldsymbol{q}}.$$

Thus ω in the dielectric constant is replaced by $(\varepsilon_{\boldsymbol{k}} - \varepsilon_{\boldsymbol{k}+\boldsymbol{q}})/\hbar$. This is clearly an approximation that assumes that the electron energies are not strongly perturbed by the scattering and that the scattering is not far from real. In fact, for the weak coupling pure superconductors ($T_c/\theta_D \ll 1$) the approximation works well. For other cases, it has to be modified and, in particular, the dynamic nature of the screening must be taken into account properly [7, 8, 9].

The properties of the dielectric constant of the electrons and the lattice can be understood on the basis of the following classical model which has been described more fully elsewhere [10]. Imagine the lattice to be a uniform jelly with uniform charge density $-\varrho_0$ in equilibrium. Imagine the electrons to form a charged fluid permeating the lattice and with charge density ϱ_0 in equilibrium. The overall system is neutral in equilibrium. In some small motion of the system we can ascribe a local displacement $\boldsymbol{u}_e(\boldsymbol{r})$ to the electron fluid and a local displacement $\boldsymbol{u}_L(\boldsymbol{r})$ to the ion jelly. The overall charge density now becomes

$$\varrho_0(1 - \operatorname{div} \boldsymbol{u}_e) - \varrho_0(1 - \operatorname{div} \boldsymbol{u}_L) = \varrho_0 \operatorname{div}(\boldsymbol{u}_L - \boldsymbol{u}_e).$$

In general, this implies the existence of an electric field $\boldsymbol{E}(\boldsymbol{r})$ governed by POISSON's equation

$$\operatorname{div} \boldsymbol{E} = 4\pi \varrho_0 \operatorname{div}(\boldsymbol{u}_L - \boldsymbol{u}_e). \tag{3}$$

In turn, the electric field causes the fluid and jelly to move according to the fluid equations of motion

$$\ddot{\boldsymbol{u}}_e = \frac{-e \boldsymbol{\nabla} p}{m \varrho_0} + \frac{e \boldsymbol{E}}{m} \tag{4}$$

$$\ddot{\boldsymbol{u}}_L = -\frac{Z e \boldsymbol{E}}{M}. \tag{5}$$

Here, e/m is the charge to mass density of the electron fluid, Ze/M is the charge to mass density of the ion jelly and p is the pressure of the electron fluid. The pressure of the ion jelly is sufficiently small to be ignored. In general p is a function of ϱ and

$$\nabla p = \frac{dp}{d\varrho} \nabla \varrho \approx -\frac{dp}{d\varrho_0} \varrho_0 \boldsymbol{\nabla} \operatorname{div} \boldsymbol{u}_e.$$

We can calculate the dependence of p on ϱ assuming that the electrons form a degenerate Fermi gas as follows. The total energy density and the charge density of the gas are given respectively by

$$E = \frac{2}{(2\pi)^3} \int d^3k\, \varepsilon_{\boldsymbol{k}}\, \theta(\varepsilon_{\boldsymbol{k}} - \varepsilon_F), \qquad \varrho_0 = \frac{2e}{(2\pi)^3} \int d^3k\, \theta(\varepsilon - \varepsilon_F), \tag{6}$$

and the pressure is given by

$$p = -\frac{\partial(EV)}{\partial V} = -E - V\frac{\partial E}{\partial V} = -E + \varrho_0 \frac{\partial E}{\partial \varrho_0}.$$

Therefore,

$$\frac{dp}{d\varrho_0} = \varrho_0 \frac{\partial^2 E}{\partial \varrho_0^2}.$$

But, from equations (6)

$$1 = \frac{\partial \varrho_0}{\partial \varrho_0} = + \frac{2e}{(2\pi)^3} \int d^3k\, \delta(\varepsilon - \varepsilon_F) \frac{\partial \varepsilon_F}{\partial \varrho_0} = 2e\, N(0) \frac{\partial \varepsilon_F}{\partial \varrho_0},$$

and

$$\frac{\partial E}{\partial \varrho_0} = \frac{2}{(2\pi)^3} \int d^3k\, \varepsilon_{\boldsymbol{k}}\, \delta(\varepsilon - \varepsilon_F) \frac{\partial \varepsilon_F}{\partial \varrho_0} = 2N(0)\, \varepsilon_F \frac{\partial \varepsilon_F}{\partial \varrho_0} = \varepsilon_F/e.$$

Hence

$$\frac{\partial^2 E}{\partial \varrho_0^2} = \frac{\partial \varepsilon_F}{e\, \partial \varrho_0} = \frac{1}{2e^2 N(0)}$$

and

$$\frac{dp}{d\varrho_0} = \frac{\varrho_0}{2e^2 N(0)}.$$

We shall write

$$v_s = \sqrt{\frac{e\, dp}{m\, d\varrho_0}} = \sqrt{\frac{\varrho_0}{2em\, N(0)}}.$$

For the free electron gas v_s is $V_F/\sqrt{3}$. With this notation the equation of motion for the electron fluid becomes

$$\ddot{\boldsymbol{u}}_e - v_s^2 \operatorname{grad} \operatorname{div} \boldsymbol{u}_e = e\boldsymbol{E}/m. \tag{7}$$

The Eqs. (3), (5) and (7) yield the linear collective oscillations. Since the system is uniform one finds oscillations with definite wavevectors $\boldsymbol{q}$. These are of two kinds, the plasma oscillations of the electron fluid with angular frequencies given by

$$\omega_p(\boldsymbol{q}) = \left(\frac{4\pi e \varrho_0}{m} + v_s^2 q^2\right)^{\frac{1}{2}}$$

and the longitudinal phonons with frequencies given by

$$\omega_{\boldsymbol{q}} = (Zm/M)^{\frac{1}{2}}\, \omega_p(0)\, v_s q / [1 + v_s^2 q^2/\omega_p(0)^2]^{\frac{1}{2}}.$$

The model is designed chiefly for dealing with the long-range Coulomb interactions and is not suitable for yielding the transverse phonons where ion-ion interactions are dominant and the lattice is important.

To calculate the dielectric constant of the system we add an external charge density $\varrho_{\boldsymbol{q}} e^{i\boldsymbol{q}\boldsymbol{r}-i\omega t}$ with wave-vector $\boldsymbol{q}$ and angular frequency ω and calculate the total resulting forced electric field. The only change in the equations is that Poisson's equation is now

$$\operatorname{div} \boldsymbol{E} = 4\pi\varrho_0 \operatorname{div}(\boldsymbol{u}_L - \boldsymbol{u}_e) + 4\pi\varrho_{\boldsymbol{q}} e^{i\boldsymbol{q}\cdot\boldsymbol{r}-i\omega t}.$$

The solution of the equations in which all fields vary as $\exp(i\boldsymbol{q}\cdot\boldsymbol{r} - i\omega t)$ yields

$$\boldsymbol{E} = -\frac{4\pi e^2 i\boldsymbol{q}\varrho_{\boldsymbol{q}}}{q^2}\,\frac{v_s^2 q^2}{\omega_p(0)^2 + v_s^2 q^2}\left(1 + \frac{\omega_q^2}{\omega^2 - \omega_q^2}\right).$$

Since the field due to the external charge in vacuo is

$$\boldsymbol{E}_0 = -\frac{4\pi e^2 i\boldsymbol{q}\,\varrho_{\boldsymbol{q}}}{q^2},$$

the dielectric constant of the medium is $\varepsilon(\boldsymbol{q}, \omega)$ where

$$\varepsilon(\boldsymbol{q}, \omega)^{-1} = \frac{v_s^2 q^2}{\omega_p(0)^2 + v_s^2 q^2}\left(1 + \frac{\omega_q^2}{\omega^2 - \omega_q^2}\right).$$

This yields the interaction between the electrons due to their interaction with the longitudinal phonons as calculated by Fröhlich [5] and Bardeen and Pines [6].

The important property of the dielectric constant is that, because of a resonance in the lattice, it possesses an anomaly where the frequency is equal to a frequency of the freely oscillating lattice. It vanishes at this frequency and is negative for lower frequencies, positive for higher frequencies. Since the Coulomb interaction is repulsive, the screened interaction is attractive for low frequencies and repulsive for high frequencies. It is the attraction which is exploited in the theory of superconductivity.

The resulting potential between the electrons falls naturally into two parts the first is

$$V_C = \frac{4\pi e^2 v_s^2}{\omega_p(0)^2 + v_s^2 q^2} \tag{8}$$

and exists even if the phonons can be ignored (in the limit $M \to \infty$). This is the screened Coulomb interaction and is repulsive. The remainder of the interaction is

$$V_{ph} = \frac{4\pi e^2 v_s^2 \omega_q^s}{(\omega_p(0)^2 + v_s^2 q^2)}\,\frac{1}{\omega^2 - \omega_q^2} \tag{9}$$

and is due to the interaction between the electrons and the phonons. This is the part which is attractive for low frequencies and which changes sign. It is clear that the electrons force the lattice to resonate at the lattice frequencies and this resonance is responsible for the change of sign.

We can expect that whenever the electrons interact with a system with a resonance frequency ω_0 there is a change of sign of the resultant interaction between the electrons at this resonance frequency. A simple extension of the previous model shows that this is so for the transverse and other phonons. The different phonon branches do not mix and one is led to an extra interaction between the electrons of the form [5, 6, 10]

$$\sum_{\boldsymbol{k},\boldsymbol{Q},\boldsymbol{K}} \frac{|g_{\boldsymbol{Q},\boldsymbol{K},\lambda}|^2 c^+_{\boldsymbol{k}+\boldsymbol{Q}\sigma} c^+_{\boldsymbol{k}'\sigma'} c_{\boldsymbol{k}'+\boldsymbol{Q}\sigma'} c_{\boldsymbol{k}\sigma}}{M[(\varepsilon_{\boldsymbol{k}}-\varepsilon_{\boldsymbol{k}-\boldsymbol{Q}})^2-\omega^2_{\boldsymbol{Q}-\boldsymbol{K}}]} \tag{10}$$

where M is the average ionic mass, $\boldsymbol{K}$ is a lattice wave-vector and $g_{\boldsymbol{Q},\boldsymbol{K},\lambda}/(M\omega_{\boldsymbol{Q}-\boldsymbol{K}})^{-\frac{1}{2}}$ is the matrix element for the scattering of an electron from the state $\boldsymbol{k}$ to $\boldsymbol{k}+\boldsymbol{Q}$ with the emission or absorption of a phonon of wave-vector $\boldsymbol{Q}-\boldsymbol{K}$, angular frequency $\omega_{\boldsymbol{Q}-\boldsymbol{K}}$. If $\boldsymbol{K}$ is not zero the scattering is an Umkapp process. The coupling g is independent of the isotopic mass. The interaction (10), as expected, changes sign at $|\varepsilon_{\boldsymbol{k}}-\varepsilon_{\boldsymbol{k}-\boldsymbol{Q}}|=\hbar\omega_{\boldsymbol{Q}-\boldsymbol{K}}$ and is attractive for $|\varepsilon_{\boldsymbol{k}}-\varepsilon_{\boldsymbol{k}-\boldsymbol{Q}}|<\hbar\omega_{\boldsymbol{Q}-\boldsymbol{K}}$.

2. Isotope Effect

This arises because of the dependence of the interaction on $\omega_{\boldsymbol{Q}}$ which itself is proportional to $M^{-\frac{1}{2}}$. If only the electron-phonon interaction is important then the only energies which can be used to measure $k_B T_c$ are the phonon energies $\hbar\omega_{\boldsymbol{Q}}$. Since these are all proportional to $M^{-\frac{1}{2}}$ it follows that

$$T_c \propto M^{-\alpha}$$

with $\alpha=\frac{1}{2}$. This was the result deduced by Fröhlich and by Bardeen, Cooper and Schrieffer [11] when they introduced their theory. Since many metals do not have $\alpha=\frac{1}{2}$ it was thought for some time that in these metals some other interaction must be important as well. However, it was pointed out, at first Bogoliubov, Tolmachev and Shirkov [12], that with a proper treatment of the Coulomb interaction α need not be $\frac{1}{2}$.

According to the BCS theory the transition temperature, T_c, of a superconductor is the highest temperature for which the equation

$$\Delta_{\boldsymbol{k}} = \sum_{\boldsymbol{k}'} V_{\boldsymbol{k},\boldsymbol{k}'}\, \Delta_{\boldsymbol{k}'} \frac{\tanh \varepsilon'/2k_B T_c}{\varepsilon'} \tag{11}$$

has a non-trivial solution, $\Delta_{\boldsymbol{k}}$. Here $V_{\boldsymbol{k},\boldsymbol{k}'}$ is the matrix element of the interaction for scattering a pair of electrons from the Bloch states $\boldsymbol{k}\uparrow$, $-\boldsymbol{k}\downarrow$ to $\boldsymbol{k}'$, $\uparrow$, $-\boldsymbol{k}'\downarrow$. If only the phonons are important $V_{\boldsymbol{K}\boldsymbol{K}'}$ has the form

$$\Delta_{\boldsymbol{k}} = \sum_{\boldsymbol{K}',\lambda} \frac{\Delta_{\boldsymbol{k}'}}{\varepsilon'} \tanh(\varepsilon'/2k_B T_c) \frac{|g_{\boldsymbol{k}-\boldsymbol{k}',\boldsymbol{K},\lambda}|^2}{M[(\varepsilon_{\boldsymbol{k}}-\varepsilon_{\boldsymbol{k}'})^2-(\hbar\omega_{\boldsymbol{k}-\boldsymbol{k}'-\boldsymbol{K},\lambda})^2]} \tag{12}$$

where there is a contribution from each phonon branch λ and all the $|g_{\boldsymbol{k}-\boldsymbol{k}',\boldsymbol{K},\lambda}|^2$ are independent of M. The sum over energies will, for $\varepsilon_{\boldsymbol{k}}$ near the Fermisurface, be confined to energies near the Fermi surface. Hence, this sum can be replaced by an integral, the density of states at the Fermi surface, $N(0)$, being introduced. Then

$$\Delta_{\boldsymbol{k}} = N(0) \int d\varepsilon' \frac{d\Omega'}{4\pi} \frac{\Delta_{\boldsymbol{k}'}}{\varepsilon'} \tanh(\varepsilon'/2k_B T_c) \sum_{\lambda} \frac{|g|^2}{[M-(\varepsilon-\varepsilon')^2-(\hbar\omega_{\boldsymbol{k}-\boldsymbol{k}',\lambda})^2]} \,.$$

If ε and ε' are replaced by x and x' where

$$x = M^{\frac{1}{2}}\varepsilon, \qquad x' = M^{\frac{1}{2}}\varepsilon',$$

$$\Delta_{\boldsymbol{k}} = N(0) \int dx' \frac{d\Omega'}{4\pi} \frac{\Delta_{\boldsymbol{k}'}}{x'} \tanh\left(\frac{x'}{2M k_B T_c}\right) \sum_{\lambda} \frac{|g|^2}{(x-x')^2 - M(\hbar\omega)^2}$$

In this equation the ionic mass appears only in the combination $M^{\frac{1}{2}}T_c$. Hence, we must have

$$T_c \propto M^{-\frac{1}{2}}.$$

To see the effect of the Coulomb interaction we look at the model of BOGOLIUBOV, TOLMACHEV and SHIRKOV [12]. The interaction terms of the Hamiltonian which are important for superconductivity are the pairing terms, and these are of the form

$$\sum_{\boldsymbol{k},\boldsymbol{k}'} V(\boldsymbol{k},\boldsymbol{k}')\, c^+_{\boldsymbol{k}\uparrow} c^+_{-\boldsymbol{k}\downarrow} c^+_{-\boldsymbol{k}'\downarrow} c_{\boldsymbol{k}'\uparrow} \,.$$

The eigenvalue equation for T_c is then

$$\Delta_{\boldsymbol{k}} = \sum_{\boldsymbol{k}'} V(\boldsymbol{k},\boldsymbol{k}') \frac{\Delta_{\boldsymbol{k}'}}{|\varepsilon'|} \tanh\left(\frac{\varepsilon'}{2k_B T_c}\right).$$

The equation above is a special case of this. Now the phonon part of $V(\boldsymbol{k}, \boldsymbol{k}')$ is well simulated by the potential V_{ph} defined by

$$\begin{aligned} V_{ph} &= -V, \text{ a negative constant, } |\varepsilon'|,\ |\varepsilon| < \hbar\omega \\ &= 0, \quad \text{otherwise.} \end{aligned} \tag{13}$$

In the model of BOGOLIUBOV, TOLMACHEV and SHIRKOV, the Coulomb part of $V(\boldsymbol{k}, \boldsymbol{k}')$ is similated by a potential V_C where

$$\begin{aligned} V_C &= V_1, \text{ a positive constant, } |\varepsilon'|,\ |\varepsilon| < \xi_c \\ &= 0, \text{ otherwise,} \end{aligned} \tag{14}$$

and ξ_c is an energy of the order of E_F. The constants V, V_1, ξ_c are all independent of ionic mass. Then

$$\begin{aligned} \Delta = \Delta_1, &\ |\varepsilon| < \hbar\omega \\ \Delta_2, &\ \hbar\omega < |\varepsilon| < \xi_c \\ 0, &\ |\varepsilon| > \xi_c. \end{aligned} \tag{15}$$

If one takes the density of electron states in energy to have the constant value $N(0)$ for all the energies for which Δ is non-zero, the Eqs. (12—15) yield the solution

$$k_B T_c = \hbar\omega \exp(-1/\varrho) \tag{16}$$

where

$$\varrho = N(0)\, V - \frac{N(0)\, V_1}{1 + N(0)\, V_1 \ln(\xi_c/\hbar\omega)}\,. \tag{17}$$

Because of the effect of the Coulomb interaction at energies more than $h\omega$ from the Fermi surface, the Coulomb interaction parameter is reduced by the factor $[1 + N(0)\, V_1 \ln(\xi_c/\hbar\omega)]$. The whole of the second term is often referred to as the Coulomb pseudopotential. It depends on the ionic mass through $h\omega$ and so changes the parameter α of the isotope effect. In fact, in this model

$$\alpha = -\frac{d \ln T_c}{d \ln M} = \frac{1}{2} + \frac{h\omega}{2\varrho^2}\frac{d\varrho}{d(\hbar\omega)} = \frac{1}{2} - \frac{1}{2\varrho^2}\left[\frac{(N(0)\, V_1)}{1 + N(0)\, V_1 \ln(\xi_c/\hbar\omega)}\right]^2. \tag{18}$$

Hence, even within this simple model, a wide range of values of α (less than $\frac{1}{2}$) is possible including, accidentally, zero. Thus even the absence of an isotope effect is not a clear indication that the phonons are not involved in the transition.

The deviations of α from $\frac{1}{2}$ clearly depend on the Coulomb pseudopotential and this depends on the properties of electrons out to the band edge. For these electrons the assumptions of the model, a constant density of states and a constant strength of interaction, are too crude and a reliable calculation must include a full treatment of the effects of band structure. Such treatments have been attempted by GARLAND [13] and SWIHART [14] and the reader is referred to their papers for details. Two other simplifications of the theory outlined here have also to be abandoned.

The first simplification we have used is to treat the interaction between the electrons as a static one and, in particular, we have replaced the dynamic dielectric constant $\varepsilon(\boldsymbol{q}, \omega)$ by a static one $\varepsilon[\boldsymbol{q}, (\varepsilon_{\boldsymbol{k}+\boldsymbol{q}} - \varepsilon_{\boldsymbol{k}})/\hbar]$. In the full treatment of the isotope effect the dynamic properties must be treated properly.

The second simplification is that the self-energy effects of the electrons have been ignored. These, however, make a significant contribution to the coupling constant ϱ and must also be included.

MCMILLAN has made a calculation of T_c for a wide variety of metals and alloys avoiding these two simplifications although his treatment of the Coulomb pseudopotential is still crude. His result for T_c is

$$T_c = \frac{\Theta}{1.45} \exp\left[-\frac{1.04\,(1+\lambda)}{\lambda - \mu^*(1 + 0.62\,\lambda)}\right] \tag{19}$$

where Θ is the Debye frequency and μ^* is the Coulomb pseudopotential given, as above, by

$$\mu^* = \frac{N(0)\, V_1}{1 + N(0)\, V_1 \ln(\xi_c/\omega_c)}\,, \tag{20}$$

and ξ_c is taken to be the band width. The parameter λ is an effective electron-phonon coupling constant and is defined by

$$\lambda = \frac{N(0)\,\langle g^2\rangle}{M\,\langle\omega^2\rangle} \tag{21}$$

where $\langle\omega^2\rangle$ as an appropriate average of the square of the phonon frequencies and g^2 is an appropriate average of the coupling constant in the interaction (10). It is assumed in the calculation that impurities are present in sufficient numbers for anisotropic effects to be unimportant. Eq. (19) has been used with success [15—18] to understand the variation of T_c amongst different groups of superconductive materials. A modified form has been used by Hertel [19]. More recently it has been generalised by Garland and Allen [20] to improve the agreement between theory and experiment. Despite this success of these equations in improving our understanding of known superconductors they have not yet been strikingly helpful in finding new superconductors. It has also been used to set theoretical upper limits on the attainable values of T_c for different kinds of material [15, 21]. However, as Matthias [22] has quite rightly pointed out, simple empirical rules have been better guides to high temperature superconductors.

3. Other Possible Mechanisms

Although it is, in principle, possible to explain the observed isotope effects assuming only the Coulomb and electron-phonon interactions, there is no reason to suppose that these are the only important interactions which could lead to superconductivity. Indeed, although the study of the electron-phonon interaction pointed physicists in the right direction for a theory, the theory of BCS does not depend strongly on this interaction. Almost any attractive interaction between the electrons could apparently lead to superconductivity and such possibilities might exist in some metals. Also any repulsive interaction could impede superconductivity. Two interactions whose influence on superconductivity have been considered are the interaction between electrons and acoustic plasmons in a narrow d-band and also the interaction between electrons and spin waves in nearly ferromagnetic materials. If either of these possibilities or any other which does not involve the lattice is alone responsible for superconductivity, then there would be no isotope effect and α would be exactly zero. In practice, it is rare to find a superconductor in which α is zero so that it seems that the electron-phonon interaction must be involved in all the known superconducting transitions. The other mechanisms can play some part and we shall consider them in turn. There seems to be no dispute that the electron-phonon and Coulomb interactions are the dominant ones in the $s-p$ metals. For metals with more complicated band structures, the transition metals, the rare earths and the actinides, it is possible that the other mechanisms play some part. The more complicated band structure itself, however, makes a direct comparison of theory with experiment quite difficult.

a) Acoustic Plasmons

The suggestion that these play a role in superconductivity has come first from Radhakrishnan [23] and independently from Fröhlich [24] and it has been developed by Rothwarf [25] and by Geilikman [26]. It has not been unequivocally demonstrated that these plasmons exist in real metals but, after a review of the experimental evidence Rothwarf concludes that they do. Fröhlich [27] and Ganguly and Wood [27a] suggest that they may help explain the phonon dispersion curves of some superconductive transition metals. The basic idea is that if we have two unfilled bands of electrons, say d-band and s-band, and if the electrons in one band (d-band) have much greater inertia than those in the other (s-band), the latter electrons will screen the former just as the electrons screen the ions and there will be a collective mode with the dispersion relation of a sound wave. Because of the frequencies involved the scattering of electrons will cause the collective modes to have a finite lifetime and this may be too short for the acoustic plasmon to be observed.

The theory of acoustic plasmons has been developed by several authors [28–30]. One can see the origin of the acoustic plasmon with a slightly modified version of the electron-lattice model. We suppose we have two charged Fermi-fluids, s and d, in which we can define displacements $\boldsymbol{u}_s$ and $\boldsymbol{u}_d$. We shall assume that the frequencies of the modes we are looking for are much greater than those of the lattice, with the result that the lattice can be assumed to be at rest. It provides a uniform charged background to ensure overall charge neutrality. We take the equilibrium charge densities of the two fluids to be ϱ_s and ϱ_d. Then the equations of motion are

$$\begin{aligned} \ddot{\boldsymbol{u}}_s - V_s^2 \operatorname{grad} \operatorname{div} \boldsymbol{u}_s &= e\boldsymbol{E}/m_s \\ \ddot{\boldsymbol{u}}_d - V_d^2 \operatorname{grad} \operatorname{div} \boldsymbol{u}_d &= e\boldsymbol{E}/m_d \end{aligned} \tag{22}$$

where m_s, m_d are the effective masses in the two bands and V_s and V_d are the sound velocities for the corresponding neutral fluids and are given by

$$V_s = \sqrt{\frac{\varrho_s}{2e\, m_s\, N_s(0)}}\,, \qquad V_d = \sqrt{\frac{\varrho_d}{2e\, m_d\, N_d(0)}}\,. \tag{23}$$

The electric field is determined from Poisson's equation,

$$\operatorname{div} \boldsymbol{E} = -4\pi\,[\varrho_s \operatorname{div} \boldsymbol{u}_s + \varrho_d \operatorname{div} \boldsymbol{u}_d - \varrho_a] = +4\pi\varrho_t \tag{24}$$

where ϱ_a is any other induced charge in the system, coming from, say, interband transitions. From the equations of motion, we find that for a motion with definite wave-vector $\boldsymbol{q}$ and angular frequency ω,

$$\begin{aligned} \operatorname{div} \boldsymbol{u}_s = \frac{e \operatorname{div} \mathbf{E}/m_s}{-\omega^2 + V_s^2 q^2} &= -\frac{4\pi e\, \varrho_t/m_s}{\omega^2 - V_s^2 q^2} \\ \operatorname{div} \boldsymbol{u}_d \qquad\qquad &= -\frac{4\pi e\, \varrho_t/m_d}{\omega^2 - V_d^2 q^2}\,. \end{aligned}$$

If, furthermore, when ϱ_s and ϱ_d can be ignored the dielectric constant is ε_0, then

$$\varrho_a = (1-\varepsilon_0)\,\varrho_t. \tag{25}$$

Hence, dividing Eq. (24) by ϱ_t, we have

$$1 = \frac{4\pi e\, \varrho_s/m_s}{\omega^2 - V_s^2 q^2} + \frac{4\pi e\, \varrho_d/m_d}{\omega^2 - V_d^2 q^2} + 1 - \varepsilon_0. \tag{26}$$

This is the dispersion equation for the collective modes of the electrons.

For a narrow d-band and broad s-band, $V_s \gg V_d$. Hence for q, ω such that

$$V_s q \gg \omega \gg V_d q \tag{27}$$

we find

$$\varepsilon_0 = -\frac{4\pi e\, \varrho_s/m_s}{V_s^2 q^2} + \frac{4\pi e\, \varrho_d/m_d}{\omega^2}\,. \tag{28}$$

As $q \to 0$, $\omega \to (\varrho_d m_s/\varrho_s m_d)^{\frac{1}{2}} V_s q$ and we have an acoustic dispersion relation corresponding to the acoustic plasmon. In this analysis we have assumed that the d-band is free-electron like and this is far from the truth. However, for sufficiently large ω, the polarisability of the d-band will have the assumed form proportional to ω^{-2} [24]. We can, therefore, regard $(4\pi\varrho_d/m_d)$ as a parameter, the constant of proportionality. This analysis leaves out of account the effects of the lifetime of the electrons and of the electron-plasmon interaction. The former of these has been discussed in what is, effectively, a free electron model by ROTHWARF [25] who concludes that the acoustic plasmon can exist at low temperatures in, for example, Nb and Nb_3Sn.

The latter, that is, the electron-plasmon interaction, leads to a finite lifetime for the plasmons through the creation of d electron-hole pairs. This creation process is forbidden by the conservation of energy and momentum if the angular frequency of the plasmon, ω_q, and its wave-vector, $\boldsymbol{q}$, are related by

$$\omega_q > v_F q, \tag{29}$$

where v_F is the Fermi velocity of d-electrons. Since the dispersion equation for the plasmons is Eq. (28), there is a wave-vector $\boldsymbol{Q}$ given by Eq. (28) and

$$\omega_Q = v_F Q, \tag{30}$$

such that plasmons with wave-vectors above this value have a finite lifetime. The equation for Q is

$$\varepsilon_0 = -\frac{4\pi e\,\varrho_s/m_s}{V_s^2 Q^2} + \frac{4\pi e\,\varrho_d/m_d}{v_F^2 Q^2}. \tag{31}$$

The dielectric constant of the material for frequencies near the plasmon frequency is obtained by adding an external charge ϱ_e, proportional to $\exp(i\boldsymbol{q}\cdot\boldsymbol{r} - i\omega t)$, to the system so that

$$\varrho_t = -\varrho_s \operatorname{div}\boldsymbol{u}_s - \varrho_d \operatorname{div}\boldsymbol{u}_d + \varrho_a + \varrho_e. \tag{32}$$

Then the dielectric constant $\varepsilon(\boldsymbol{q}, \omega)$ is defined so that, in the forced solution of the equations of motion,

$$\varrho_t = \varrho_e/\varepsilon(\boldsymbol{q}, \omega). \tag{33}$$

Using the expressions for $\boldsymbol{u}_s$, $\boldsymbol{u}_d$, ϱ_a in terms of ϱ_t that we found earlier, we now find from Eq. (32) that

$$\varepsilon(\boldsymbol{q}, \omega) = \varepsilon_0 + \frac{4\pi e\,\varrho_0/m_s}{V_s^2 q^2} - \frac{4\pi e\,\varrho_d/m_d}{\omega^2} = \frac{4\pi e\,\varrho_d}{m_d \omega_q^2}\,\frac{\omega^2 - \omega_q^2}{\omega^2}\,(V_s q \gg \omega \gg V_d q). \tag{34}$$

This has a form similar to that arising from the interaction of the electrons with the ions. To make the analogue of the BCS one-parameter model of superconductivity, FRÖHLICH and ROTHWARF cut-off the interaction at the wave-vector Q given by Eq. (31). Then the Debye frequency in the BCS formulae is replaced by ω_Q. From (34) the strength of the interaction is

$$\frac{V(q)\, m_d\, \omega_q^2}{4\pi e\,\varrho_d} \tag{35}$$

where $V(\boldsymbol{q})$ is the Coulomb potential. In order that this interaction should yield a transition temperature comparable with that due to the electron-phonon interaction it is necessary that these quantities should be comparable with their counterparts. It is also necessary for Q to be greater than the Fermi wave-vector for the s-band so that the scattering of pairs of electrons to form the superconductive state can take place across the Fermi surface. Otherwise the amount of phase space available for the interaction would be severely reduced. So far there have been no realistic calculations to obtain the effect of acoustic plasmons on T_c. There is strictly no need to cut-off the interaction at the frequency ω_Q. Above this frequency the dielectric constant is complex but one can set up an integral equation for the gap parameter analogous to that used by McMillan to deal with the electron-phonon interaction. So far this has not been done.

b) Spin Fluctuations

The spin fluctuations arise from the exchange interaction between electrons. Hence there is no simple classical or semiclassical theory of their effect on superconductivity. There are, however, certain resemblances between phonons and spin fluctuations which are best brought out by reference to diagrams related to perturbation theory.

There are two effects of phonons which are important for superconductivity. One is their effect on the self-energy of the electrons and is illustrated in Fig. 1. This effect exists in the normal state as well and leads to a change in the mass of electrons. The second effect is the repeated scattering of a pair of electrons which results in the formation of Cooper pairs. This is illustrated in Fig. 2.

It is the interaction in this diagram which we calculated in Section 1.

The spin fluctuations assume importance when the metal is nearly ferromagnetic. Then the propagator of an electron-hole pair, as illustrated in Fig. 3, becomes large as the total momentum and energy of the pair tends to zero. In this case, the contribution to the susceptibility of the diagrams of Fig. 4 which contain the electron-hole propagator are also large. For a simple model in which the Coulomb interaction is treated as a δ-function interaction, $V_c\delta(\boldsymbol{r}-\boldsymbol{r}')$, the static spin susceptibility [31, 32] is

$$\chi=\chi_0/[1-N(0)\,V_c];$$

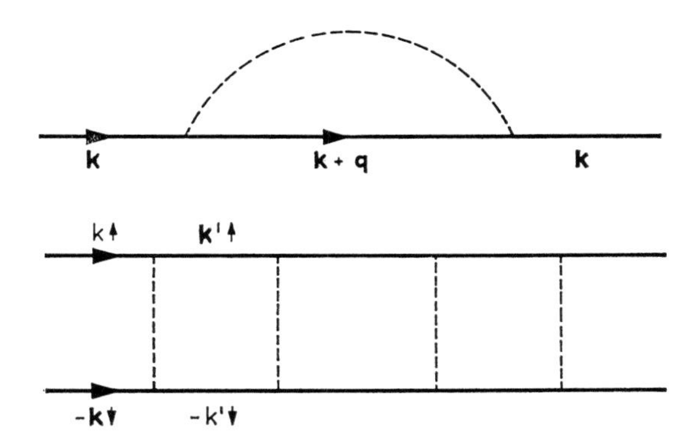

Fig. 1.
A Feynman diagram which contributes to the self-energy of an electron due to its interaction with phonons. The solid lines represent electron propagators, the dashed lines phonon propagators

Fig. 2.
A typical ladder Feynman diagram showing the repeated scattering through the exchange of phonons of a pair electrons with total momentum and spin of zero

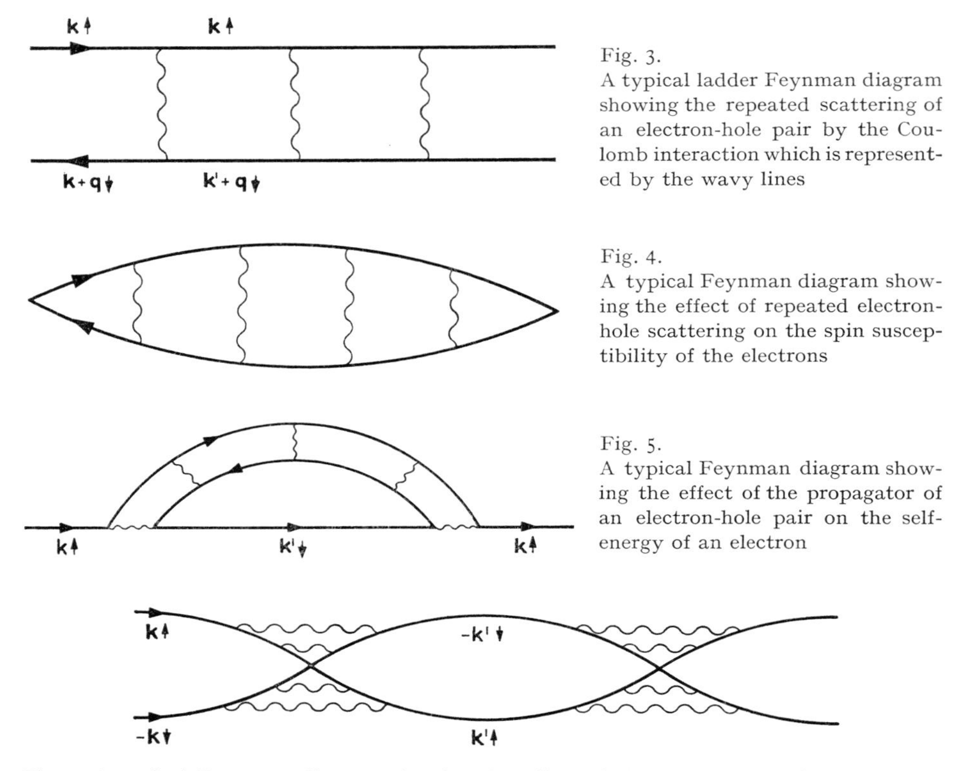

Fig. 3. A typical ladder Feynman diagram showing the repeated scattering of an electron-hole pair by the Coulomb interaction which is represented by the wavy lines

Fig. 4. A typical Feynman diagram showing the effect of repeated electron-hole scattering on the spin susceptibility of the electrons

Fig. 5. A typical Feynman diagram showing the effect of the propagator of an electron-hole pair on the self-energy of an electron

Fig. 6. A typical Feynman diagram showing the effect of the propagators of electron-hole pairs on the repeated scattering of a pair of electrons with total momentum and spin of zero

here χ_0 is the susceptibility when the Coulomb interaction between the electrons is ignored, $N(0)$ is the density of electron states at the Fermi level and V_c is of the order of the Coulomb energy of electrons. If $0<[1-N(0)V_c]\ll 1$, the system has a greatly enhanced spin susceptibility. It is for this kind of metal that the spin fluctuations are likely to be important.

The propagator of an electron-hole pair as illustrated in Fig. 3 is similar to the propagator of a phonon. For example, it leads to a change in the self-energy of the electrons through diagrams of the form shown in Fig. 5 [33—35], and to a scattering of electrons as shown in Fig. 6. The latter diagram shows very clearly the exchange nature of the resultant interaction between the electron pairs. As a result of this exchange the interaction tends to be repulsive and so inhibits the formation of superconductors. This is also to be expected from the fact that, as the enhanced diamagnetic susceptibility shows, the interaction tends to align spins whereas, in the superconductive state, the paired spins are antiparallel*. From this point of view, the effect of spin fluctuations on superconductivity was

* This interaction could be important for producing Cooper pairs with parallel spin in (say) relative p-states, but there is no evidence for this pairing in known superconductors.

first pointed out by Doniach [36]. The diagrammatic approach was developed by Berk and Schrieffer [33, 34]. It is possible to add the effects of the spin fluctuations as shown in Fig. 6 to that of the phonons to obtain a more general equation for T_c [34, 37]. Since, however, there have, as yet, been no detailed calculations based on such an equation we shall not write it out here. We emphasize two results which should generally follow if the spin fluctuations are important. One is that the exponent α of the isotope effect would be altered and the other is that the interaction tends to lower T_c and this decrease should correlate with an enhanced susceptibility. There is some evidence that this occurs in 4d H.C.P. transition metal alloys [38] and in lanthanum intermetallic compounds [39].

c) Excitons

c) The search for room temperature superconductors has led to a discussion of the possibility of superconductivity being caused by high energy excitations, particularly excitons. According to the BCS theory the transition temperature is proportional to the energy of the excitation that mediates the interaction between the electrons. If one can raise this energy without weakening the strength of the interaction one can raise T_c. As excitons in insulators have energies measured in eV, if they can be exploited, a room temperature superconductor can be found. This is the basis of the first proposals advanced by Little [40] and Ginzburg [41]. The difficulty in realizing such a possibility is that the excitons usually exist only outside the metal and coupling between the electrons and excitons are likely to be weak. In order to have such a coupling it is necessary to consider metallic samples which have at least one small dimension. Little proposed a chain organic molecule with readily polarisable side-chains as a possible system for this kind of superconductivity, but there is no evidence for any superconductivity in such systems. Indeed, it has been argued that it is impossible for superconductivity to arise in the way proposed [42]. The system proposed by Ginzburg is an insulator-metal-insulator sandwich. Such systems can be superconductors but none which exhibits Ginzburg's mechanism or any high temperature superconductivity has been found*. Geilikman [43] has considered the possibility of high energy excitations in metals with some localised electrons, either intrinsic or introduced by alloying. Again, although this is a possibility, in principle, no high temperature superconductors of this kind have been found.

4. Summary

It is clear from the measured isotope effects of known superconductors that in nearly all of these the electron-phonon mechanism originally proposed by Fröhlich plays a part. Although in the early days of the BCS theory it was thought that this mechanism would lead to a value of α equal to $\frac{1}{2}$, it is now realised that it can lead to almost any value, depending on the band structure and the strength of the Coulomb interaction, and that, in principle, it may be the sole mechanism responsible for superconductivity in all known superconductors. Probably, the value of $\alpha = -2$ in U is the hardest to explain in this way.

* *Note added in proof:* Recently, D. Allender, J. Bray and J. Bardeen [Phys. Rev. B7, 1020 (1973)] have discussed in detail a particular model for an excitonic mechanism for superconductivity.

Other mechanisms which can lead to an electron-electron interaction have also been proposed for causing or inhibiting superconductivity. However, so far, none has definitely been shown to be of importance. Different isotope effects for T_c, $\Delta(0)$ and H_c can indicate another mechanism. However, a varying density of states near the Fermi surface in conjunction with the usual mechanism can also lead to such different isotope effects. We thus have the paradoxical situation (or is it a paradox?) that although any attractive interaction between electrons could, in principle, be responsible for superconductivity only that originally proposed by FRÖHLICH has definitely been established and that one could be the cause in all superconductors.

References

1. FRÖHLICH, H.: Phys. Rev. **79**, 845 (1950).
2. FRÖHLICH, H., PELZER, H., ZIENAU, S.: Phil. Mag. **41**, 221 (1950).
3. MAXWELL, E.: Phys. Rev. **78**, 477 (1950).
4. SCHAFROTH, M. R.: Helv. Phys. Acta **24**, 645 (1951).
5. FRÖHLICH, H.: Proc. Roy. Soc. (London). Ser. A **215**, 291 (1952).
6. BARDEEN, J., PINES, D.: Phys. Rev. **99**, 1140 (1955).
7. ELIASHBERG, G. M.: Zh. Eksperim. i Teor. Fiz. **38**, 966 (1960). [English tranl.: Soviet Phys. JETP **11**, 696 (1960).]
8. NAMBU, Y.: Phys. Rev. **117**, 648 (1960).
9. SCHRIEFFER, J. R.: Theory of Superconductivity, Chap. 7. New York: Benjamin 1964.
10. RICKAYZEN, G.: Theory of Superconductivity, Chap. 3. New York: Wiley 1965.
11. BARDEEN, J., COOPER, L. N., SCHRIEFFER, J. R.: Phys. Rev. **108**, 1175 (1957).
12. BOGOLIUBOV, N. N., TOLMACHEV, V. V., SHIRKOV, D. V.: A New Method in the Theory of Superconductivity. USSR: Acad. Sci. USSR 1958; translation, New York: Consultants Bureau 1959.
13. GARLAND, J. W.: Phys. Rev. Letters **11**, 114 (1963); — Phys. Rev. **153**, 460 (1967).
14. SWIHART, J. C.: IBM J. Res. Develop. **6**, 14 (1962).
15. MCMILLAN, W. L.: Phys. Rev. **167**, 331 (1968).
16. HOPFIELD, J. J.: Superconductivity. Proceedings of the International Conference on the Science of Superconductivity, Stanford, U.S.A., 1969, Ed. Chilton, F., p. 41. Amsterdam: North-Holland 1971.
17. GASPARI, G. D., GYORFFY, B. L.: Phys. Rev. Letters **28**, 801 (1972).
18. DYNES, R. C.: Solid State Commun. **10**, 615 (1972).
19. HERTEL, P.: Z. Phys. **248**, 272 (1971).
20. GARLAND, J. W., ALLEN, P. B.: Superconductivity. Proceedings of the International Conference of the Science of Superconductivity, Stanford, U.S.A., 1969, Ed. Chilton, F., p. 669. Amsterdam: North-Holland 1971.
21. COHEN, M. L., ANDERSON, P. W.: AIP Conference Proceedings on Superconductivity in *d*- and *f*-Band Metals, Rochester, New York, U.S.A., 1971.
22. MATTHIAS, B. T.: Superconductivity. Proceedings of the International Conference of the Science of Superconductivity, Stanford, U.S.A., 1969, Ed. Chilton, F., p. 69. Amsterdam: North-Holland 1971.
23. RADHAKRISHNAN, V.: Phys. Letters **16**, 247 (1965).
24. FRÖHLICH, H.: J. Phys. C **1**, 1 (1968).
25. ROTHWARF, A.: Phys. Rev. B **2**, 3560 (1970).
26. GEILIKMAN, B. T.: Fiz. Tverd. Tela **12**, 1881 (1970). [Translation: Soviet Phys.—Solid State **12**, 1497 (1971).
27. FRÖHLICH, H.: Phys. Letters **35** A, 325 (1971).

27a. GANGULY, B. N., WOOD, R. F.: Phys. Rev. Letters **28**, 681 (1972).

28. PINES, D.: Can. J. Phys. **34**, 1379 (1956).
29. NOZIÈRES, P., PINES, D.: Phys. Rev. **109**, 1062 (1958).
30. ABRIKOSOV, A. A.: Zh. Eksperim. i Teor. Fiz. **39**, 1797 (1960).

31. WOLFF, P. A.: Phys. Rev. **120**, 814 (1960).
32. IZUYAMA, T., KIM, D. J., KUBO, R.: J. Phys. Soc. Japan **18**, 1025 (1963)
33. BERK, N. F., SCHRIEFFER, J. R.: Bull. Am. Phys. Soc. **11**, 78 (1966).
34. BERK, N. F., SCHRIEFFER, J. R.: Phys. Rev. Letters **17**, 433 (1966).
35. DONIACH, S., ENGLESBERG, S.: Phys. Rev. Letters **17**, 750 (1966).
36. DONIACH, S.: Proceedings of the Manchester Many-Body Conference, Sept. 1964 (unpublished).
37. GLADSTONE, G., JENSEN, M. A., SCHRIEFFER, J. R.: Superconductivity, Ed. Parks, R. D., Chap. 13. New York: Marcel Dekker 1969.
38. RIBLET, G., JENSEN, M. A.: Superconductivity. Proceedings of the International Conference of the Science of Superconductivity, Stanford, U.S.A., 1969, Ed. Chilton, F. p. 622. Amsterdam: North-Holland 1971.
39. TOXEN, A. M., GAMBINO, R. J., VAN DER HOEVEN, B. J.: Superconductivity. Proceedings of the International Conference of the Science of Superconductivity, Stanford, U.S.A., 1969, Ed. Chilton, F., p. 626. Amsterdam: North-Holland 1971.
40. LITTLE, W. A.: Phys. Rev. A**134**, 1416 (1964).
41. GINZBURG, V. L.: Contemporary Phys. **9**, 355 (1968).
42. KUPER, C.: Phys. Rev. **150**, 189 (1966).
43. GEILIKMAN, B. T.: Zh. Eksperim. i Teor. Fiz. **48**, 1194 (1965). [Translation: Soviet Phys. JETP **21**, 796 (1965).]

Transition Temperatures of Superconductors

Bernd T. Matthias

University of California, San Diego*, Institute for Pure and Applied Physical Sciences, La Jolla, California 92037 and Bell Telephone Laboratories, Murray Hill, New Jersey 07974

In 1950, Fröhlich and Bardeen laid the foundation for the theory of superconductivity. Today, after more than 20 years and many errors, my own included, the crucial obstacle in raising the superconducting transition temperature is substantially the same one that was foreseen by them.

This obstacle is the instability—or metastability—of the crystal lattice for compounds having high transition temperatures. Since 1950, the theory progressed by giving explanations for isotope effects, Cooper pairs, tunneling, the Josephson effect, and type II superconductors. But as far as superconducting transition temperatures are concerned, and any way to raise them, there has not been any theoretical progress whatsoever. If one were willing to take into account as a negative progress, erroneous hypotheses and predictions based on BCS for higher superconducting transition temperatures, the proposals of Little, McMillan, Ginzburg, Parmenter, or Gor'kov would be in the forefront since they have been, without exception, spectacularly unsuccessful. Only antigravity of magnetic monopoles can compete with this negative progress which is perhaps more so here than in any other field, as there has never been a single positive experimental result yet verifying any of the theoretical speculations.

It has now been 15 years since the first quantitative theory, namely BCS, was introduced. Unfortunately, this theory with all the good it has done has also given impetus to all of the above mentioned fiascos. This is too bad, as BCS in itself is such a beautiful theory. But, the trend set forth by Little, Ginzburg and associates should not obscure the fact that today we think we really do know how to gradually raise the transition temperature and what the limitations are.

By our method, which is strictly experimental and empirical, it has now become more and more evident that high transition temperatures and d-electrons are inseparable. Again, it was Fröhlich who first pointed to the underlying reason [1]. For several years, after the role of d-electrons for high transition temperatures had become established empirically, partially formed ideas of a "magnetic ionic interaction" causing superconductivity had crossed my mind. I tried to forget phonons entirely. However, this proved to be inconsistent since, even then, the lattice always had to be involved. Until the analogy with the piezoelectric effect

* Research in La Jolla sponsored by the Air Force Office of Scientific Research, under AFOSR contract number AFOSR/F-44620-72-C-0017.

became obvious, the difficulties were both of a logical as well as of a sematic nature. This all disappeared—at least in my mind—when the formal analogy to the piezoelectric effect began to impress itself. The piezoelectric effect is the property of certain crystal classes (those which lack a center of inversion) in which mechanical deformation and electric polarization are a function of one another, and thus become intimately connected.

The two major contributors to the piezoelectric effect are the ionic or lattice polarization, and the electronic or core polarization. For a given elastic deformation, these two kinds of polarizations frequently have the opposite sign and while they are two entirely separate mechanisms, they are still rigorously connected with each other. It is for this reason that it was impossible to consider them separately until a piezoelectric effect could be found in a crystal with only one kind of ion, namely an element. Here the lattice of ionic polarization could be neglected simply because the charge on all ions is the same. This finally happened when the piezoelectricity of selenium single crystals was discovered [3]*. Now, the effect was due *only* to the electronic polarizability. In magnitude, it turned out to be very similar to the ionic polarizability calculated for ionic crystals.

In superconductivity, the electron-lattice interaction in a metal will again encounter both polarizabilities. The ionic polarizability will depend mostly upon lattice structure and atomic mass while the electronic polarizability will be dominantly influenced by the actual electronic configuration of the individual atom. For that reason it was possible to predict superconductivity in the periodic system from the position of the isolated atom. It also enabled us to predict superconducting isotope effects different from the original $T_c = 1/\sqrt{M}$. The ultimate prediction—many years before it was found—was the opposite sign of this effect for lanthanum and uranium. In the latter, it was finally discovered experimentally [4] and turned out to be $T_c \sim M^2$.

So far, so good, but simply assuming an electronic polarizability was not of too much use for the electron theory of metals. For a change, we needed a theoretical proof.

As mentioned earlier, it was again FRÖHLICH who finally pointed out the real meaning of all this [1, 2]. He showed that the electronic polarizability was decisively determined by the plasma frequencies in the metal, and thus it was for the first time accessible to experiment. Any real increase in superconducting transition temperatures now will have to rely on and use these insights. At the same time, his is the first theoretical insight as to why superconductivity can be predicted from the nature of the isolated atom.

References

1. FRÖHLICH, H.: J. Phys. C **1**, 544 (1968); also Phys. Letters **26** A, 169 (1968).
2. FRÖHLICH, H.: Phys. Letters **35** A, 325 (1971).
3. GOBRECHT, H., HAMISCH, H., TAUSEND, A.: Z. Physik **148**, 209 (1957).
4. FOWLER, R. D., LINDSAY, J. D. G., WHITE, R. W., HILL, H. H., MATTHIAS, B. T.: Phys. Rev. Letters **19**, 892 (1967).

* At very high pressures, selenium becomes superconducting. However, there does not seem to be any immediate connection between superconductivity and the piezoelectric effect as the pressure induced superconductivity of selenium is presumably due to a polymorphic transition.

The Effect of Stress on the Superconductivity of Tin

F. R. N. NABARRO and BARBARA D. ROTHBERG

University of the Witwatersrand, Department of Physics, Johannesburg/
South Africa

A. Introduction

1. Superconductors Under Stress. The phenomenon of superconductivity was first satisfactorily explained by BARDEEN, COOPER and SCHRIEFFER [1], who exploited FRÖHLICH's idea [2], [3] that the coupling of electrons and phonons in a metal would lead to an attractive term in the interaction between two electrons. Clearly, any procedure which allows the parameters of the metal to be continuously varied will provide a sensitive test of the theory. One such procedure is the addition of alloying elements, but this may introduce complications by turning a "clean" into a "dirty" metal. The application of elastic stress is free from this disadvantage; moreover, the stress can be applied and removed reversibly while the specimen is in the superconducting state.

The simplest stress to apply is an isotropic hydrostatic pressure. If the metal is cubic, the resulting strain is also isotropic, but for a metal of lower symmetry a hydrostatic stress alters the ratios of the crystal axes, while leaving the symmetry of the crystal unchanged. Hydrostatic stress does not produce plastic deformation. More information can be obtained by applying a uniaxial tensile or compressive stress. The resulting shear strains may or may not alter the symmetry of the crystal. The practical difficulty is that pure metal crystals readily deform plastically under very small stresses. Single crystal whiskers may remain elastic up to strains of 1 or 2 per cent. Unfortunately, their magnetic properties in the superconducting state show large temperature-dependent size effects, and this complicates the interpretation of the observations. Extensive measurements of the critical temperature T_c of strained whiskers of tin and its alloys have been made by a group at Clemson University, South Carolina [4–7], and one of us [8] has made measurements in a magnetic field.

2. The Theoretical Interpretation. Since superconductivity is controlled by an interaction between electrons and phonons, and elastic strain is likely to affect both the Fermi surface of the electrons and the phonon spectrum, it is interesting to discover whether the changes in the superconducting parameters depend predominantly on the Fermi surface or predominantly on the phonon spectrum. In particular, we may attempt to relate strong or unusual stress dependences to special features of the Fermi surface or of the phonon spectrum. The two principal parameters which can be measured are T_c and the critical magnetic field at zero temperature, H_0.

Probably the most detailed theoretical study of the effect of pressure on the superconducting properties of metals, including tin, is that of CARBOTTE and VASHISHTA [9]. Their analysis "suggests that changes in the electronic band structure under pressure may not be as important as changes in the lattice dynamics when considering changes in superconducting properties under a volume contraction". Similarly, ZAVARITSKII et al. [10], who studied electron tunnelling under pressure, found that the decrease in electron-phonon coupling under pressure "is primarily the result of the variation of the lattice oscillation spectrum", while OTT [11], in discussing the change in length of gallium single crystals at the superconducting transition, considers only the influence of lattice vibrations. On the other hand, the effect of pressure on the superconductivity of indium and its alloys has been explained in detail by considering singularities in the Fermi surface [12–14], while HAVINGA and his colleagues have in a series of papers (e.g. [15] and [16]) successfully explained the properties of superconducting alloys by saying that when the Fermi surface approaches a Brillouin zone boundary the electron wave function contains more than one orthogonalized plane wave. This leads to strong umklapprozesse, and hence to strong electron-phonon coupling.

It seems likely that studies under directional stresses may clarify the position.

B. The Effect of Hydrostatic Pressure

3. The Basic Formulae. We base our analysis on three results of the weak-coupling isotropic B.C.S. theory, neglecting Coulomb repulsion between the electrons:

$$3.5\, k_B T_c = 2\Delta_0 \tag{3.1}$$

$$H_0^2/8\pi = \tfrac{1}{2} N(0)\,\Delta_0^2 \tag{3.2}$$

$$\Delta_0 = 2\hbar\omega_D \exp\left(-1/N(0)\,V\right). \tag{3.3}$$

Here k_B is BOLTZMANN's constant, Δ_0 is the energy gap at $T=0$, $N(0)$ is the density of states at the Fermi surface for electrons of one spin, ω_D is the Debye temperature and V is the electron-electron coupling parameter. The value of $N(0)\,V$ for tin is about 0.25 [17]. Both T_c and H_0 depend on the strain parameter ε.

Eliminating Δ_0 between (3.1) and (3.2), we find

$$N(0) = H_0^2/(3.5)^2\, \pi k_B^2 T_c^2. \tag{3.4}$$

If the Fermi surface does not change appreciably on straining the material, $N(0)$ is constant, and H_0/T_c is constant ("double similarity"). Then

$$d \ln H_0/d \ln T_c = 1. \tag{3.5}$$

If, as another extreme case, the predominant effect of strain arises from a change in $N(0)$, we obtain from (3.1), (3.2) and (3.3)

$$\frac{d \ln T_c}{d\varepsilon} = \frac{1}{N(0)\,V}\,\frac{d \ln N(0)}{d\varepsilon} \tag{3.6}$$

and

$$\frac{d \ln H_0}{d\varepsilon} = \left(\frac{1}{2} + \frac{1}{N(0)\,V}\right) \frac{d \ln N(0)}{d\varepsilon}, \qquad (3.7)$$

so that $d \ln H_0/d \ln T_c = \frac{1}{2} N(0)\,V + 1 \simeq 1.125.$ (3.8)

A determination of $d \ln H_0/d \ln T_c$ under strain can distinguish directly between these cases, but it is clear that rather precise measurements are required.

In general we have

$$d \ln T_c = d \ln \omega_D + \left(N(0)\,V\right)^{-1} \left(d \ln N(0) + d \ln V\right). \qquad (3.9)$$

From (3.1) and (3.2) we obtain

$$d \ln H_0 = d \ln T_c + \tfrac{1}{2}\, d \ln N(0). \qquad (3.10)$$

The three parameters ω_D, $N(0)$ and V can be regarded as functions of the independent variables M, the isotopic mass, and v, the volume. If we consider first the isotope effect (e.g. MAXWELL [18] and LOCK, PIPPARD and SHOENBERG [19]), then M varies while v is kept constant, and it is found experimentally that $T_c \propto M^{-\beta}$, where, for tin, $\beta = 0.47$. We have $\partial \ln \omega_D/\partial \ln M = -\frac{1}{2}$, and can use equations (3.9) and (3.10) as follows: The observation of double similarity gives

$$\partial \ln H_0/\partial \ln M = \partial \ln T_c/\partial \ln M,$$

which by (3.10) verifies the expected relation $\partial \ln N(0)/\partial \ln M = 0$. From equation (3.9)

$$\begin{aligned}\partial \ln V/\partial \ln M &= N(0)\,V(\partial \ln T_c/\partial \ln M - \partial \ln \omega_D/\partial \ln M)\\ &= N(0)\,V(-\beta + \tfrac{1}{2}) \gtrsim 0.007.\end{aligned}$$

We can also use equations (3.9) and (3.10) to analyze the variation with volume. The quantity $-\partial \ln \omega_D/\partial \ln v$ is by definition the Grüneisen constant, which has been tabulated at room temperature (GSCHNEIDNER [20]) and has been found by WHITE [21] not to vary much with temperature for copper. From equation (3.10) we have

$$\partial \ln N(0)/\partial \ln v = 2(\partial \ln H_0/\partial \ln v - \partial \ln T_c/\partial \ln v) \qquad (3.11)$$

and from (3.9):

$$\begin{aligned}\partial \ln V/\partial \ln v = N(0)\,V(\partial \ln T_c/\partial \ln v - \partial \ln \omega_D/\partial \ln v)\\ - \partial \ln N(0)/\partial \ln v.\end{aligned} \qquad (3.12)$$

Since in weak-coupling BCS theory, $N(0)$ is proportional to the electronic specific heat γ, which is proportional to $(H_0/T_c)^2$ (SERAPHIM and MARCUS [22]), our analysis should be equivalent to that of SCHIRBER and SWENSON [23].

4. The Experimental Results and their Analysis. The most reliable values of dT_c/dp for polycrystalline tin are probably those of JENNINGS and SWENSON [24] who covered the range from 0 to 10 kiloatmospheres, obtaining

$$T_c = 3.732 - 4.95 \times 10^{-5}\, P + 3.9 \times 10^{-10}\, P^2$$

where T is in K and P in atmospheres. The curvature of their plot of T_c vs. P can be eliminated by extrapolating to zero degrees absolute BRIDGMAN's data

[25] on the variation with pressure of the compressibility of tin. The results of JENNINGS and SWENSON are conveniently expressed as $\partial \ln T_c/\partial \ln v = 7.51 \pm 0.25$.

The parameter H_0 must be obtained by extrapolation; the variation of H_c with T is well known for bulk tin at zero pressure. Measurements were made by SCHIRBER and SWENSON [23] at temperatures between 1.15 K and 3.7 K and at pressures up to 2800 atm. using polycrystalline tin. They obtained

$$\partial \ln H_0/\partial \ln v = 8.30 \pm 0.25.$$

Using the room temperature value of the Grüneisen constant for Sn [20] in the absence of other data, we have

$$-\partial \ln \omega_D/\partial \ln v = 2.27 \pm 0.2. \tag{4.1}$$

From Eqs. (3.11) and (3.12) we then find

$$\partial \ln N(0)/\partial \ln v = 1.6 \pm 1.0 \tag{4.2}$$

and

$$\partial \ln V/\partial \ln v = 0.8 \pm 1.1. \tag{4.3}$$

It is obvious that existing measurements are not accurate enough to support the present analysis. We can only conclude that ω_D and $N(0)$ vary more with v than does V. The free electron value for $\partial \ln N(0)/\partial \ln v$ is two-thirds and it seems likely that the actual value is appreciably greater than this. Our value of

$$\partial \ln N(0)\, V/\partial \ln v = 2.4$$

agrees with those derived by GLADSTONE et al. [26] and LEVY and OLSEN [27] from the same data. Our value of $\partial \ln N(0)/\partial \ln v = 1.6 \pm 1.0$ agrees only moderately well with SCHIRBER and SWENSON's value of $\partial \ln \gamma/\partial \ln v = 1.2 \pm 0.4$, derived from the same data.

C. Uniaxial Tension

5. Resistivity in the Normal State. We expect in general that dilatation or uniaxial extension will soften the phonon spectrum, and so increase the resistivity ϱ. Indeed, in simple univalent metals [28], moderate pressure always decreases the resistivity. Large reductions in volume, of the order $-\delta v/v \gtrsim 0.2$, increase the energy gap at the Brillouin zone boundary, distort the Fermi surface, introduce umklapprozesse, and lead to increases in the resistivity. The first unexpected observation of the effect of uniaxial tension on the resistivity of single crystals of tin was made by ALLEN [29] in 1937. While tension along the tetragonal [001] axis produces the expected increase in resistivity, tension along [100] up to strains of the order of 0.4 per cent decreases the resistivity (and indeed the resistance) of the sample. These results were confirmed by DAVIS [4] and ourselves.

It seems possible to interpret the observations in terms of the model of KLEMENS et al. [30], which was successfully applied by CASE and GUETHS [31] to explain the temperature dependence of the anisotropy of the electrical resistivity of tin. The model is valid in the temperature range in which phonon scattering dominates the residual resistivity, but still deviates electrons only through small angles. Small pockets of the Fermi surface do not contribute appreciably to the current, because the electrons on them suffer frequent umklapprozesse. The current is carried principally by large areas of the Fermi surface which lie roughly perpendicular to the direction of the current. CASE and GUETHS adopt a very simplified model in which the largest of these areas are taken to bound electrons overlapping {100} faces of the zone. In fact the most prominent large flat areas of the Fermi surface as determined by GOLD and PRIESTLEY [32] and by WEISZ [33], appear in the reduced zone representations as ribbons of holes in the fourth zone, which give rise to the orbits labelled by GOLD and PRIESTLEY as ς. It is clear from their Fig. 6 that these large flat areas of the Fermi surface are parallel to {100}. It is easier, in discussing the effect of strain, to use the extended zone representation, in which this "hole" surface becomes a set of electron surfaces overlapping zone boundaries, as visualized by CASE and GUETHS. Extension along [100] of a crystal carrying a current along [100] causes the zone boundaries to shrink away from the (100) sheets of the Fermi surface, thus increasing their area, reducing the strength of umklapprozesse and decreasing the resistivity, as is observed. [In the reduced zone representation, the zone boundary shrinks towards the hole surface, but this surface, which has been mapped from the fourth zone by a translation through three reciprocal lattice vectors, moves towards the origin faster than does the zone boundary.]

The experiments on whiskers allow strains of the order of 1–2 per cent. For [001] whiskers a linear increase in resistivity ϱ, with $d \ln \varrho / d \ln \varepsilon \simeq 8.8$ is observed. We analyze the large value of $d \ln \varrho / d\, \varepsilon_{001}$ by saying that 1 per cent elongation along [001] produces an increase in volume of $(S_{33}+2S_{13})/S_{33}=0.51$ per cent. With a Grüneisen constant of 2.27, this would, in an isotropic material, produce an increase in ω_D of 1.16 per cent. At low temperatures, $\varrho \propto \omega_D^4$, so that the greatest expected increase in ϱ is only 4.6 per cent. Straining parallel to [001] must therefore increase the electron-phonon coupling, unless it should produce a large decrease in the effective number of carriers.

For [100] whiskers tested at various temperatures the observations show a wide scatter, because the bulk residual resistivity, thermal resistivity and boundary scattering can all be important. All observations show a strongly non-linear dependence of resistivity on strain, the resistance reaching a minimum at strains of between 0.3 and 1.6 per cent, though the resistivity is still decreasing slowly. It is remarkable that this non-linearity is pronounced at strains as small as 0.4 per cent, whereas the non-linearity in the pressure dependence of the resistivity of simple metals becomes strong only at strains of the order of 10 per cent. If, as a representative value, we take the curve of $\ln \varrho$ as a function of strain ε to deviate from a straight line by 5×10^{-3} for $\varepsilon = 1$ per cent, we find $d \ln \varrho / d \varepsilon^2 = 50$. This presumably arises because neither hydrostatic pressure on a cubic crystal nor [001] tension on a tetragonal crystal alters the symmetry, while [100] tension reduces the symmetry of a tetragonal crystal to orthorhombic. We therefore concentrate our attention on the shearing of (001) planes. Because [001] is initially a fourfold axis, the effects of shear in (001) planes must be even functions of the shear strain (e.g. PIPPARD [34]). The pockets of the Fermi surface, which are very sensitive to strain, contribute little to the resistivity, and so we enquire whether there may be features of the phonon spectrum which are sensitive to [100] tension, but less so to [001] tension. In particular, we look for deformations which, if finite, would represent a transformation of the lattice into that of another stable structure, the relative stability of the new structure being increased by the applied stress.

The known stable phases of tin, in addition to white β-tin, are grey α-tin and Sn III, which is formed under high pressures. Grey tin may in principle be formed from white tin by a simple lattice deformation [35], but the deformations involved are very large (105 per cent elongation along the c axis), and the possibility of such a transformation will not appreciably affect the vibrations of the white tin lattice. The lattice of Sn III, which is simple body-centred tetragonal, may be formed from that of white tin by sliding [001] rods of atoms in [001] directions. This is a high-pressure phase and will not be stabilized by tension.

Finally, we consider the possibility that face-centred cubic tin might be stable under negative pressures. Since the ratio of the density of white tin to its atomic weight is 0.061, while the corresponding ratio for f.c.c. lead is 0.055, it is not unreasonable to consider that f.c.c. tin might be stable under negative pressures. The $\{100\}$ planes of white tin do not deviate much from a close-packed hexagonal structure (Fig. 1 a). If these planes re-arrange into a close-packed form, an f.c.c. structure may be formed in several ways by sliding these planes over one another in [010] directions. If such a latent transformation influences the elastic properties of white tin, the evidence will be that large slidings of [100] planes over one another will require less energy than is estimated by extrapolating according to Hooke's Law from the energy required to produce small relative displacements. Strong anomalies in the diffuse background in X-ray diffraction by white tin have in fact been observed [36], and correspond roughly with the present picture. The observations are interpreted in terms of transverse shearing waves propagating in all [Okl] directions, with displacements along [100]. The authors suggest an explanation in terms of static metastable inclusions of grey tin, but such an explanation does not now seem plausible.

Any strain which increases the volume will increase the relative stability of f.c.c. tin, and so soften these modes of vibration. It remains to explain why [001] tension, which makes the configuration of $\{100\}$ planes more nearly close packed, is only moderately effective, while [100] tension is very effective, although the final adjustment to an f.c.c. structure involves a contraction along the [100] axis normal to the active $\{100\}$ planes. The answer is probably simply that the $\{100\}$ planes slide more freely over one another when they are further apart; it is the intermediate configuration of maximum energy, not the final f.c.c. configuration, which controls the frequency of the vibrations.

6. The Superconducting Critical Temperature. In agreement with the general principles expounded by Fröhlich [2], the influence of uniaxial stress on the critical temperature T_c largely parallels its influence on the normal resistivity. We attribute the differences which are observed to the fact that portions of the Fermi surface in strong interaction with Brillouin zone boundaries, which have little influence on the normal resistivity because they are short-circuited by the larger sheets of the Fermi surface, have a larger influence on the superconducting properties (e.g. [12]).

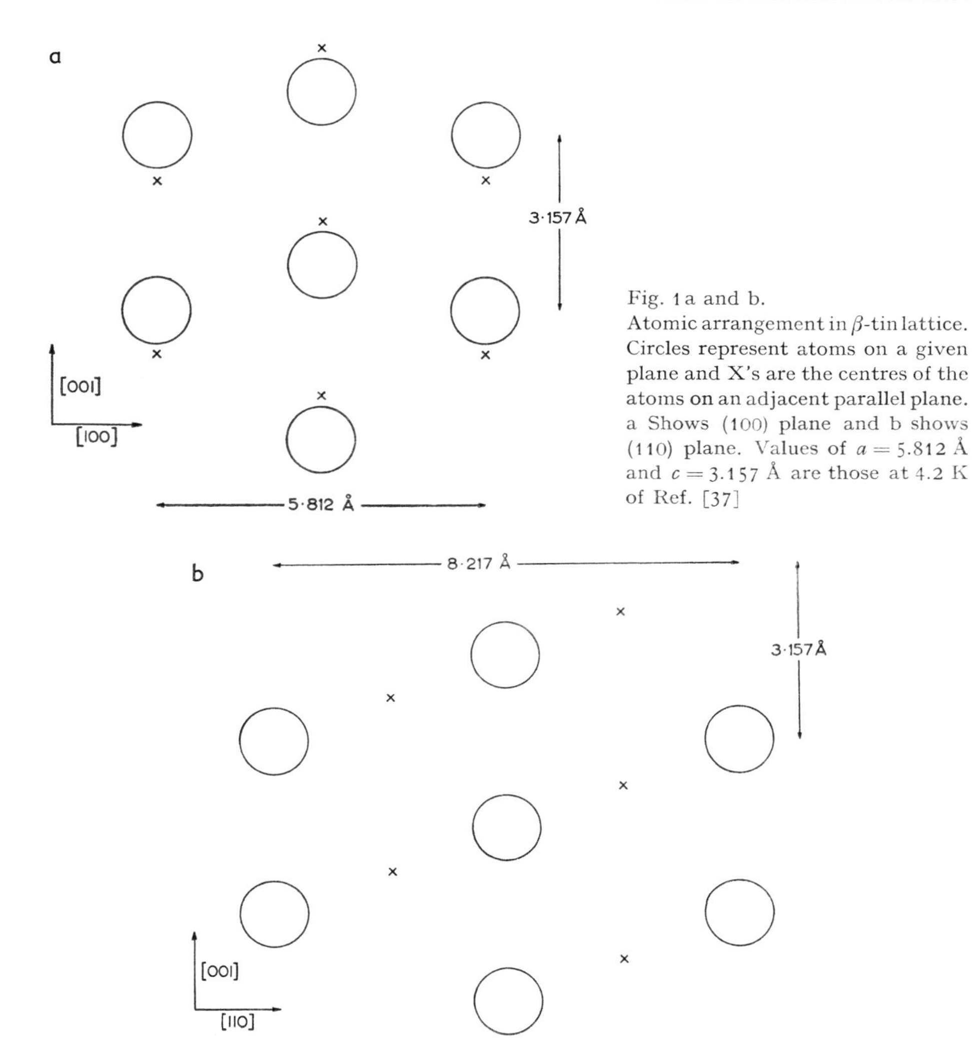

Fig. 1a and b.
Atomic arrangement in β-tin lattice. Circles represent atoms on a given plane and X's are the centres of the atoms on an adjacent parallel plane. a Shows (100) plane and b shows (110) plane. Values of $a = 5.812$ Å and $c = 3.157$ Å are those at 4.2 K of Ref. [37]

Since T_c is a scalar, the linear change of T_c with axial stress σ must be a function of the angle θ between the tensile axis and the tetragonal axis of the form

$$\frac{d \ln T_c}{d\sigma_\theta} = P + Q \cos^2 \theta. \tag{6.1}$$

The most accurate experiments appear to be those of COOK [7], our own observations and those of DAVIS agreeing reasonably well with his. Fig. 2a shows the measurements made by B.D.R. [8] of ΔT_c as a function of elastic strain for whiskers of four orientations. However, the variable experimentally imposed is strain, not stress, and COOK used the inaccurate elastic constants of BRIDGMAN to convert his observations. We have recomputed COOK's values using the elastic constants determined at low temperatures by RAYNE and CHANDRASEKHAR [37]. Fig. 2b shows the verification of Eq. (6.1).

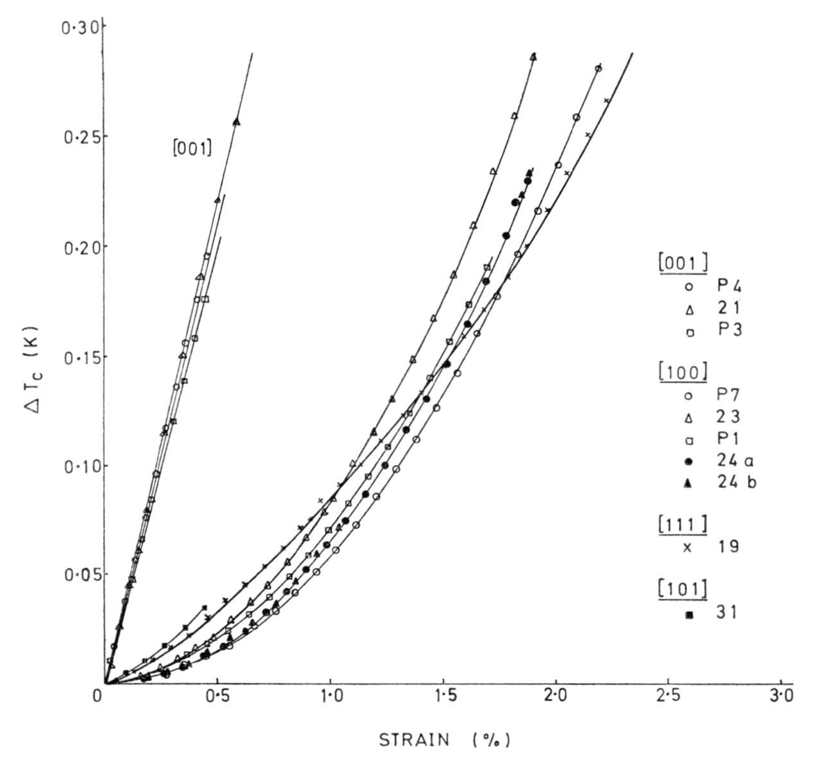

Fig. 2a. Change in critical temperature with elastic strain, for whiskers of four axial orientations (Ref. [8])

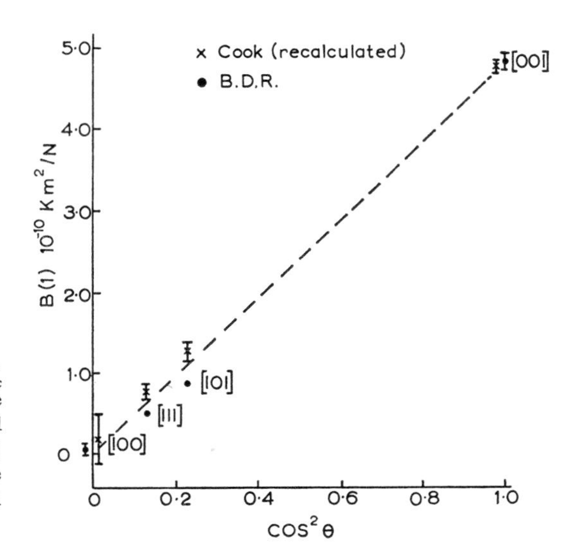

Fig. 2b. Variation of $B(1)$ such that $\Delta T_c = B(1)\,\sigma + B(2)\,\sigma^2$, with $\cos^2\theta$, where θ is the angle between the crystal axis and the [001] direction (Ref. [8] and Ref. [7]). For clarity, values have been displaced sideways at $\theta = 0$ and $\theta = \pi/2$

In terms of strain ε, the linear dependence of T_c on ε varies between

$$d \ln T_c/d\varepsilon_{100} = 0.16 \quad \text{and} \quad d \ln T_c/d\varepsilon_{001} = 11.$$

These values may be compared with DAVIS's values for the normal resistivity at 77 K of $d \ln \varrho/d\varepsilon_{100} = -0.5$ and $d \ln \varrho/d\varepsilon_{001} = 8.8$. We have attempted to explain the small negative value of $d \ln \varrho/d\varepsilon_{100}$, and a small value of $d \ln T_c/d\varepsilon_{100}$ is to be expected if $N(0)\,V$ does not vary much with strain. The large linear increase in T_c with [001] strain is compatible with either of the suggested explanations of the large linear increase of ϱ with the same strain. Either the phonon spectrum is softened, or the Fermi surface is brought into strong interaction with a Brillouin zone boundary.

While T_c is a linear function of ε_{001} up to strains of nearly 1 per cent, T_c, like ϱ, has a large quadratic dependence on ε_{100}, with $d \ln T_c/d\varepsilon_{100}^2 \simeq 170$ from COOK's data and $\simeq 175$ from our own. We again attribute this dependence to a softening of the vibrational spectrum, partly because the features of the Fermi surface which seem most likely to contribute to large quadratic effects do not have the necessary geometrical configuration. Since the observed changes in T_c are large, the features responsible must be fairly large. The obvious candidates are the four pipes of holes parallel to [001] in the third zone. These pipes (like the sheets of holes in the fourth zone) lie in $\langle 110 \rangle$ directions from the origin. Tension along [100] affects them all equally, and no larger quadratic effects would be expected from [100] tension than from [001] tension, which produces no observable quadratic effects. If these pipes (or the sheets in the fourth zone) were responsible for quadratic effects, the quadratic effects produced by tension along [110], which expands two of the pipes and contracts the other two, should be much greater. Whiskers are not available in this orientation, but COOK [7] and B.D.R. [8] each tested one [111] whisker, which should show a similar effect. The graph of T_c as a function of ε_{111} deviates recognisably from a parabola, but the curvature is always less than that for a [100] whisker.

It is perhaps remarkable that although stress along [001] produces a strictly linear change in T_c, and stress along $\langle 100 \rangle$ produces large quadratic effects, hydrostatic stress, which is simply the sum of equal stresses along [001], [100] and [010], produces an essentially linear change. In fact JENNINGS and SWENSON [24] showed that if T_c is plotted against volume rather than against pressure the graph is linear as far as $\delta v/v = -0.015$, whereas uniaxial tension shows large quadratic effects at $\varepsilon_{100} = 0.005$. This implies that the quadratic effects depend directly on the change of symmetry produced by shears in the base plane. COOK's value $dT_c/d\sigma_{001} + 2dT_c/d\sigma_{100}$ for the sum of the three linear dependences, recomputed with the elastic constants of [37], is 5.08×10^{-5} K/bar, in good agreement with JENNINGS and SWENSON's value $dT_c/dp = -4.9 \times 10^{-5}$ K/bar.

It still remains unsatisfactory that the shears which are observed to produce large quadratic effects represent the sliding of $\{110\}$ planes over one another, while our discussion considers the sliding of $\{100\}$ planes over one another. The $\{110\}$ planes in white tin are also approximately hexagonal close packed (Fig. 1 b), but the shears required to produce an f.c.c. structure are in approximately $\langle 1\bar{1}2 \rangle$ directions, which does not agree even roughly with PRASAD and WOOSTER's interpretation of their observations.

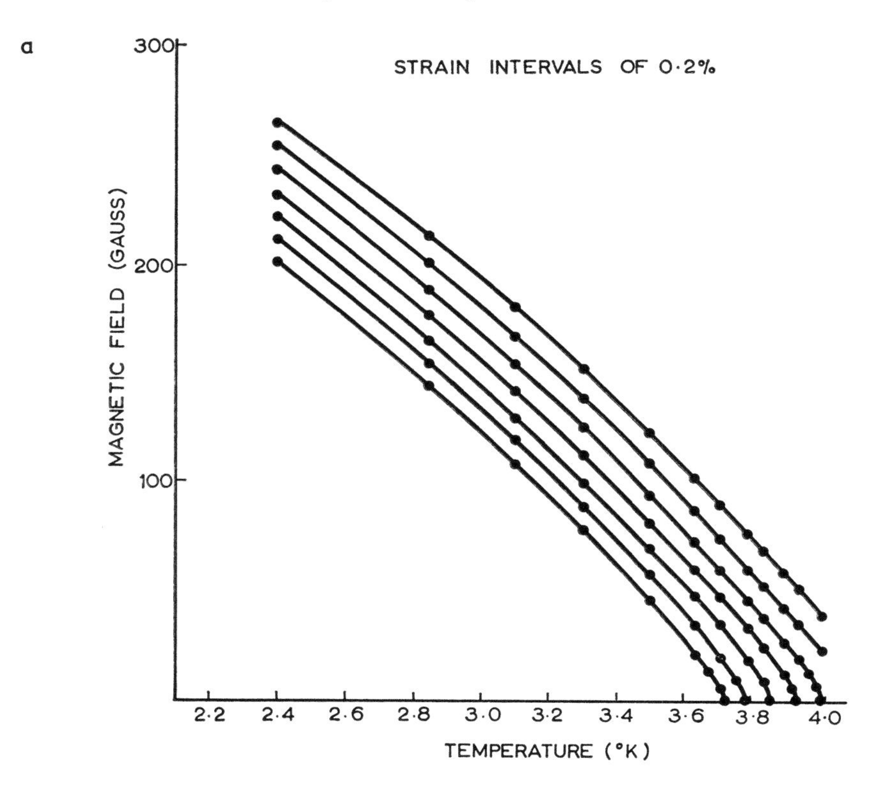

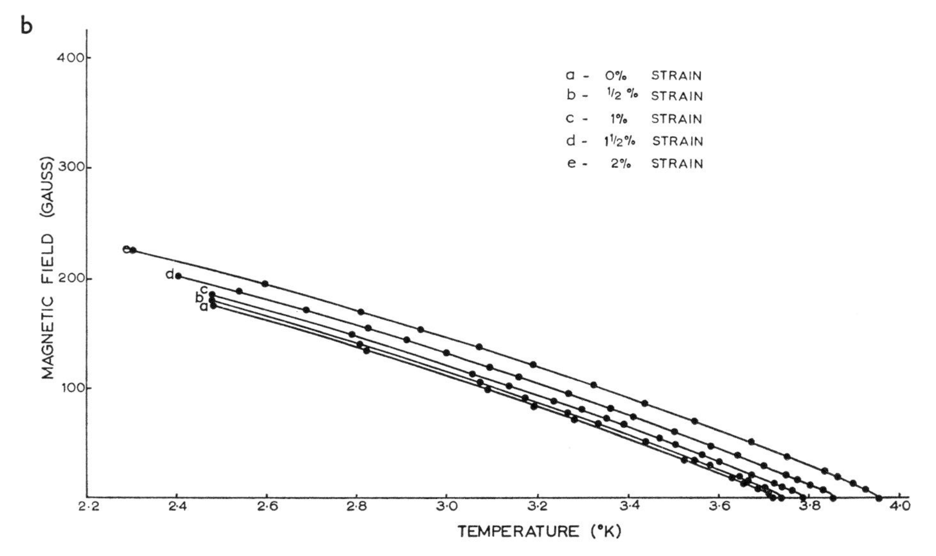

Fig. 3a and b. Critical magnetic field $H_c(\varepsilon, T)$ as a function of strain ε and temperature T for whiskers of two axial orientations. a Shows curves for [001] whisker and b shows curves for [100] whisker (Ref. [8]). In both cases the zero-strain curve is that on the extreme left

7. Measurements of H_0 Under Uniaxial Stress. Preliminary measurements of H_0 under uniaxial stress have been made by B.D.R. [8]. Fig. 3a shows the linear shift of the curve of $H_c(T)$ with strain of a whisker of [001] orientation, and Fig. 3b shows the quadratic shift for a [100] whisker. If these curves could be extrapolated accurately to $T=0$ to give H_0, and a correction made for the small size of the specimens, the analysis of §§ 3 and 4 could be applied to determine the changes of $N(0)$ and V produced by uniaxial stress. Unfortunately the temperature dependence of the size effect has so far prevented accurate extrapolation.

D. The Influence of Alloying

8. Whiskers of Indium Alloys and Whiskers of Tin Alloys. While measurements on whiskers of indium alloys [14] showed changes in the dependence of T_c on stress which supported a model in which the change of T_c with stress is associated with singularities in the Fermi surface, COOK's observations [7] on whiskers of tin alloys showed that the graphs of ΔT_c as a function of elongation for various alloys could be superposed with remarkable accuracy on the graphs for pure tin whiskers of the same orientations. These observations seem to exclude the possibility that the changes of T_c with stress in pure tin could be caused by changes in relatively small features of the Fermi surface.

E. Conclusions

Even the measurements under hydrostatic pressure are not accurate enough to distinguish clearly between changes in the Fermi surface and changes in the electron-phonon interaction. The measurements on whiskers under uniaxial tension give accurate information on T_c, but, so far, only very rough estimates of the changes of H_0 with strain. However, it seems fairly clear that the influence of the phonon spectrum is more important than that of the Fermi surface, and there is direct evidence of anomalies in the lattice vibrations of tin.

We acknowledge valuable discussions with D. S. MCLACHLAN, D. G. PETTIFOR and M. J. STEPHEN, and are especially grateful to M. J. SKOVE, who has kept us fully informed of the work in progress at Clemson University.

References

1. BARDEEN, J., COOPER, L. N., SCHRIEFFER, J. R.: Phys. Rev. **108**, 1175 (1957).
2. FRÖHLICH, H.: Phys. Rev. **79**, 845 (1950).
3. FRÖHLICH, H.: Proc. Roy. Soc. (London), Ser. A **223**, 296 (1952).
4. DAVIS, J. H.: Dissertation Clemson University 1966.
5. DAVIS, J. H., SKOVE, M. J., STILLWELL, E. P.: Solid State Comm. **4**, 597 (1966).
6. WATLINGTON, C. L.: Thesis Clemson University 1969.
7. COOK, J. W.: Dissertation Clemson University 1971.
8. ROTHBERG, B. D.: Thesis University of the Witwatersrand Johannesburg 1972.
9. CARBOTTE, J. P., VASHISHTA, P.: Can. J. Phys. **49**, 1493 (1971).
10. ZAVARITSKII, N. V., ITSKEVICH, E. S., VORONOVSKII, A. N.: Zh. Eksperim. Teor. Fiz. **60**, 1408 (1971) = Soviet Phys. JETP **33**, 762 (1971).
11. OTT, H.: Solid State Comm. **9**, 2225 (1971).
12. HIGGINS, R. J., KAEHN, H. D.: Phys. Rev. **182**, 649 (1969).
13. MAKAROV, V. I., VOLYNSKII, I. YA.: Zh. Eksperim. Teor. Fiz. **57**, 3 (1969) = Soviet Phys. JETP **30**, 1 (1970).

14. Overcash, D. R., Skove, M. J., Stillwell, E. P.: Phys. Rev. B **3**, 3765 (1971).
15. Havinga, E. E.: Phys. Letters **26** A, 244 (1968).
16. Havinga, E. E., Maaren, M. H. van: Low Temperature Physics LT12. Academic Press of Japan 1971.
17. Meservey, R., Schwartz, B. B.: Superconductivity (R. D. Parks, Ed.), p. 355. New York: Dekker 1969.
18. Maxwell, E.: Phys. Rev. **86**, 235 (1952).
19. Lock, J. M., Pippard, A. B., Shoenberg, D.: Proc. Cambridge Phil. Soc. **47**, 811 (1954).
20. Gschneidner, K.: Solid State Physics **16**, 275 (1964).
21. White, G. K.: Cryogenics **1**, 1 (1961).
22. Seraphim, D. P., Marcus, P. M.: IBM J. Res. Develop. **6**, 94 (1962).
23. Schirber, J. E., Swenson, C. A.: Phys. Rev. **127**, 72 (1962).
24. Jennings, L. D., Swenson, C. A.: Phys. Rev. **112**, 31 (1958).
25. Bridgman, P. W.: The Physics of High Pressure. London: G. Bell and Sons 1949.
26. Gladstone, G., Jensen, M. A., Schrieffer, J. R.: Superconductivity (R. D. Parks, Ed.), p. 771. New York: Dekker 1969.
27. Levy, M., Olsen, J. L.: Physics of High Pressure and the Condensed Phase (A. F. van Itterbeek, Ed.), p. 525. Amsterdam: North Holland Publishing Co. 1965.
28. Dugdale, J. S.: Science **134**, 77 (1961).
29. Allen, M.: Phys. Rev. **52**, 1246 (1937).
30. Klemens, P. G., Baarle, C. van, Gorter, F. W.: Physica **30**, 1470 (1964).
31. Case, S. K., Gueths, J. E.: Phys. Rev. B **2**, 3843 (1970).
32. Gold, A. V., Priestley, M. G.: Phil. Mag. **5**, 1089 (1960).
33. Weisz, G.: Phys. Rev. **149**, 504 (1966).
34. Pippard, A. B.: Phil. Mag. **46**, 1115 (1955).
35. Hall, E. O.: The Mechanism of Phase Transformations in Metals, p. 87. London: The Institute of Metals 1956.
36. Prasad, S. C., Wooster, W. A.: Acta Cryst. **9**, 35 (1956).
37. Rayne, J. A., Chandrasekhar, B. S.: Phys. Rev. **120**, 1658 (1960).

Aspects of Condensation in ^{4}He II

F. W. Cummings

University of California, Dept. of Physics, Riverside, Calif. 92502/U.S.A.

I. Introduction

The primary aim of the present article is to review certain aspects of the theoretical problem of ^{4}He II, liquid helium below the lambda transition $T = T_\lambda \cong 2.2$ °K. We shall not be concerned in what follows with the low temperature thermodynamic properties which make virtually no reference to superfluidity; these aspects seem now to be adequately understood [1]. Superfluidity has a profound effect on few of the thermally excited modes, and does not affect the correlation energy appreciably. Instead we give our attention here to certain aspects related to the appearance of a macroscopically occupied "coherent" mode, which is generally believed to be the most relevant property of the superfluid or superconducting state. In particular, a method for the experimental determination of $\varrho_c(T)$, the density of the condensate as a function of the temperature, will be discussed, as well as a connection between $\varrho_c(T)$ and the superfluid density $\varrho_s(T)$ of the two fluid model.

Mathematically, the presence of a macroscopically occupied mode is expressed by the condition, first put forward by Penrose [2], that the first order reduced density matrix, defined by

$$\Omega_1(\underset{\sim}{x}'; \underset{\sim}{x}'') = N \sum_\lambda \omega_\lambda \int \Psi_\lambda(\underset{\sim}{x}', \underset{\sim}{x}_2, \ldots \underset{\sim}{x}_N, t)\, \Psi_\lambda^*(\underset{\sim}{x}'', \underset{\sim}{x}_2, \ldots x_N, t)\, d\underset{\sim}{x}_2 \ldots d\underset{\sim}{x}_N \quad \text{(I-1)}$$

have an eigenvalue of order N, the total number of particles. Being hermitian, Ω_1, possesses the diagonal spectral decomposition

$$\Omega_1(\underset{\sim}{x}'; \underset{\sim}{x}'') = n_0\, \varphi_0(\underset{\sim}{x}')\, \varphi_0^*(\underset{\sim}{x}'') + \sum_{\alpha \neq 0} n_\alpha\, \varphi_\alpha(\underset{\sim}{x}')\, \varphi_\alpha(\underset{\sim}{x}''), \quad \text{(I-2)}$$

where the mode functions $\varphi_\alpha(\underset{\sim}{x}, t)$ have the normalization

$$\frac{1}{V} \int \varphi_\alpha^*(\underset{\sim}{x})\, \varphi_\beta(\underset{\sim}{x})\, d\underset{\sim}{x} = \delta_{\alpha\beta}, \quad \text{(I-3)}$$

and the eigenvalue n_0 is $\sim N$. The zero subscript denotes the condensed mode. The second term on the right of (I-2) behaves "normally" in the sense that it vanishes for the two points $\underset{\sim}{x}'$ and $\underset{\sim}{x}''$ taken a distance apart greater than the coherence length l_1, i.e.

$$\Omega_1 = \varphi^*(\underset{\sim}{x}'')\, \varphi(\underset{\sim}{x}') + \Lambda_1(\underset{\sim}{x}'; \underset{\sim}{x}''), \quad \text{(I-4)}$$

and

$$\Lambda_1(\underset{\sim}{x}'; \underset{\sim}{x}'') \to 0 \quad \text{for} \quad |\underset{\sim}{x}' - \underset{\sim}{x}''| > l_1. \tag{I-5}$$

In ^{4}He II, the distance l_1 is about 4 Å [3]. The density of the system is

$$\varrho(\underset{\sim}{x}) = \Omega_1(\underset{\sim}{x}; \underset{\sim}{x}) = \varrho_c(\underset{\sim}{x}) + \varrho_d(\underset{\sim}{x}), \tag{I-6}$$

where $\varrho_c(\underset{\sim}{x})$ is the density of the condensate $\varphi^*(\underset{\sim}{x})\,\varphi(\underset{\sim}{x})$ and $\varrho_d(\underset{\sim}{x})$ is the density associated with Λ_1, and often called the "depletion density". A perfect Bose gas at $T = 0$ °K has zero depletion, but the presence of the strongly repulsive "cores" in real helium gives a zero point kinetic energy of magnitude comparable to the potential energy at $T = 0$ °K, and thus the depletion part of $\Omega_1(\underset{\sim}{x}';\underset{\sim}{x}'')$ is appreciable even down to absolute zero temperature. Theoretical estimates of the condensate fraction ϱ_c/ϱ give 8 percent [4], and 11 percent [3].

The thermodynamic parameter $\varrho_s(T)$, the superfluid density, has a simple behavior as a function of temperature and can be simply measured by a variety of methods [5]; at $T = 0$ °K, $\varrho_s = \varrho$, the total density. It is, of course, ϱ_s which enters the phenomenological two fluid model, whereas the condensate density ϱ_c is the quantity which plays a fundamental role in microscopic theories.

The ideas reviewed in this article regarding the possibility of measuring the condensate density by measurement of the liquid structure factor, as well as an equation explicity relating ϱ_c to ϱ_s, are largely based on earlier ideas of Fröhlich, who first pointed out, for example, that the assumption of the existence of a macroscopically occupied mode in Ω_1, equations (I-4) and (I-5), has some very interesting implications for the second order reduced density matrix, Ω_2, whose diagonal elements are related to the liquid structure factor by a fourier transform. It is these implications which we discuss in the following section.

II. Form for Ω_2 when ODLRO Present in Ω_1

The aim of this section is largely a review of work by Fröhlich [6] and Terreaux [7] regarding the form assumed by the second order reduced density matrix $\Omega_2(\underset{\sim}{x}', \underset{\sim}{y}'; \underset{\sim}{x}'', \underset{\sim}{y}'')$ as a consequence of the appearance of the large eigenvalue in Ω_1 or as Yang [8] has termed it, the appearance of "Off-Diagonal-Long-Range-Order," or "ODLRO". By definition,

$$\Omega_2(\underset{\sim}{x}', \underset{\sim}{y}'; \underset{\sim}{x}'', \underset{\sim}{y}'') = N(N-1) \sum_\lambda \omega_\lambda \int \Psi_\lambda(\underset{\sim}{x}', \underset{\sim}{y}', \underset{\sim}{x}_3 \ldots \underset{\sim}{x}_N, t)\, \Psi_\lambda^*(\underset{\sim}{x}'', \underset{\sim}{y}'', \underset{\sim}{x}_3 \ldots \underset{\sim}{x}_N, t) \cdot d\underset{\sim}{x}_3 \ldots d\underset{\sim}{x}_N, \tag{II-1}$$

and also possesses a large eigenvalue, of order N^2 in Bose systems, as a consequence of the presumed large eigenvalue in Ω_1. Besides the conditions placed on Ω_2 due to the presence of ODLRO in Ω_1, there are several other conditions which Ω_2 may be easily seen to satisfy. Two of these conditions follow simply as a result of the definitions (II-1) and (I-1), that is

$$\begin{aligned} \Omega_2(\underset{\sim}{x}', \underset{\sim}{y}'; \underset{\sim}{x}'', \underset{\sim}{y}'') &= \Omega_2(\underset{\sim}{y}', \underset{\sim}{x}'; \underset{\sim}{x}'', \underset{\sim}{y}'') = \Omega_2(\underset{\sim}{y}', \underset{\sim}{x}'; \underset{\sim}{y}'', \underset{\sim}{x}'') \\ &= \Omega_2^*(\underset{\sim}{y}'', \underset{\sim}{x}''; \underset{\sim}{y}', \underset{\sim}{x}'), \end{aligned} \tag{II-2}$$

which is a consequence of the Bose symmetry of Ψ, and the trace condition

$$\int \Omega_2(\underset{\sim}{x}'\, \underset{\sim}{y}; \underset{\sim}{x}'', \underset{\sim}{y})\, d\underset{\sim}{y} = (N-1)\, \Omega_1(\underset{\sim}{x}'; \underset{\sim}{x}''). \tag{II-3}$$

The next two conditions express the physical nature of the system we are considering, that is, a liquid consisting of Bose particles which interact via two body potentials which have a strongly repulsive "core". The "asymptotic" condition, correct to order $1/N$,

$$\Omega_2(\underset{\sim}{x}', \underset{\sim}{y}'; \underset{\sim}{x}'', \underset{\sim}{y}'') \to \Omega_1(\underset{\sim}{x}'; \underset{\sim}{x}'')\, \Omega_1(\underset{\sim}{y}'; \underset{\sim}{y}''), \tag{II-4}$$

when the pair $(\underset{\sim}{x}', \underset{\sim}{x}'')$ is a distance greater than l_2 ("correlation length") from the pair $(\underset{\sim}{y}', \underset{\sim}{y}'')$, expresses the fact that we are dealing with a liquid, down to $T=0$ °K. The correlation length l_2 in ^{4}He II is about 10 Å, and is roughly speaking, the distance over which the liquid is capable of responding like a solid, e.g. sustaining shear. The "core" condition is expressed by

$$\Omega_2(\underset{\sim}{x}', \underset{\sim}{y}'; \underset{\sim}{x}'', \underset{\sim}{y}'') \to 0 \tag{II-5}$$

when $|\underset{\sim}{x}' - \underset{\sim}{y}'| < 2r_0$ or $|\underset{\sim}{x}'' - \underset{\sim}{y}''| < 2r_0$, where r_0 is the core radius. The four conditions we will impose on Ω_2 so far do not involve ODLRO, which is assumed to set in for $T < T_\lambda$. There are two further conditions, which assume Eqs. (I-4) and (I-5). The first simply combines Eqs. (I-4), (I-5) and (II-4) in the limit that all four points in Ω_2 are further apart than l_2, that is

$$\Omega_2 \to \varphi^*(\underset{\sim}{x}'')\, \varphi^*(\underset{\sim}{y}'')\, \varphi(\underset{\sim}{x}')\, \varphi(\underset{\sim}{y}') \tag{II-6}$$

for all points further apart than l_2.

The second condition imposed by "ODLRO" in Ω_1 follows from the exact equation of motion for Ω_1, namely,

$$\begin{aligned} i\hbar \frac{\partial}{\partial t} \Omega_1(\underset{\sim}{x}'; \underset{\sim}{x}'') = &-\frac{\hbar^2}{2m} (\nabla^2_{\underset{\sim}{x}'} - \nabla^2_{\underset{\sim}{x}''})\, \Omega_1(\underset{\sim}{x}'; \underset{\sim}{x}'') \\ &+ \int [V(\underset{\sim}{x}' - \underset{\sim}{y}) - V(\underset{\sim}{x}'' - \underset{\sim}{y})]\, \Omega_2(\underset{\sim}{x}', \underset{\sim}{y}; \underset{\sim}{x}'', \underset{\sim}{y})\, d\underset{\sim}{y}. \end{aligned} \tag{II-7}$$

Use of Eq. (I-4), (I-5) in (II-7) in the limit that $\underset{\sim}{x}'$ and $\underset{\sim}{x}''$ are taken greater than l_1 apart leads to an exact equation of motion for the "macroscopic wave function" $\varphi(\underset{\sim}{x}, t)$ since in this limit Eq. (II-7) separates into a function of $\underset{\sim}{x}'$ plus a function of $\underset{\sim}{x}''$. This results in

$$i\hbar \frac{\partial}{\partial t} \varphi(\underset{\sim}{x}', t) = -\frac{\hbar^2}{2m} \nabla^2_{\underset{\sim}{x}'} \varphi(\underset{\sim}{x}', t) + \lim_{|\underset{\sim}{x}' - \underset{\sim}{x}''| > l_1} \int \frac{V(\underset{\sim}{x}' - \underset{\sim}{y}) \Omega_2(\underset{\sim}{x}', \underset{\sim}{y}; \underset{\sim}{x}'', \underset{\sim}{y})}{\varphi^*(\underset{\sim}{x}'')}\, d\underset{\sim}{y}, \tag{II-8}$$

where it is required that the last term on the right of equation (II-8) be a function of $\underset{\sim}{x}'$ alone.

The conditions imposed above by Eqs. (II-2), (II-3), (II-4), (II-5), (II-6), and (II-8) taken together are quite restrictive and allow determination of a form for Ω_2,

$$\begin{aligned} \Omega_2(\underset{\sim}{x}', \underset{\sim}{y}'; \underset{\sim}{x}'', \underset{\sim}{y}'') = &\mathcal{S}_1(\underset{\sim}{x}', \underset{\sim}{y}'; \underset{\sim}{x}'', \underset{\sim}{y}'')\, \varphi^*(\underset{\sim}{x}'')\, \varphi^*(\underset{\sim}{y}'')\, \varphi(\underset{\sim}{x}')\, \varphi(\underset{\sim}{y}') \\ &+ \mathcal{S}_2(\underset{\sim}{x}', \underset{\sim}{y}'; \underset{\sim}{x}'', \underset{\sim}{y}'')\, [\varphi^*(\underset{\sim}{x}'')\, \varphi(\underset{\sim}{x}') \Lambda_1(\underset{\sim}{y}'; \underset{\sim}{y}'') + \varphi^*(\underset{\sim}{x}'')\, \varphi(\underset{\sim}{y}') \Lambda_1(\underset{\sim}{x}'; \underset{\sim}{y}'') \\ &+ \varphi^*(\underset{\sim}{y}'')\, \varphi(\underset{\sim}{y}') \Lambda_1(\underset{\sim}{x}'; \underset{\sim}{x}'') + \varphi^*(\underset{\sim}{y}'')\, \varphi(\underset{\sim}{x}') \Lambda_1(\underset{\sim}{y}'; \underset{\sim}{x}'')] + \Lambda_2. \end{aligned} \tag{II-9}$$

The "screening functions" $\mathcal{S}_1$ and $\mathcal{S}_2$ are similar in that they are required to satisfy the conditions, in view of Eqs. (II-5) and (II-8),

$$\mathcal{S}_{1,2} \to 1 \quad \text{for} \quad |\underset{\sim}{x}' - \underset{\sim}{y}'| \quad \text{or} \quad |\underset{\sim}{x}'' - \underset{\sim}{y}''| > l_1, \tag{II-10}$$

and

$$\mathcal{S}_{1,2} \to 0 \quad \text{for} \quad |\underset{\sim}{x}' - \underset{\sim}{y}'| \quad \text{or} \quad |\underset{\sim}{x}'' - \underset{\sim}{y}''| < 2r_0. \tag{II-11}$$

The last term in (II-9), Λ_2, is required to satisfy all the conditions which would be satisfied by an Ω_2 describing a normal liquid, that is, without ODLRO. That is to say, Λ_2 is required to satisfy a condition analogous to the trace condition,

$$\int \Lambda_2(\underset{\sim}{x}', \underset{\sim}{y}; \underset{\sim}{x}'', \underset{\sim}{y})\, d\underset{\sim}{y} = (N_d - 1)\Lambda_1(\underset{\sim}{x}'; \underset{\sim}{x}''), \tag{II-12}$$

where

$$N_d = \int \varrho_d(\underset{\sim}{x})\, d\underset{\sim}{x}, \tag{II-13}$$

as well as the analogous asymptotic condition,

$$\Lambda_2 \to \Lambda_1(\underset{\sim}{x}'; \underset{\sim}{x}'')\Lambda_1(\underset{\sim}{y}'; \underset{\sim}{y}'') \tag{II-14}$$

for $(\underset{\sim}{x}', \underset{\sim}{x}'')$ a distance l_2 away from the pair $(\underset{\sim}{y}', \underset{\sim}{y}'')$. Λ_2 is thus required to vanish whenever a primed variable is a distance greater than the coherence length away from a doubly primed variable, as would, for example, the second order reduced density matrix of liquid argon. Clearly also Λ_2 obeys the Bose symmetry conditions as well as the core condition. Thus we see that we may consider Λ_2 as containing only "normal" correlations, and is what is left after all "ODLRO effects" have been extracted from Ω_2. Clearly also, $\Omega_2 \to \Lambda_2$ when $\varphi \to 0$ for

$$T \geqq T_\lambda \cong 2.2\ °\mathrm{K}.$$

It is perhaps helpful to point out that McMillan [3] has evaluated $\mathcal{S}_1$, the "condensate screening" in Eq. (II-9), in the "bulk" (translation-rotational symmetry) situation at $T = 0$ °K. In a first principle's variational calculation, he approximated $\Psi(\underset{\sim}{x}_1, \underset{\sim}{x}_2 \ldots \underset{\sim}{x}_N)$ by a "Bijl" type product

$$\Psi = \prod_{i \neq j} f(\underset{\sim}{x}_i - \underset{\sim}{x}_j)$$

and assumed a two parameter function $\exp\{-(a/r)^b\}$ for $f(r)$. It was then found that

$$\begin{aligned}\Omega_2(\underset{\sim}{x}', \underset{\sim}{y}'; \underset{\sim}{x}'', \underset{\sim}{y}'') &\to \langle \psi^+(\underset{\sim}{x}'')\, \psi^+(\underset{\sim}{y}'') \rangle \langle \psi(\underset{\sim}{x}')\, \psi(\underset{\sim}{y}') \rangle \\ &= m(|\underset{\sim}{x}'' - \underset{\sim}{y}''|)\, m(|\underset{\sim}{x}' - \underset{\sim}{y}'|),\end{aligned} \tag{II-15}$$

when $(\underset{\sim}{x}'', \underset{\sim}{y}'')$ was taken very far from $(\underset{\sim}{x}', \underset{\sim}{y}')$, greater than l_2, say. In the Bijl approximation, $m(r)$ is given by

$$m(r) = f(r)\, \Omega_1(r), \tag{II-16}$$

and $f(\mathrm{r})$ and $\Omega_1(r)$ are both given by the numerical calculation of reference [3]. Now referring back to Eq. (II-9) in the bulk situation we have, in the same limit as Eq. (II-15)

$$\Omega_2(\underset{\sim}{x}', \underset{\sim}{y}'; \underset{\sim}{x}'', \underset{\sim}{y}'') \to S_1(\underset{\sim}{x}' - \underset{\sim}{y}')\, S_1(\underset{\sim}{x}'' - \underset{\sim}{y}'')\, \varrho_c^2, \tag{II-17}$$

where

$$\mathcal{S}_1(\underset{\sim}{x}', \underset{\sim}{y}'; \underset{\sim}{x}'', \underset{\sim}{y}'') = S_1(\underset{\sim}{x}' - \underset{\sim}{y}')\, S_1(\underset{\sim}{x}'' - \underset{\sim}{y}''). \tag{II-18}$$

Comparison of Eqs. (II-17) and (II-18) with (II-15) and (II-16) gives

$$S_1(r) = m(r)/\varrho_c = f(r)\, \Omega_1(r)/\varrho_c. \tag{II-19}$$

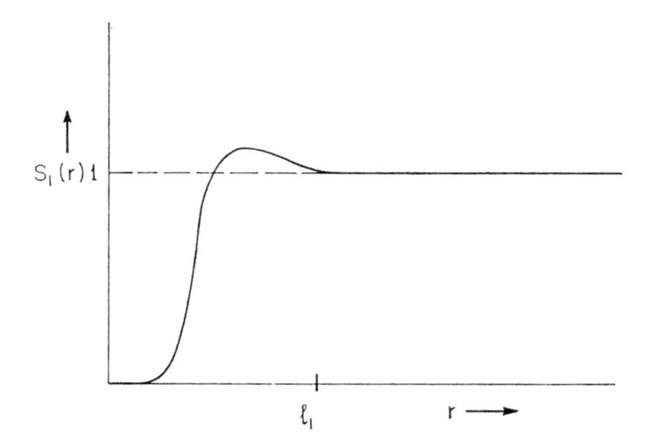

Fig. 1.
The screening function $S_1(r)$ is shown at $T = 0$ °K, taken from the values given by McMillan's calculation. The "anomolous pair function" is given by $\varrho_c [S_1(r) - 1]$.

Taking $f(r)$ and Ω_1 from reference [3], we have shown the screening function S_1 in figure 1. The "anomolous pair function" $m(r) - \varrho_c$ is also plotted in reference [3] [and there compared with the Bogoliubov weak coupling theory] and via Eq. (II-19) is thus equal to $\varrho_c [S_1(r) - 1]$.

III. An Exact Formula for $\varrho_c(T)$

Having discussed the implications of ODLRO in Ω_1 on the form of Ω_2, this form is now used to obtain a simple exact expression for $\varrho_c(T)$ [10]. To this end, we examine the diagonal elements of Eq. (II-9) in the bulk limit,

$$\Omega_2 = \Omega_2(\underset{\sim}{x}' \underset{\sim}{y}'; \underset{\sim}{x}', \underset{\sim}{y}') = \Omega_2(r).$$

We also examine $\Omega_2(r)$ in the region $l_1 < r < l_2$ where Eq. (II-9) reduces to

$$\Omega_2(r) = \varrho_c^2 + 2\varrho_c \varrho_d + \varrho_d^2 \tilde{g}(r). \tag{III-1}$$

The normal part of $\Omega_2(r)$, in view of the asymptotic condition, Eq. (II-14), and the "core" condition, $\Lambda_2(r) \to 0$ for $r < 2r_0$, has been written as

$$\Lambda_2(r) \equiv \varrho_d^2 \tilde{g}(r) \tag{III-2}$$

where $\tilde{g}(r) \to 0$ for $r < 2r_0$, and $\tilde{g}(r) \to 1$ for $r > l_2$. Thus $\tilde{g}(r)$ is the "pair distribution function" for the "normal" part of Ω_2. Defining the measurable pair distribution function as usual by

$$\Omega_2(r) = \varrho^2 g(r) \tag{III-3}$$

allows (II-1) to be written as

$$\varrho^2 [g(r, T) - 1] = \varrho_d^2 [\tilde{g}(r, T) - 1], \tag{III-4}$$

or

$$\varrho_c(T) = \varrho \left\{ 1 - \left[\frac{g(r, T) - 1}{\tilde{g}(r, T) - 1} \right]^{\frac{1}{2}} \right\}. \tag{III-5}$$

g and $\tilde{g}$ are to be thought of as functions of both pressure and temperature as well as r, but in what follows we may envision the pressure as being held constant. Our claim is that (III-5) is an exact expression to order $1/N$.

In the rest of this section we wish to discuss some speculative, hopefully plausible assumptions regarding $\tilde{g}(r, T)$, which will have the merit of being able to be checked experimentally. The function $\tilde{g}(r, T)$, representing the structure of the "normal" part of Ω_2, is not directly measurable, but we wish to suggest possible connections between $\tilde{g}$ and experimentally observable quantities, and thus to have an expression for $\varrho_c(T)$ which is verifiable.

Eq. (III-5) is remarkable in requiring the ratio $(g-1)$ to $(\tilde{g}-1)$ be independent of r. Any legitimate form assumed for $\tilde{g}-1$ must share this property, and "r independence" provides a check.

It is important to stress the point that the statement of ODLRO in Ω_1, Eqs. (I-4), (I-5) and (I-6), demands that we describe *each* helium atom as being partly in the condensate and partly in the localized depletion, since $\Omega_1(\underset{\sim}{x}'; \underset{\sim}{x}'')$ is a function of the coordinates $\underset{\sim}{x}'$ and $\underset{\sim}{x}''$ of a *single* particle. This is to be contrasted to the statement, sometimes seen, that ϱ_c is proportional to the number of helium atoms in the condensate and ϱ_d to the number in the depletion. The first, and correct, description leads to a picture of bulk helium for $T < T_\lambda$ as consisting of a uniform condensate in which "lumps" (for lack of a better word) are imbedded, the lumps being localized to within a radius l_1. Since the total density remains almost constant as the temperature is varied below T_λ, the lumps will "melt" into the condensate as temperature is lowered, for $T < T_\lambda$.

In a scattering experiment, neutrons or x-rays, the scattering will take place only from the lumps and not from the uniform condensate, in the same way that an e.m. wave will scatter only from the bubbles and imperfections in traversing an otherwise uniform plate of glass (except for forward scattering). The scattering experiment will reflect, via a fourier transform, the relative probability of finding a second particle a distance r from a first, i.e. the information contained in $\Omega_2(r)$. As we *lower* the temperature, in the region $T < T_\lambda$, we may expect, then, due to the growth of the condensate and consequent loss of density of the lumps, that the scattering become less intense. We might thus expect, based on these intuitive notions, the following temperature sequence upon measurement of $g(r, T)$: for temperatures *in the region* $T > T_\lambda$, the maxima of $g(r, T)$ are expected to increase with decreasing temperature to reflect the increasing relative localization and decreasing potential (and also kinetic) energy; the helium atoms act as if they were trying to approach the solid state, the ultimate classical end point of which would be when all helium atoms are at rest at fixed lattice points. At some point in temperature, however, this process is halted by intervention of noticable quantum effects; the uncertainty principle steps in and demands that energy loss through increasing localization be accompanied by an increase in the zero point kinetic energy. We may suppose that the point at which relative localization becomes an unacceptable way to give up energy to the heat bath, is the lambda transition, $T = T_\lambda$. At this point the maxima of $g(r, T)$ have their greatest amplitude (although $g(r, T)$ is not necessarily differentiable with temperature at T_λ). Below T_λ, nature must find a way other than increasing localization (reflected in increasing maxima of $g(r)$) for the system to lose energy. For $T < T_\lambda$ it may be reasonable to suppose that the function $\tilde{g}(r, T)$ representing the relative structure of the "*lumps*" *alone*, remain very close to its value at T_λ, when $\tilde{g}(r, T_\lambda) = g(r, T_\lambda)$. This conjecture is illustrated schematically in Fig. 2.

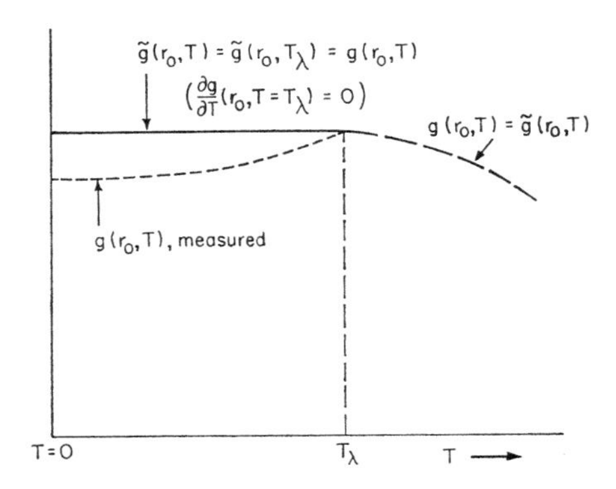

Fig. 2.
Conjecture for the behavior of a particular maximum at $r_0 > l_1$ of $g(r, T)$ and $\tilde{g}(r, T)$, shown above and below T_λ; the solid line represents one plausible conjecture for the behavior of $\tilde{g}(r_0, T)$.

A plot of a given maxima (say) of $g(r, T)$ vs. temperature is shown schematically in Fig. 2. Clearly a reversal in trend of the value of the maxima of $g(r, T)$ occurs at $T = T_\lambda$, from increasing to decreasing as T is decreased. The curves of figure 2 do not represent actual measurements of g, but conjecture only; however the trend reversal does agree with the scant experimental evidence to date [10, 11].

Let us imagine, first of all, that both g and $\tilde{g}$ may be represented for T in the vicinity of T_λ, for $T < T_\lambda$ as a linear function of $(T - T_\lambda)$

$$g(r, T) \cong g(r, T_\lambda) + \frac{\partial g}{\partial T}(r, T_\lambda)(T - T_\lambda), \tag{III-6}$$

$$\tilde{g}(r, T) \cong g(r, T_\lambda) + \frac{\partial \tilde{g}}{\partial T}(r, T_\lambda)(T - T_\lambda). \tag{III-7}$$

Then Eq. (III-5) becomes, in the vicinity of T_λ,

$$\varrho_c(T)/\varrho = \left\{\left[\frac{\partial \tilde{g}}{\partial T}(r, T_\lambda) - \frac{\partial g}{\partial T}(r, T_\lambda)\right] \Big/ 2\left[g(r, T_\lambda) - 1\right]\right\}(T - T_\lambda). \tag{III-8}$$

Eqs. (III-6) and (III-7) with (III-5) then give ϱ_c as linear in $T - T_\lambda$, in agreement with scaling law arguments [9]. TAYLOR has discussed in this volume the relation of ϱ_c to ϱ_s near T_λ, i.e.,

$$\varrho_s(T) = \frac{2m}{\hbar^2} B(T)\, \varrho_c(T), \tag{III-9}$$

and this allows us to relate the bracket in Eq. (III-8) to the parameter B of the "Landau-Ginzburg" equation for ^{4}He II. Taking the scaling law results

$$B = \beta (T_\lambda - T)^{-1/3}, \tag{III-10}$$

and

$$\varrho_s = \alpha (T_\lambda - T)^{2/3} \tag{III-11}$$

then gives in Eq. (III-8) with (III-9),

$$\left[\frac{\partial g}{\partial T}(r, T_\lambda) - \frac{\partial \tilde{g}}{\partial T}(r, T_\lambda)\right] \Big/ \left(g(r, T_\lambda) - 1\right) = \frac{\hbar^2 \alpha}{m \beta \varrho}. \tag{III-12}$$

Everything in (III-12) can be measured except $\frac{\partial \tilde{g}}{\partial T}$ and β.

Consistent with our discussion about Eq. (III-6) we may suppose that $\tilde{g}$ is a function which remains constant in the vicinity of T_λ, that is, that

$$\frac{\partial \tilde{g}}{\partial T}(T=T_\lambda^-)=\frac{\partial \tilde{g}}{\partial T}(T=T_\lambda^+)=0.$$

With this assumption, (III-12) becomes

$$\frac{\partial}{\partial T}\ln(g-1)\bigg|_{T=T_\lambda}=\frac{\hbar^2\alpha}{m\beta\varrho} \qquad \left(\text{if } \frac{\partial \tilde{g}}{\partial T}\bigg|_{T_\lambda}=0\right). \tag{III-13}$$

If the left hand side of Eq. (III-13) should turn out experimentally to depend on r, this would be evidence that $\partial\tilde{g}/\partial T$ is non zero, but in fact a function of r. If this should turn out to be independent of r, then Eq. (III-13) gives the previously phenomenological constant B in the expansion of the free energy

$$F=F_n-\frac{\hbar^2}{2m}B\int \varphi^*\nabla^2\varphi\, d\underset{\sim}{x}+c\int\varphi^*\varphi\, d\underset{\sim}{x}+\cdots \tag{III-14}$$

in terms of experimentally determined quantities.

It would seem further reasonable to presume that

$$\tilde{g}(r,T)=\tilde{g}(r,T_\lambda)=g(r,T_\lambda)\begin{Bmatrix}0\leqq T\leqq T_\lambda,\\ l_1<r<l_2\end{Bmatrix} \tag{III-15}$$

the argument being simply that since $\tilde{g}(r,T)$ represents the structure of the lumps, there seems to be nothing further to be gained as T decreases below T_λ from the energetic point of view by changing the relative position of two lumps. With this assumption, we may then write Eq. (III-5) as [10, 11]

$$\varrho_c(T)/\varrho=1-\left[\frac{g(r,T)-1}{g(r,T_\lambda)-1}\right]^{\frac{1}{2}}, \tag{III-16}$$

if $\tilde{g}(r,T)=\tilde{g}(r,T_\lambda)$, $T<T_\lambda$.
Eq. (III-16) allows determination of ϱ_c as a function of temperature, and at the same time provides a check on its validity: the function under the square root must be independent of r, $l_1<r<l_2$, and in particular, the zeroes of $g-1$ must be independent of temperature.

Existing scant experimental data seems to somewhat substantiate the above assumption, Eq. (III-16), but only more extensive experiments measuring $g(r,T)$ for many values of T above and below T_λ will provide an answer. The existing data [10] gives a value for ϱ_c of about ten percent at about one degree kelvin, plus or minus five percent. The zero crossings are constant to within about three percent. More extensive measurements are presently underway at the National Bureau of Standards, Washington D. C.

We close this article with a brief mention of an interesting formula recently put forth by Hyland and Rowlands [12], and also obtained by a different method by Haug and Weiss [13]. This formula relates the macroscopically defined two fluid model density ϱ_s and the microscopically defined condensate density ϱ_c, that is

$$\varrho_s=\varrho_c\left[\frac{a\varrho+Ts}{a\varrho_c+Ts}\right]. \tag{III-17}$$

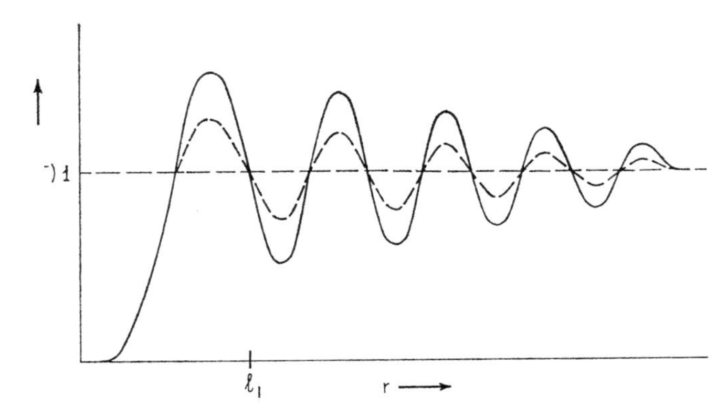

Fig. 3. Schematic of the pair distribution function shown for two different temperatures: the dashed line may represent $T \approx 4$ °K and the solid curve then represents $T \approx 2.3$ °K, $T > T_\lambda$. As temperature decreases below T_λ, $g(r, T)$ will return to a curve similar to the dashed curve, in the sense that its amplitude for $T < T_\lambda$, will decrease.

a is an integral involving the screened two-body potential $V(r)$ and depends, perhaps weakly, on temperature. Ts is the temperature times the entropy density The above formula has the virtue that it exhibits the correct limiting behavior as $T \to 0$, when $\varrho_s = \varrho$, as well as when $T \to T_\lambda$, when $\varrho_s \approx \varrho_c \to 0$. However, as discussed earlier, the equation $\varrho_s \approx \varrho_c$, $T \lesssim T_\lambda$, while being in accord with the original Landau assumption concerning the meaning of the order parameter, appears to be at variance with the scaling law results of JOSEPHSON, and WONG, and is discussed by ANDREW TAYLOR in this volume, namely that $\varrho_s \sim (T_\lambda - T)^{2/3}$ and $\varrho_c \sim (T_\lambda - T)$. The source of the discrepancy is not understood and seems to merit further study.

References

1. FEENBERG, E.: Am. J. Phys. **38**, 684 (1970).
2. PENROSE, O.: Phil. Mag. **42**, 1373 (1951).
3. MCMILLAN, W. L.: Phys. Rev. **138** A, 442 (1966).
4. PENROSE, O., ONSAGER, L.: Phys. Rev. **104**, 577 (1956).
5. WILKS, J.: The properties of Liquid and Solid Helium. Clarendon Press Oxford 1967.
6. FRÖHLICH, H.: J. Phys. Soc. Japan **26**, (Suppl.) 189 (1969); — Phys. Kondens. Materie **9**, 350 (1969) and also four lectures delivered at Summer Semester of 1969 at Institute of Theor. Phys., Univ. of Stuttgart, available from editor R. Hübner.
7. TERREAUX, C.: Collective Phenomena **1**, 54 (1972).
8. YANG, C. N.: Rev. Mod. Phys. **34**, 694 (1962).
9. WONG, V. K.: Phys. Letters **27** A, 269 (1968).
10. CUMMINGS, F. W., HYLAND, G., ROWLANDS, G.: Phys. Kondens. Materie **12**, 90 (1970).
11. CUMMINGS, F. W.: Macroscopic Wave Functions. In: Proc. of Bienniel I.U.P.A.P. Conference on Statistical Mechanics, 1971, ed. by Stuart Rice. Chicago: Chicago Univ. Press 1972.
12. HYLAND, G., ROWLANDS, G.: J. Low Temp. Phys. **7**, 271 (1972).
13. HAUG, H., WEISS, K.: Physica **59**, 29 (1972).

The Derivation of Time-Dependent Ginzburg-Landau Equations in Superfluids

A. W. B. TAYLOR

University of Liverpool, Dept. of Theoretical Physics, Liverpool/England

Abstract

The general structure of the equations of motion for the order parameter in superfluid helium and in superconductors are derived by application of conventional techniques in statistical mechanics. These equations of motion are valid at all temperatures and not only near the lambda point or critical point. It is shown how, near the lambda point and critical point, these equations can be reduced to the usual Ginzburg-Landau equations.

1. Introduction

Prof. FRÖHLICH has pointed out [1, 2] that the equations for macroscopic physics do not require for their derivation a complete solution of the underlying many body problem. The general structure of macroscopic equations can often be deduced from general considerations of the microscopic problem. The role of the microscopic theory is then focussed on finding expressions for the parameters which may appear in such macroscopic equations (e.g. transport coefficients). Prof. FRÖHLICH [1] illustrated this in the case of the hydrodynamic Navier-Stokes equations showing how the structure of these equations is largely a matter of the symmetries of the physical system and not of the detailed microscopic structure.

Macroscopic equations which are proving to be extremely useful in superconductors and in superfluid helium are the time dependent Ginzburg-Landau equations. These are equations of motion for the order parameter in superfluid helium [3, 4, 5] and in superconductors [6, 7, 8] (analogous to the Landau equations or the Bloch-Bloembergen equation in ferromagnetic relaxation theory). In superfluid helium the order parameter is the statistical average $\langle\psi^\dagger(x)\rangle$, $\psi^\dagger(x)$ being the spinless Boson creation operator. In superconductors it is the average $\langle\psi_\uparrow^\dagger(x)\,\psi_\downarrow^\dagger(x)\rangle$, $\psi_\sigma^\dagger(x)$ being the Fermion creation operator (with spin).

In superconductors, where a highly successful detailed microscopic theory is available, the time dependent Ginzburg-Landau equation has been derived in full detail from the microscopic theory [6, 7, 8]. These derivations do not distinguish between the general structure of the equation and the explicit expressions for the parameters in the equation but produce both in a single derivation. As a consequence of the exigencies of such a calculation the resultant equation is valid only near the critical temperature.

In this paper we apply conventional techniques of statistical mechanics [9, 10, 11] to the problem of deriving a time dependent equation of motion for the order parameters in superfluid helium and in superconductors. In superfluid helium (with zero normal component velocity) the resultant equation (§ 2) is

$$(i\hbar\,\partial/\partial t+\mu)\,\langle\psi^{\dagger}\rangle=\eta\,\delta F/\delta\langle\psi\rangle \tag{1.1}$$

and for the superfluid current density

$$\nabla\cdot\boldsymbol{J}=i\{\langle\psi\rangle\,\delta F/\delta\langle\psi\rangle-\text{c.c.}\}/\hbar \tag{1.2}$$

where μ is a local chemical potential, η is a complex number the imaginary part of which is a transport coefficient and F is the free energy density. In the absence of dissipation one immediately recovers the well known results

$$\langle\psi\rangle\sim\exp i\mu t/\hbar,\qquad \delta F/\delta\langle\psi\rangle=0 \tag{1.3}$$

Since the derivation relies only on the existence of $\langle\psi\rangle$ and does not invoke any specific microscopic model or approximation it does not produce any expressions for η or F. This is the task of the microscopic theory. *The advantage of this general approach is that the Eqs. (1.1) and (1.2) are valid for all temperatures $T\leqq T_\lambda$ and not just near T_λ as the conventional equations are.*

Near T_λ one can apply to (1.1) and (1.2), (§ 3), recent considerations on the structure of F near T_λ [12, 13]. The usual time dependent Ginzburg-Landau equation is then immediately obtained together with the result that the superfluid density σ_s depends upon temperature T as $(1-T/T_\lambda)^{\frac{2}{3}}$.

In the case of a superconductor (§ 4) the resultant equation of motion is

$$(i\hbar\,\partial/\partial t+2\mu)\,\langle\psi_\uparrow^{\dagger}\psi_\downarrow^{\dagger}\rangle=\eta\,\delta F/\delta\langle\psi_\downarrow\psi_\uparrow\rangle \tag{1.4}$$

and for the superfluid current

$$\nabla\cdot\boldsymbol{J}=2mi\{\langle\psi_\uparrow^{\dagger}\psi_\downarrow^{\dagger}\rangle\,\delta F/\delta\langle\psi_\uparrow^{\dagger}\psi_\downarrow^{\dagger}\rangle-\text{c.c.}\}/\hbar. \tag{1.5}$$

Again these equations are valid for all values of temperature and magnetic field and not just near the critical point. In contrast to the superfluid helium we have a well established microscopic theory [14] and to find $\delta F/\delta\langle\psi_\downarrow\psi_\uparrow\rangle$ near the critical point (§ 5) one need only refer to the appropriate sections of the textbooks. This gives immediately the standard Ginzburg-Landau equations.

The conventional technique of statistical mechanics [9, 10, 11] used can be described as follows. The density matrix ϱ is written as

$$\varrho=\varrho_o+\Delta_\varrho \tag{1.6}$$

where ϱ_o is a local equilibrium destribution and Δ_ϱ describes the deviation from local equilibrium. The local equilibrium destribution is written in the form

$$\varrho_o=\exp-\sum_i a_i A_i/\text{Tr}\exp-\sum_i a_i A_i \tag{1.7}$$

where the a_i are various operators and the A_i are intensive quantities determined by the conditions

$$\operatorname{Tr} \varrho\, a_i = \operatorname{Tr} \varrho_o\, a_i. \tag{1.8}$$

The choice of the a_i depends upon the system being studied. In the case of a normal fluid one would choose energy density, number density, momentum density. For a boson superfluid one has only to add ψ and $\psi^\dagger$ to this list and for a superconductor $\psi_\uparrow^\dagger \psi_\downarrow^\dagger$ and $\psi_\downarrow \psi_\uparrow$.

The deviation from local equilibrium, Δ_ϱ, is found by solving the Liouville equation for ϱ under the assumption that in the infinite past $\varrho = \varrho_o$. Taking Δ_ϱ linear in the thermodynamic forces, one substitutes this Δ_ϱ into

$$-i\hbar \frac{\partial}{\partial t} \langle a_i \rangle = \operatorname{Tr} \varrho_o [H, a_i] + \operatorname{Tr} \Delta_\varrho [H, a_i] \tag{1.9}$$

to obtain an equation of motion for $\langle a_i \rangle$ including dissipative terms (second term on right) with dissipative coefficient given by Kubo formulae.

2. Boson Superfluid

Superfluid helium is described as a collection of bosons interacting through a two body potential.

The Hamiltonian is

$$H = \frac{\hbar^2}{2m} \int dx\, \nabla\psi^\dagger(x) \cdot \nabla\psi(x) + \frac{1}{2} \int dx\, dy\, V(x-y)\, \psi^\dagger(x)\, \psi^\dagger(y)\, \psi(y)\, \psi(x) \tag{2.1}$$

where $\psi^\dagger(x)$, $\psi(x)$ are boson creation and annihilation operators and V is the spherically symmetric two body potential. We use the conventional distinction between a normal system and a superfluid system, that in the superfluid system the expectation value $\langle \psi(x) \rangle$ is non zero. This allows the familiar definition of the condensate density σ_c and the curl free superfluid velocity $\boldsymbol{v}_s$, as

$$\langle \psi(x) \rangle = \sqrt{\sigma_c} \exp -i\theta, \qquad \boldsymbol{v}_s = \hbar \nabla \theta / m. \tag{2.2}$$

The Liouville equation for the density matrix ϱ

$$i\hbar\, \partial\varrho | \partial t = [H, \varrho] \tag{2.3}$$

leads to the exact equation of motion

$$-i\hbar \frac{\partial}{\partial t} \langle \psi^\dagger(x) \rangle = -\frac{\hbar^2}{2m} \nabla^2 \langle \psi^\dagger(x) \rangle + \int dy\, V(x-y) \langle \psi^\dagger(x)\, \psi^\dagger(y)\, \psi(y) \rangle. \tag{2.4}$$

The local equilibrium density matrix is now taken to be

$$\varrho_o = \exp -Q / \operatorname{Tr} \exp -Q, \tag{2.5}$$

$$\begin{aligned} Q = \int dx \frac{1}{kT(x,t)} \{ & \hat{\varepsilon}(x) - \mu(x,t)\, \hat{\sigma}(x) - \boldsymbol{v}_n(x,t) \cdot \hat{\boldsymbol{J}}(x) \\ & -\alpha(x,t)\, \psi(x) - \alpha^*(x,t)\, \psi^\dagger(x) \} \end{aligned} \tag{2.6}$$

where $\hat{\varepsilon}$ is the energy density operator, $\hat{\sigma}$ is the particle density operator and $\hat{\boldsymbol{J}}$ is the current density operator defined by

$$\hat{\varepsilon}(x) = \frac{\hbar^2}{2m} \nabla\psi^\dagger(x) \cdot \nabla\psi(x) + \frac{1}{2} \int dy\, V(x-y)\, \hat{\sigma}(x)\hat{\sigma}(y) \tag{2.7}$$

$$\hat{\sigma}(x) = \psi^\dagger(x)\psi(x) \tag{2.8}$$

$$\hat{\boldsymbol{J}}(x) = -i\hbar\{\psi^\dagger(x)\nabla\psi(x) - \text{c.c.}\}/2. \tag{2.9}$$

The local temperature T, local chemical potential μ, "normal" component velocity $\boldsymbol{v}_n$ and the complex intensive quantity α are defined by the conditions

$$\varepsilon(x, t) \equiv \mathrm{Tr}\,\varrho\,\hat{\varepsilon}(x) = \mathrm{Tr}\,\varrho_o\,\hat{\varepsilon}(x) \tag{2.10a}$$

$$\sigma(x, t) \equiv \mathrm{Tr}\,\varrho\,\hat{\sigma}(x) = \mathrm{Tr}\,\varrho_o\,\hat{\sigma}(x) \tag{2.10b}$$

$$\boldsymbol{J}(x, t) \equiv \mathrm{Tr}\,\varrho\,\hat{\boldsymbol{J}}(x) = \mathrm{Tr}\,\varrho_o\,\hat{\boldsymbol{J}}(x) \tag{2.10c}$$

$$\langle\psi^\dagger(x)\rangle \equiv \mathrm{Tr}\,\varrho\,\psi^\dagger(x) = \mathrm{Tr}\,\varrho_o\,\psi^\dagger(x) \tag{2.10d}$$

$$\langle\psi(x)\rangle \equiv \mathrm{Tr}\,\varrho\,\psi(x) = \mathrm{Tr}\,\varrho_o\,\psi(x). \tag{2.10e}$$

The total entropy is defined by

$$S_{\mathrm{tot}} = -k\,\mathrm{Tr}\,\varrho_o \log\varrho_o \tag{2.11}$$

and so

$$\delta S_{\mathrm{tot}} = \int dx\,\{\delta\varepsilon/T - \mu\,\delta\sigma/T - \boldsymbol{v}_n \cdot \delta\boldsymbol{J}/T - \alpha\,\delta\langle\psi\rangle/T - \alpha^*\,\delta\langle\psi^\dagger\rangle/T\} \tag{2.12}$$

where δ denotes an arbitrary infinitesimal variation.

It is not obvious from this that there exists a local entropy density $S(x, t)$ satisfying

$$\delta S = \delta\varepsilon/T - \mu\,\delta\sigma/T - \boldsymbol{v}_n \cdot \delta\boldsymbol{J}/T - \alpha\,\delta\langle\psi\rangle/T - \alpha^*\, d\langle\psi^\dagger\rangle/T. \tag{2.13}$$

However, if the right hand side of (2.13) can be shown to be a perfect differential then an $S(x, t)$ satisfying (2.13) does exist. By virtue of (2.11) and (2.12) this $S(x, t)$, whatever it may be, gives the total entropy when integrated and so we can refer to it as the local entropy density.

It is shown in the Appendix that the right hand side of (2.13) is indeed a perfect differential if the correlation functions and cross-correlation functions of $\hat{\varepsilon}$, $\hat{\sigma}$, $\hat{\boldsymbol{J}}$, ψ and $\psi^\dagger$ all have a finite spatial range and that the intensive quantities $1/T$, μ/T, $\boldsymbol{v}_n/T$, α/T and α^*/T are effectively constant over the distances of the order of this finite correlation distance.

Thus, we assume that the intensive quantities have sufficiently slow spatial variation. This is not a special property of superfluid systems but presumably a basic requirement of the physical meaning of local equilibrium within the framework of non-equilibrium thermodynamics. The problem of proving the existence of a local thermodynamic identity such as (2.13) exists in the case of a normal fluid, which is, of course, a special case of the superfluid ($\alpha = \alpha^* = 0$).

In passing we point out that if α is written in the form

$$\alpha(x, t) = -i\hbar\nabla \cdot \{\sigma_s(\boldsymbol{v}_s - \boldsymbol{v}_n)\}/2\langle\psi\rangle \tag{2.14}$$

then the thermodynamic identity can be shown to become

$$T\delta S = \delta\varepsilon - \mu\,\delta\sigma - \boldsymbol{v}_n\cdot\delta\boldsymbol{J} - \sigma_s(\boldsymbol{v}_s - \boldsymbol{v}_n)\,\delta\boldsymbol{v}_s \tag{2.15}$$

which is the thermodynamic identity postulated in the two fluid model of superfluid helium, σ_s being the density of the superfluid component. It has been shown elsewhere [15] that the choice (2.14) also leads to the usual two fluid model expressions for the various fluxes, e.g.

$$\boldsymbol{J} = \sigma\boldsymbol{v}_n + \sigma_s(\boldsymbol{v}_s - \boldsymbol{v}_n) \tag{2.16}$$

$$= \sigma_n\boldsymbol{v}_n + \sigma_s\boldsymbol{v}_s \tag{2.17}$$

where $\sigma_n = \sigma - \sigma_s$. However, we shall not use here the explicit expression (2.14) for α.

We now define a free energy density F * by

$$F = \varepsilon - TS - \mu\sigma. \tag{2.18}$$

From (2.13), it is straightforward to show that

$$\delta F = -S\,\delta T + \sigma\,\delta\mu + \alpha\,\delta\langle\psi\rangle + \alpha^*\,\delta\langle\psi^\dagger\rangle \tag{2.19}$$

so that

$$\alpha(x,t) = \left.\frac{\delta F}{\delta\langle\psi(x)\rangle}\right|_{T,\mu}. \tag{2.20}$$

Now, the identity

$$\mathrm{Tr}\,\varrho_o[Q,\psi^\dagger(x)] = 0 \tag{2.21}$$

yields, on evaluating the commutator,

$$\begin{aligned} -\frac{\hbar^2}{2m}\nabla^2\langle\psi^\dagger(x)\rangle_o + \int dy\,V(x-y)\,\langle\psi^\dagger(x)\,\psi^\dagger(y)\,\psi(y)\rangle_o \\ = \mu\langle\psi^\dagger(x)\rangle_o + \frac{i\hbar}{2}(\nabla\cdot\boldsymbol{v}_n)\langle\psi^\dagger(x)\rangle_o + i\hbar\,\boldsymbol{v}_n\cdot\nabla\langle\psi^\dagger(x)\rangle_o \\ -\alpha(x) + \frac{\hbar^2}{2m}T\left(\nabla\frac{1}{T}\right)\cdot\nabla\langle\psi^\dagger(x)\rangle_o \\ +\frac{1}{2}\int dy\,\boldsymbol{v}(x-y)\left\{1-\frac{T(x)}{T(y)}\right\}\langle\psi^\dagger(x)\,\psi^\dagger(y)\,\psi(y)\rangle_o \end{aligned} \tag{2.22}$$

where the subscript o denotes that the averages are taken with the density matrix ϱ_o of (2.6).

Since T is assumed to have slow spatial variation we can neglect the last two terms in (2.22). This gives, on comparison with Eq. (2.4),

$$\mathrm{Tr}\,\varrho_o[H,\psi^\dagger(x)] = \alpha - \left(\mu + \frac{i\hbar}{2}\nabla\cdot\boldsymbol{v}_n + i\hbar\,\boldsymbol{v}_n\cdot\nabla\right)\langle\psi^\dagger\rangle \tag{2.23}$$

as the equation of motion for $\langle\psi^\dagger\rangle$ in local equilibrium.

* If the temperature T is uniform, the total free energy is

$$F_{\mathrm{tot}} = \int dx\,F = -kT\log Tr\exp\{-Q - \int dx\,(\alpha\langle\psi\rangle + \mathrm{c.c.})|T\}$$

and

$$\alpha(x,t) = \delta F_{\mathrm{tot}}/\delta\langle\psi(x)\rangle$$

the differentiation being a functional differentiation.

To find the contribution of Δ_ϱ, the derivation of the density matrix from local equilibrium, to the equation of motion

$$-i\hbar \frac{\partial}{\partial t} \langle \psi^\dagger(x) \rangle = \mathrm{Tr}\, \varrho_o [H, \psi^\dagger(x)] + \mathrm{Tr}\, \Delta_\varrho [H, \psi^\dagger(x)]. \tag{2.24}$$

we apply conventional techniques [9, 10, 11] of statistical mechanics. These provide Δ_ϱ linear in the deviation from equilibrium* and give

$$-i\hbar \frac{\partial}{\partial t} \langle \psi^\dagger(x) \rangle = \mathrm{Tr}\, \varrho_o [H, \psi^\dagger(x)] + i \int dx' \int_0^\infty d\tau\, K(x', \tau)\, \alpha(x - x', t - \tau) \tag{2.25}$$

where the correlation function K is given by a Kubo formula

$$K(x, \tau) = \int_0^{1/kT} d\lambda \langle e^{iHt/\hbar} [\psi^\dagger(x), H] e^{-iHt/\hbar} e^{-\lambda Q} [H, \psi(0)] e^{\lambda Q} \rangle_o \tag{2.26}$$

in which the average (and Q) is taken with $\alpha = 0$, i.e. over the normal equilibrium state.

If α has sufficiently slow spatial and temporal variation we can take it outside the integral in (2.25) to give

$$-i\hbar\, \partial \langle \psi^\dagger \rangle / \partial t = \mathrm{Tr}\, \varrho_o [H, \psi^\dagger(x)] + i\gamma\alpha \tag{2.27}$$

where

$$\gamma = \int dx' \int_0^\infty d\tau\, K(x', \tau) \tag{2.28}$$

is a real, positive transport coefficient.

Substituting (2.20) and (2.23) in (2.27) we obtain

$$\left(i\hbar \frac{\partial}{\partial t} + \mu + \frac{i\hbar}{2} \nabla \cdot \boldsymbol{v}_n + i\hbar \boldsymbol{v}_n \cdot \nabla\right) \langle \psi^\dagger(x) \rangle = \eta\, \delta F / \delta \langle \psi(x) \rangle \tag{2.29}$$

where

$$\eta = 1 - i\gamma.$$

To find the current $\boldsymbol{J}$ we evaluate the commutator in the identity

$$\mathrm{Tr}\, \varrho_o [Q, \hat{\sigma}(x)] = 0 \tag{2.30}$$

to obtain

$$\nabla \cdot (\boldsymbol{J} - \sigma \boldsymbol{v}_n)/T = i(\alpha \langle \psi \rangle - \mathrm{c.c.})/\hbar T. \tag{2.31}$$

Neglecting derivatives of T, this yields

$$\nabla \cdot (\boldsymbol{J} - \sigma \boldsymbol{v}_n) = i(\alpha \langle \psi \rangle - \mathrm{c.c.})/\hbar \tag{2.32}$$

$$= i\left\{\langle \psi \rangle \frac{\delta F}{\delta \langle \psi \rangle} - \mathrm{c.c.}\right\}/\hbar. \tag{2.33}$$

Eq. (2.29) is our time dependent equation of motion for the order parameter and (2.33) the accompanying equation for the current.

* There are terms in $\Delta\varrho$ linear in $\nabla 1/T$, $\nabla \mu/T$, α and α^*. Only the terms linear in α and α^* gives a non-zero contribution to $\mathrm{Tr}\, \Delta\varrho [H, \psi^\dagger]$, the Kubo coefficients of the other terms vanishing due to the various symmetry properties of the equilibrium density matrix. This is an example of the Cune principle in non-equilibrium thermodynamics [16] that fluxes and forces of different symmetries do not couple.

3. The Equations Near T_λ

Taking $\boldsymbol{v}_n = 0$ for simplicity, Eqs. (2.29) and (2.33) are

$$\left(i\hbar \frac{\partial}{\partial t} + \mu\right) \langle \psi^\dagger \rangle = \eta\, \delta F / \delta \langle \psi \rangle \tag{3.1}$$

$$\nabla \cdot \boldsymbol{J} = i\{\langle \psi \rangle \delta F / \delta \langle \psi \rangle - \text{c.c.}\}. \tag{3.2}$$

These equations are valid for all temperatures $T \leqq T_\lambda$, the lambda temperature of liquid helium. However, near T_λ we can take $\langle \psi^\dagger \rangle$ as small and expand the total free energy in terms of $\langle \psi^\dagger \rangle$ as [12]

$$F_{\text{tot}} = F_n - B \int dx \langle \psi^\dagger \rangle \nabla^2 \langle \psi \rangle - c \int dx |\langle \psi^\dagger \rangle|^2 + D \int dx |\langle \psi^\dagger \rangle|^4 / 2 \tag{3.3}$$

where F_n is the free energy of the normal state and B, C, D are real parameters.

This gives

$$\left(i\hbar \frac{\partial}{\partial t} + \mu\right) \langle \psi^\dagger \rangle = \eta (-B \nabla^2 + C + D |\langle \psi^\dagger \rangle|^2) \langle \psi^\dagger \rangle \tag{3.4}$$

$$\boldsymbol{J} = \sigma_s \boldsymbol{v}_s \tag{3.5}$$

where

$$\sigma_s = 2m B \sigma_c / \hbar^2 \tag{3.6}$$

verifying that the superfluid density σ_s is not equal to the condensate density

$$\sigma_c = |\langle \psi^\dagger \rangle|^2.$$

Scaling theory and semi-microscopic arguments show that [12]

$$C \sim (1 - T/T_\lambda) \qquad B \sim (1 - T/T_\lambda)^{-\frac{1}{3}} \tag{3.7}$$

and D can be taken as constant.

Since the equilibrium solution is

$$\sigma_c = \frac{C}{D} \sim (1 - T/T_\lambda) \tag{3.8}$$

we obtain to lowest order in $\boldsymbol{v}_s$

$$\sigma_s \sim (1 - T/T_\lambda)^{\frac{2}{3}} \tag{3.9}$$

in agreement with experiment [17].

4. Superconductors

In superconductors the appropriate Hamiltonian is

$$\begin{aligned} H = \sum_\sigma \int dx\, \psi_\sigma^\dagger(x) \left(-i\hbar \nabla + \frac{e}{c} \boldsymbol{A}\right)^2 \psi_\sigma(x)/2m \\ + \frac{1}{2} \int dx\, dx \sum_{\sigma,\sigma'} V(x-y)\, \psi_\sigma^\dagger(x)\, \psi_{\sigma'}^\dagger(y)\, \psi_{\sigma'}(y)\, \psi_\sigma(x) \end{aligned} \tag{4.1}$$

where $\boldsymbol{A}(x)$ is the vector potential and the creation and annihilation operators now satisfy Fermi anticommutation relations. The conventional distinction between superconductors and normal metals is that in the former the expectation value $\langle \psi_\uparrow^\dagger \psi_\downarrow^\dagger \rangle$ is non-zero. Analogous to (2.2) we can define a condensate density

σ_c and a curl free superfluid velocity $\boldsymbol{v}_s$ by

$$\langle\psi_\uparrow^\dagger(x)\,\psi_\downarrow^\dagger(x)\rangle = \sqrt{\sigma_c}\exp - i\theta, \qquad \boldsymbol{v}_s = \hbar\nabla\theta/m. \tag{4.2}$$

From the Liouville equation the exact equation of motion for the order parameter is

$$\begin{aligned} -i\hbar\frac{\partial}{\partial t}\langle\psi_\uparrow^\dagger(x)\,\psi_\downarrow^\dagger(x)\rangle = &-\frac{\hbar^2}{2m}\{\langle\nabla^2\psi_\uparrow^\dagger(x)\,\psi_\downarrow^\dagger(x)\rangle + \langle\psi_\uparrow^\dagger(x)\,\nabla^2\psi_\downarrow^\dagger(x)\rangle\} \\ &+\left\{\frac{e^2}{mc^2}\boldsymbol{A}^2 - \frac{i\hbar e}{mc}\nabla\cdot\boldsymbol{A} + \frac{i\hbar e}{mc}\boldsymbol{A}\cdot\nabla\right\}\langle\psi_\uparrow^\dagger(x)\,\psi_\downarrow^\dagger(x)\rangle \\ &+2\int dy\, V(x-y)\sum_{\sigma'}\langle\psi_\sigma^\dagger(y)\,\psi_\sigma(y)\,\psi_\uparrow^\dagger(x)\,\psi_\downarrow^\dagger(x)\rangle. \end{aligned} \tag{4.3}$$

In analogy with (2.5) the local equilibrium density matrix is taken to be

$$\varrho_o = \exp - Q/\mathrm{Tr}\exp - Q \tag{4.4}$$

with

$$\begin{aligned} Q = \int dx\{\hat{\varepsilon}(x) - \mu(x,t)\,\hat{\sigma}(x) - \alpha(x,t)\,\psi_\uparrow^\dagger(x)\,\psi_\downarrow^\dagger(x) \\ -\alpha^*(x,t)\,\psi_\downarrow(x)\,\psi_\uparrow(x)\}/kT(x,t) \end{aligned} \tag{4.5}$$

where the energy density operator $\hat{\varepsilon}(x)$ now includes the magnetic terms and the density operator $\hat{\sigma}(x)$ is the total density operator, i.e. including a sum over spins.

The intensive quantities T, μ, α are determined by the equations analogous to (2.10), i.e.

$$\varepsilon(x,t) \equiv \mathrm{Tr}\,\varrho\,\hat{\varepsilon}(x) = \mathrm{Tr}\,\varrho_o\,\hat{\varepsilon}(x) \tag{4.6a}$$

$$\sigma(x,t) \equiv \mathrm{Tr}\,\varrho\,\hat{\sigma}(x) = \mathrm{Tr}\,\varrho_o\,\hat{\sigma}(x) \tag{4.6b}$$

$$\mathrm{Tr}\,\varrho\,\psi_\uparrow^\dagger(x)\,\psi_\downarrow^\dagger(x) = \mathrm{Tr}\,\varrho_o\,\psi_\uparrow^\dagger(x)\,\psi_\downarrow^\dagger(x) \tag{4.6c}$$

$$\mathrm{Tr}\,\varrho\,\psi_\downarrow(x)\,\psi_\uparrow(x) = \mathrm{Tr}\,\varrho_o\,\psi_\downarrow(x)\,\psi_\uparrow(x). \tag{4.6d}$$

By similar arguments to the boson case one can show that there is a free energy density F such that

$$\alpha(x,t) = \delta F/\delta\langle\psi_\downarrow(x)\,\psi_\uparrow(x)\rangle. \tag{4.7}$$

By evaluating the commutator in the identity

$$\mathrm{Tr}\,\varrho_o[Q, \psi_\uparrow^\dagger(x)\,\psi_\downarrow^\dagger(x)] \tag{4.8}$$

and assuming T sufficiently slow we obtain the analogous equation to (2.23),

$$\mathrm{Tr}\,\varrho_o[H, \psi_\uparrow^\dagger(x)\,\psi_\downarrow^\dagger(x)] = 2\mu\langle\psi_\uparrow^\dagger(x)\,\psi_\downarrow^\dagger(x)\rangle + \alpha(x,t)\{1-\sigma(x,t)\} \tag{4.9}$$

as the equation of motion for the order parameter in local equilibrium.

The contribution of Δ_ϱ, the deviation from local equilibrium, to the equation of motion,

$$-i\hbar\frac{\partial}{\partial t}\langle\psi_\uparrow^\dagger(x)\,\psi_\downarrow^\dagger(x)\rangle = \mathrm{Tr}\,\varrho_o[H, \psi_\uparrow^\dagger(x)\,\psi_\downarrow^\dagger(x)] + \mathrm{Tr}\,\Delta_\varrho[H, \psi_\uparrow^\dagger(x)\,\psi_\downarrow^\dagger(x)] \tag{4.10}$$

is again found by simple application of conventional statistical mechanical techniques. To first order in the deviation from equilibrium we find

$$\begin{aligned} -i\hbar \frac{\partial}{\partial t} \langle \psi_\uparrow^\dagger(x)\, \psi_\downarrow^\dagger(x) \rangle \\ = \operatorname{Tr} \varrho_o [H, \psi_\uparrow^\dagger(x)\, \psi_\downarrow^\dagger(x)] + i \int dx' \int_0^\infty d\tau\, K(x', \tau)\, \alpha(x - x', t - \tau) \end{aligned} \tag{4.11}$$

where the correlation function K is now given by the Kubo formula

$$K(x, \tau) = \hbar \int_0^{1/kT} d\lambda \langle e^{iH\tau/\hbar} [\psi_\uparrow^\dagger(x)\, \psi_\downarrow^\dagger(x), H]\, e^{-iH\tau/\hbar} \cdot e^{-\lambda Q} [H, \psi_\downarrow(0)\, \psi_\uparrow(0)]\, e^{\lambda Q} \rangle_o \tag{4.12}$$

in which the average (and Q) are taken with $\alpha = 0$ i.e. over the equilibrium state of a normal system in the presence of a magnetic field.

If α has sufficiently slow spatial and temporal variation we can take it outside the integral in (3.11) to give

$$-i\hbar \frac{\partial}{\partial t} \langle \psi_\uparrow^\dagger(x)\, \psi_\downarrow^\dagger(x) \rangle = \operatorname{Tr} \varrho_o [H, \psi_\uparrow^\dagger(x)\, \psi_\downarrow^\dagger(x)] + i\gamma\, \alpha(x, t) \tag{4.13}$$

where

$$\gamma = \int dx' \int_0^\infty d\tau\, K(x', \tau) \tag{4.14}$$

is a real, positive transport coefficient.

Substituting (3.7) and (3.9) the equation of motion of the order parameter is

$$\left(i\hbar \frac{\partial}{\partial t} + 2\mu\right) \langle \psi_\uparrow^\dagger(x)\, \psi_\downarrow^\dagger(x) \rangle = \eta\, \delta F / \delta \langle \psi_\downarrow(x)\, \psi_\uparrow(x) \rangle \tag{4.15}$$

where

$$\eta = (\sigma - 1 - i\gamma).$$

The current $\boldsymbol{J}$ is given by

$$\boldsymbol{J}(x, t) = \operatorname{Tr} \varrho_o \hat{\boldsymbol{J}}(x) + \operatorname{Tr} \Delta_\varrho \hat{\boldsymbol{J}}(x) \tag{4.16}$$

where the current operator is

$$\hat{\boldsymbol{J}}(x, t) = \sum_\sigma \left\langle \psi_\sigma^\dagger(x) \left(-i\hbar\nabla + \frac{e}{c}\boldsymbol{A}\right) \psi_\sigma(x) \right\rangle \Big/ 2 + \text{c. c.} \tag{4.17}$$

The second term on the right is the dissipative current linear in the temperature gradient and electromagnetic field, the coefficients being given by the usual Kubo formulae for a normal system. The first term on the right is the superfluid current and is found by evaluating the commutator in the identity

$$\operatorname{Tr} \varrho_o [Q, \hat{\sigma}(x)] = 0. \tag{4.18}$$

We find (c.f. (2.31)) for this current

$$\nabla \cdot \boldsymbol{J}_s(x, t) = 2mi \{\alpha^* \langle \psi_\uparrow^\dagger \psi_\downarrow^\dagger \rangle - \text{c.c.}\}/\hbar \tag{4.19}$$

$$= 2mi \left\{ \langle \psi_\uparrow^\dagger \psi_\downarrow^\dagger \rangle \frac{\delta F}{\delta \langle \psi_\uparrow^\dagger \psi_\downarrow \rangle} - \text{c.c.} \right\} \Big/ \hbar. \tag{4.20}$$

Eq. (3.15) is the hydrodynamic equation of motion for the order parameter and (3.20) the accompanying equation for the superfluid current.

5. The Equations Near T_c

According to the microscopic theory [14] of superconductors the free energy is given by

$$F = -kT \log \operatorname{Tr} \exp - H_{\mathrm{eff}}/kT \tag{5.1}$$

where (omitting the Hartree potential for simplicity)

$$\begin{aligned} H_{\mathrm{eff}} = \sum_{0} \int d x \, \psi_\sigma^\dagger [(\boldsymbol{p} - e\boldsymbol{A}/c)^2/2m - \mu] \psi_\sigma \\ + V \int dx [\langle \psi_\uparrow^\dagger(x) \psi_\downarrow^\dagger(x) \rangle \psi_\downarrow(x) \psi_\uparrow(x) + \mathrm{c.c.}] \\ - V \int dx \langle \psi_\uparrow^\dagger(x) \psi_\downarrow^\dagger(x) \rangle \langle \psi_\downarrow(x) \psi_\uparrow(x) \rangle \end{aligned} \tag{5.2}$$

and where $V(x-x') = V\delta(x-x')$ has been assumed.

Therefore,

$$\delta F / \delta \langle \psi_\downarrow \psi_\uparrow \rangle = \operatorname{Tr} e^{-H_{\mathrm{eff}}/kT} V \psi_\uparrow^\dagger \psi_\downarrow^\dagger / \operatorname{Tr} e^{-H_{\mathrm{eff}}/kT} - V \langle \psi_\uparrow^\dagger \psi_\downarrow^\dagger \rangle. \tag{5.3}$$

Near the critical temperature T_c the first term on the right is expanded in terms of the order parameter to yield

$$\begin{aligned} \delta F / \delta \langle \psi_\downarrow \psi_\uparrow \rangle = B [i\hbar \nabla - 2e\boldsymbol{A}/c]^2 \langle \psi_\uparrow^\dagger \psi_\downarrow^\dagger \rangle \\ - C \langle \psi_\uparrow^\dagger \psi_\downarrow^\dagger \rangle + D \langle \psi_\uparrow^\dagger \psi_\downarrow^\dagger \rangle |\langle \psi_\uparrow^\dagger \psi_\downarrow^\dagger \rangle|^2 \end{aligned} \tag{5.4}$$

with explicit expressions for B, C, D in terms of the microscopic parameters. In contrast to the superfluid helium only $C \sim (1 - T/T_c)$ is temperature dependent.

The time dependent Ginzburg-Landau equation valid near T_c is

$$\left(i\hbar \frac{\partial}{\partial t} + 2\mu\right) \langle \psi_\uparrow^\dagger \psi_\downarrow^\dagger \rangle = \eta \{ B [i\hbar \nabla - 2e\boldsymbol{A}/c]^2 - C + D |\langle \psi_\uparrow^\dagger \psi_\downarrow^\dagger \rangle|^2 \} \langle \psi_\uparrow^\dagger \psi_\downarrow^\dagger \rangle. \tag{5.5}$$

In the absence of dissipation the right hand side is zero, giving the usual time independent Ginzburg-Landau equation.

The superfluid current given by Eq. (4.20) is

$$\boldsymbol{J} = \sigma_s \{ \boldsymbol{v}_s - 2e\boldsymbol{A}/mc \} \tag{5.6}$$

with

$$\sigma_s = 4m^2 B |\langle \psi_\uparrow^\dagger \psi_\downarrow^\dagger \rangle|^2. \tag{5.7}$$

References

1. Fröhlich, H.: Physica **37**, 215 (1967).
2. Fröhlich, H.: Proc. Inter. Conf. on Stat. Mech. (1968), p. 189 (publ. as Suppl. to J. Phys. Soc. Japan **26**).
3. Ginzburg, V. L., Pitaevskii, L. P.: Soviet. Phys. JETP **7**, 858 (1958).
4. Vyama, H.: Progr. Theoret. Phys. (Kyoto) **45**, 25 (1971); **46**, 34 (1971).
5. Usui, T.: Progr. Theoret. Phys. (Kyoto) **41**, 1603 (1969).
6. Schmid, A.: Phys. Kondens. Materie **5**, 302 (1966).
7. Abrahams, E., Tsuneto, T.: Phys. Rev. **152**, 416 (1966).
8. Ebisawa, H., Fukuyama, H.: Progr. Theoret. Phys. **46**, 1042 (1971).
9. Kubo, R.: Lectures in Theoret. Phys. (Boulder) **1**, 120 (1958).
10. Mori, H.: Phys. Rev. **115**, 298 (1959).
11. Robertson, B.: Phys. Rev. **160**, 175 (1967); **166**, 206 (1968) (Erratum).
12. Wong, V. K.: Phys. Letters **27** A, 269 (1968).

13. AMIT, D. J.: J. Phys. Chem. Solids **131**, 1099 (1970).
14. SAINT JAMES, D., SARMA, G., THOMAS, E. J.: Type 11 Superconductivity. London: Pergamon 1969.
15. TAYLOR, A. W. B., EVANS, J. W.: Collective Phenomena **1**, 37 (1972).
16. DE GROOT, S. R., MAZUR, P.: Non-Equilibrium Thermodynamics. Amsterdam: North-Holland 1962.
17. CLOW, J. R., REPPY, J. D.: Phys. Rev. Letters **16**, 887 (1966).

Appendix

The local equilibrium density matrix is written as

$$\varrho_o = \exp -Q/\mathrm{Tr}\exp -Q \tag{A1}$$

with

$$Q = \sum_i \int dx\, a_i(x)\, A_i(x,t) \tag{A2}$$

where

$$a_1 = \hat{\varepsilon}, \quad a_2 = \hat{\sigma}, \quad a_3 = \hat{\boldsymbol{J}}, \quad a_4 = \psi, \quad a_5 = \psi^\dagger \tag{A3}$$

$$A_1 = 1/T, \quad A_2 = -\mu/T, \quad A_3 = -\boldsymbol{v}_n/T, \quad A_4 = -\alpha/T, \quad A_5 = -\alpha^*/T. \tag{A4}$$

We wish to show that

$$\sum_i A_i(x,t)\,\delta\langle a_i(x)\rangle_o \tag{A5}$$

is a perfect differential.

Now

$$\delta\langle a_i(x)\rangle_o = \mathrm{Tr}\,\delta\varrho_o\, a_i(x) \tag{A6}$$

$$= -\sum_j \int dx' \langle a_j(x')\, a_i(x)\rangle_c\, \delta A_j(x',t) \tag{A7}$$

where the subscript c denotes a symmetrised correlation average,

$$\langle BC\rangle_c = \langle BC + CB\rangle/2 - \langle B\rangle_o\langle C\rangle_o \tag{A8}$$

for any two operators B and C.

If the correlation average in (A 7) has a finite spatial range and if the intensive quantities A_j are effectively constant over distances of the order of this correlation range then we can replace $\delta A_j(x',t)$ in (A 7) by $\delta A_j(x,t)$ giving

$$\delta\langle a_i(x)\rangle = -\sum_j \int dx' \langle a_j(x')\, a_i(x)\rangle_c\, \delta A_j(x,t) \tag{A9}$$

$$= -\sum_j \int dx' \langle a_j(x-r)\, a_i(x)\rangle_c\, \delta A_j(x,t). \tag{A10}$$

Now

$$\langle a_j(x-r)\, a_i(x)\rangle_c = \langle a_j(x-r/2)\, a_i(x+r/2)\rangle_c' \tag{A11}$$

the dash denoting that the density matrix used in the right hand side is the spatially shifted one,

$$\varrho_o' = \exp\{-ir\int dx\,\hat{\boldsymbol{J}}(x)/2\hbar\}\,\varrho_o\,\exp\{+ir\int dx\,\hat{\boldsymbol{J}}(x)/2\hbar\} \tag{A12}$$

$$= \exp -\sum_i \int dx\, A_i(x)\, a_i(x+r/2)/\mathrm{Tr}\exp -Q \tag{A13}$$

$$= \exp -\sum_i \int dx\, A_i(x-r/2)\, a_i(x)/\mathrm{Tr}\exp -Q \tag{A14}$$

since the correlation average in (A 11) is zero for r greater than the correlation range and the A_i are effectively constant over this distance we can replace $A_i(x-r/2)$ by $A_i(x)$ for use in (A 11), i.e. we can use $\varrho_o' = \varrho_o$ for substitution in (A 11).

Therefore,

$$\langle a_j(x-r)\, a_i(x)\rangle_c = \langle a_j(x-r/2)\, a_i(x+r/2\rangle_c \tag{A 15}$$

and therefore

$$\delta\langle a_i(x)\rangle/\delta A_j(x) = -\int dr \langle a_j(x-r/2)\, a_i(x+r/2)\rangle \tag{A 16}$$

$$= \delta\langle a_j(x)\rangle/\delta A_i(x). \tag{A 17}$$

Eq. (A 17) is a necessary and sufficient condition that

$$\sum_i \langle a_i(x)\rangle \gamma A_i(x) \tag{A 18}$$

be a perfect differential. There exists a quantity $\Omega(x, t)$ such that

$$\delta\Omega = \sum_i \langle a_i(x)\rangle\, \delta A_i(x). \tag{A 19}$$

Therefore, defining the local entropy density by

$$S(x, t) = -\Omega + \sum_i \langle a_i(x)\rangle A_i(x, t) \tag{A 20}$$

we have

$$\delta S = \sum_i A_i \delta\langle a_i\rangle. \tag{A 21}$$

We have thus established the existence of a local entropy density.

Scattering of ^{4}He Atoms by the Surface of Liquid ^{4}He

C. G. KUPER

Technion-Israel Institute of Technology, Department of Physics, Haifa/Israel

Dedication

When mentioning the age of a friend, a Jew will traditionally add the prayer "עד מאה ועשרים"—"may he live to be 120". Prof. H. FRÖHLICH has reached the statutory age of retirement from his Chair, but in his approach to physics he remains so young that his significant contributions will continue for many years to come, and—as in the past—will be a source of inspiration to the new generation of physicists. I have great pleasure in dedicating this paper to Professor FRÖHLICH, and wishing him "עד מאה ועשרים".

Abstract

A classical hydrodynamical model is proposed which leads to inelastic scattering of helium atoms by the free surface of liquid helium. The qualitative features are in agreement with preliminary experimental results of EDWARDS' group. The model predicts also that the scattering of rotons impinging on the surface from below will be elastic or nearly elastic.

1. Introduction

The kinetics of the evaporation and condensation of helium has recently become an active field of research. KING and JOHNSTON [1] have attempted to measure the energy distribution of the atoms which evaporated into a vacuum from liquid ^{4}He at a temperature of 0.6 K. The distribution was reported to be approximately Maxwellian, but with a mean energy $\sim$1.6 K.

A related topic which is interesting both theoretically and experimentally is the accommodation coefficient of helium, i.e. the probability that a gaseous atom impinging on the surface will be captured by the liquid phase. There is some rather indirect experimental evidence [2] that this probability is of order unity, together with one qualitative observation that tends to support a much smaller value [3]. Very recently ECKARDT, EDWARDS, GASPARINI and SHEN (EEGS) [4] designed an elegant experiment to measure the accommodation coefficient directly, but the preliminary results display some totally unexpected features. In this paper, their experiment will be analyzed in terms of a new model.

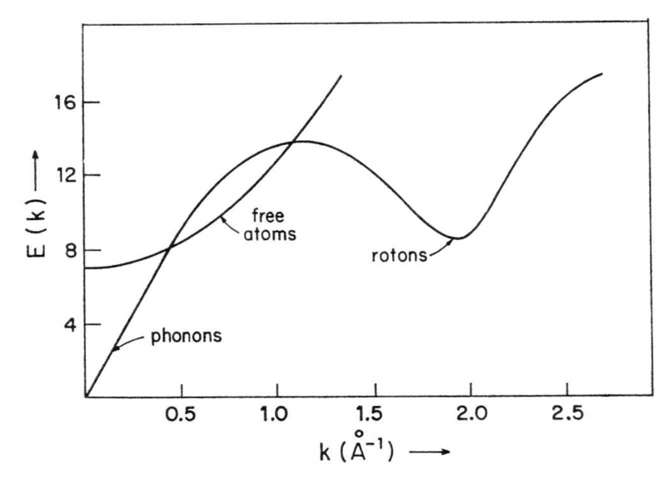

Fig. 1. Comparison of the parabolic free-atom spectrum with the Landau phonon-roton spectrum

2. The Tunnelling Model

Theoretical treatments of the problem of evaporation or condensation have been based on a model [5] in which the important microscopic processes are (a) those in which an atom colliding with the surface is absorbed, and one quasiparticle (phonon or roton) is created with conservation of energy, and (b) the inverse process: a quasiparticle is annihilated and an atom evaporated. In the simplest version of this model, the difference in shape between the free-particle spectrum

$$E = L + p^2/2m \tag{1}$$

and the Landau roton-phonon spectrum (Fig. 1) creates an acoustic mismatch, with consequent strong reflection except for those few quasiparticles states for which E and $\boldsymbol{p}$ satisfy Eq. (1) to within the natural linewidth of the Landau spectrum. In Eq. (1), L is the chemical potential per atom of the liquid; i.e. the binding energy of an atom in the liquid, and E represents the energy of a free atom relative to an atom in the "condensate". In the more general formulation of WIDOM *et al.* and ANDERSON [6] the model assumes a "tunnelling" interaction which connects atomic states with quasiparticle states. In other words the transition matrix element becomes a parameter of the theory.

The WIDOM-ANDERSON model was proposed to account qualitatively for the King-Johnston effect [1]. At 0.6 K, most of the thermal excitations in the liquid phase are phonons, whose energy is not sufficient to allow tunnelling. But the few rotons which are present have energies in excess of 8.65 K. Since $L = -7.15$ K per atom, all rotons *can* tunnel, and moreover will leave a residue of at least 1.5 K as kinetic energy of the evaporated atom. Thus roton tunnelling will give a threshold of 1.5 K in the spectrum of the evaporated atoms; on top of this sharp peak is a weak continuous spectrum arising from phonons with energies greater than 7.15 K (Fig. 2). The model explains the rather high energy of the evaporated atoms, but the King-Johnston experiment does not exhibit the

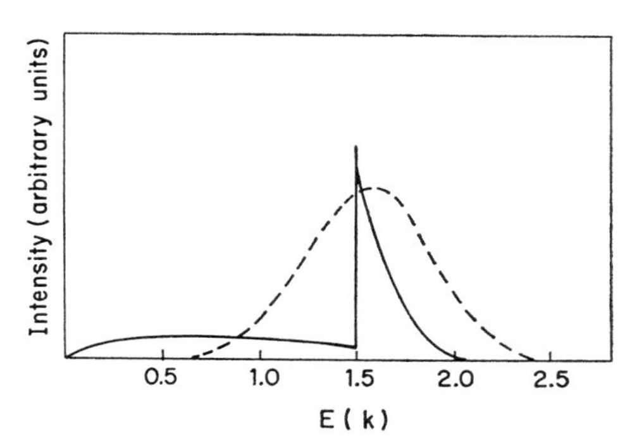

Fig. 2.
Comparison of the King-Johnston distribution [1] of evaporated ^{4}He atom energies (broken curve) with the predictions [6] of WIDOM et al. and of ANDERSON (full curve). Not only is the vertical scale arbitrary, but also the relative magnitudes of the phonon background and the roton peak

predicted threshold. This discrepancy will be discussed briefly in § 6, but has not been resolved. The validity of the King-Johnston experiment has been questioned, but if the effect is genuine, it clearly demonstrates an inadequacy of the theory.

3. The Experiment of Edwards' Group

On the assumption that the tunnelling model is valid, EEGS [4] designed an experiment which would give the accommodation coefficient directly. The experiment consists in allowing an atomic beam, of ^{4}He atoms at $\sim$2 K, to impinge on the free surface of the cold liquid (at $\sim$0.05 K) at a moderately well-defined angle and observing the scattered-beam intensity as a function of angle (Fig. 3). If—as predicted by the tunnelling model—the reflection is specular (i.e. $\Theta_f = \Theta_i$), then the intensity of the reflected beam will differ from the incident-beam intensity by the atoms which have been lost into the liquid. The incident beam could be calibrated by a control experiment in which the cell contained no liquid. They produce the atomic beam by means of a heat pulse which burns the Rollin film off a suitably placed surface. The magnitude of the pulse is chosen so as not to raise the local temperature above about 2 K. Hence the atoms in the beam do not have enough energy to create a roton (or a high-energy phonon) without being captured by the liquid; the fact that an atom is seen in the reflected beam establishes that it has not produced any excitation more energetic than 2 K, i.e. than a phonon of wave number $\sim$0.1 Å^{-1}.

The preliminary results of EEGS were quite unexpected. They find that (a) the reflection is largely non-specular, but is markedly peaked towards glancing angles of emergence Θ_f and (b) the distribution of Θ_f is insensitive to the angle Θ_i of incidence. This result seems to make the one-excitation tunnelling model untenable and supports a conjecture by PEIERLS [7] that multiple-soft-phonon emission should be important. In this paper a classical hydrodynamical model is used to study the energy loss through soft-phonon radiation. The model appears able to account for the inelastic scattering observed by EDWARDS' group.

The geometry of the EEGS experiment almost suggests several variants, which EDWARDS has expressed his intention to try. Either the heater or the detector, or both, could be below the surface; one could thus study the production of rotons by an incident atomic beam, the evaporation of atoms by an incident roton beam, or the pure scattering of rotons by the free surface. The last of these possibilities is amenable to theoretical treatment by the model of § 4. The analysis, which exhibits an interesting new feature, is given in § 5. The other two possible beam experiments involve *both* one-excitation tunnelling and the classical field; their analysis is therefore more complicated, and will not be attempted here.

4. Scattering of Low-Energy Atoms by the Free Surface

In the EEGS experiment, the incident beam consists of atoms whose kinetic energy is ~ 2 K. This fact implies that any atom detected in the reflected beam could not have produced any excitation of energy $\gtrsim 2$ K. In particular, no phonons of wave number in excess of ~ 0.1 Å nor rotons could be produced. But soft phonons and surface quanta ("ripplons") can be produced in large numbers*. Sufficient for the present argument is the fact that in the case of non-capture, *all the permitted excitations are irrotational.* Then the tangential component $p_i \sin \Theta_i$ of the incident momentum p_i is certainly preserved in the scattering, and plays no role in exciting the liquid [8]:

$$p_i \sin \Theta_i = p_f \sin \Theta_f. \tag{2}$$

On the other hand, the normal component $p_i \cos \Theta_i$ can and will change (even in specular reflection it will change sign). But the motion of the liquid which it produces must possess axial symmetry. It is as if a loudspeaker diaphragm covers a small part of the liquid surface, and the loudspeaker receives an impulsive signal.

The total impulse received on the surface is

$$P = p_i \cos \Theta_i + p_f \cos \Theta_f. \tag{3}$$

In general this impulse will generate both sound and surface waves, and the hydrodynamical equations couple these modes of motion in a complicated way. For the present work, it will be sufficient to assume (a) that the amplitudes are sufficiently small for the usual linearized hydrodynamical equations of acoustic theory to be valid, and (b) that macroscopic hydrodynamics can be extrapolated down to atomic distances without gross error. Under these assumptions, the amplitude of every Fourier component of both the radiated sound and surface wave is proportional to the impulse P, and hence the total energy radiated is proportional to P^2. Conservation of energy requires that

$$p_i^2/2m - p_f^2/2m = \alpha P^2/2m, \tag{4}$$

where $\alpha < 1$ is a dimensionless coefficient. In principle α can be calculated, but it is sensitive to the details of the model (e.g. the duration of the pulse and the size of the loudspeaker diaphragm). In Appendix A, the acoustic contribution to α is estimated, with plausible values for the model parameters; α is shown to be of order unity.

---

* If the atom is captured, then of course a roton *can* be produced, but the atom is then absent from the reflected beam.

Combining Eqs. (2) and (4) gives:

$$p_i^2 \cos^2 \Theta_i - p_f^2 \cos^2 \Theta_f = \alpha P^2. \quad (5)$$

Solving (3) and (5) for P and Θ_f, and using (2) to eliminate p_f, we have:

$$P = 2 p_i \cos \Theta_i / (1 + \alpha), \quad (6)$$

and

$$\tan \Theta_f = \{(1+\alpha)/(1-\alpha)\} \tan \Theta_i. \quad (7)$$

We see that for all α in the range $0 < \alpha \leqq 1$, the angle of reflection is greater than the angle of incidence, in qualitative agreement with EDWARDS' results. Moreover, the theory predicts that Θ_f is independent of p_i. This statement can be tested experimentally, but I do not known if EDWARDS *et al.* have yet obtained any data relating to it.

However, Eq. (7) *does* state that Θ_f depends on Θ_i, in apparent contradiction to the EEGS finding that Θ_f is insensitive to Θ_i. I believe that the discrepancy is probably instrumental, arising from the limited collimation of the incident beam. If $\alpha \simeq \frac{1}{2}$, then (7) gives $\Theta_f = \tan^{-1}(3 \tan \Theta_i)$, and already for $\Theta_i \sim 45°$, Θ_f is close to 75°. The angular distribution is strongly biased towards angles near 90°, and hence is much less sensitive than it would be for specular reflection. However, for small angles, $\Theta_f \sim 3\Theta_i$, and this should show up if the beam can be defined well enough. If the preliminary suggestion that Θ_f does not depend on Θ_i should be strictly confirmed, it would be very hard to explain.

5. Scattering of Rotons by the Free Surface

Consider one of the—as yet untried—variations of the EEGS experiment: a beam of rotons impinges on the underside of the free surface, and is scattered. The geometry is that of Fig. 3, with the sign of the z-axis reversed. But first let us ask whether it is possible to produce a beam of rotons. A local heat pulse, raising the temperatuie of a small region to a temperature ~ 2 K, will produce a significant concentration of rotons, but of course it will also produce phonons. However, all

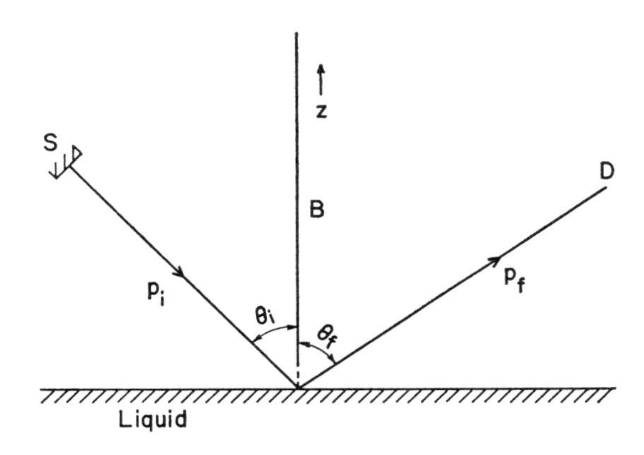

Fig. 3. The geometry of the EEGS experiment [4]. A heat pulse applied to the source S generates the beam, and B is a screen perpendicular to the free surface which prevents atoms from S arriving directly at the detector D

the phonons travel at the speed of sound (2.4×10^4 cm sec^{-1}), while the speed of a roton depends on its energy. At 2 K, the mean thermal velocity of rotons is 1.2×10^4 cm sec^{-1} (see Appendix B). Hence if the detector is switched on after the purely acoustic part of the pulse has already reached it, only rotons will be detected. Thus it is possible in principle to discriminate against phonons and to observe pure roton scattering.

A roton always carries sufficient energy to free an atom into the vapour and leave an energy residue ~ 1.5 K, which can be divided between the kinetic energy of the free atom and any radiated soft phonons and ripplons. But there is never enough energy available to free an atom *and* preserve the roton. Hence the roton-scattering picture looks very similar to the scattering of free atoms. Eqs. (2) and (3) hold without change. However, Eq. (4) must be replaced by

$$(p_i - p_0)^2/2m^* - (p_f - p_0)^2/2m^* = \alpha P^2/2m. \tag{8}$$

Here p_0 and m^* are the Landau roton parameters (and Δ has cancelled), while α is the same parameter as entered in (4) if we assume that the acoustic and surface-wave radiation depend on the impulse only and not on the structure of the incident particle. It is convenient to define

$$\gamma = \alpha m^*/m = \alpha/6. \tag{9}$$

In Appendix B it is shown that, for thermal rotons at 2 K,

$$|\bar{p} - p_0| / p_0 \sim 7 \times 10^{-2}. \tag{10}$$

Hence, from (8)

$$\gamma P^2/p_0^2 = (p_i - p_0)^2/p_0^2 - (p_f - p_0)^2/p_0^2 \lesssim 5 \times 10^{-3},$$

i.e.

$$P/p_0 < 0.07/\sqrt{\gamma}. \tag{11}$$

But from (3)

$$p_i \cos\Theta_i < P < 2p_i \cos\Theta_i, \tag{12}$$

whence

$$P/p_0 > \cos\Theta_i. \tag{13}$$

If α has indeed the same value as for atomic scattering, then $\gamma \sim 10^{-1}$, and Eqs. (11) and (13) are contradictory except for large angles $\Theta_i \gtrsim 80°$. The contradiction can be removed only by requiring $\alpha_{\text{rot}} \ll \alpha_{\text{atom}}$, i.e. either roton scattering must be much more nearly elastic than atomic scattering, or the above mechanism is forbidden, and roton scattering is truly elastic.

6. Discussion

In § 4 we have seen that a classical acoustic model gives inelastic scattering of ^{4}He atoms by the surface of cold liquid ^{4}He, in general agreement with the preliminary findings of EEGS [4]. Moreover, in § 5 it has been shown that roton scattering, under the same assumptions, will be almost elastic; an experimental test of this prediction would be valuable.

However, although this model has succeeded in explaining the EEGS result, it has done so at the expense of the KING-JOHNSTON [1] result, since our model seems to leave no room for the WIDOM-ANDERSON [6] mechanism. The difficulty is that the present theory is completely classical.

If the one-excitation tunnelling matrix element were really known as a function of the wave vector $\boldsymbol{k}$—and if it did indeed exhibit an "infrared divergence"—it would be natural to attempt a BLOCH-NORDSIECK [9] transformation in order to separate the one-roton part of the scattering (or tunnelling) from the many-soft-phonon part. We would then hope that the residual roton-one-atom term could account for the King-Johnston 1.6 K peak, while the inelastic scattering of EEGS is represented by the classical "splash" of phonons and rippons. It is harder to see what could cause the broadness of the King-Johnston peak, since even in this model it would appear as a threshold at 1.5 K. If the King-Johnston effect is genuine, the theory is necessarily incomplete.

Lacking the necessary detailed knowledge of the k-dependent matrix element, we are not in a position to perform the Bloch-Nordsieck transformation explicitly. The present model represents an implicit guess as to the form of the Hamiltonian after the transformation. But we have neglected the Bloch-Nordsieck cut-off—we might guess that the classical acoustic and ripplon field has a cut-off wave vector k_c above which the Widom-Anderson quantum treatment is appropriate.

Acknowledgements

I am indebted to Prof. D. O. EDWARDS for stimulating my interest in this problem, for many lively discussions, and for letting me have the preliminary experimental results before publication. I thank Prof. S. ROSENBLAT for an illuminating comment on the generation of surface waves, and Prof. SHULAMIT ECKSTEIN for drawing my attention to the argument of reference [8].

References

1. KING, J. A., JOHNSTON, W. D.: Phys. Rev. Letters **16**, 1191 (1966).
2. ATKINS, K. R., ROSENBAUM, B., SEKI, H.: Phys. Rev. **113**, 751 (1959).
   OSBORNE, D. V.: Proc. Phys. Soc. (London) **80**, 103, 1343 (1962).
3. GRIMES L. G., JACKSON, L. C.: Advan. Physics **7**, 1435 (1959).
4. EDWARDS, D. O.: Private communication.
   ECKARDT, J., EDWARDS, D. O., GASPARINI, F. M., SHEN, S. Y.: Abstract submitted to the 13th Internat. Conference on Low Temperature Physics (Boulder, Colorado, 1972).
5. TILLEY, J.: Ph. D. Thesis, University of St. Andrews (1965); see also Ref. [6].
6. WIDOM, A.: Phys. Letters **29** A, 96 (1969).
   HYMAN, D. S., SCULLY, M. O., WIDOM, A.: Phys. Rev. **186**, 231 (1969).
   WIDOM, A., HYMAN, D. S.: Phys. Rev. A **2**, 818 (1970).
   ANDERSON, P. W.: Phys. Letters **29** A, 563 (1969).
7. PEIERLS, R. E.: Private communication.
8. Eq. (2) remains true under much more general conditions—see LANDAU, L. D., and LIFSCHITZ, E. M.: Fluid Mechanics, p. 253. London: Pergamon Press 1959.
9. BLOCH, F., NORDSIECK, A.: Phys. Rev. **52**, 54 (1937).
10. e.g. MORSE, P. M.: Vibration and Sound, 2nd ed., p. 345. New York: McGraw-Hill 1948.

Appendix A

Acoustic Radiation from an Impulse on a Piston in an Infinite Baffle

The centre of the piston will be taken as the origin of a system of polar coordinates, with the polar axis normal to the surface. A time-dependent acceleration $A(t)$ of the piston generates a pressure wave [10]:

$$p = \frac{\varrho s}{2\pi r \sin\Theta} \int_{t-(r+a\sin\Theta)/s}^{t-(r-a\sin\Theta)/s} d\tau\, A(t) \{a^2 \sin^2\Theta - [r - s(t-\tau)]^2\}^{\frac{1}{2}}. \tag{A1}$$

Here s is the speed of sound and ϱ is the fluid density. Assuming that the amplitude is small, we may put $\varrho = \varrho_0$ in (A1). The velocity potential ϕ is

$$\phi = \varrho_0^{-1} \int_{-\infty}^{t} p\, dt. \tag{A2}$$

If the piston is suddenly set in motion at time $t=0$ and stopped at $t=b/s$, i.e. $A(t) \propto \delta(t) - \delta(t-b/s)$, where $b > 2a$, the integrals may be evaluated easily to give

$$\phi = 0, \qquad \mu < -1 \tag{A3.1}$$

$$\phi = (C/r)\{\tfrac{1}{2}\pi + \arcsin\mu + \mu\sqrt{(1-\mu^2)}\}, \qquad |\mu| < 1, \tag{A3.2}$$

$$\phi = \pi C/r, \qquad 1 < \mu < (b/a\sin\Theta) - 1, \tag{A3.3}$$

$$\phi = (C/r)\{\tfrac{1}{2}\pi - \arcsin\mu' - \mu'\sqrt{(1-\mu'^2)}\}, \qquad |\mu'| < 1, \tag{A3.4}$$

$$\phi = 0 \qquad \mu' > 1, \tag{A3.5}$$

where

$$\mu = (ts - r)/a\sin\Theta, \tag{A4.1}$$

$$\mu' = \mu - b/a\sin\Theta. \tag{A4.2}$$

For simplicity let us study the qualitatively similar but more convenient potential (see Fig. 4):

$$\phi = (C/r)\{0,\ 1+\mu,\ 2,\ 1-\mu',\ 0\} \tag{A5}$$

in an obvious notation; the five alternatives refer respectively to the five domains of $ts - r$ defined in Eqs. (A3).

The radial velocity v_r is

$$v_r = -\partial\phi/\partial r = \phi/r - (C/ar\sin\Theta)\{0,\ 1,\ 0,\ -1,\ 0\}, \tag{A6}$$

while the tangential velocity has a higher $(1/r)$-power dependence, and may be neglected for large r.

The z-component of the total momentum is

$$\begin{aligned} P &= \varrho_0 \int d^3r\, v_r \cos\Theta \\ &= 2\pi C \varrho_0 (b + \tfrac{4}{3}a), \end{aligned} \tag{A7}$$

since the second term in (A6) contributes only in domains (2) and (4), and these contributions cancel; the ϕ/r term contributes in domains (2), (3) and (4).

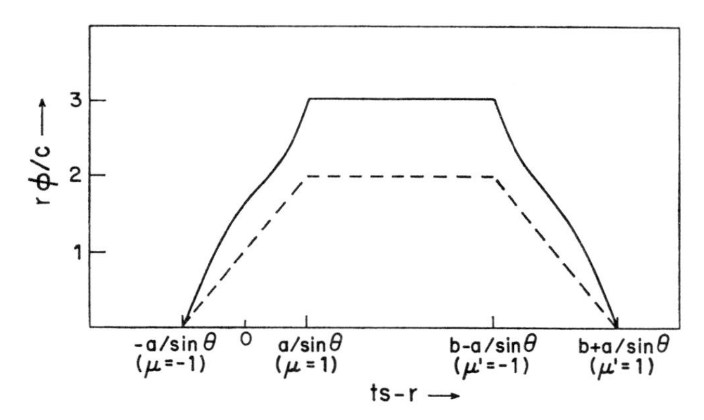

Fig. 4. Velocity potential in the sound wave generated by an impulse. Full curve: Eq. (A3); broken curve: Eq. (A5)

In a small-amplitude sound wave, the total energy is double the kinetic energy, so that the energy is

$$\begin{aligned} E &= \varrho_0 \int d^3 r\, v_r^2 \\ &= 8\pi^2 C^2 \varrho_0 / a \end{aligned} \tag{A8}$$

where the dominant contributions arise from the regions (2) and (4) of large velocity and the ϕ/r term of (A6) has been neglected.

Hence the acoustic part of α, as defined in Eq. (4), is

$$\alpha_{ac} = 2Em/P^2 = 4m/\varrho_0 a\,(b + \tfrac{4}{3}a)^2. \tag{A9}$$

Taking $b = 2a \sim 10$ Å (the de Broglie wavelength of a 2 K ⁴He atom), (A9) gives $\alpha_{ac} \sim 0.3$.

Appendix B

Some Parameters of Thermal Rotons

The Landau roton spectrum

$$E = \Delta + (p - p_0)^2 / 2m^* \tag{B1}$$

gives a velocity

$$v = \partial E / \partial p = (p - p_0)/m^*. \tag{B2}$$

The thermal expectation value is

$$\bar{v} = \int d^3 p\, v\, e^{-E/kT} / \int e^{-E/kT} d^3 p \tag{B3}$$

$$= \int_0^\infty (v + p_0/m)^2 v\, dv \exp(-\tfrac{1}{2} m^* v^2/kT) / \int_0^\infty (v + p_0/m)^2 dv \exp(-\tfrac{1}{2} m^* v^2/kT) \tag{B3'}$$

However, the parabolic form (B1) is in any case valid only for $p - p_0 \ll p_0$; we may therefore neglect v compared with p_0/m in the density-of-states factor, to give

$$\bar{v} = \int_0^\infty v \exp(-\tfrac{1}{2}m^* v^2/kT)\,dv \Big/ \int_0^\infty \exp(-\tfrac{1}{2}m^* v^2/kT)\,dv = (2kT/\pi m^*)^{\frac{1}{2}} \tag{B4}$$

or

$$|\bar{p} - p_0| = (2m^* kT/\pi)^{\frac{1}{2}}. \tag{B4'}$$

Similarly

$$\bar{E} = \Delta + \tfrac{1}{2}kT. \tag{B5}$$

For $T \simeq 2$ K, (B4) and (B5) give

$$\left.\begin{array}{l} \bar{E} - \Delta \simeq 1\,\mathrm{K}; \quad \bar{v} \simeq 1.2 \times 10^4\,\mathrm{cm\,sec^{-1}}, \\ |\bar{p} - p_0|/p_0 \simeq 0.07. \end{array}\right\} \tag{B6}$$

Chapter III. Dielectric Theory

Historically, one of the earliest physical examples of collective behaviour was the investigation of dielectric materials. Many excellent contributions have been presented (Mossotti, 1850; Clausius, 1879; Onsager, 1936; Kirkwood, 1939; Debye, 1929; Fröhlich, 1949) and these attempts have lost their actuality to a lesser extent than in other regions. Hence, the famous book of Fröhlich (1949, 1958), which summarizes these methods, is still very much in use and the reader is referred to it. On the one hand, these older methods have been a bridge for the creation of modern methods in microscopic many-body physics ("dielectric description"). On the other hand, because of their unrestrictive nature they have stimulated specifications, extensions and comparisons with other methods. The four contributions in this chapter are examples of this sort.

Some Comments on the Dynamical Dielectric Constants

RYOGO KUBO

University of Tokyo, Department of Physics

In his celebrated book *Theory of Dielectrics* [1], published in 1949, Professor FRÖHLICH gave an elegant treatment of a statistical theory of dielectric constants. In particular, the general theorems derived in § 7 are transparent and very useful. Extensions of this treatment to dynamic cases have been studied by several authors [2] along the line of the linear response theory developed by the present author [3]. Here a few remarks will be made on this subject from the viewpoint of application of the fluctuation-dissipation theorem.

The theory of dielectrics is complicated by the long-range nature of the dipolar interaction. Dielectric constants are material constants, but the shape of the samples has to be considered in the course of a calculation. Let us first review a few typical situations which are usually assumed in deriving theoretical formulae of dielectric constants. They are:

a) a spherical sample in a homogeneous external field,

b) a spherical sample embedded in a medium of the same material, having the form of a thick plate placed in a parallel plate condenser,

c) a plate with an infinite extension but a finite thickness placed in a parallel plate condenser,

d) an infinite sample of the material on which an inhomogeneous distribution of fictitious charges is superposed.

In each of these cases we define the interaction of the system with external sources of electric field to causes dielectric polarization. The interaction energy is written as

$$H_{\text{ext}} = -\eta E_0 M. \tag{1}$$

where M is the electric moment of the sample and η is a factor determined by the geometry and the dielectric constant. If the field E_0 is oscillatory with the frequency ω, the response of the system, that is the induced polarization of the dielectrics, may be expressed as

$$P = M/V = \chi(\omega)\,\eta E_0 \tag{2}$$

where V is the volume and $\chi(\omega)$ is the admittance. It is given by the linear response theory in the form,

$$\chi(\omega) = (VkT)^{-1}\{\langle M; M\rangle - i\omega \int_0^\infty \langle M(0); M(t)\rangle \exp(-i\omega t)\, dt\}, \tag{3}$$

or

$$\chi(\omega) = (V k T i\omega)^{-1} \int_0^\infty \langle J(0); J(t) \rangle \exp(-i\omega t)\, dt \tag{4}$$

where

$$J(t) = \dot{M}(t) \tag{5}$$

is the current associated with the electric moment. In the expressions (3) and (4), the notation $\langle X; Y \rangle$ implies a canonical correlation function [4] of relevant quantities defined by

$$\langle X; Y \rangle = \beta^{-1} \int_0^\beta \mathrm{Tr}\,\{e^{-\beta(\mathscr{H}-\psi)}\, e^{\lambda\mathscr{H}} X e^{-\lambda\mathscr{H}} Y\}\, d\lambda$$

in terms of the equilibrium density matrix, $\exp\beta(\psi - \mathscr{H})$, of the system at temperature $T = 1/k\beta$ in the absence of external field. The canonical correlation takes care of quantum effects but is reduced to the ordinary correlation in the classical limit.

Now the polarization induced by the external source can be given by quasi-statics in each case in the form

$$P = \mu(\varepsilon)\, E_0$$

where ε is the complex dielectric constant at the frequency ω. Equating (2) to (6), we obtain the equation

$$\mu(\varepsilon) = \eta\, \chi(\omega) \tag{6}$$

to determine the dielectric constant. Explicit forms of this equation will be discussed below.

Case a). The external sourse is the charge on the condenser plate producing a homogeneous field E_0. Thus in Eq. (1) we have

$$\eta = 1.$$

Polarization in the sphere is uniform and is given by

$$P = \frac{3}{4\pi}\, \frac{\varepsilon - 1}{\varepsilon + 2}\, E_0, \tag{7}$$

so that Eq. (6) is

$$\frac{3(\varepsilon - 1)}{\varepsilon + 2} = 4\pi\, \chi_a(\omega). \tag{8}$$

Case b). The system is a spherical sample placed in a spherical cavity in a dielectrics of the same material. The condenser produces a uniform displacement E_0 and the cavity field

$$G = 3 E_0 (2\varepsilon + 1),$$

so that

$$\eta = 3/(2\varepsilon + 1). \tag{9}$$

Uniform polarization in the system is

$$P = (1 - \varepsilon^{-1})\, E_0 / 4\pi. \tag{10}$$

Thus Eq. (6) takes the form

$$\frac{(\varepsilon - 1)(2\varepsilon + 1)}{3\varepsilon} = 4\pi\, \chi_b(\omega). \tag{11}$$

Case c). The charge on the condenser produces a uniform displacement E_0 and polarization

$$P = (1 - \varepsilon^{-1}) E_0 / 4\pi.$$

Thus Eq. (6) is

$$1 - \varepsilon^{-1} = 4\pi \chi_c(\omega). \tag{12}$$

Case d). The fictitious charge distribution $\varrho_e(r, t)$ is assumed to be harmonic in space and time, producing the displacement,

$$E_{0q} \exp(iqr + i\omega t)$$

and the polarization,

$$P_q = \{1 - \varepsilon(q, \omega)^{-1}\} E_{0q} / 4\pi.$$

Thus Eq. (6) now reads

$$1 - \varepsilon(q, \omega)^{-1} = 4\pi \chi_d(q, \omega). \tag{13}$$

In this case we note that the admittance function χ_d can also be written as

$$\chi_d = -q^{-2} \int_0^\infty \langle (\varrho_{-q}(0), \varrho_q(t)) \rangle \exp(-i\omega t)\, dt \tag{14}$$

where $\varrho_q(t)$ is the Fourier component of the induced charge distribution and the bracket in the integrand means a Poisson bracket. Eq. (14) is known as the Nozieres-Pines formula [5].

In the above expressions of the dielectric constant ε, Eqs. (8), (11), (12), and (13), the admittance $\chi(\omega)$ is defined in terms of correlation functions of the moment or the current in the sample of dielectrics with a given geometry. The nature of fluctuation in fact depends on the shape of the sample.

Fluctuation of the electric moment or the polarization may be described by a generalized Langevin equation [4, 6],

$$\dot{M}(t) = -\int_{t_0}^{t} \gamma(t - t')\, M(t')\, dt' + J_r(t) \equiv J_s(t) + J_r(t) \qquad (t \geqq t_0). \tag{15}$$

The current $J(t) = \dot{M}(t)$ is here divided into the systematic part and the random part. The systematic current $J_s(t)$ is driven by the moment M itself and depends on its memeory, and the random current $J_r(t)$ is the rest of current which fluctuates in a random way. Eq. (15) may be written in a somewhat different manner as

$$\dot{M}(t) = -\int_{-\infty}^{t} \gamma(t - t')\, M(t')\, dt' + J_r'(t) \qquad -\infty < t < \infty \tag{16}$$

in which the memory extends to the infinite past. Eq. (16) looks more adequate to express a stationary random process of $M(t)$ in thermal equilibrium. But Eq. (15) is equivalent to it and is in some sense more convenient. It represents also a stationary process because the initial time t_0 is arbitrary. The random current $J_r(t)$ is not totally independent of $M(t)$, but is assumed to satisfy the condition

$$\langle M(t_0); J_r(t) \rangle = 0, \qquad t > t_0.$$

Eq. (15) gives by this condition the correlation function of $M(t)$, the Fourier-Laplace transform of which is given by

$$\int_0^\infty \langle M_0(t_0); M(t_0+t)\rangle \exp(-i\omega t)\, dt = \langle M; M\rangle/(i\omega + \gamma[\omega])^{-1} \tag{17}$$

where $\gamma[\omega]$ is defined by

$$\gamma[\omega] = \int_0^\infty \gamma(t) \exp(-i\omega t)\, dt.$$

Using Eq. (17) one can show from Eq. (15) that

$$\gamma[\omega] = \langle M; M\rangle^{-1} \int_0^\infty \langle J_r(t_0); J_r(t_0+t)\rangle \exp(-i\omega t)\, dt \tag{18}$$

and also that

$$\gamma_t[\omega] = i\omega\gamma[\omega]/(i\omega + \gamma[\omega]) \tag{19}$$

where

$$\gamma_t[\omega] = \langle M; M\rangle \int_0^\infty \langle J(t_0); J(t_0+t)\rangle \exp(-i\omega t)\, dt. \tag{20}$$

Eq. (18) is the second fluctuation-dissipation theorem, a generalized form of the Nyquist theorem, which states in this case that the kernel of the retarded current is determined by the correlation function of the random current. Eq. (19) tells us that the correlation function of the total current is related to that of the random current in a simple manner.

The same results are obtained from Eq. (16) if we assume that the random current $J_r'(t)$ satisfies the fluctuation-dissipation theorem (18). Then the power spectrum of $M(t)$ becomes

$$I_M(\omega) = \langle M; M\rangle \operatorname{Re}\gamma[\omega]/|i\omega + \gamma[\omega]|^{-2}, \tag{21}$$

which is equivalent to Eq. (17) since $\gamma[\omega]$ is analytic in the lower half-plane of the complex variable ω. Thus, Eqs. (15) and (16) are equivalent to each other.

By Eqs. (3) and (4), we find that

$$\gamma_t[\omega] = i\omega\chi(\omega)/\chi(0) \tag{22}$$

so that we can write the admittance $\chi(\omega)$, Eq. (3), as

$$\chi(\omega) = \chi(0)\,\gamma[\omega]/(i\omega + \gamma[\omega]). \tag{23}$$

In the presence of an external field, the Langevin Eqs. (15) or (16) has to be modified to include the effect of the external force. Eq. (16) will be generalized to

$$\dot{M}(t) = \int_{-\infty}^{t} \{-\gamma(t-t')\, M(t') + \alpha(t-t')\, E_0(t')\}\, dt' + J_r(t). \tag{24}$$

The external field is assumed to produce a retarded current and the random current to be independent of the external field, which is right in the framework of the linear response theory. The moment M induced by periodic external force E_0 is given by

$$M = E_0\alpha[\omega]/(i\omega + \gamma[\omega]) \tag{25}$$

which must be identified with the expression (2), so that we have

$$\int_0^\infty \alpha(t) \exp(-i\omega t)\, dt = \gamma[\omega]\, \eta(\varepsilon(\omega)). \tag{26}$$

Mori [7] has shown that a dynamic equation of a physical quantity can be cast into the form of Eq. (15) by introducing a proper projection. Applying his prescription, an explicit form of $J_r(t)$ in Eq. (15) can be obtained from the microscopic equation of motion. This, in principle, allows one to calculate $\gamma[\omega]$, Eq. (18), which will give the admittance $\chi(\omega)$, Eq. (23), and hence the dielectric constant $\varepsilon(\omega)$ by Eq. (6). The only complication here, which is not at all easy to resolve, is that the evolution of $J_r(t)$ thus defined is not real dynamics but is rather an artificial modification of it.

We shall not attempt here to pursue this approach any further for calculating any realistic models of dielectrics, but we rather limit ourselves to commenting a few general points.

It is generally accepted as obvious that the complex dielectric constant $\varepsilon(\omega)$ satisfies the Kramers-Kronig relation. One may wonder, however, if it is really so obvious, because $\varepsilon(\omega)$ is determined as a solution of such an equation as Eq. (6) where μ and η are given functions of ε while $\chi(\omega)$ as defined by Eq. (3) or (4) is analytic in the lower half-place of complex ω and satisfies the Kramers-Kronig relation. The mapping defined by Eq. (6) must conserve this property, which seems not too obvious. There is no simple answer to this seemingly puzzling question. However, we note that Eqs. (12) and (23) for case c) give the relations

$$1 - \varepsilon(\omega)^{-1} = 4\pi\chi_c(0)\, \gamma_c(\omega)/(i\omega + \gamma_c[\omega]), \tag{27}$$

and

$$1 - \varepsilon(0)^{-1} = 4\pi\chi_c(0), \tag{28}$$

and therefore,

$$\varepsilon(\omega) = (i\omega + \gamma_c[\omega])/(i\omega + \varepsilon(0)^{-1}\gamma_c[\omega]). \tag{30}$$

where $\gamma_c[\omega]$ is defined by Eq. (18) for the fluctuating current in case c). The real part of $\gamma_c[\omega]$ is seen to be positive for ω in the lower complex plane where $\gamma_c[\omega]$ is analytic. Then the imaginary part of the denominator of the expression (30) is proved never to vanish in the lower half-plane of ω so that the function $\varepsilon(\omega)$ is analytic in the same region and satisfies the Kramers-Kronig relation.

Eq. (3) gives

$$\langle M; M\rangle = (1 - \varepsilon(0)^{-1})\, kTV/4\pi$$

and so Eq. (18) is written as

$$\gamma_c[\omega] = \{4\pi/(1 - \varepsilon(0)^{-1})\, kTV\} \int_0^\infty \langle J_r(t_0); J_r(t_0 + t)\rangle_c \exp(-i\omega t)\, dt. \tag{31}$$

If the material is not really a dielectrics but is a conductor, the static dielectric constant $\varepsilon(0)$ is infinite. Then Eqs. (30) and (31) are reduced to

$$\varepsilon(\omega) = 1 + 4\pi\sigma/i\omega \tag{32}$$

and

$$\sigma(\omega) = (kTV)^{-1} \int_0^\infty \langle J_r(t_0); J_r(t_0 + t)\rangle_c \exp(-i\omega t)\, dt \tag{33}$$

where $\sigma(\omega)$ is the conductivity and is expressed in terms of the correlation function of fluctuating current. It must be borne in mind that the part of the current induced by polarization is eliminated in the expression (33). More generally, Eq. (13) can be written as

$$\varepsilon(q, \omega) - 1 = (4\pi/i\omega kT) \int_0^\infty \langle j_{r,-q}(0); j_{rq}(t)\rangle_d \exp(-i\omega t)\, dt \tag{34}$$

for the dielectric constant $\varepsilon(q, \omega)$. The fluctuating current $j_{rq}(t)$ is defined in the same way as that in Eq. (33).

Eq. (32) is also obtained by considering case a). We find from Eq. (3) by putting $\varepsilon(0) = \infty$ that

$$1 = 4\pi\chi_a(0)/3$$

or

$$\langle M; M\rangle_a = 3kTV/4\pi \tag{35}$$

so that Eq. (18) becomes

$$\gamma_a[\omega] = 4\pi\sigma(\omega)/3 \tag{36}$$

where the conductivity σ is defined by the same expression as (32), the suffix c being replaced by a. Eq. (8) gives

$$\frac{\varepsilon - 1}{\varepsilon + 2} = \frac{4\pi\sigma/3}{i\omega + 4\pi\sigma/3} \tag{37}$$

which yields Eq. (32).

Eq. (11) is not useful for conductors for the obvious reason that polarization of a sphere in a conducting medium has no restoring force. It is useful, however, for a dielectrics, as was clearly noticed by FRÖHLICH. On the right-hand side of Eq. (11) the admittance function χ may be written as

$$\chi(\omega) = (N/V)\,\bar{\chi}(\omega) \tag{38}$$

where

$$\bar{\chi}(\omega) = \frac{1}{kT}\left\{\langle m; M\rangle - i\omega \int_0^\infty \langle m(0); M(t)\rangle e^{-i\omega t}\, dt\right\} \tag{39}$$

represents the admittance of a dielectric sphere embedded in the same medium when an oscillatory force acts on one of the elementary dipoles. Or, in other words,

$$\bar{\chi}(0) = \langle m; M\rangle/kT \tag{40}$$

is the induced polarization of the sphere when a static force of unit strength acts on an elementary dipole in the sphere and

$$\langle m(0); M(t)\rangle/kT \tag{41}$$

represents relaxation of this induced polarization when the force is removed at time $t = 0$. In a classical case the correlation function in (41) can be written as

$$\langle m(0); M(t)\rangle = \int P_e(m)\, dm \,\langle M(t)\rangle_m$$

where $P_e(m)$ is the probability distribution to find a dipole at the value m and $\langle M(t)\rangle_m$ is the average of the total moment M of a large sphere containing the dipole m when the value m is specified at $t=0$. In a quantum-mechanical case this interpretation is not quite justified and so we interprete (40) or (41) as response or relaxation. By an electrostatic calculation, Fröhlich showed that the moment $\langle M(0)\rangle_m$ is independent of the size of the sphere so long as it is large enough. The same is true for $\langle M(t)\rangle_m$ in the framework of quasi-statics where $\varepsilon(\omega)$ is used instead of $\varepsilon(0)$. It will also be true for the response or relaxation (41).

Glarum, Cole, and a few other investigators [2] have discussed the complex dielectric constants of polar liquids along the lines here briefly described. There remain, however, some ambiguities concerning the way the autocorrelation and cross-correlation functions are mutually related. The present status of the theory seems thus not to be quite satisfactory. It should be possible to develop a more rigorous theory without introducing any ad hoc assumptions.

Reference

1. Fröhlich, H.: Theory of Dielectrics. Oxford: Clarendon Press 1949.
2. Glarum, S. H.: J. Chem. Phys. **33**, 1371 (1960).
   Cole, R. H.: J. Chem. Phys. **42**, 637 (1965).
   Scaife, B. K. P.: Proc. Phys. Soc. **84**, 616 (1964).
   Fatuzzo, E., Mason, P. R.: Proc. Phys. Soc. **90**, 729, 741 (1967).
3. Kubo, R.: J. Phys. Soc. Japan **12**, 570 (1957).
4. Kubo, R.: Rept. Progr. Phys. **29**, Part 1, 255 (1966).
5. Nozières, P., Pines, D.: Nuovo Cimento **9**, 470 (1958).
6. Kubo, R., in: Many-body Theory, ed. R. Kubo. Tokyo: Shokabo, and New York: Benjamin 1966.
7. Mori, H.: Progr. Theoret. Phys. (Kyoto) **33**, 423 (1965).

Vibrational Absorption and Electronic Deformation in Solids

B. SZIGETI

J. J. Thomson Physical Laboratory, University of Reading, England

This is a review of work by the author relating to the contribution of electronic deformation to the vibrational polarization and absorption of solids, including work carried out in collaboration with others.

1. Introduction

The fundamental theory of ionic crystals, developed mainly by M. BORN, was based essentially on a model of rigid, non-overlapping and spherical charged atoms. After the advent of wave mechanics and the development of the theory of the covalent bond, it came to be realized that in covalent materials like diamond the bond is formed by sharing electrons, and it was also accepted that in many "weakly ionic" materials the bonds are partially covalent and that for such materials the picture of spherical ions cannot be exact. But for a long time not very much was known about the amount of covalency or electron sharing in the various weakly ionic solids. The strength of ionicity or covalency was judged mainly by qualitative criteria, such as chemical behaviour, ionic conductivity, and the strength of the infrared absorption.

In many cases the infrared evidence contradicted the indications by the other criteria. For instance, materials like SiO_2, SiC, GaAs and other III–V compounds exhibit strong infrared absorption, yet all the other criteria indicate that they have predominantly covalent character. More recently wave mechanical calculations also pointed to an essentially covalent state in such materials. Nonetheless, for such materials approximations based on a picture of spherical, although polarizable, ions are still frequently used, and it is often thought that the strong infrared absorption supports such a view.

In fact, however, the evidence available today indicates fairly clearly that materials like GaAs, SiC or SiO_2 are to a large extent covalent, and it follows that for such materials, in processes where the valency electrons are important, the model of spherical atoms or ions cannot even serve as the starting point for an approximation. It is known that even in the case of the alkali halides electronic deformation has an important rôle in the lattice polarization and absorption, and in the mainly covalent materials such deformations probably represent the dominating factor. As a result, strong infrared absorption is not evidence of ionic character.

The present paper reviews some aspects of previous work by the author relating to the contributions of electronic deformation to the lattice polarization and absorption, including work carried out in collaboration with others. Apart from the discussion of impurity absorption in Section 7, the review will be restricted to diatomic cubic crystals.

2. Static Dielectric Constant

We start by recalling that the macroscopic electric field E at any point in a solid is the spatial average value of the field over a small macroscopic region near that point. The actual value of the field is of course subject to strong variations from point to point, partly because of the strong variations of the field near a nucleus or electron, and also because of thermal fluctuations. We can look upon the macroscopic field as consisting of those Fourier components of the total field whose wavelengths are long compared with atomic size.

Inside a material the macroscopic field can be split into two parts [1]

$$E = E_e + E_i \tag{1}$$

where the "external field" E_e is due to sources outside the material and the "internal field" E_i is the electrostatic field due to the polarisation in the material. The vibrating particles in the material may also emit a radiation field, but as this radiation is not a function of the coordinates of the particles but of their time derivatives, it is convenient to include all radiation fields in the external field E_e.

In a diatomic cubic crystal, let q denote the displacement of the two kinds of atoms relative to each other. As long as the frequency of the external field is low compared with the electronic transition energies, the electronic wave functions follow q and E_e adiabatically. The polarization P is thus a unique function of q and E_e and we can write, in first order

$$P = \left(\frac{\partial P}{\partial E_e}\right)_q E_e + \left(\frac{\partial P}{\partial q}\right)_{E_e} q. \tag{2}$$

The internal field E_i is a function of P and therefore does not appear explicitly in this equation. The external force is represented by E_e, not by the total macroscopic field E. If q is in thermal equilibrium with the external field, then

$$P = E_e \left[\left(\frac{\partial P}{\partial E_e}\right)_q + \left(\frac{\partial P}{\partial q}\right)_{E_e} \frac{\partial q}{\partial E_e}\right]. \tag{3}$$

P and q of course represent macroscopic averages over a small region in the same way as E or E_e do. Since cubic crystals are electrically isotropic, P and q have the same direction as E_e. If E_e varies in space then q and P, and hence also E_i, vary with the same wave length so that (3) is valid everywhere in the material.

In Eq. (3), the first term in the square brackets is the purely electronic polarization, i.e. the polarization that would take place if the lattice were fixed. This is the only polarization that occurs at optical frequencies where the lattice cannot follow the field. The second term is the polarization due to the lattice displacements and includes that part of the electronic polarization which follows the lattice displacements adiabatically. The first term arises from mixing-in higher electronic states and the second from mixing-in other vibrational states.

Eq. (3) gives the response of the crystal to the external field E_e. Since any radiation emitted by the crystal is included in E_e, it follows that the states and the vibrational frequencies of the crystal in terms of which we describe the response, must be those appropriate to the *crystal Hamiltonian without the radiation field*. In contrast, the so-called "polariton" description, which has become popular in recent years, uses the states appropriate to the combined Hamiltonian of the crystal and the radiation field (cf. HUANG [2]). Such a description would not be very convenient for the present discussion.

It is well known that when the polarization is uniform in the material, for a given value of P the internal field E_i depends on the shape of the specimen. On the other hand, if the wavelength of P is short compared with the size of the specimen but long compared with atomic diameter, then E_i is different according to whether the wave is longitudinal or transverse. In particular, for a longitudinal wave

$$E_i = -4\pi P \tag{4a}$$

for a transverse wave

$$E_i = 0 \tag{4b}$$

and for a uniformly polarized sphere

$$E_i = -\frac{4\pi}{3} P \tag{4c}$$

(FRÖHLICH and MOTT [3]; LYDDANE and HERZFELD [4]).

Since E_i contributes to the restoring force, it follows that the response to a given E_e and hence the various derivatives occurring in Eq. (3) depend on shape or on wave form. However, instead of q and E_e we may of course also describe the state of the crystal in terms of the variables q and E. In anology to Eq. (3) we then have

$$P = E\left[\left(\frac{\partial P}{\partial q}\right)_E \frac{dq}{dE} + \left(\frac{\partial P}{\partial E}\right)_q\right]. \tag{5}$$

The derivatives occurring in this equation are independent of shape or waveform, since at a given frequency the total macroscopic field uniquely determines the polarization. This is because in any small region of the crystal the particles only register the total field E and do not know which part of E is exerted by other regions inside the crystal and which part of E comes from outside the crystal.

On the other hand, the right-hand-side of Eq. (5) is not written as a sum of independent contributions from the vibrational and the purely electronic transitions, but also contains cross-terms between these. This is because both terms contain E as a factor, and in general both types of polarization contribute to E_i and hence to E. This fact has lead to mistakes in a number of cases. In contrast, the two terms in Eq. (3) do represent independent contributions to the polarization.

The situation is simplified in the case of transverse waves, because for these according to Eqs. (4b) and (1) $E_i = 0$, hence $E_e = E$, so that the two terms in

Eq. (5) are then independent. In fact, for transverse waves Eqs. (3) and (5) are identical.

The dielectric constant ε is defined by the relation

$$\varepsilon - 1 = 4\pi P/E. \tag{6}$$

In view of what has been said, the ratio P/E and hence ε are of course independent of shape or wave form, although in general they depend on frequency. Let ε_0 denote the "optical" dielectric constant, i.e. the dielectric constant at such frequencies where the lattice can no longer follow the field while the purely electronic polarization is still in equilibrium with E. Since at such frequencies $q=0$, from Eq. (5) we have

$$\varepsilon_0 - 1 = 4\pi(\partial P/\partial E)_q. \tag{7}$$

In first order, in a lattice wave the dipole moment per atom pair is proportional to q. In the absence of an external field, we denote the factor of proportionality by e^*. This quantity depends on the internal field created by the displacement, since that field may polarize the electron shells of the ions and thus contribute to the dipole moment. In particular for a transverse wave, where $E_i=0$, we denote the factor of proportionality by e_t^*. Thus if N is the number of atom pairs per unit volume, $P=N e_t^* q$. Because for a transverse wave $E_i=0$, $\partial P/\partial q$ in this case is equal to $(\partial P/\partial q)_E$. Hence

$$\left(\frac{\partial P}{\partial q}\right)_E = N e_t^*.$$

e_t^* is usually called the "transverse effective charge" of the ions. Using Eqs. (5) and (7), for low frequencies Eq. (6) can now be written

$$\varepsilon_{st} = \varepsilon_0 + 4\pi N e_t^* \frac{dq}{dE} \tag{8}$$

where ε_{st} denotes the static dielectric constant.

In a transverse lattice wave V the potential energy per atom pair can be written

$$V = \tfrac{1}{2} m \omega_t^2 q^2$$

where m is the reduced mass of an atom pair and ω_t the frequency of a transverse vibration. If there is also an electric field E, V contains the additional term $-E e_t^* q$, since $e_t^* q$ is that part of the dipole moment per atom pair which is connected with the lattice displacement. We then have

$$V = \tfrac{1}{2} m \omega_t^2 q^2 - E e_t^* q. \tag{9}$$

The corresponding equation would not be so simple for cases other than a transverse wave, because in other cases $E_i \neq 0$ so that a part of E contributes to the restoring force and hence to ω.

In equilibrium $\partial V/\partial q = 0$. Hence from Eq. (9) the equilibrium value of q is given by

$$q = E \frac{e_t^*}{m\omega_t^2}.$$

Differentiating with respect to E and inserting the result in Eq. (8) we get

$$\varepsilon_{st} = \varepsilon_0 + 4\pi N \frac{e_t^{*2}}{m\omega_t^2}. \tag{10}$$

In this equation ε_{st} has been expressed in terms of the properties of the transverse lattice waves, because as we have seen the conditions are simplest for transverse waves. Moreover, ω_t is directly obtained from experiment, as it is the frequency at which electromagnetic radiation is absorbed.

However, it is also of interest to consider a uniform optical displacement in a solid of spherical shape, i.e. a lattice displacement where at any instant q has the same value everywhere in the spherical specimen. According to Eq. (4c) in this case $E_i \neq 0$ so that both the restoring force and the polarization of the electron shells differ from their values for a transverse wave. It is not clear how far a uniform optical displacement approximates a normal vibration of a spherical solid, but in any case we can define a "frequency" ω_s and an effective charge e_s^* by the relations

$$V = \tfrac{1}{2} m\omega_S^2 q^2$$
$$P = N e_S^* q$$

where V is the potential energy per ion pair and P the polarization for a uniform optical displacement in a sphere, in the absence of an external field. Using purely macroscopic arguments, LYDDANE, SACHS and TELLER [5] showed that ω_S and e_S^* are connected with ω_t and e_t^* by the relations

$$\begin{aligned} e_t^* &= \frac{\varepsilon_0 + 2}{3} e_S^* \\ \omega_t^2 &= \frac{\varepsilon_0 + 2}{\varepsilon_{st} + 2} \omega_S^2. \end{aligned} \tag{11}$$

Using the first of these relations, Eq. (10) can be written

$$\varepsilon_{st} = \varepsilon_0 + 4\pi N \left(\frac{\varepsilon_0 + 2}{3}\right)^2 \frac{e_S^{*2}}{m\omega_t^2} \tag{12}$$

(SZIGETI [1]). We have replaced e_t^* by e_S^* because, as we shall see, in many cases e_S^* is more directly connected with the porperties of the ions. It is worth stressing that the only approximation involved in Eqs. (10) and (12) is that the dipole moment is assumed to be of first order and the potential energy of second order in the displacement, and in addition it is also assumed that semiconducting effects can be neglected. Otherwise Eqs. (10) and (12) are free of all assumptions and they are quite generally valid for all diatomic cubic crystals. They can of course be generalized for non-cubic and for multiatomic crystals.

An expression for the total vibrational absorption can of course be easily obtained from the preceeding relations. If ε''_ω denotes the imaginary part of the dielectric constant at frequency ω, then according to the Kramers-Kronig relations we have

$$\varepsilon_{st} - 1 = \frac{2}{\pi} \int_0^\infty \frac{\varepsilon''_\omega}{\omega} d\omega$$

(cf. for instance Fröhlich [6]). ε''_ω measures the absorption per cycle and is of course different from zero only inside the absorption regions. Since the purely electronic and the vibrational transitions contribute independently, we can separate this relation into two parts

$$\varepsilon_0 - 1 = \frac{2}{\pi} \int_{\text{p.e.}} \frac{\varepsilon''_\omega}{\omega}\, d\omega$$

and

$$\varepsilon_{st} - \varepsilon_0 = \frac{2}{\pi} \int_{\text{vib}} \frac{\varepsilon''_\omega}{\omega}\, d\omega \tag{13}$$

where the first integral goes over the high frequency or purely electronic absorption bands and the second over the vibrational absorption region.

Comparison with Eq. (10) gives for the integrated vibrational absorption

$$\omega_t^2 \int_{\text{vib}} \frac{\varepsilon''_\omega}{\omega}\, d\omega = 2\pi^2 N \frac{e_t^{*2}}{m}. \tag{10b}$$

We remark that since $N e_t^* = (\partial P/\partial q)_E$, using standard relations the following alternative formula can also be derived for the integrated vibrational absorption

$$c \int_{\text{vib}} \varrho_\omega \alpha_\omega\, d\omega = 2\pi^2 N \frac{e_t^{*2}}{m} \tag{10c}$$

where c denotes the velocity of light in vacuum, ϱ_ω is the refractive index and α_ω the conventional absorption coefficient at frequency ω. As $c\varrho\alpha = \omega\varepsilon''$, the identity of Eqs. (10b) and (10c) is easily seen for the case when the absorption is a narrow line at ω_t. In fact, however, the left-hand-sides of the two equations remain equal to each other to a very good approximation even when the vibrational absorption is fairly broad.

3. Effective Field and Clausius-Mossotti Relation

We now recall briefly the assumptions on which the Clausius-Mossotti relation is based.

If one can assume that the constituent particles of a material, i.e. the ions, atoms or molecules, do not overlap and only interact with each other through a classical electrostatic field, then μ the dipole moment of a particle is uniquely determined by the field acting on it. In first order we can then write

$$\mu = \alpha E_{\text{eff}} \tag{13}$$

where the factor of proportionality α is called the polarizability of the particle. E_{eff} the "effective field" is the total field less the field exerted by the particle, since the particle does not act upon itself. As is well known, in electrically isotropic materials, which include cubic crystals, for non-overlapping spherical particles the effective field is given by

$$E_{\text{eff}} = E + \frac{4\pi}{3} P. \tag{14}$$

If all the particles are of the same kind and their number per unit volume is N, then $P = N\mu$. Eq. (6) and the last two equations thus result in the Clausius-Mossotti relation

$$\frac{4\pi}{3} N\alpha = \frac{\varepsilon - 1}{\varepsilon + 2} \tag{15}$$

where both ε and α are, of course, frequency dependent. Eqs. (14) and (15) are also valid for diatomic cubic crystals if we define $\alpha = \alpha_1 + \alpha_2$, where α_1 and α_2 are the polarizabilities of the two atoms or ions and in this case N in Eq. (15) denotes the number of such pairs per unit volume.

It is worth stressing that the effective field introduced here is the field acting on a *whole atom* or ion or molecule. This is a different quantity from the effective field acting on *an electron*, which has become of considerable interest in band theory calculations. For instance in the case of ionic crystals, the field acting on an electron includes the field exerted by the nucleus and by the other electrons of its own ion, and that field can therefore not be combined with the ionic polarizability to give a dipole moment.

We also note from Eq. (14) that the Clausius-Mossotti effective field is expressed in terms of the macroscopic quantities E and P, and thus represents an average over the effective fields acting on all the atoms in a small region.

Clearly the assumptions on which the Clausius-Mossotti theorem is based are for many materials not fullfilled. This fact is frequently ignored in the literature and even today the theorem is often applied to materials where it has no basis whatever.

Basically, the assumption in Eq. (13) implies that α is a property of the particle and is independent of its surroundings. Hence in changes of density the right-hand-side of Eq. (15) should vary proportionally to N and thus to the density. For many molecular materials whose molecules have no permanent dipole moments this has been found to be true; in fact for several such materials it has been found that $(\varepsilon - 1)/N(\varepsilon + 2)$ has practically the same value in the gaseous, liquid and solid phases (cf. for instance FRÖHLICH [6]). This is because the Van der Waals binding forces represent a weak perturbation on the molecules, and so does of course the electric field. Both perturbations can therefore be treated in first order, hence they superpose independently and α is thus not affected by the binding forces.

Turning to the non-molecular solids, we note that the inter-atomic or inter-ionic restoring forces in a lattice displacement arise mainly from non-classical short-range forces and certainly cannot be expressed in the foım of a classical effective field. Moreover, since the restoring force is due to the inter-ionic forces, the lattice polarizability is a property of the crystal and not of the individual ions. The Clausius-Mossotti theorem has just no bearing whatever on the "vibrational polarization" i.e. on the polarization due to the lattice displacements.

As far as the "purely electronic" polarization is concerned, i.e. that part of the polarization which occurs independently of the lattice displacements, we must distinguish between the strongly ionic and the covalent crystals. In a covalent material the electrons are shared between neighbouring atoms; this is the case, for instance, for diamond, and to an appreciable extent also for materials like GaAs (cf. Figs. 1 and 2). In this case the assumptions on which Eqs. (13) and (14)

Fig. 1.
From DAWSON [7]. Measured valency electron charge density in diamond. The contour lines represent a conventional "difference" Fourier. Hence the contour lines in this figure do not indicate absolute magnitudes of valency electron charge density, while those in Fig. 2 do. The four black dots indicate the nuclear positions

Fig. 2. From WALTER and COHEN [8]. Calculated valence-electron-density contour map (in units of e per primitive cell) for GaAs in the (1, — 1, 0) plane. The orientation of the plane (dashed lines) with respect to the primitive cell is shown in the inset. The core radii for Ga and As are 0.23 and 0.18 of the Ga-As distance. The radii are those of spheres containing 80% of the outermost shell of core electrons

are based are completely invalid. Even in a purely electronic polarization the valency electrons of neighbouring atoms interact essentially with non-classical forces, and their interaction can therefore certainly not be described in terms of a classical electric field; moreover, since the valency electrons are shared between neighbouring atoms, the atoms do not represent individual polarizable units. As in the vibrational case, the polarizability is a property of the crystal and not of the individual atoms, and the Clausius-Mossotti theorem cannot even serve as a basis for an approximation.

This conclusion is borne out by the experimentally observed variation of ε_0 with density and elastic strains in such crystals: in many cases, the observed changes differ both in order of magnitude and in sign from what the Clausius-Mossotti theorem would predict (cf. [9, 10]).

The situation is probably quite different for the purely electronic polarization of the strongly ionic crystals. The ions in such materials can probably be approximated by spheres which overlap only very little. For such a model, we can assign a separate electronic dipole moment to each ion, and the electric field acting on each ion is then given by the Clausius-Mossotti field (14). But because the interionic forces are not weak we can no longer expect that α is independent of these forces. In other words, at high frequencies where only purely electronic polarization takes place we may expect that Eqs. (13) and (14) are valid, but that α changes somewhat in an elastic compression or expansion since the interionic forces change when the lattice parameter is altered. For the alkali halides the experimentally found variation of ε_0 in compression and other elastic deformations does, on the whole, agree with these assumptions [10].

Although there are important arguments against representing the alkali halides as spherical ions without appreciable sharing of electrons, all in all the reasons for accepting the spherical picture as a first approximation are rather strong. This view seems to be shared by the majority of physicists working in this field.

4. Strongly Ionic Crystals

If for the strongly ionic crystals the picture of non-overlapping spheres were exact, then the electronic polarization should be uniquely determined by the Clausius-Mossotti Eqs. (13)–(15), provided that α and ε are replaced by α_0 and ε_0, the values for optical frequencies. These relations should also determine that part of the electronic polarization which occurs as a result of a lattice displacement. In view of Eq. (14), in the absence of an external field $E_{\text{eff}}=E_i+4\pi P/3$; it is thus seen from Eq. (4) that while E_{eff} does not vanish for a transverse wave, it does vanish for the uniform polarization of a spherical specimen. Therefore, according to the Clausius-Mossotti equations, in the uniformly polarized sphere the ions should be displaced rigidly, without polarization of their electron shells, and e_S^* should therefore be equal to the static charge on the ions. Thus if e_0 denotes the elementary charge, e_S^* should be equal to e_0 for monovalent ions, to $2e_0$ for divalent ions, etc.

On the other hand, e_S^* can be obtained from Eq. (12) since all the other quantities in that equation can be experimentally measured. For most alkali halides the values of e_S^*/e_0 so obtained are between 0.7 to 0.8 (Szigeti [1]). Values of e_S^*/e_0, calculated from Eq. (12) but using recent measurements of ε_{st}, ε_0 and ω_t [11], are reproduced in Table 1 for a number of ionic crystals.

The measured data thus differ significantly from the predictions based on non-overlapping spherical ions. This however need not mean that in alkali halides the static charge on the ions is significantly less than e_0 and that there is therefore a strong covalent contribution to the bond; it is much more likely that for the alkali halides the assumption of spherical ions with charge e_0 is *almost* correct, but the ions overlap slightly and when they are displaced relative to each other not only do they get polarized by the effective electrostatic field but also by non-classical short-range forces. The latter act mainly in the overlap region and the dipole moment which they produce in a lattice displacement causes e_S^* to differ from e_0 (Szigeti [13]).

An alternative explanation for the difference between e_S^* and e_0 would be to assume that the valency electrons have a fairly large probability for being near the alkali ions. In this case the static charge could be less than e_0 and, what is

Table 1. All the figures for ε_s, ε_0 and ω_t represent recent measurements by LOWNDES and MARTIN [11]. Cf. also LOWNDES [12].

	290 °K				2 °K			
	ε_{st}	ε_0	ω_t(cm^{-1})	e_S^*/e_0	ε_{st}	ε_0	ω_t (cm^{-1})	e_S^*/e_0
LiCl	11.86	2.75	203.0	0.79	10.83	2.79	221.0	0.77
NaCl	5.90	2.33	164.0	0.77	5.45	2.35	178.0	0.76
KCl	4.84	2.17	142.0	0.81	4.49	2.19	151.0	0.79
RbCl	4.89	2.18	116.5	0.83	4.53	2.20	126.0	0.81
LiBr	13.23	3.16	173.0	0.73	11.95	3.22	187.0	0.73
NaBr	6.27	2.60	134.0	0.74	5.78	2.64	146.0	0.73
KBr	4.90	2.35	114.0	0.78	4.52	2.38	123.0	0.75
RbBr	4.86	2.34	87.5	0.80	4.51	2.36	94.5	0.78
CsBr	6.66	2.78	73.5	0.82	6.39	2.83	78.0	0.82
NaI	7.28	3.01	115.5	0.73	6.62	3.08	123.5	0.71
KI	5.09	2.63	102.0	0.74	4.69	2.67	109.5	0.72
RbI	4.91	2.58	75.5	0.77	4.55	2.61	81.5	0.74
CsI	6.54	3.02	62.0	0.78	6.29	3.09	65.6	0.77
LiF	9.0	1.92	305.0	0.81	8.50	1.93	318.0	0.80
NaF	5.08	1.74	246.5	0.83	4.73	1.75	262.0	0.82
KF	5.50	1.85	194.0	0.91	5.11	1.86	201.5	0.88
RbF	6.48	1.93	158.0	0.95	5.99	1.94	163.0	0.92
CsF	8.08	2.16	127.0	0.95	7.27	2.17	134.0	0.93
AgCl	11.14	3.92	105.5	0.69	9.51	3.97	120.0	0.68
AgBr	12.44	4.62	79.5	0.68	10.61	4.68	91.6	0.67
TlCl	32.6	4.76	63.0	0.87	37.6	5.00	60.4	0.86
TlBr	30.4	5.34	47.9	0.84	35.1	5.64	47.2	0.84

probably more important, the Clausius-Mossotti equations would be invalid so that e_S^* would significantly differ from the static charge. Such a picture is rather unlikely, even though there is no definite evidence which would allow us to reject it completely. It is of interest to note however that if this second explanation is correct then a considerable larger part of the dipole moment is due to electronic deformation induced by non-classical effects than is the case for the first explanation.

If we accept the first explanation, namely that in the alkali halides the ions are almost spherical, we note that probably not much overlap is needed in order to account for the discrepancy between e_S^* and e_0. This is because the deformation in a displacement does not depend on the overlap forces in equilibrium, but on their *derivative* with respect to the displacement, and the relative importance of the derivative of such short-range forces is always greater than that of the forces themselves.

If it is true that in the alkali halides the difference between e_S^* and e_0 is due mainly to overlap effects, then we may expect that the shift of charge due to these effects is not uniform over the ions but is largest in the overlap regions. On the basis of this assumption it was possible to derive a relation between ε_s, ε_0, ω_t and the compressibility β, namely

$$\frac{1}{\beta} = \frac{R^2}{3u} \frac{\varepsilon_{st}+2}{\varepsilon_0+2} m\,\omega_t^2 \tag{16}$$

where R is the nearest neighbour distance and u the volume occupied by an ion pair (SZIGETI [13]). For the alkali halides this relation agrees with experiment within about 10%, and this seems to support the view that in the alkali halides the deformation due to the short-range forces occurs mainly near the overlap regions. For a further discussion, see SZIGETI [13]; BORN and HUANG [14] pp. 111–116; JONES *et al.* [15].

Some years after Eqs. (12) and (16) were derived, phonon dispersion curves started to be measured and it was found that these are also easiest to explain if one assumes electronic deformations due to short-range effects. However, the most widely used force constant model in lattice dynamics, namely the "shell modell", assumes that the deformation due to the short-range forces can be well approximated by a single dipolar term, representing uniform shift of charge over the whole atom in the same way as the polarization induced by an electric field (cf. COCHRAN [16]). Yet it should be noted that while the perturbing Hamiltonian due to a macroscopic electric field is uniform over the whole atom, this is certainly not the case for the perturbation due to overlap effects and therefore one would not really expect that the two perturbations should produce the same deformation.

The shell model certainly gives a very simple description of the electronic deformations, and on the basis of this model it is usually possible to calculate force constants which agree well with the measured lattice frequencies. This in itself is however no proof that the shell model represents a good approximation to the actual deformations, since very many other force constant models would agree with the measured frequencies equally well [17].

5. Weakly Ionic Crystals

Table 1 shows that for all the alkali halides e_S^* is less than e_0. Since it is presumably mainly the negative ions which are deformed, the fact that $e_S^* < e_0$ indicates that in an optical lattice displacement the negative ions cannot carry their electron shells completely with them. This is what one would expect for a strongly ionic material: the short-range forces try to reduce the overlap and hence force back the electrons of the negative ions as they penetrate into the positives.

For materials which are less strongly ionic than the alkali halides the interpretation of the difference between e_S^* and $z e_0$ is not so simple (z = valency). For such materials we cannot exclude the possibility of some covalency, i.e. some sharing of the valency electrons, in which case the static charge is less than $z e_0$. On the other hand, the sharing may for some materials increase and for others decrease with interionic distance, hence in an optical lattice displacement these covalency effects may increase the dipole moment for some materials and decrease it for others. In addition, if there is some electron sharing then the Clausius-Mossotti Eqs. (13) and (14) are no longer correct.

In view of these various effects the magnitude of e_S^* cannot be looked upon as a unique function of ionic character. For instance, as Table 1 shows, TlCl and TlBr have a larger e_S^* than NaCl, yet we know from other criteria that they are less ionic than NaCl. Since $e_t^* = e_S^* (\varepsilon + 2)/3$, comparison of the data for the thallium halides and the alkali halides also shows that e_t^* and hence the infra-red absorption would be a still less suitable criterion for ionic character.

With increasing electron sharing the ions cease to be individually polarizable units and the Clausius-Mossotti Eqs. (13) and (14) lose all validity. It is not possible to retain Eq. (13) and to replace the Clausius-Mossotti effective field (14) by some other classical effective field, partly because when the electrons are shared the polarizability of an individual atom has no meaning, and also because the interaction between neighbouring atoms ceases to be classical altogether. The whole interaction between neighbours then comes into the range of non-classical effects.

In general, one may expect that the more pronounced the electron sharing, the more profoundly the wave functions will be altered when neighbouring atoms are displaced relative to each other. Hence, although in the predominantly covalent diatomic materials (e.g. GaAs, SiC, etc.) the static charge may be small, the electronic deformation produced in a displacement by short-range effects may be expected to be much larger than in the alkali halides. If this is correct, then in the predominantly covalent materials e_S^* or e_t^* bear no relation to the static charge or to the ionicity of the bond, nor does of course the magnitude of the infra-red absorption, as this is related to e_t^* [cf. Eq. (12b)]. These matters will be discussed in more detail in the next sections.

6. Survey of Covalent Crystals

In Table 2, values for ε_0 and the effective charges e_S^* and e_t^* have been collected for a number of mainly covalent crystals. By "mainly covalent" we mean that the binding is due mainly to electron sharing.

Selenium and tellurium are of course not diatomic, nor are they cubic, but they have been included in the table because of the special interest that attaches to them. These two materials exhibit a large vibrational absorption even though all their atoms are chemically identical. Their crystal structure is trigonal, with three atoms in a unit cell. The crystal symmetry is such that the electronic deformation in a displacement can produce a dipole moment, although in the undisplaced configuration the posibility of a static charge on the atoms is excluded.

We note that for the covalent materials listed in Table 2 the values of e_S^* are smaller than for the alkali halides, but nevertheless they are still quite large; at the same time, the values of e_t^* and hence the vibrational absorption are in almost every case larger than for the alkali halides. Other covalent diatomic crystals which have not been included in the table but for which measurements are available behave in a similar manner.

Since theoretical considerations [22] indicated that the static charge in the III–V compounds or in silicon carbide is probably small, BURSTEIN [23] concluded that the polarization in these materials is due only to a very small extent to the static charge and to a much larger extent to the electronic deformation induced by short-range effects. In selenium and tellurium, where the static charge is zero, the polarization can be caused only by short-range effects. The long-range internal field can of course make a further contribution, but this field can only arise if a polarization has already been created by the short-range interactions.

Table 2. Covalent materials

		ε_0	ε_S^*	e_t^*
GaP	[18]	8.46	0.58	1.51
GaAs	[19]	10.9	0.43	2.20
GaSb	[19]	14.4	0.30	1.81
InP	[19]	9.6	0.60	2.63
InAs	[19]	12.2	0.56	2.65
InSb	[19]	15.7	0.34	2.48
AlSb	[19]	10.2	0.48	2.16
SiC (zinc blende structure)	[20]	6.7	0.94	2.73
$Se_{\parallel c}$	[21]	12.8	0.13	0.7
$Se_{\perp c}$	[21]	7.7	0.32	1.35
$Te_{\parallel c}$	[21]	39.5	0.14	1.8
$Te_{\perp c}$	[21]	22.9	0.42	2.5

Note. Selenium and tellurium have a trigonal structure. There is one infrared active mode with the polarization P parallel to the trigonal c-axis, and two doubly degenerate modes (i.e. four altogether) with P perpendicular to the c-axis. For the case $\perp c$, e_t^* in the table is the sum of the two $\perp c$ modes, and e_S^* for $\perp c$ is the sum of the e_S^* of the same two modes. Hence the values of e_t^* and e_S^* given for $\perp c$ have no direct physical meaning, but they are more useful for comparison purposes than the individual effective charges of the two modes would be.

Notwithstanding these matters, it is still frequently assumed that the vibrational absorption is a measure of ionicity and of static charge. In the next sections we discuss work by LEIGH and SZIGETI which contradicts this view. The results of that work indicate that in the mainly covalent materials the lattice polarization and hence the vibrational absorption are determined to a large extent by short-range interactions *even if the static charge is not small;* we can expect no clear relation between the effective charge and the static charge, and the two need not even have the same sign.

LEIGH and SZIGETI's conclusions are based on an analysis that could only be applied directly to the vibrational polarization and absorption induced by a charged impurity in an unpolar crystal. However, the nature of the arguments is such that one can expect the conclusions to be valid also for pure covalent crystals.

7. Charged Impurity in an Unpolar Crystal

7.1. General Considerations

In Sections 7.1 to 7.5 we discuss the papers by LEIGH and SZIGETI [24, 25, 7, 26] relating to the absorption induced by a charged impurity in an unpolar cubic crystal. We use the word "unpolar" to indicate that the crystal exhibits no first-order vibrational absorption in the pure state, e.g. diamond, solid argon, etc. The procedure used in the papers is valid for any charged substitutional impurity in any cubic unpolar crystal, but LEIGH and SZIGETI only considered impurities in a covalent unpolar crystal, i.e. in diamond, silicon or germanium. In particular B^- and P^+ impurities in silicon were discussed in detail as the infra-red absorption spectra of these impurities had been investigated experimentally by ANGRESS, GOODWIN and SMITH [27]. Towards the later stages of the theoretical work experimental spectra also became available for B^- and As^+ impurities in silicon (ANGRESS, GOODWIN and SMITH [28]).

It is believed that impurities like B^-, P^+ and As^+ all have four valency electrons and each is bound to its four silicon neighbours by four covalent bonds. Presumably the valency electrons are not shared equally between the impurity atom and its silicon neighbours, so that some of the impurity charge spreads over to the neighbouring atoms. But LEIGH and SZIGETI stressed that even if all the static charge were localized on the impurity atoms, the vibrations of the neighbouring silicon atoms would still cause absorption, partly because of polarization by the short-range interactions between these atoms and the impurity and partly due to a dielectric interaction which extends considerably beyond the nearest neighbours of the impurity. This dielectric interaction follows from the fact that the charged impurity exerts an electric field which polarizes the electrons in its surroundings. The polarization has opposite signs on opposite sides of the impurity atom and therefore in the equilibrium configuration the total polarization vanishes. But, due to the Raman and the photoelastic effects, the polarizability may change in a lattice displacement. If the displacement is not symmetric with respect to the foreign atom then the polarizability changes by different amounts on opposite sides and the total polarization induced by the foreign atom no longer vanishes. Vibrations which have this property are therefore infra-red active even if in the course of these vibrations the charged foreign atom is not displaced at all.

It was shown that both this dielectric effect and also the short-range interactions between the impurity and its neighbours result in considerable absorption. The calculations showed that even if all the charge is on the foreign atom, and even if only the dielectric effect is taken into account, the total absorption by the vibrations of the *uncharged* host atoms is larger than the absorption by the vibration of the charged impurity itself, unless the impurity is much lighter than the host atoms. This conclusion is in agreement with the experimentally measured spectra. A further analysis also showed that the absorption by the impurity itself is determined more by its short-range interactions than by its static charge.

7.2. Apparent Charge and Integrated Absorption

For a pure diatomic cubic crystal, in Eq. (2) we expressed the polarization P as a function of the optical displacement q and of the external field E_e. As we remarked afterwards, for certain purposes it is more convenient to use q and the total field E as independent variables rather than q and E_e. Instead of Eq. (2) we can then write

$$P = \left(\frac{\partial P}{\partial q}\right)_E q + \left(\frac{\partial P}{\partial E}\right)_q E. \tag{16}$$

If we want to take the effect of impurities into account, the dipole moment does not only depend on q, but in the most general case it may depend on the displacements of all the atoms.

Let $u_{n\xi}$ denote the displacement of the nth atom in one of the three Cartesian directions; ξ stands for x or y or z. Further, let M be the total dipole moment of the crystal, and M_x P_x and E_x the x-components of M, P and E. Thus, if v is the volume, then $M_x = vP_x$. For the general case with impurities, instead of Eq. (16) we can then write

$$M_x = \sum_{n,\xi} \left(\frac{\partial M_x}{\partial u_{n\xi}}\right)_{E_x} u_{n\xi} + \left(\frac{\partial M_x}{\partial E_x}\right)_u E_x \tag{17}$$

with similar expressions for M_y and M_z. In the last term the suffix u means that the derivative is to be formed with all the displacements kept zero. The expression written in this form is valid not only for macroscopic cubic symmetry but also for other structures, provided that x, y and z are chosen to coincide with the principal directions of the macroscopic dielectric tensor. We have written Eq. (17) for M rather than for P, since clearly $(\partial M/\partial u_{n\xi})_E$ is independent of the volume of the crystal while $(\partial P/\partial u_{n\xi})_E$ is not.

It is convenient to define a tensor η_n by the relation

$$\eta_{n\xi;x} = \left(\frac{\partial M_x}{\partial u_{n\xi}}\right)_E \tag{18}$$

with similar expressions for $\eta_{n\xi;y}$ and $\eta_{n\xi;z}$. We call the tensor η_n the *apparent charge* of the nth atom since, as we shall see, it determines the total absorption by the vibrations of that atom.

If the nth atom is at the centre of cubic or tetrahedral symmetry then η_n is a scalar, but even in a cubic crystal in the neighbourhood of an impurity the cubic symmetry is removed and for atoms near an impurity η_n is therefore a second-rank tensor.

As mentioned in Section 2, for a specimen that has the shape of a slab the internal field E_i always vanishes parallel to the long edges, whatever the polarization. In the language of macroscopic physics, this is because the surface charges are too far away to exert a field. Hence if we consider a specimen in the shape of a slab, with the x-direction lengthwise, then in the absence of an external field E_x is zero for all internal displacements. If M_x (slab) denotes the x-component of the dipole moment of such a slab, instead of Eq. (18) the apparent charges can therefore be written

$$\eta_{n\xi;x} = \frac{\partial M_x\,(\text{slab})}{\partial u_{n\xi}}\,. \tag{19}$$

This is a useful definition of η_n as for many problems it is convenient to consider a slab-shaped crystal [24, 29].

If we wished to use also $\eta_{n\xi;y}$ and $\eta_{n\xi;z}$ then these would have to be defined with respect to a slab whose long edges are oriented accordingly. For cubic crystals it is sufficient to calculate macroscopic properties in the x-direction only.

Let $\alpha_\omega(x)$ be the conventional absorption coefficient and $\varrho_\omega(x)$ the refractive index for light polarized in the x-direction and with frequency ω. Using well-known relations, one can derive the following expression for the integrated vibrational absorption

$$\int\limits_{\text{vib}} \varrho_\omega(x)\,\alpha_\omega(x)\,d\omega = \frac{2\pi^2}{vc} \sum_n \frac{1}{m_n} (\eta^2_{nx;x} + \eta^2_{ny;x} + \eta^2_{nz;x}) \tag{20}$$

where m_n is the mass of the nth atom, c the velocity of light and v the volume of the crystal. The integral on the left extends over the whole vibrational spectrum and the sum on the right over all the atoms.

If the crystal has over-all cubic symmetry then the quantities on the left are independent of direction, hence the sum on the right must also be independent of how the x-direction is chosen, even though some of the individual terms may depend on it. For cubic crystals we can therefore write [22]

$$\int_{\text{vib}} \varrho_\omega \alpha_\omega \, d\omega = \frac{2\pi^2}{v c} \sum_n \frac{1}{m_n} \left(\eta^2_{n\,x;\,x} + \eta^2_{n\,y;\,x} + \eta^2_{n\,z;\,x}\right) \tag{21}$$

Relations similar to Eq. (21) had been derived before by several other authors, but it was not pointed out that the derivatives represented by the η are to be taken with E kept constant, or for a slab, as defined in Eqs. (18) and (19).

On the right-hand-side of Eq. (21), the nth term represents the total absorption by the vibrations of the nth atom. In general, the $\eta_{n\,\xi;x}$ are rather complicated quantities, as they include the effects of the adiabatic electronic deformation induced by the displacement of the nth atom. As a rule, this deformation is not localized on the nth atom but may extend to some distance from it. Nevertheless, the η have the advantages that they do not depend on the force constants and can therefore in principle be calculated from static considerations. Such a calculation will be carried out for a special case in the next subsections.

It is of interest to compare Eq. (21) with the formulae for pure diatomic cubic crystals which were discussed in section 2. In a pure diatomic cubic crystal each atom is at the centre of cubic symmetry, so each η_n is a scalar. As there are only two kinds of atoms, η_n can only have two values, say η_1 and η_2. Since in a transverse optical wave the internal field E_i is zero, it follows from Eq. (16) and from the definition of e_t^* in Section 2 that η_1 and η_2 are equal to e_t^* and to $-e_t^*$, respectively. Remembering that m the reduced mass of an atom pair is defined by $1/m = (1/m_1) + (1/m_2)$, it is readily seen that for this case the last equation reduces to Eq. (10c).

7.3. *Macroscopic Calculation of Apparent Charges*

We may expect that for low impurity concentration the effects of the various impurity atoms add linearly, and it is therefore sufficient to consider the effect of a single foreign atom. We therefore consider a valency crystal like diamond or silicon; the crystal has the shape of a slab with the x-direction parallel to four of the long edges, and it contains a *single substitutional foreign atom with charge e.*

As explained in Section 7.1., the charged foreign atom causes the vibrations of the host atoms in its neighbourhood to absorb. In other words the foreign atom induces apparent charges on the host atoms even if their true charges are zero. This is due partly to short-range effects and partly to the dielectric interaction explained in Section 7.1. The short-range effects are restricted to near neighbours of the impurity while the dielectric effect extends rather further. In this section we shall consider the latter.

For atoms sufficiently far from the impurity, the apparent charges due to the dielectric interaction can be calculated exactly from the macroscopic photoelastic and Raman effects. Such a purely macroscopic calculation leads to

$$\eta_{nx;x} = \frac{\varepsilon^2}{4\pi N}\left[p_{11}\frac{\partial F_x}{\partial x} + p_{44}^0\left(\frac{\partial F_y}{\partial y} + \frac{\partial F_z}{\partial z}\right)\right]_{\boldsymbol{r}_n} \tag{22a}$$

$$\eta_{ny;x} = \frac{\varepsilon^2}{4\pi N}\left[p_{12}\frac{\partial F_x}{\partial y} + p_{44}^0\frac{\partial F_y}{\partial x} \pm 2\beta F_z\right]_{\boldsymbol{r}_n} \tag{22b}$$

$$\eta_{nz;x} = \frac{\varepsilon^2}{4\pi N}\left[p_{12}\frac{\partial F_x}{\partial z} + p_{44}^0\frac{\partial F_z}{\partial x} \pm 2\beta F_y\right]_{\boldsymbol{r}_n} \tag{22c}$$

(LEIGH and SZIGETI [24]). N denotes the number of atoms per unit voulme and ε the dielectric constant, both for the pure host crystal. Since the host crystal is unpolar, ε has practically the same value in the near infrared as at low frequencies. β is the "Raman constant" of the host crystal, which for the diamond structure was defined by the relation

$$\varepsilon^2\beta = \frac{\partial\varepsilon_{xy}}{\partial q_z} = \frac{\partial\varepsilon_{yx}}{\partial q_x} = \frac{\partial\varepsilon_{zx}}{\partial q_y} \tag{23}$$

where ε_{xy} etc. are the off-diagonal terms in the dielectric tensor and q_x, q_y and q_z denote the x, y and z-displacements of the two sub-lattices relative to each other. p_{11} and p_{12} are two of the conventional photoelastic constants of the host crystal, while p_{44}^0 is related to the third photoelastic constant p_{44} in a manner to be discussed later. For the present, we may assume that p_{44}^0 is equal to the measured constant p_{44}.

In Eqs. (22) the vector $\boldsymbol{r}_n$ denotes the position of the nth atom with respect to the impurity, and F is the electric field exerted by the impurity with charge e. The suffic $\boldsymbol{r}_n$ after the square brackets indicates that for F and its derivatives the values at $\boldsymbol{r}_n$ are to be taken. The field $\boldsymbol{F}$ at $\boldsymbol{r}_n$ is given by

$$\boldsymbol{F}(\boldsymbol{r}_n) = e\,\boldsymbol{r}_n/\varepsilon\, r_n^3. \tag{24}$$

From this relation the various components of $\boldsymbol{F}$ and their derivatives are easily obtained.

In Eqs. (22b) and (22c), the upper or lower sign in front of β is to be used according to whether the nth atom and the impurity are on the same or on opposite sub-lattices.

As the expressions (22) are based on a macroscopic calculation they are exact only if the nth atom is far from the impurity atom, in which case the η_n are of course very small anyway. Nonetheless, as is usual for dielectric interactions, we may assume that the expressions are approximately valid even for atoms near the impurity. In particular, if all the charge is on the impurity atom itself, we may probably assume that Eqs. (22) are approximately valid from second neighbours outward but they must be invalid for nearest neighbours of the impurity. Nearest neighbours share their valency electrons with the impurity, hence their dielectric interaction can certainly not be described on a macroscopic basis, and in addition they are also bound to interact strongly through non-classical short-range effects.

If the foreign atom shares some of its charge with its neighbours, in a cubic structure the centre of the total charge is still at the centre of the impurity atom. In that case Eq. (23) and hence Eqs. (22) are still valid at sufficient distances, but we can then not expect Eqs. (22) to be valid for second neighbours. For the time being, however, we shall continue to consider the limiting case when all the static charge is on the foreign atom itself.

Eq. (24) shows that $F(\boldsymbol{r}_n)$ is proportional to $1/r_n^2$, while its derivatives to $1/r_n^3$. It thus follows that the photoelastic terms in Eqs. (22) fall off more rapidly with distance from the foreign atom than the Raman terms. In fact, it turns out that in the $\eta^2_{n\xi;x}$ from second neighbours outward the photoelastic terms are negligible compared with the Raman terms, and Eqs. (22) then give

$$\eta^2_{n\,x;\,x} \simeq 0 \tag{25a}$$

$$\eta^2_{n\,y;\,x} = \left(\frac{\varepsilon^2 \beta}{2\pi N}\right)^2 (F_z^2)_{\boldsymbol{r}_n} \tag{25b}$$

$$\eta^2_{n\,z;\,x} = \left(\frac{\varepsilon^2 \beta}{2\pi N}\right)^2 (F_y^2)_{\boldsymbol{r}_n}\,. \tag{25c}$$

Table 3 refers to a silicon crystal containing one B^- and one P^+ impurity, and it gives estimates for the absorption by the vibration of the impurity atoms, by their nearest neighbours and by further neighbours. The absorption of the foreign atoms was taken from the measured absorption of the local mode of boron [24, 27], and it was assumed that the B^- and P^+ impurities have the same apparent charges, so that the difference in their absorption is only due to the different atomic masses. This is not a reliable approximation, but because of its much larger mass the absorption by the phosphorous is in any case much smaller than the absorption by the boron and hence the error in the approximation is not very important.

The figure for the absorption of the nearest neighbours was obtained by calculating their contribution to Eq. (20) from Eqs. (22). For the reasons given, these figures are quite unreliable and they have therefore been put in brackets, nonetheless they give an idea of the magnitude one may expect. The third row in the table, which gives the absorption by all neighbours beyond the first, was obtained by inserting Eq. (24) in (21) and summing over the appropriate terms.

Table 3 shows that the absorption by the atoms beyond first neighbours is remarkably large. All the figures in the table are taken from the paper by LEIGH and SZIGETI [24], who for β used the value $|\beta| = 0.2/r_1$ (r_1 = nearest neighbour distance). This value was taken from a measurement by RUSSEL [30], done inside the dispersion region, while the value actually needed is the one at frequencies below the dispersion. Since then RALSTON and CHANG [31] measured the Raman scattering intensity for silicon below the dispersion, and from their data we calculate $|\beta| \simeq 0.28/r_1$ (the measurements do not give the sign of β). With this new value, the absorption of atoms beyond first neighbours would be even larger than according to Table 3. We have not adjusted the figures in the third row to this new value of β. In contrast, the figures for the absorption by the foreign atoms are somewhat too high in Table 3, as they did not take into account the fact that the neighbouring silicon atoms also make a contribution to the absorption by the local mode of the boron.

For P^+ impurity, whose mass is similar to the host atoms, the table shows that even with the old value of β the combined absorption of the host atoms beyond first neighbours is larger than the absorption by the charged foreign atom itself. With the new value of β this effect would be still more pronounced. Since the

Table 3. Estimated absorption due to the vibrations of the various atoms, in the case of one B^- and one P^+ impurity atom in silicon, if all the actual charge is on the foreign atoms

To get the absorption, each number in the table has to be multiplied by $2\pi^2 e^2/p c v m_{\mathrm{Si}}$ ($v =$ volume per foreign atom of either kind; m_{Si}, m_{B} and m_{P} denote the atomic masses).

	B^-	P^+
Foreign atom	$2.6 = \left(\frac{m_{\mathrm{Si}}\eta_0^2}{m_{\mathrm{B}} e^2}\right)$	$0.9 = \left(\frac{m_{\mathrm{Si}}\eta_0^2}{m_{\mathrm{P}} e^2}\right)$
Four nearest neighbours	(1.0)	(1.0)
All further neighbours	1.4	1.4
Total for one B^- and one P^+	8.3	

absorption by the host atoms beyond first neighbours was calculated purely from macroscopic physics, the results demonstrate that even on the basis of macroscopic physics we find that the vibrations of totally uncharged atoms can absorb, and their combined absorption can be larger than the absorption by the charged atom which produces the effect.

In view of what has been said, the figures in Table 3 are far from accurate because of the uncertainties both in the theory and in the experiments. Nonetheless, it seems safe to conclude that for an equal amount of B^- and P^+ impurities in silicon the absorption by the vibrations of the silicon atoms is at least as strong as the absorption by the two kinds of impurity atoms taken together. This conclusion is borne out by the measured infrared spectra, as can be seen by the following argument.

As is well known, since the boron atom is considerably lighter than the silicon atom the boron impurity in silicon exhibits a "local mode" of vibration whose frequency is higher than all the lattice frequencies of silicon. The local mode consists essentially of the vibrations of the B^- while the silicon atoms are approximately stationary. On the other hand, in the "lattice modes" or "band modes" of silicon, the displacement of the B^- relatively to its neighbours is rather small. In contrast, since the atomic masses of phosphorous and silicon are very similar, all the vibrations of the P^+ impurity take place within the band modes of silicon.

Dawber and Elliott [32] derived relations from which one can calculate the ratio of the vibrational amplitude of B^- in the local mode to its integrated amplitude in the band modes. From this, and assuming that only the vibrations of the two impurities absorb, for B^- and P^+ impurities in silicon they derived a formula which predicts the ratio of the absorption by the local mode to the integrated absorption by the band modes. In fact, Angress *et al.* [27, 28] found that the percentage absorbed by the band modes is about 2.5 times larger than predicted by Dawber and Elliott, which can be only explained by the fact that the vibrations of the host atoms, taken together, absorb about 1.5 times as much as the foreign atoms. The experimental spectra of B^- and As^+ impurities in silicon lead to a similar result [28]. The experiments are thus in agreement with the prediction that the vibrational absorption by all the host atoms taken together is larger than the vibrational absorption by the charged foreign atoms.

7.4. Polarization Induced by the Displacement of a Spherical Region Around a Charged Foreign Atom

We again consider an unpolar crystal containing one charged impurity atom. Since the two alternative definitions of the η given by Eqs. (18) and (19) respectively, are equivalent, if we use the definition (18) the expressions obtained for the η in the last section are valid independently of the shape of the crystal.

Eq. (21) shows that the integrated absorption depends on the sum of the squares of the apparent charges. Therefore, in view of the results of the last section, the Raman terms in the absorption by the nth atom are proportional to $1/r_n^4$ and the photoelastic terms fall off even more rapidly with r_n. Hence only atoms in the neighbourhood of the foreign atom make a significant contribution to the absorption.

The contributions of distant atoms can however not be neglected when linear sums over the η_n are considered. In particular, in this section we shall be interested in sums over the diagonal terms $\eta_{nx;x}$. Eqs. (22a) and (23) show that $\eta_{nx;x}$ falls off as $1/r_n^3$. Since the number of atoms within distance r_n from the impurity is proportional to r_n^3, it follows that in the sum $\Sigma_n \eta_{nx;x}$ the contributions of quite distant atoms may be important.

Using the results of the last section, it is not difficult to prove [25] that

$$\sum_n^{(r)} \eta_{nx;x} = 0 \tag{26}$$

where the sum goes over all atoms at a given distance $|r|$ from the impurity. The values of the individual $\eta_{nx;x}$ are in general of course not zero, even though they may be small. Eq. (26) is exact as long as r is small compared with the dimensions of the crystal but large enough for Eq. (22) to be valid. Further, if A denotes a spherical region with the foreign atom at its centre, LEIGH and SZIGETI [25] have proved that

$$\sum_n^{(A)} \eta_{nx;x} = e\,S \tag{27}$$

where

$$S = 1 + \tfrac{1}{3}\,\varepsilon\,(p_{11} + 2p_{44}^0).$$

As before, ε, p_{11} and p_{44}^0 are the values for the pure crystal. In Eq. (27) the sum goes over all the atoms inside the spherical region A, including the foreign atom itself. e denotes the charge of the foreign atom.

Eq. (27) again is exact as long as A is small compared to the size of the crystal but large enough for Eqs. (22) to be balid for the atoms *outside* A. Within these limits the size of A is irrelevant, as can also be seen from Eq. (26).

As remarked earlier, if the foreign atom shares some of its charge with its neighbours Eqs. (22) and (23) still remain correct at sufficiently large distances from the impurity, and it follows therefore that the validity of Eqs. (26) and (27) is not affected either.

Before discussing the consequences of these relations we have to explain the meaning p_{44}^0. The measured photoelastic constant p_{44} is defined by

$$p_{44} = -\frac{1}{\varepsilon^2}\,\frac{d\,\varepsilon_{xy}}{d\,S_{xy}}$$

Table 4

	S'	$\frac{\varepsilon+2}{3}$
Diamond	-0.13	2.6
Silicon	-0.24	4.7
Germanium	-0.13	6.0

where S_{xy} denotes an xy-strain. In the diamond lattice an xy-strain induces a displacement of the two sub-lattices relative to each other in the z-direction. Denoting this "optical" lattice displacement by q_z, the last equation can be written

$$p_{44} = -\frac{1}{\varepsilon^2}\left(\frac{\partial \varepsilon_{xy}}{\partial S_{xy}}\right)_{q_z} - \frac{1}{\varepsilon^2}\left(\frac{\partial \varepsilon_{xy}}{\partial q_z}\right)_{S_{xy}} \frac{d q_z}{d S_{xy}} \tag{28}$$

p_{44}^0 is defined as the photoelastic constant for a pure strain, without the optical displacement. That is

$$p_{44}^0 = -\frac{1}{\varepsilon^2}\left(\frac{\partial \varepsilon_{xy}}{\partial S_{xy}}\right)_{q_z}. \tag{29}$$

Let S' denote the value which S would have if p_{44}^0 were equal to the measured constant p_{44}. Table 4 contains values of S' for unpolar crystals taken from a critical review of the measured data by LEIGH and SZIGETI [26]. These differ somewhat from the values used in LEIGH and SZIGETI's earlier paper [25] but are believed to be based on more reliable measurements. As we shall see, the most important fact is that although the measured values of the photoelastic constants exhibit a large experimental scatter, there is no doubt that S' is much less than unity.

To get S itself, i.e. to correct for the difference between p_{44}^0 and p_{44}, we need to estimate the second term is Eq. (28). That term involves the quantity $\partial \varepsilon_{xy}/\varepsilon^2 \partial q_z$, which according to Eq. (23) is equal to the Raman constant β. As already mentioned, although the magnitude of β has been measured for diamond [33, 34] and silicon [31] so far it has not been possible to obtain its sign. In addition, the magnitude of dq_z/dS_{xy} is subject to an appreciable experimental uncertainty. However, in the arguments which follow we shall only use the fact that for diamond and silicon S is substantially less than $+1$, and this seems certain.

For diamond it was shown [26] that the experimental values give $|S-S'| \simeq 0.06$, and since $S \simeq -0.13$, even allowing for the experimental uncertainties it is clear that S is much less than unity.

Concerning silicon, as we said earlier the recent measurement of the Raman scattering intensity by RALSTON and CHANG [31] gives $|\beta| \simeq 0.28 r_1$ (r_1=nearest neighbour distance). Using the probable value of dq_z/dS_{xy} [26] we get $|S-S'| \simeq 0.6$, so that the approximate value of S is either -0.85 or $+0.35$, depending on the sign of β. Notwithstanding the uncertainties involved, it thus remains practically certain that also in the case of silicon S is substantially less than $+1$.

Eq. (27) is of interest both in its macroscopic and its microscopic aspects. We shall discuss the macroscopic aspects first, and in order to fix our ideas we shall assume that the specimen considered has the shape of a slab as described in connection with Eq. (19), and that A is a macroscopic spherical region, i.e. it contains very many atoms. The charged foreign atom is of course at the centre of A. The x-direction is parallel to four of the long edges of the slab.

We now consider a situation where A is displaced rigidly in the x-direction by a small amount X with respect to the rest of the crystal. Since every atom inside the region A is displaced by X, Eqs. (19) and (27) show that $M(X)$ the dipole moment induced by the displacement is given by

$$M(X) = S e X. \tag{30}$$

This is clearly true if A is displaced rigidly and all the atoms outside A remain at rest. It follows from Eq. (26) that Eq. (30) remains correct if A is displaced rigidly by the amount X while outside A the displacements decrease from X to 0 gradually with $|r|$ as long as the displacements are a unique function of $|r|$, i.e. of the distance from the foreign atom.

As is well known, if the Clausius-Mossotti theorem were valid then the displacement of the macroscopic spherical region A should create a dipole moment $M(X) = \frac{1}{3}(\varepsilon+2) e X$, i.e. it follows from Eq. (30) that S would be equal to $(\varepsilon+2)/3$. The values of $(\varepsilon+2)/3$ for diamond, silicon and germanium are shown in Table 4. As we said earlier, although the actual value of S may differ appreciably from S, it is nevertheless certain for diamond and practically certain for silicon that S is considerably less than $+1$, and hence it differs from the Clausius-Mossotti value by a large factor. In germanium the value of $S - S'$, and hence of S, is not known.

We recall that Eq. (27) is free of assumptions, hence it is valid independently of whether or not the Clausius-Mossotti theorem is correct. It can be shown that if the latter theorem were correct then the photoelastic constants would satisfy the relation

$$p_{12} + 2 p_{44}^{0} = (\varepsilon - 1)/\varepsilon$$

so that S would be equal to $(\varepsilon+2)/3$.

This complete failure of the Clausius-Mossotti theorem for the displacement considered is of considerable interest, as it used to be thought that the theorem is valid for macroscopic displacements. It is also worth noting that if the effective field were the Drude field then for the displacement considered we would have $M(X) = eX$, i.e. S would be unity. In diamond and silicon the actual value of S and hence of $M(X)$ is thus much less than would be the case for either the Clausius-Mossotti or the Drude field. This means that even for the macroscopic displacement considered, the concept of a classical effective field fails completely; this failure is due to the non-electrostatic short-range effects which act at the internal surface between the region A and its surroundings and which change the electronic polarizability at the internal surface, thus changing the dipole moment induced by the electrostatic field of the foreign atom. It is rather interesting that these short-range effects can take place at a macroscopic distance from the charged foreign atom.

We also note that according to the experimental data S' is negative, and it is quite possible that S itself is also negative. There is of course no basic reason why this should not be so. If the whole crystal were displaced rigidly by the amount X the dipole moment would of course be eX. But because the region A is displaced relative to the rest of the crystal an additional electronic dipole moment results due to the cause just discussed, and the fact that $S < +1$ means that this electronic dipole moment has opposite sign to eX. If S is negative that merely implies that the electronic dipole moment is not only opposite to eX but in addition its magnitude is larger than eX.

7.5. *Polarization by Short-Range Effects*

The apparent charges of atoms near an impurity are second-rank tensors, because the presence of the foreign atom removes the normal symmetry of the diamond lattice. But the foreign atom itself is at the centre of tetrahedral symmetry so that its apparent charge, which we shall denote by η_0, is a scalar.

To start with we shall consider the same limiting case as in the previous sections, namely the case when all the static charge e is on the foreign atom.

It follows from crystal symmetry that the $\eta_{nx;x}$ for the four nearest neighbours are equal. We denote this quantity by $\eta_{1x;x}$. Moreover, if the macroscopic expressions (22) were valid for the nearest neighbours, then $\eta_{1x;x}$ would be zero, although $\eta_{1y;x}$ and $\eta_{1z;x}$ would have fairly large values. However, the macroscopic expression for $\eta_{1x;x}$ vanishes only bacause those derivatives of the electrostatic field F which occur in Eq. (22a) happen to vanish at the positions of the nearest neighbours, and it is important to note that there is no symmetry reason at all why any short-range contributions to $\eta_{1x;x}$, which of course are not included in (22a), should vanish.

We now consider the microscopic implications of Eq. (27). If the impurity interacted even with its nearest neighbours according to the macroscopic relations (22), then it would follow from Eq. (26) that the contributions of all the host atoms to the left-hand-side of Eq. (27) would cancel. That equation would then reduce to

$$\eta_0 = S e. \tag{31}$$

However, as we said before, we can obviously not expect that the macroscopic relations should be valid for nearest neighbours, but they probably represent a good approximation from second neighbours outward as long as all the static charge e is on the foreign atom. The contributions of the nearest neighbours to Eq. (27) then do not vanish, but those of all the other host atoms within A cancel. As a result Eq. (27) does not reduce to (31) but to

$$\eta_0 + 4\eta_{1x;x} = S e. \tag{32}$$

Although S is determined to an appreciable extent by short-range effects, the expression for S as given after Eq. (27) was obtained from macroscopic considerations and contains only macroscopic quantities. Moreover, S is a property of the host crystal, and for a given host crystal Se is therefore uniquely determined by the impurity charge e. In contrast, the difference between η_0/e and S, which is expressed by Eq. (32), is due to short-range interactions between the impurity and its neighbours, such as bond-deformation or charge transfer accompanying a displacement, and also to the fact that the electrostatic interaction between the impurity and its nearest neighbours is not expressed correctly by the macroscopic relations. These short-range effects are obviously not determined uniquely by the charge e, even their sign may be different for different impurities and the difference $(\eta_0/e) - S$ may therefore have very different values for different impurity atoms.

Without any short range effects η_0 would be equal to Se and S would be equal to $(\varepsilon+2)/3$ (or possibly to 1 if one wants to argue that in semiconductors like silicon the Drude field is appropriate). In fact, S is much less than $(\varepsilon+2)/3$ (and also substantially less than 1) because of short-range effects at macroscopic distances from the impurity. On the other hand, the difference between η_0 and Se is due to the short-range effects mentioned in the last paragraph, which act in the immediate neighbourhood of the impurity. These latter type of short-range effects, which are not proportional to e, must be expected to be at least as important as those which determine S. These matters indicate that η_0 is determined more by the various short-range effects than by the static charge e. It may of course happen that the various short-range effects partially cancel, and in particular it may occur that η_0 is nearer to e than Se is, but this does not reduce the importance of the short-range effects.

It is worth noting that just as there is no basic reason why S should not be negative, similarly it is also quite possible that in some cases η_0 has opposite sign to e.

So far we have not made use of the measured infrared absorption. However, our conclusion concerning the importance of the short-range effects could be tested if S and η_0 were known from experiment, because then $\eta_0 - Se$ would be determined and $\eta_{1x;x}$ could also be obtained from Eq. (32). When LEIGH and SZIGETI derived [9] that equation they thought that $S \simeq S'$. Since the experimental local mode absorption of B^- in silicon [27], after making an approximate allowance [9] for the small contribution of the silicon atoms to this mode, gives for B^- $|\eta_0| \simeq 0.8$, and since S' is much less than 1, it was concluded that $|Se|$ is much smaller than $|\eta_0|$. From this it followed that η_0 is determined overwhelmingly by the short-range effects not included in Se and that therefore η_0 is practically independent of the static charge e.

As explained in Section (7.4), for diamond it is true that $S - S'$ is small, and hence S is small in comparison to unity. Hence for an impurity in diamond the apparent charge η_0 is indeed determined principally by those short-range effects which are not included in Se.

For silicon, however, $S - S'$ is not small and, as we said in Section 7.4, the probable value of S is either $+0.35$ or -0.85, according to the sign of the Raman constant β which is not known. Thus $|Se|$ is not small compared to the η_0 of B^- impurity in silicon, and η_0 is thus determined partly by Se and partly by those short-range effects which are not included in Se. Since S itself is determined to a considerable extent by short-range effects, the earlier conclusion that η_0 is determined more by the short-range effects than by the static charge agrees with the experimental data of B^- in silicon. Because of the experimental uncertainties attaching to the values of both S and of η_0, it is not quite certain that for B^- in silicon Se and η_0 are substantially different, but because of the arguments presented earlier it would be very surprising if they were not.

We have only considered the limiting case when all the static charge is on the foreign atom, but the conslusion that η_0 is determined to a large extent by short-range effects remains valid *a fortiori* if part of the static charge spreads over to the neighbours of the impurity. In the first place, if the charge spreads then the importance of the static charge on the impurity itself is obviously reduced; moreover, what is probably more important, the spreading of the charge means that the electron wave functions on the neighbours are profoundly altered by the presence of the impurity and hence the short-range interactions of the impurity extend further than they would do if the charge were not shared.

The conclusion that there is no clear relation between the static charge of an impurity and its vibrational absorption finds its most striking experimental confirmation in the fact that the local mode absorption for the *uncharged* C impurity in silicon [35] is about five times as large as for the charged B^- impurity is silicon [27]. In this connection, we also recall our earlier conclusion that uncharged host atoms in the neighbourhood of an impurity can exhibit appreciable absorption, and that this is also in agreement with experiment.

The analysis discussed in Sections 7.3–7.5 cannot be applied to pure crystals, as it utilizes the electrostatic field which a charged impurity exerts. Nonetheless, we can expect that the main features of the results are valid also for the pure covalent crystals, since the type of bond in covalent crystals like GaAs, AlP or SiC is not substantially different to the B–Si or P–Si bonds in the impurity case. The results thus imply that not only in the impurity case, but also in the case of the pure covalent crystals the vibrational absorption depends to a large extent on electronic polarization by short-range effects, and therefore strong absorption need not indicate an appreciable static charge or ionic character.

Prior to the work described in Sections 7.1–7.5 a number of physicists, including the present author [36], had already taken the view that strong infrared absorption is possible without a large static charge, or in fact without any static charge at all. Burstein and his collaborators [23, 37] had concluded that in the covalent materials, where the static charge is small, the lattice polarization is due mainly to electronic deformation induced by short-range effects. Their conclusion was based on the fact that covalent materials whose static charge was believed to be small exhibited a large vibrational absorption in the infrared, and also on the assumption that in these materials the effective field which represents long-range interactions is approximately equal to the macroscopic field. The experimental data on the absorption by selenium and tellurium [21, 38, 39], where the static charge is definitely zero, were published after the work described here was finished.

The arguments described in Sections 7.1–7.5 are of interest for several reasons. They show that considerations of a different nature to those used before, and which make use of the photoelastic data but not of the measured infrared absorption, also indicate that for the lattice polarization and vibrational absorption of covalent materials the deformation by short-range effects is more important than the static charge. Moreover, what has not been suggested previously, these considerations indicate that in the case of mainly covalent bonds this conclusion remains true even if the static charge is fairly large. The results of course also contributed to the understanding of the infrared absorption by impurity centres, in particular by demonstrating the importance of the vibrational absorption of host atoms in the neighbourhood of an impurity. Finally, it was shown that at least in some covalent crystals, some long-range dielectric interactions are far outside the range predicted by either the Clausius-Mossotti or the Drude effective field.

Concerning the importance of short-range effects for the lattice polarization and absorption, we recall that even in the case of the alkali halides the contribution of these effects to the effective charge e_S^* amounts to about $0.25\,e_0$. It is thus not really surprising if in the covalent materials they represent the dominating factor.

8. Covalency and Static Charge

It has been seen that for the mainly covalent materials the magnitude of the static charge is not very important in determining the strength of the vibrational absorption. In fact, one can probably go further and conclude that for the covalent materials the static charge is not a proper physical concept and can only be defined in an arbitrary manner.

The basis of this statement can be understood from the properties of the density distribution of the valency electrons of covalent crystals. Fig. 2 on p. 154 shows such a density distribution for GaAs, calculated recently by WALTER and COHEN [8]. The figure shows that the charge density has appreciable values all the way along the line connecting two neighbouring atoms and exhibits a *maximum* between the atoms. It may seem that the mathematical properties of the density map suggest an obvious way in which to draw the boundaries between the atoms, and if we do this then the charge within the boundaries determines by definition the static charge of the atoms. But from the physical point of view such a boundary is completely artificial; it is quite arbitrary to say that the valency electron charge density on one side of this "boundary" belongs to one atom and on the other side to the other atom, when in reality the electrons on both sides of the "boundary" are shared between the atoms. There is clearly no true boundary between the atoms and in fact the atoms are not separate units.

This is very different from the case of the alkali halides, where the charge distribution around each ion is almost spherical and the charge density almost vanishes between the ions. In that case the ions can be considered as separate units and the static charge of each ion has a proper meaning. If an ion is displaced in an alkali halide crystal, in a first approximation it is reasonable to assume that it carries its static charge with it; the deformation of the electron shells represents only a correction, albeit possibly a large one.

In contrast, in the case of the lattice vibrations of a covalent crystal like GaAs, not even in a first approximation can we think of the atoms as being displaced rigidly. A consideration of Fig. 2 shows that such a picture of rigid displacements would lead to complete nonsense, and this again points to the fact that in the covalent crystals the atoms are not separate or well defined units, and we cannot attribute a definite charge separately to each atom.

References

1. Szigeti, B.: Trans. Faraday Soc. **45**, 155 (1949).
2. Huang, K.: Proc. Roy. Soc. (London), Ser. A **208**, 352 (1951).
3. Fröhlich, H., Mott, N. F.: Proc. Roy. Soc. (London), Ser. A **171**, 496, (1939).
4. Lyddane, R. H., Herzfeld, K. F.: Phys. Rev. **54**, 846 (1938).
5. Lyddane, R. H., Sachs, R. G., Teller, E.: Phys. Rev. **59**, 673 (1941).
6. Fröhlich, H.: Theory of Dielectrics, 2nd Ed. Oxford: Clarendon Press 1958.
7. Dawson, B.: Proc. Roy. Soc. (London), Ser. A **298**, 280 (1967).
8. Walter, J. P., Cohen, M. L.: Phys. Rev. Letters **26**, 17 (1971).
9. Leigh, R. S., Szigeti, B.: Proc. Irvine Conf. on Localized Excitations in Solids (Editor R. F. Wallis), p. 159. New York: Plenum Press 1968.
10. Aggarwal, K. G., Szigeti, B.: J. Phys. C **3**, 1097 (1970).
11. Lowndes, R. P., Martin, D. H.: Proc. Roy. Soc. (London), Ser. A **164**, 167 (1969).
12. Lowndes, R. P.: Thesis, Queen Mary College, University of London (1967).
13. Szigeti, B.: Proc. Roy. Soc. (London), Ser. A **204**, 51 (1950).
14. Born, M., and Huang, K.: Dynamical Theory of Crystal Lattices. Oxford: Clarendon Press 1956.
15. Jones, G. O., Martin, D. H., Mawer, P. A., Perry, C. H.: Proc. Roy. Soc. (London), Ser. A **261**, 10 (1961).
16. Cochran, W.: Rept. Progr. Phys. **26**, 1 (1963).
17. Leigh, R. S., Szigeti, B., Tewary, V. K.: Proc. Roy. Soc. (London), Ser. A **320**, 505 (1971).
18. Kleinman, D. A., Spitzer, W. G.: Phys. Rev. **118**, 110 (1960).
19. Hass, M., Henvis, B. W.: J. Phys. Chem. Solids **23**, 1099 (1962).
20. Spitzer, W. G., Kleinman, D. A., Frosch, C. J.: Phys. Rev. **113**, 133 (1959).
21. Lucovsky, G., Keezer, R. C., Burstein, E.: Solid State Comm. **5**, 439 (1967).
22. Slater, J. C.: Quantum Theory of Molecules and Solids II, Chap. 4. New York: McGraw-Hill Book Co. 1961.
23. Burstein, E.: in "Lattice Dynamics, Proceedings of the International Conference held at Copenhagen Aug. 5.—9., 1963". ed. R. F. Wallis. Pergamon Press 1965.
24. Leigh, R. S., Szigeti, B.: Proc. Roy. Soc. (London), Ser. A **301**, 211 (1967).
25. Leigh, R. S., Szigeti, B.: Phys. Rev. Letters **19**, 566 (1967).
26. Leigh, R. S., Szigeti, B.: J. Phys. C: Solid State Phys. **3**, 782 (1970).
27. Angress, J. F., Goodwin, A. R., Smith, S. D.: Proc. Roy. Soc. (London), Ser. A **287**, 64 (1965).
28. Angress, J. F., Goodwin, A. R., Smith, S. D.: Proc. Roy. Soc. (London), Ser. A **308**, 111 (1968).
29. Szigeti, B.: Proc. Roy. Soc. (London), Ser. A **258**, 377 (1960).
30. Russell, J. P.: Appl. Phys. Letters **6**, 223 (1965).
31. Ralston, J. M., Chang, R. K.: Phys. Rev. B **2**, 1858 (1970).
32. Dawber, P. G., Elliot, R. J.: Proc. Phys. Soc. (London) **81**, 453 (1963).
33. Anastassakis, E., Iwasa, S., Burstein, E.: Phys. Rev. Letters **17**, 1051 (1966).
34. Angress, J. F., Cooke, C., Maiden, A. J.: J. Phys. C **1**, 1769 (1968).
35. Newman, R. C., Willis, J. B.: J. Phys. Chem. Solids **26**, 373 (1965).
36. Szigeti, B.: J. Phys. Chem. Solids **24**, 225 (1963).
37. Brodsky, M., Burstein, E.: Bull. Am. Phys. Soc. **7** (II), 214 (1962).
38. Geick, R., Schröder, M., Stuke, J.: Phys. Stat. Sol. **24**, 99 (1967).
39. Grosse, P., Lutz, M., Richter, W.: Solid State Comm. **5**, 99 (1967).

Electrically Induced Stresses in Dielectric Fluids

B. K. P. Scaife

Engineering School, Trinity College, Dublin/Ireland

1. Introduction

The macroscopic theory of electrically induced stresses can fairly be said to be complete [1–4]. Some work has been done on the microscopic theory of such phenomena [5]. Despite the well established nature of the basic theory of electrostrictive phenomena one still finds disagreement in the literature [6], as to how the basic theory should be applied. Furthermore it is not unusual to find considerable confusion as to precisely what is predicted by theory in any particular case.

In an attempt to clarify the essential features of electrostriction and in an attempt to refute a criticism of the basic theory [6], we shall consider in some detail the stresses induced in a spheroidal non-conducting liquid droplet polarized by an axial electric field. We shall also discuss the deformation of a bubble in a non-conducting dielectric fluid.

The hydrostatic pressure, p_E, induced in a spheroidal droplet by an axial external field, E_0, is easily calculated by thermodynamics with the result that [5],

$$\mathsf{p}_E = -\frac{1}{2} E_0^2 \left(\frac{\partial \gamma_3}{\partial V}\right)_T, \tag{1.1}$$

where γ_3 is the polarizability, along the axis, of the spheroid of volume V. Whether p_E is positive or negative will depend upon two factors, namely, the precise shape of the spheroid and the density dependence of the relative permittivity of the fluid. To what degree the fluid is compressible is of no consequence.

2. Deformation of a Spheroidal Droplet

We shall address ourselves here to the investigation of the surface stresses induced in a spheroidal drop of non-conducting liquid when placed in an external electric field $\vec{E}_0 = E_0 \vec{i}_3$ parallel to the axis of symmetry, Fig. 1.

The uniform potential gradient within the fluid, $-\vec{E}_i$, is related to $\vec{E}_0$ by the expression

$$\vec{E}_i = \vec{E}_0 / [1 + \lambda_3(\varepsilon - 1)], \tag{2.1}$$

in which ε is the relative permittivity of the fluid and λ_3 is the depolarizing factor for the axis of symmetry of the spheroid. For a prolate spheroid, of eccentricity e [3b],

$$\lambda_3 = [(1 - e^2)/2e^3]\{\ln[(1+e)/(1-e)] - 2e\}.$$

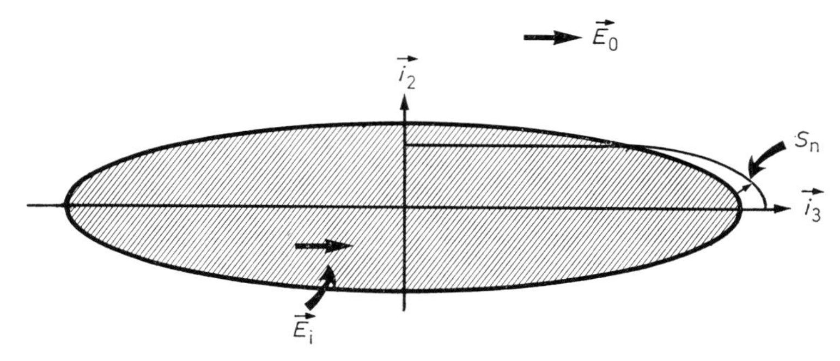

Fig. 1. Shewing the distribution of normal stress, S_n, at the surface of a spheroidal droplet of non-conducting dielectric fluid polarized by a uniform external field, $\vec{E}_0$, parallel to the axis

For a sphere $e=0$ and $\lambda_3=(1/3)$; as $e\to 1$ the spheroid degenerates into a long, needle-shaped drop for which $\lambda_3\to 0$.

Classical electrostatic theory [1–4], indicates that the normal, outward component of surface stress, S_n, satisfies the equation

$$S_n=\tfrac{1}{2}\varepsilon_0(\varepsilon-1)\left[\varepsilon E_{in}^2+E_{it}^2\right]-\tfrac{1}{2}\varepsilon_0 E_i^2\varrho(\partial\varepsilon/\partial\varrho)_T, \tag{2.2}$$

in which ε_0 is the absolute permittivity of free space, $E_{in}=E_i\cos\psi$, $E_{it}=E_i\sin\psi$, ψ is the angle between the outward normal and i_3 and ϱ is the density of the fluid.

For purposes of exposition we shall assume that the dielectric fluid is non-polar and that the dependence of ε on ϱ is determined by the Clausius-Mossotti formula. Consequently [7a],

$$\varrho(\partial\varepsilon/\partial\varrho)_T=(\varepsilon-1)(\varepsilon+2)/3,$$

which means that

$$S_n=\tfrac{1}{2}\varepsilon_0(\varepsilon-1)^2E_i^2\left[\cos^2\psi-\tfrac{1}{3}\right]. \tag{2.3}$$

An indication as to how S_n varies over the surface is shewn in Fig. 1. Notice that S_n is proportional to $(\varepsilon-1)^2$ and that as $\psi\to 0$ it is positive, whereas when $\psi\to(\pi/2)$ S_n is negative. Eq. (2.3) clearly indicates that a spheroidal drop will tend to become less oblate, or more prolate, depending upon the original shape.

The term $-\tfrac{1}{2}\varepsilon_0E_i^2\varrho(\partial\varepsilon/\partial\varrho)_T$ plays an essential rôle in determining the sign of S_n. Indeed, without it S_n would be outwards at all points of the surface.

It is instructive to obtain an explicit expression for the hydrostatic pressure, p_E, induced in the spheroid. To do so we make use of Eq. (1.1) which means that we require an expression for γ_3. From Eq. (2.1) it follows that

$$\gamma_3=\varepsilon_0(\varepsilon-1)V/[1+\lambda_3(\varepsilon-1)], \tag{2.4}$$

therefore

$$\mathsf{p}_E=-\tfrac{1}{2}\varepsilon_0(\varepsilon-1)^2E_i^2\left[\lambda_3-\tfrac{1}{3}\right]. \tag{2.5}$$

As has been shewn before [5], p_E vanishes for a sphere ($\lambda_3=\tfrac{1}{3}$). Thus the effect of an axial electric field is to induce a hydrostatic compression ($\mathsf{p}_E<0$) in an oblate spheroid ($\lambda_3>\tfrac{1}{3}$) and a tension ($\mathsf{p}_E>0$) in a prolate spheroid ($\lambda_3<\tfrac{1}{3}$).

For a dipolar liquid which obeys ONSAGER's equation [7b], and for which the relative permittivity at very high frequencies is unity ($n^2 = 1$),

$$\varrho(\partial\varepsilon/\partial\varrho)_T = \varepsilon(\varepsilon - 1)(2\varepsilon + 1)/(2\varepsilon^2 + 1), \tag{2.6}$$

Eqs. (2.3) and (2.5) become

$$S_n = \tfrac{1}{2}\varepsilon_0(\varepsilon - 1)^2 E_i^2\{\cos^2\psi - [1/(2\varepsilon^2 + 1)]\}, \tag{2.7}$$

and

$$\mathsf{p}_E = -\tfrac{1}{6}\varepsilon_0(\varepsilon - 1)^2 E_i^2\{\lambda_3 - [1/(2\varepsilon^2 + 1)]\}, \tag{2.8}$$

respectively.

3. Deformation of Spheroidal Bubble

The case of a spheroidal bubble is only slightly more complicated than the previous case. In this instance the normal stress, T_n, directed from the fluid into the bubble is, for a non-polar dielectric, given by the relation

$$T_n = \tfrac{1}{2}\varepsilon_0(\varepsilon - 1)^2 G_i^2\{\cos^2\psi\,[(\varepsilon^2 + 2)/3\varepsilon^2] - \tfrac{1}{3}\} \tag{3.1}$$

in which the electric field, G_i, within the bubble has the value

$$\vec{G}_i = \varepsilon\vec{E}_0/[\varepsilon - \lambda_3(\varepsilon - 1)]$$

where $-\vec{E}_0$ is the potential gradient far away from the bubble. Eq. (3.1) clearly indicates that a spheroidal bubble will tend to become more oblate. The situation is illustrated in Fig. 2. An interesting feature of Eq. (3.1) is the fact that the change in sign of T_n occurs at a much smaller angle than in the previous case. It must, however, be borne in mind that Eq. (3.1) applies only to non-polar dielectric fluids which obey the Clausius-Mossotti equation. Even for a polar liquid it so happens that the general features of the expression for T_n are still valid. In fact the equation corresponding to Eq. (3.1) for a dipolar medium reads

$$T_n = \tfrac{1}{2}\varepsilon_0(\varepsilon - 1)^2 G_i^2\,[\cos^2\psi - \tfrac{1}{3}]\,[3/(2\varepsilon^2 + 1)].$$

4. The Calculation of the Stress Tensor, σ_E

We shall study the electrically induced stresses when a fluid dielectric spheroid, of volume V, is placed in a uniform external field, $\vec{E}_0$, parallel to the axis of the spheroid. Our attention will be confined to that part of the stress tensor which is quadratic in E_0 and therefore our results will be valid only for small values of E_0.

Each of the N molecules within the volume V will carry an electric dipole moment $\vec{m}_j$ ($j = 1, 2, 3, \ldots, N$), which may, or may not, be permanent and which depends upon internal molecular coordinates $\vec{r}_{j1}, \vec{r}_{j2}, \ldots, \vec{r}_{jM}$, such that

$$\vec{m}_j = \sum_{k=1}^{M} e_k\vec{r}_{jk},$$

where e_k is one of the constituent charges within the molecule.

The radius vector $\vec{R}_j$ will specify the position of the centre of mass of molecule j. The right-handed Cartesian components of a vector will be indicated by a Greek suffix which can take on the values 1, 2 or 3 (denoting the x, y and z

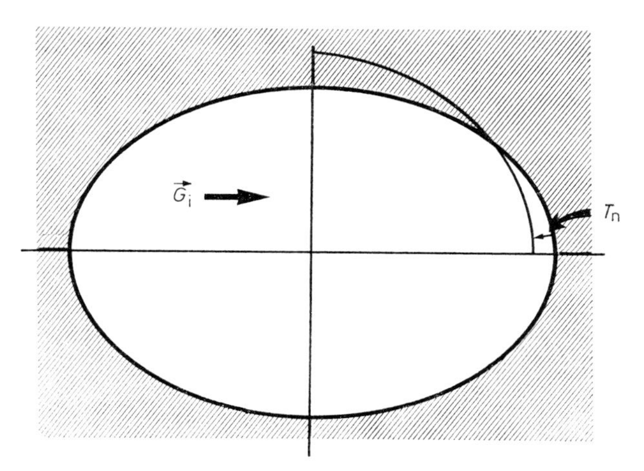

Fig. 2. Shewing the distribution of normal stress, T_n, at the surface of a bubble in a non-conducting dielectric fluid polarized by an axial electric field

directions); a repeated Greek suffix is to be summed, thus $m_{j\alpha} m_{j\alpha} = m_{j1}^2 + m_{j2}^2 + m_{j3}^2 = |\vec{m}_j|^2$. Unit vectors in the x, y and z directions are denoted by $\vec{i}_1$, $\vec{i}_2$ and $\vec{i}_3$ respectively.

The stress tensor, $\boldsymbol{\sigma}$, of a fluid in equilibrium, may be expressed in dyadic form as follows:

$$\boldsymbol{\sigma} = \vec{i}_1 \vec{i}_1 \sigma_{11} + \vec{i}_2 \vec{i}_2 \sigma_{22} + \vec{i}_3 \vec{i}_3 \sigma_{33},$$

or as a matrix

$$\begin{bmatrix} \sigma_{11} & 0 & 0 \\ 0 & \sigma_{22} & 0 \\ 0 & 0 & \sigma_{33} \end{bmatrix}$$

All the off diagonal elements $\sigma_{\alpha\beta}$ $(\alpha \neq \beta)$ must vanish in a fluid in equilibrium. The stress tensor in the absence of an external field, $\boldsymbol{\sigma}_0$, will change to $\boldsymbol{\sigma}_0 + \boldsymbol{\sigma}_E$ following the application of the external field. Our calculation of σ_{E11}, σ_{E22} and σ_{E33} is a simple extension of the method adopted by CHAPMAN and COWLING [8]. For simplicity we shall ignore all intermolecular interaction except dipole-dipole interaction.

For a given set of molecular coordinates, $\vec{r}_{i1}$, $\vec{r}_{i2}$, ..., and $\vec{r}_{k1}$, $\vec{r}_{k2}$, ..., the force, $\vec{F}_{ik}$, exerted on molecule i by molecule k may be expressed as:

$$F_{ik\alpha} = -\frac{\partial}{\partial R_{i\alpha}} \left(-m_{i\beta} E_{i\beta}^{(k)}\right). \tag{4.1}$$

$\vec{E}_i^{(k)}$ is the electric field, acting on molecule i, produced by molecule k. In fact

$$4\pi\varepsilon_0 E_{i\beta}^{(k)} = \{[3(R_{k\beta} - R_{i\beta})(R_{k\gamma} - R_{i\gamma})/R_{ik}^5] - (\delta_{\beta\gamma}/R_{ik}^3)\} m_{k\gamma},$$

in which ε_0 is the absolute permittivity of free space and $\vec{R}_{ik} = \vec{R}_k - \vec{R}_i$.

Eq. (4.1) may also be written

$$F_{ik\alpha} = -\frac{\partial}{\partial R_{i\alpha}} u_{ik},$$

where $u(\vec{r}_{i1}, \vec{r}_{i2}, \ldots; \vec{r}_{k1}, \vec{r}_{k2}, \ldots, \vec{r}_{kM}; \vec{R}_{ik})$, is the energy of the dipole-dipole interaction between molecules i and k.

Let us choose the centre of the spheroid as the origin of the coordinates $\vec{R}_j$. The contribution to σ_{33} from dipole-dipole forces may be determined as follows. We shall calculate the force in the direction $\vec{i}_3$ (parallel to the spheroid axis and the external field $\vec{E}_0 = \vec{i}_3 E_0$) transmitted across a small elemental surface $\mathrm{d}\vec{S} = \vec{i}_3 \mathrm{d}S$ situated at the centre of the system, Fig. 3. Assuming that the molecular number density $n = N/V$ is uniform throughout the system, we conclude that the number of molecules in the element of volume $\mathrm{d}\vec{R}_i$ will be $n \mathrm{d}\vec{R}_i$. The solid angle subtended by $\vec{i}_3 \mathrm{d}S$ at $\vec{R}_i$ is $\mathrm{d}S \cos\Theta_i / R_i^2$ and, therefore, with $\vec{R}_k$ antiparallel to $\vec{R}_i$, the molecules within the volume element $(\mathrm{d}S \cos\Theta_i/R_i^2)\, R_{ik}^2\, \mathrm{d}R_{ik}$ interact, through the area $\mathrm{d}S$, with the molecules within $\mathrm{d}\vec{R}_i = R_i^2\, \mathrm{d}R_i \sin\Theta_i\, \mathrm{d}\Theta_i\, \mathrm{d}\Phi_i$. Thus between these two volume elements there are

$$n^2 (\mathrm{d}S \cos\Theta_i/R_i^2)\, R_{ik}^2\, \mathrm{d}R_{ik} (R_i^2\, \mathrm{d}R_i \sin\Theta_i\, \mathrm{d}\Theta_i\, \mathrm{d}\Phi_i)$$

pairs of molecules which interact with one another across $\mathrm{d}S$. The average of F_{ik3} over all the molecular coordinates will be denoted by $\overline{F_{ik3}}$ and is a function of R_{ik}. The average force, in the $\vec{i}_3$ direction, transmitted across $\mathrm{d}S$ between the two volume elements is

$$\mathrm{d}S\, n^2\, \mathrm{d}R_i \cos\Theta_i\, \overline{F}_{ik3}(\vec{R}_{ik})\, R_{ik}^2\, \mathrm{d}R_{ik} \sin\Theta_i\, \mathrm{d}\Theta_i\, \mathrm{d}\Phi_i.$$

Clearly R_i can vary between 0 and R_{ik} while R_{ik} remains fixed. Therefore the mean force transmitted across $\vec{i}_3\, \mathrm{d}S$ in the $\vec{i}_3$ direction due to all pairs of molecules $\vec{R}_{ik}$ apart is

$$\mathrm{d}S\, n^2\, R_k \cos\Theta_i\, \overline{F}_{ik3}(\vec{R}_{ik})\, \mathrm{d}\vec{R}_k,$$

with the understanding that $\vec{R}_i$ is now zero. By a similar procedure for $\vec{i}_1\, \mathrm{d}S$ and $\vec{i}_2\, \mathrm{d}S$ and by integrating over *half* the volume of the system, we find that

$$\boldsymbol{\sigma} = \tfrac{1}{2} \int n^2 [\vec{i}_1 \vec{i}_1 R_{k1} \overline{F}_{ik1} + \vec{i}_2 \vec{i}_2 R_{k2} \overline{F}_{ik2} + \vec{i}_3 \vec{i}_3 R_{k3} \overline{F}_{ik3}]\, \mathrm{d}\vec{R}_k.$$

Because u_{ik} is a function of R_{ik},

$$\frac{\partial u_{ik}}{\partial R_{i\alpha}} = -\frac{\partial u_{ik}}{\partial R_{k\alpha}}$$

and so the expression for the stress tensor can be simplified to read

$$\boldsymbol{\sigma} = \frac{1}{2} \int n^2 \left[\vec{i}_1 \vec{i}_1 R_{k1} \overline{\frac{\partial u_{ik}}{\partial R_{k1}}} + \vec{i}_2 \vec{i}_2 \overline{\frac{\partial u_{ik}}{\partial R_{k2}}} + \vec{i}_3 \vec{i}_3 \overline{\frac{\partial u_{ik}}{\partial R_{k3}}}\right] \mathrm{d}\vec{R}_k,$$

in which $\overline{(\partial u_{ik}/\partial R_{k\alpha})}$ is the average value obtained after averaging $\partial u_{ik}/\partial R_{k\alpha}$ over all molecular coordinates. This average will be a function of the external field because*

$$\overline{\left(\frac{\partial u_{ik}}{\partial R_{k\alpha}}\right)} = \frac{\int \frac{\partial u_{ik}}{\partial R_{k\alpha}} \exp\left\{-\beta\left[\left(\sum_{j=1}^{N} u_{0j}\right) + \left(\sum_{m>n}^{N} u_{mn}\right) - \vec{M}\cdot\vec{E}_0\right]\right\} \mathrm{d}\vec{r}_{11}\, \mathrm{d}\vec{r}_{12} \ldots \mathrm{d}\vec{r}_{NM}}{\int \exp\left\{-\beta\left[\left(\sum_{j=1}^{N} u_{0j}\right) + \left(\sum_{m>n}^{N} u_{mn}\right) - \vec{M}\cdot\vec{E}_0\right]\right\} \mathrm{d}\vec{r}_{11}\, \mathrm{d}\vec{r}_{12} \ldots \mathrm{d}\vec{r}_{NM}},$$

* A short-range repulsive potential has been omitted for simplicity.

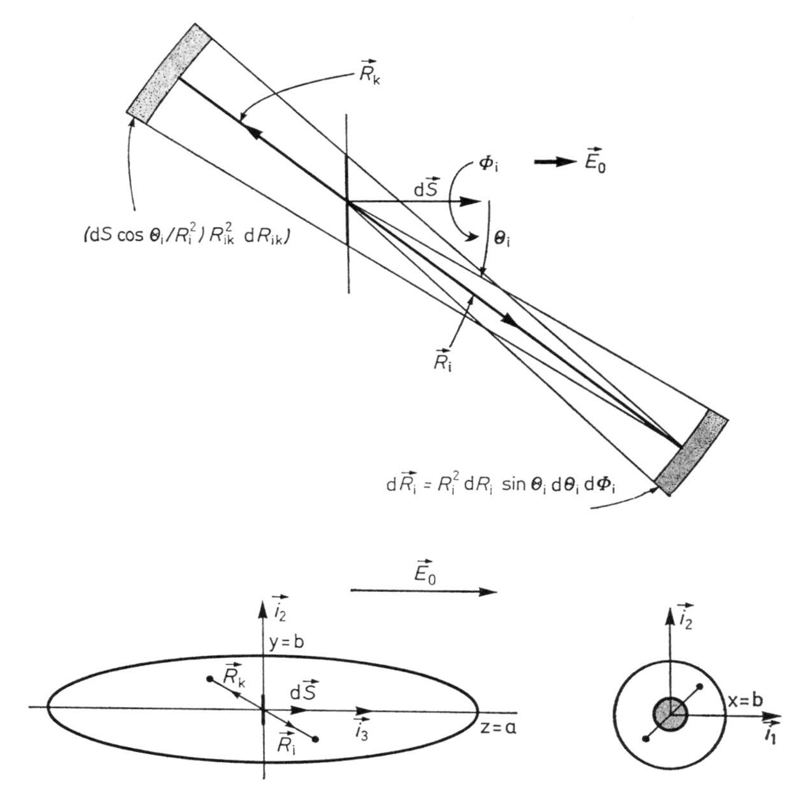

Fig. 3. Illustrating the geometry involved in the calculation of the stress transmitted across the surface element $d\vec{S}$

in which, as usual, $\beta = (\boldsymbol{k}\,T)^{-1}$, $\vec{M} = \sum_{j=1}^{N} \vec{m}_j$ is the total dipole moment of the system and u_{0j} is the internal energy of molecule j.

By retaining only terms which are quadratic in E_0, we conclude that

$$\overline{\left(\frac{\partial u_{ik}}{\partial R_{k\alpha}}\right)} = \left\langle \frac{\partial u_{ik}}{\partial R_{k\alpha}} \right\rangle_0 + \frac{\beta^2 E_0^2}{2} \left\langle \frac{\partial u_{ik}}{\partial R_{k\alpha}} M_3^2 \right\rangle_0 - \frac{\beta^2 E_0^2}{2} \left\langle \frac{\partial u_{ik}}{\partial R_{k\alpha}} \right\rangle_0 \langle M_3^2 \rangle_0,$$

where $\langle \cdots \rangle_0$ indicates an average over all molecular coordinates in the absence of an external field.

The field dependent part, $\boldsymbol{\sigma}_E$, of the stress tensor may thus be written

$$\boldsymbol{\sigma}_E = \frac{1}{4}\, \beta^2 E_0^2 n^2 \int \left\{ \vec{i}_\alpha\, \vec{i}_\alpha \left[\left\langle R_{k\alpha} \frac{\partial u_{ik}}{\partial R_{k\alpha}} M_3^2 \right\rangle_0 - \left\langle R_{k\alpha} \frac{\partial u_{ik}}{\partial R_{k\alpha}} \right\rangle_0 \langle M_3^2 \rangle_0 \right] \right\} d\vec{R}_k.$$

The pressure tensor $\mathbf{p}$ is merely the negative of the stress tensor. Therefore the field dependent part, $\mathbf{p}_E$, of the pressure tensor, $\mathbf{p}$, is simply $-\boldsymbol{\sigma}_E$. The hydrostatic pressure, p_E, induced by the external field is one third of the spur of $\mathbf{p}_E$, that is

$$\mathrm{p}_E = -\tfrac{1}{3}(\sigma_{11} + \sigma_{22} + \sigma_{33})$$

which implies that

$$\mathsf{p}_E = -\frac{1}{12}\beta^2 E_0^2 n^2 \int \left[\left\langle \vec{R}_k \cdot \frac{\partial u_{ik}}{\partial \vec{R}_k} M_3^2 \right\rangle_0 - \left\langle \vec{R}_k \cdot \frac{\partial u_{ik}}{\partial \vec{R}_k} \right\rangle_0 \langle M_3^2 \rangle_0 \right] \mathrm{d}\vec{R}_k. \quad (4.2)$$

Attention has already been drawn to the fact [5], that for a spherical specimen containing a non-polar dielectric (i.e. a dielectric described by the Clausius-Mossotti relation) p_E, the electrically induced hydrostatic pressure, vanishes. This means that, for a sphere, the integral in Eq. (4.2) vanishes for those substances for which $\langle \vec{m}_j \rangle_0$ vanishes. One can most easily verify this by considering an assembly of three-dimensional harmonic oscillators.

The correctness of Eq. (4.2) may be checked by deriving it directly from Eq. (1.1) for p_E, which was based on thermodynamics.

A simple generalization of the Fröhlich [7c], susceptibility formula leads one to the following expression for the polarizability, γ_3, of a spheroid along its axis of symmetry:

$$\gamma_3 = \beta \langle M_3^2 \rangle_0$$

Eq. (1.1) may now be written

$$\mathsf{p}_E = -\tfrac{1}{2} E_0^2 \beta (\partial \langle M_3^2 \rangle_0 / \partial V)_T, \quad (4.3)$$

in which

$$\langle M_3^2 \rangle_0 = \frac{\int M_3^2 \exp\left\{-\beta\left[\left(\sum_{j=1}^{N} u_{0j}\right) + \left(\sum_{m>n}^{N} u_{mn}\right)\right]\right\} \mathrm{d}\vec{r}r_{11} \ldots \mathrm{d}\vec{r}_{NM}}{\int \exp\left\{-\beta\left[\left(\sum_{j=1}^{N} u_{0j}\right) + \left(\sum_{m>u}^{N} u_{mn}\right)\right]\right\} \mathrm{d}\vec{r}r_{11} \ldots \mathrm{d}\vec{r}_{NM}}$$

Differentiation of this expression with respect to V can be carried out by a procedure used by Green [9], in a similar connection. To this end it is convenient to assume that the spheroid has semiaxes (b, b, a) with eccentricity $e = [1-(b/a)^2]^{\frac{1}{2}}$ and volume $V = 4\pi a b^2/3$. We now introduce reduced variables $R'_{i1} = R_{i1}/b$, $R'_{i2} = R_{i2}/b$, $R'_{i3} = R_{i3}/a$. Let f be a function of $\vec{R}_i$, then

$$\frac{\partial f}{\partial V} = \frac{\partial f}{\partial R_{i\alpha}} \frac{\partial R_{i\alpha}}{\partial V}.$$

Now

$$(\partial R_{i1}/\partial V) = R'_{i1}(\partial b/\partial V) = R'_{i1} b/3V = R_{i1}/3V,$$

and, similarly,

$$(\partial R_{i\alpha}/\partial V) = R_{i\alpha}/3V.$$

Therefore

$$(\partial f/\partial V) = (R_{i\alpha}/3V)(\partial f/\partial R_{i\alpha}) = (\vec{R}_i/3V) \cdot (\partial f/\partial \vec{R}_i),$$

and hence

$$(\partial \langle M_3^2 \rangle_0/\partial V) = -(\beta/3V)\langle M_3^2 \sum_{m>n}^{N} [\vec{R}_m \cdot (\partial u_{mn}/\partial \vec{R}_m) + \vec{R}_n \cdot (\partial u_{mn}/\partial \vec{R}_n)] \rangle_0$$

$$+ (\beta/3V)\langle M_3^2 \rangle_0 \langle \sum_{m>n}^{N} [\vec{R}_m \cdot (\partial u_{mn}/\partial \vec{R}_m) + \vec{R}_n \cdot (\partial u_{mn}/\partial \vec{R}_n)] \rangle_0.$$

Recalling that $(\partial u_{mn}/\partial \vec{R}_m) = -(\partial u_{mn}/\partial_n \vec{R})$ and that the summation over the N molecules may be replaced by a volume integral and that

$$\sum_{j=1}^{N} \to \int n \, d\vec{R}$$

we see that,

$$(\partial \langle M_3^2 \rangle_0 / \partial V) = -(\beta n^2/12V) \int [\langle \vec{R}_k \cdot (\partial u_{ik}/\partial \vec{R}_k) M_3^2 \rangle_0 - \langle \vec{R}_k \cdot (\partial u_{ik}/\partial \vec{R}_k) \rangle_0 \langle M_3^2 \rangle_0] \, d\vec{R}_k, \tag{4.4}$$

with the understanding that $\vec{R}_i = 0$. It follows immediately that Eqs. (4.3) and (4.4) are in agreement with Eq. (4.2).

5. Discussion and Conclusions

Eq. (2.3), for the outward, normal, stress, S_n, at the surface of a non-polar, non-conducting dielectric spheroidal droplet placed in an axial field, provides a useful way of shewing how easily our physical intuition can lead us astray. This equation clearly indicates that a prolate droplet will tend to elongate along its symmetry axis. Let us for a moment try to understand how this stretching arises from intermolecular interaction. The effect of the external field is to induce in each and every one of the molecules a dipole moment the average value of which will be the same for each and every molecule and its direction will be parallel to the external field. At first sight we might try to analyse this situation by asking ourselves how a line of axially polarized particles would interact. Obviously such an assembly would tend to contract. Conversely, if the line of particles were polarized at right angles to the axis the system would tend to expand. But this is the opposite to what occurs in a polarized prolate spheroid. We conclude therefore that the simple model just discussed is quite inadequate to analyse the problem.

Happily there is a simple way of reaching an understanding of Eq. (2.3). It depends upon a comparison of the energy of a molecule just outside the surface of the fluid with that of a molecule just inside the surface. Whatever stress may exist at the surface it will be in such a direction as to tend to move molecules from regions where the energy is less negative to regions where it is more negative. If we may assume the validity of the Lorentz expression [7a], for the local field, namely $(\varepsilon + 2) E_i/3$, we can easily compare molecular energies on either side of the fluid surface at the pole and at the equator. At the pole the ratio of the energy outside to that inside is $9\varepsilon^2/(\varepsilon+2)^2$ and at the equator the ratio is $9/(\varepsilon+2)^2$. Since the first expression always exceeds unity and the second is always less than unity, we conclude that there is an outward stress at the pole and an inward stress at the equator, in agreement with Eq. (2.3).

It will be noticed that all the expressions for stress involve the factor $(\varepsilon - 1)^2$. This guarantees that such stresses will be proportional to N^2, indicating that electrostriction can only occur as a result of molecular interaction. We may also remark that the failure to include the term in $\varrho(\partial\varepsilon/\partial\varrho)_T$ in Eq. (2.2), necessarily leads to relations for stress which are proportional to N rather than to N^2.

The calculation of the stress-tensor in terms of dipole-dipole interaction clearly demonstrates the fact that stresses may exist even when the scalar hydrostatic pressure, p_E, vanishes.

From what has been said here the conclusion is inescapable that a spheroidal bubble in a non-conducting fluid tends to become more oblate when subjected to an axial electric field.

References

1. POCKELS, F.: Encyklopädie der mathematischen Wissenschaften, Vol. 5, Part 2, p. 350. Leipzig: Teubner 1906.
2. JEANS, J. H.: The Mathematical Theory of Electricity and Magnetism, 5th Ed., Chap. 6, p. 140. Cambridge: University Press 1925.
3a. STRATTON, J. A.: Electromagnetic Theory, Chap. 2, p. 139. New York: McGraw-Hill Book Company Inc. 1941.
3b. STRATTON, J. A.: Electromagnetic Theory, Chap. 3, p. 214. New York: McGraw-Hill Book Company Inc. 1941.
4. PANOFSKY, W. K. H., PHILLIPS, M.: Classical Electricity and Magnetism, Chap. 6, p. 86. Cambridge, Mass.: Addison-Wesley Publishing Company, Inc. 1955.
5. Scaife, B. K. P.: Proc. Phys. Soc. (London) B **69**, 153 (1956).
6. GARTON, C. G., KRASUCKI, Z.: Proc. Roy. Soc. (London), Ser. A **280**, 211 (1964).
7a. FRÖHLICH, H.: Theory of Dielectrics, 2nd Ed., Chap. 2, p. 26. Oxford: Clarendon Press 1958.
7b. FRÖHLICH, H.: Theory of Dielectrics, 2nd Ed., Cap. 2, p. 35. Oxford: Clarendon Press 1958.
7c. FRÖHLICH, H.: Theory of Dielectrics, 2nd Ed., Chap. 2, p. 36. Oxford: Clarendon Press 1958.
8. CHAPMAN, S., COWLING, T. G.: The Mathematical Theory of Non-Uniform Gases, 2nd Ed., Chap. 16, p. 285. Cambridge: University Press 1952.
9. GREEN, H. S.: The Molecular Theory of Fluids, Chap. 2, p. 52. Amsterdam: North-Holland Publishing Company 1952.

Dielectric Constant of Alkali Halide Crystals Containing OH^- Substitutional Impurities

J. H. SIMPSON

National Research Council, Ottawa/Canada

Abstract

FRÖHLICH's generalized theory of the static dielectric constant is applied to a KCl crystal containing OH^- substitutional impurities. Dipolar interaction between nearest neighbours is considered and local ordering is taken into account using short-range order theory. The results agree well with experiment except in the very low temperature region. The simplicity of the theoretical model and the long relaxation time in this temperature region (which affects the accuracy of the experimental results) probably both contribute to the disagreement at low temperatures.

1. Introduction

One of the many subjects in which H. FRÖHLICH has made important fundamental contributions is the theory of the dielectric constant. An important advance in this field was his generalized theory of the static dielectric constant whose development was virtually complete about twenty-five years ago. A measure of the pioneering quality of this work is obtained from the literature which indicates that appreciable numbers of applications of the theory did not appear for some 15 years.

A promising field of application of dielectric-constant theory is the study of the properties of polar defects in ionic crystals. The system of this type that has been most thoroughly investigated, experimentally and theoretically, consists of OH^- ions substituded for anions in the KCl lattice. Absorption measurements under intense electric fields have indicated [1] that the OH^- dipoles orient themselves along the (100) axes of the crystal and that their directions can be changed by application of an electric field. Dielectric constant and loss measurements [2] on crystals containing varying amounts of OH^- indicate that the dipoles change orientation readily at temperatures of the order of 1 °K for fields of frequency as high as 10000 hertz. As a result of orientation under low field strengths (15 V/cm) the dielectric constant shows a pronounced peak at temperatures ranging from about 0.3 to 3 °K, depending on the density of impurities.

Various theoretical models have been used in attempts to explain this behaviour. Probably the earliest is that of ZERNICK [3] who assumes spontaneous polarization (ferroelectricity) in a model in which the interaction of a pair of dipoles is treated explicitly while the remaining interactions are treated by PIRENNE's modification of ONSAGER's theory. KLEIN [4] assumes that long-range order does not exist and approximates dipolar interaction by the scalar $(1/r^3)$ potential. LAWLESS [5] assumes antiparallel ordering between dipole pairs using the tensorial dipole interaction potential,—an approach, which seems intuitively to be the most satisfying. However his calculations do not agree with the experimental data and he invokes non-equilibrium processes to account, qualitatively, for the differences.

In view of this discrepancy it seems worthwhile to enquire whether a model exhibiting local ordering of a larger number of dipoles would give improved results. For a quasi-random arrangement of impurities of the type discussed here any cooperative effect must be essentially the "ordering of neighbours" described by BETHE [6]. The technique of applying the general theory can be modelled on an approach outlined by FRÖHLICH in § 8 of Ref. [7].

2. A Simple Model for a Crystal containing Dipolar Impurities

As a preliminary step towards a theory of the dielectric constant of dispersed substitutional impurities we consider a model in which classical* impurity dipoles are coordinated locally on a simple cubic lattice of lattice parameter $a = N^{-1/3}$ where N is the quantity of impurities which are submerged in a continuum of dielectric constant ε_0. (For OH^- in KCl, N is typically 10^{19}/cc and $\varepsilon_0 = 4.35$.) This is an idealized adaptation of the actual quasi-random distribution of impurities but, in view of the lack of structure in the curve of short-range order *vs* temperature, the lack of averaging over possible dipole positions may not show as a serious defect in the calculated characteristic. It is assumed that interaction between impurities is purely dipolar and, in this initial approach, interaction between nearest neighbours only is considered. The dipoles are further assumed to have two positions only, 180° apart, and to be subject to a special type of antiferroelectric ordering which arises from the tensorial nature of the dipolar interaction. This ordering, which was shown by SAUER [9] to give minimum energy for a crystal whose linear dimensions are roughly equal, is shown in Fig. 1. The dipoles in a single column are parallel to each other and directed along the line joining them; adjacent columns are antiparallel. BETHE's theory [6] is used in the determination of the effect of ordering on the dielectric constant.

While these approximations are quite severe, they are probably not as drastic as appears at first sight. For example, a model using dipolar forces with nearest neighbour interaction only was used by BERLIN and THOMSEN [10] and shown by LAX [11] to lead to the same equations as one in which long range interaction is included, except that the critical temperature and field are somewhat modified. The two-position model is an abstraction of the actual experimental situation in which the OH^- ions can orient themselves along any one of the six directions of a cubic lattice and may be expected to over-emphasize the ordering effect at low

* Quantum effects discussed by BAUR and SALZMAN [8] and by LAWLESS [5], which are important at the lower end of the temperature range considered here, will not be included in this initial approach.

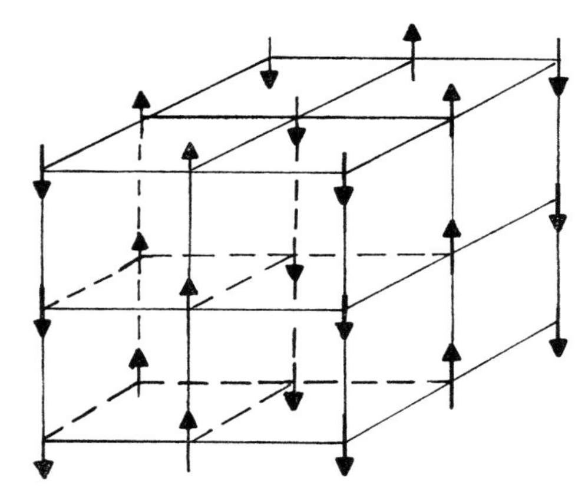

Fig. 1.
Antiferroelectric ordering of a crystal of approximately cubic dimensions

temperatures. For the present problem the six-direction model would be equivalent to treating a system containing six types of molecule and five possible activation energies. Such complications will not be considered but it may be noted in passing that a treatment of a two dimensional square net, consisting of four kinds of atom with non-directional short-range forces [12] leads to a transition temperature that is 20% lower than that of a similar net containing two types of atom. For the present case, in which a transition *per se* does not exist, it seems likely that the effect on the temperature at the maximum of the dielectric constant will be less than that obtained when a sharp transition occurs but no definite prediction is possible.

3. Calculation of the Dielectric Constant

For the calculation of dielectric constant we use the equation

$$(\varepsilon - \varepsilon_0)(2\varepsilon + \varepsilon_0)/\varepsilon = (4\pi N_0 \mu/k T_0)(T_0/T)(\overline{\boldsymbol{m}\boldsymbol{m}^*/\mu^2}) \tag{1}$$

where ε is the dielectric constant of a crystal containing N impurity dipoles per *cc* of dipole moment μ, ε_0 is the dielectric constant of the pure crystal, $\boldsymbol{m}$ is the electric dipole moment of a cell containing q dipoles and $N_0 = N/q$ is the number of such cells per *cc*, $\boldsymbol{m}^*$ is the moment of a spherical region polarized by $\boldsymbol{m}$ and $\overline{\boldsymbol{m}\boldsymbol{m}^*}$ is the average value of $\boldsymbol{m}\boldsymbol{m}^*$, taking into account all possible configurations each weighted according to its probability. The quantity $\overline{\boldsymbol{m}\boldsymbol{m}^*}/\mu^2$ is therefore a pure number. T is the absolute temperature, whose value is T_0 at the peak of the curve of dielectric constant *vs* temperature.

Eq. (1) is derived from Eq. (7.39) of Fröhlich's book [7], on the assumption that the dielectric constant of the pure crystal can be introduced empirically in the same way as n^2 in that equation. Each impurity (OH^-) ion is thus considered to contribute to the induced dielectric constant of the crystal in the same way as an anion (Cl^-) of the crystal and in addition to contribute polarization because of its dipolar nature. Since the number of impurity ions is less than 0.1% of the number of crystal anions the change of crystal dielectric constant ε_0 resulting from this assumption is negligible.

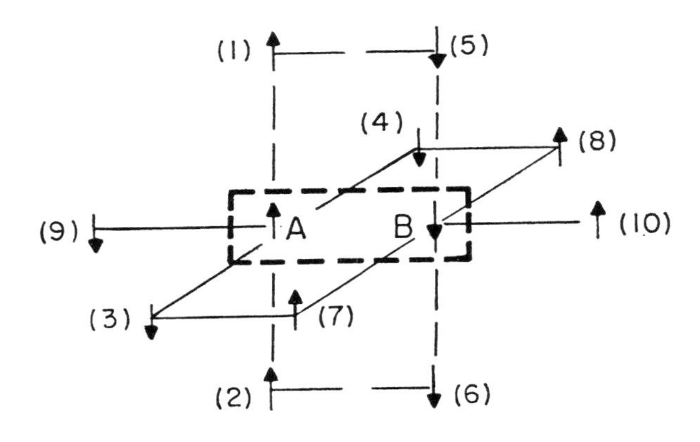

Fig. 2.
Central cell (dashed line) used in dielectric constant calculation and its immediate surroundings (shown perfectly ordered)

The dipole moment μ is the external dipole moment, called μ_e by FRÖHLICH and for the present case is given by Eq. A 2.27 of Ref. [7] with $\varepsilon_1 = \varepsilon_2 = \varepsilon_0$. Thus $\mu = (\varepsilon_0 + 2)\mu_v/3$ where μ_v is the dipole moment of the OH^- ion in vacuum. If Eq. (1) is expressed in terms of the internal moment μ_i it becomes identical with Eq. (B 1.39) of [7] with ε_∞ replaced by ε_0.

In applying Eq. (1) we consider a cell containing two dipoles A and B whose directions in the perfectly ordered state would be opposite, as shown in Fig. 2. We treat the interaction of the cell with its neighbours by BETHE's first approximation to the short range order. However for the present case, in which the interaction is assumed to be dipolar we must consider two types of interaction, depending on whether the line joining two neighbouring dipoles is parallel or perpendicular to the dipole direction. In the former case the interaction energy is $\mp 2\mu^2/a^3$ and in the latter, $\pm \mu^2/a^3$, where the upper sign applies to parallel and the lower sign to anti-parallel dipoles. We now put $V = \mu^2/a^3$ and $x = \exp(V/kT)$ and define the above interactions as type (1) and type (2) respectively. If we call the direction of each dipole in Fig. 2 its "right" (R) direction, we obtain the partition functions (PF's) shown in Table 1.

Table 1. Partition functions for the two types of dipolar interaction

Dipole pair configuration	Partition function	
	Type (1) interaction	Type (2) interaction
RR	$1/x^2$	$1/x$
$RW = WR$	x^2	x
WW	$1/x^2$	$1/x$

If A has z_1 neighbours of type (1) and z_2 of type (2), the PF when n_1 neighbours of type (1) are W, n_2 neighbours of type (2) are W and A is R becomes

$$r(n_1, n_2) = C \binom{z_1}{n_1} x^{2n_1} \cdot x^{-2(z_1 - n_1)} \binom{z_2}{n_2} x^{n_2} \cdot x^{-(z_2 - n_2)} \qquad (2)$$

where C is a normallizing constant. A similar expression for $w(n_1, n_2)$ is obtained when A is W. The total probability that A is R becomes

$$r = C\left[x^{-2z_1}\sum_{n_1=0}^{z_1}\binom{z_1}{n_1}x^{4n_1}\right]\left[x^{-z_2}\sum_{n_2=0}^{z_2}\binom{z_2}{n_2}x^{2n_2}\right] \tag{3}$$

or

$$r = C\,(x^{-2}+x^2)^{z_1}(x^{-1}+x)^{z_2}. \tag{4}$$

The probability that A is W is similarly determined and gives the same expression so that $r = w = \frac{1}{2}$ as expected for a system showing short range order only.

To determine $\overline{\boldsymbol{m}\boldsymbol{m}^*}/\mu^2$ we now follow a procedure analogous to that in Ref. [13] for the case in which the interaction between the cell of Fig. 2 and its immediate surroundings is taken into account. The probabilities and dipole moments for the various configurations of the cell are given in Table 2.

Table 2. Moments and probabilities for the central cell

Cell configuration	Moment $(=m)$	Probability of configuration
$A-R,\ B-R$	0	$p(r, r) \quad = 1/2(1+x^2)$
$A-R,\ B-W$	2μ	$p(r, w) \quad = x^2/2(1+x^2)$
$A-W,\ B-R$	-2μ	$p(w, r) \quad = p(r, w)$
$A-W,\ B-W$	0	$p(w, w) = p(r, r)$

A typical set of probabilities and moments (when $A-R$, $B-W$) for two of the ions contributing to $\boldsymbol{m}^*$ is shown in Table 3.

Table 3. Moments and probabilities for two first shell ions when $A-R$, $B-W$

First shell configuration	Moment (= part of $\boldsymbol{m}^*$)	Probability of configuration
$(1)-R,\ (5)-R$	0	$x^2(1/x)(1/x^2) = (1/x)$
$(1)-R,\ (5)-W$	2μ	$x^2 \cdot x \cdot x^2 = x^5$
$(1)-W,\ (5)-R$	-2μ	$(1/x^2)\,x\,(1/x^2) = 1/x^3$
$(1)-W,\ (5)-W$	0	$(1/x^2)(1/x)\,x^2 = 1/x$

The average moment of the dipoles (1) and (5) is therefore

$$2\mu(x^8-1)/[x^2(1+x^6)+(1+x^2)].$$

When similar calculations are carried out for all of the dipoles of Fig. 2 we obtain the following expression

$$\frac{\overline{\boldsymbol{m}\boldsymbol{m}^*}}{\mu^2} = \frac{2x^2 r}{1+x^2}\left[4+\frac{8(1-x^8)}{(1+x^2)+x^2(1+x^6)}-\frac{8(1-x^4)}{3+x^4}-\frac{4(1-x^2)}{1+x^2}\right] \tag{5}$$

$$= \frac{8x^2}{1+x^2}\left[\frac{x^2}{1+x^2}+\frac{1-x^8}{(1+x^2)+x^2(1+x^6)}-\frac{1-x^4}{3+x^4}\right]. \tag{6}$$

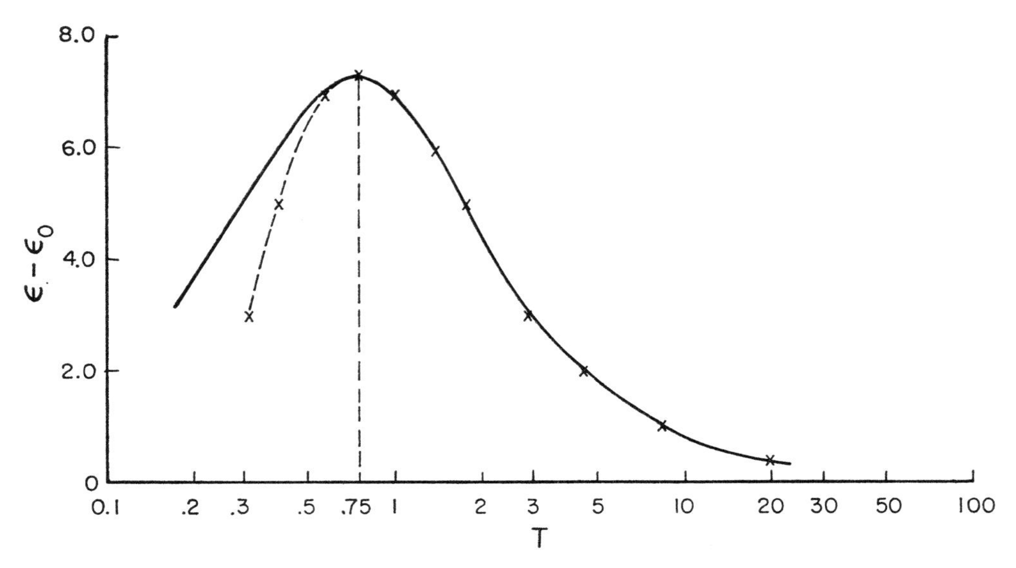

Fig. 3. Variation of $\varepsilon - \varepsilon_0$ with temperature for a KCl crystal containing a density (determined chemically) of 6.7×10^{19} OH^- ions per cc. Full line: experimental results; Points: theory according to Eq. (7)

This assumes that dipoles external to those shown in Fig. 2 are not polarized by the central cell and that their average moment is not appreciably affected by interaction with the outer dipoles of Fig. 2. The external dipole moments therefore cancel and the mean square moment of the spherical region in FRÖHLICH's theory is given by Eq. (6). Using this equation and the relation $x = \exp(V/kT)$ we may obtain a curve for the excess dielectric constant $(\varepsilon - \varepsilon_0)$ as a function of temperature. Thus, from Eq. (1) we obtain

$$4(\varepsilon - \varepsilon_0) = -(3\varepsilon_0 - Ay) + [(3\varepsilon_0 - Ay)^2 + 8\varepsilon_0 Ay]^{\frac{1}{2}} \tag{7}$$

with

$$A = 4\pi N \mu^2 / k T_0 \quad \text{and} \quad y = (T_0/T)\,(\overline{\boldsymbol{m}\,\boldsymbol{m}^*}/\mu^2). \tag{8}$$

This expression may be fitted to a typical experimental curve such as that of sample 3 in Ref. [2] to give the result shown in Fig. 3. Here the magnitude of A has been adjusted so that the theoretical and experimental curves coincide at the peak value (at which $\varepsilon - \varepsilon_0 = 7.27$ and $T = T_0 = 0.75$ °K).

The agreement between experimental and theoretical results is seen to be excellent at higher temperatures. The fact that the experimental values of dielectric constant exceed those of the theory at temperatures below T_0 may be ascribed to the following causes:

(i) The drop in dielectric constant as the temperature is lowered is probably more rapid for a two-position than for a six-position model.

(ii) The relaxation time may be such that the system cannot reach equilibrium at the applied frequency—as has been suggested by LAWLESS [5]. The dielectric loss peak measured in this region by KÄNZIG et al. [2] seems to confirm this conclusion.

It remains to determine whether the magnitude of the dielectric constant and the position of the peak on the temperature scale conform to the chemically determined impurity density and to the value of the dipole moment determined from other measurements. From the value of x at the peak of the theoretical curve for $(\varepsilon - \varepsilon_0)$ (when $T = 0.75$ °K) it is found that the magnitude of the interaction energy between parallel dipoles such as B and (10) in Fig. 2 is $V = -7.858 \times 10^{-17}$ ergs. The theoretical expression for this quantity, when the effect of crystal polarization is included macroscopically, is

$$V = N\mu^2/\varepsilon_0 \tag{9}$$

whence $N\mu^2 = 34.18$ if N is expressed in multiples of $10^{19}/cc$ and μ is in Debye units.

From Eq. (7) it is found that the value of A required to fit the magnitude of the dielectric constant at its maximum is 11.5 and this leads to a second value $N\mu^2 = 35.47$. The agreement between values of $N\mu^2$ obtained from the transition temperature and from the magnitude of the dielectric constant is thus so good as to appear fortuitous. For the curves of samples 1, 2 and 4 of [2] the ratio of the second to the first of these values of $N\mu^2$ has the values 0.70, 0.87 and 0.93 respectively—all of which are reasonable in view of the simplicity of the model.

An estimate of N can be obtained by using a value of μ obtained from the literature. From KUHN and LÜTY [1], we obtain $\mu_v = 1.83$ Debye and $\mu = 3.87$ Debye, leading, for the sample of Fig. 3, to $N = 2.37 \times 10^{19}$/cc. This is 35% of the OH^- concentration of 6.7×10^{19}/cc (which is accurate to within $\pm 20\%$) obtained from PH measurements [2]. For the other samples mentioned above the corresponding figures are 134%, 90% and 93% for samples 1, 2 and 4 respectively. The high value for sample 1 probably arises from the inaccuracy of the chemical measurements but it is difficult to account for the low value for sample 3 unless it is related to the carbonate content of the crystal. Apart from this discrepancy it seems that most of the dipoles contribute to the dielectric constant in contradiction to the results of SHEPHERD and FEHER [14] obtained from specific heat measurements assuming non-interacting dipoles.

The value of μ_v obtained by KUHN and LÜTY from the statistical mechanics of the dipole alignment process, and used in making the above estimates, agrees approximately with the recognized value from dielectric measurements on other substances as noted in [1]. A different experimental value (5.9 Debye units) was obtained by SHEPHERD and FEHER. The deviation between these values for the dipole moment, as well as the above discrepancy in the numbers of dipoles, might possibly be explained by the application of short-range order theory to the specific heat calculation.

4. A Simple Approximation

The approximation used in the proceding section gives quite reasonable results but it is conceivable that satisfactory agreement with experiment could be obtained using a simpler model. To test this possibility an approximation was tried in which the spherical region was assumed to consist of dipoles A and B only and BETHE's order of neighbours was used to determine the required probabilities. The right hand side of Eq. (6) then becomes $2x/(1+x)$. The resulting curve for $(\varepsilon-\varepsilon_0)$ is very similar to that of Fig. 3 but the value of $N\mu^2$ obtained from the dielectric constant is 64.0 in poor agreement with that (34.2) obtained from the dipolar interaction energy. It must therefore be concluded that the interaction of the nearest neighbours with those of the central cell plays an important part in deciding the magnitude of the dielectric constant and the position of its peak on the temperature scale.

5. Conclusions

The results of this paper allow us to draw the following conclusions concerning the dielectric constant of the KCl-OH^- system:

(i) FRÖHLICH generalized theory of the static dielectric constant forms the basis of a powerful method of analysing such systems.

(ii) The dielectric constant of quasi-random systems of this type is determined primarily by short-range ordering between neighbouring dipoles and, as suggested by BROUT [15], the maximum of the dielectric constant results from competition between the breakup of local order and the T^{-1} law characteristic of classical paramagnetic and paraelectric systems.

(iii) The main divergence between the experimental data and the theoretical predictions of this paper occurs in the low temperature region. It does not arise from the omission of quantum effects [5] as these would increase the disparity. An important effect is the increase in dipolar relaxation time at these temperatures but the simplicity of the model may also be a factor. One would expect the trend towards perfect local order, and thus the decrease in dielectric constant with temperature, to take place more slowly in a six-position system than in the two-position system considered here.

Acknowledgments

The author is indebted to B. A. KETTLES who programmed these calculations and to Dr. E. W. FENTON for several useful discussions.

References

1. KUHN, W., LÜTY, F.: Solid State Comm. **2**, 281 (1964).
2. KÄNZIG, W., HART, H. R., ROBERTS, S.: Phys. Rev. Letters **13**, 543 (1964).
3. ZERNICK, W.: Phys. Rev. **139** A, 1010 (1965); **158**, 562 (1967).
4. KLEIN, M. W.: Phys. Rev. **141**, 489 (1966).
5. LAWLESS, W. N.: Phys. Kondens. Materie **5**, 100 (1966); — Phys. Rev. Letters **17**, 1048 (1966).
6. BETHE, H.: Proc. Roy. Soc. (London), Ser. A **150**, 552 (1935).
7. FRÖHLICH, H.: Theory of Dielectrics, Second Ed. Oxford: Clarendon Press 1957.
8. BAUR, M. E., SALZMAN, W. R.: Phys. Rev. Letters **16**, 701, **17**, 159 (1966).
9. SAUER, J. A.: Phys. Rev. **57**, 142 (1940).
10. BERLIN, T. H., THOMSEN, J.: J. Chem. Phys. (USA) **20**, 1368 (1952).
11. LAX, M.: J. Chem. Phys. (USA) **20**, 1351 (1952).
12. ASHKIN, J., TELLER, E.: Phys. Rev. **64**, 178 (1943).
13. SIMPSON, J. H.: Canad. J. Phys. **29**, 163 (1951).
14. SHEPHERD, I., FEHER, G.: Phys. Rev. Letters **15**, 194 (1965).
15. BROUT, R.: Phys. Rev. Letters **14**, 175 (1965).

Chapter IV. Reduced Density Matrices

In recent years the usefulness of reduced density matrices has been stressed for obtaining rigorous general derivations of macroscopic properties from microscopic considerations without being confronted with the detailed solution of the N-body problem. Pioneering efforts in this field have been undertaken by YANG (1962) and FRÖHLICH (1967). Specific attention has been given to the concept of "macroscopic wave functions" and to the particular property of "off-diagonal long-range order" (ODLRO) in many-particle systems. This property is assumed to play a predominant role in superconductors, superfluids and lasers and, very recently, also in biological systems. It is believed that ODLRO may reveal new kinds of thermodynamic phases which are characterized by a kind of "phase order", and in such a phase one or more degrees of freedom may be highly excited above thermal equilibrium (cf. the laser).

The three papers in this chapter utilize reduced density matrices as a common tool. The first of them (TERREAUX and LAL) exploits a still hypothetical mathematical form which in a specific problem (van der Waals cohesion) leads to reasonable results and may become a useful theoretical device. In the second (HYLAND and ROWLANDS) some steps are taken towards the derivation of a generalized Ginzburg-Landau equation, i.e. on the basis of reasonable assumptions a general GL-equation is written down. The third paper (WEISS and HAUG) gives a derivation of the basic conservation laws for a superfluid of the ODLRO species.

Hypothetical Infinite Series Expressions of Reduced Density Matrices*

C. TERREAUX and P. LAL

University of Liverpool, Department of Theoretical Physics,
Liverpool/England

Abstract

Reduced density matrices Ω_n are assumed to be expressible in terms of two matrices Ω_i and Ω_j of lower order, $i+j=n$, and of an infinite number of translation invariant factors describing the relevant correlations between Ω_i and Ω_j. The compatibility of such a hypothesis with the hierarchical set of equations of motion of reduced density matrices is not investigated here. It is shown however that the form of the various terms of the hypothetical series of Ω_n is then determined by the symmetry properties of the reduced density matrices and by the requirement of unambiguous formal description of the translation invariance of the correlation factors occurring in the series. The present investigation is mainly concerned with Ω_2, whose hypothetical series expression is shown to lead to a generalized form of the van der Waals cohesion forces.

1. Introduction

A number of general physical properties valid for systems consisting of a great number of particles, such as e.g. the Josephson effect, or the propagation of sound waves in a fluid, must be derivable from microphysics in a manner which does not require detailed solutions of the N body problem. One set of such general features are connected with the reduced density matrices Ω_n. This defines FRÖHLICH's programme of research where detailed treatments of the N body problem are put into their proper place, which is the calculation of magnitude of various parameters entering in the formulation of macroscopic properties. It leads to a hierarchy of various approaches to problems.

In his original paper on that topic, FRÖHLICH [1] showed that the basic equations of hydrodynamics expressed in terms of a density and a velocity field fellow exactly from the equation of motion of the first reduced density matrix Ω_1, provided the force density is not specified. A specification of the force density leading to the Navier-Stokes equations was obtained by assuming the system to be near some equilibrium state, in addition to the use of the exactly valid symmetry

* With best wishes to Prof. H. FRÖHLICH for a happy and fruitful retirement.

properties of the local limit of the second reduced density matrix Ω_2, which enters in the general expression of the force density. The assumption about near equilibrium state was introduced to validate a pseudo-expansion of Ω_2 in terms of velocity, of deviation from the equilibrium macrodensity and of translation invariant correlation factors which were left unspecified.

The present paper arises from an attempt to specify as well as generalize such a treatment of Ω_2 by describing the interrelations between the various terms of the series Ω_2, no assumption related to equilibrium state being used. The treatment fits in the context of the programme above. In the Section 2 we shall assume that Ω_2 can be expressed in terms of two Ω_1's and of an infinite number of translation invariant factors describing the relevant correlations between the two Ω_1's. Whether such an assumption is or is not compatible with the equation of motion of Ω_2 will not be investigated here. It is expected to bear some relationship with diatomic molecular formation. Without actually solving anything we shall find however that the form of the various terms of such a hypothetical series Ω_2 is then determined by the symmetry properties of Ω_2 and by the requirement of unambiguous formal description (i) of the translation invariance of the correlation factors of Ω_2 and (ii) of the transfer of the action of the rest of the system on such correlation factors. These factors are linked to the reduced density matrices of higher order as all Ω_n are; they can be estimated by approximate methods, however the exact determination of each one of them involves a task as impracticable as the detailed solving of the N body problem itself. In the present paper we shall not be concerned with approximate determinations of these translation invariant correlation factors of Ω_2.

In conjunction with a development procedure the hypothetical series expression of Ω_2 will be used to derive a corresponding unambiguously defined expression of the Navier-Stokes force. It includes in particular the van der Waals cohesion forces term.

The Section 3 comments about the generalization of the above treatment to any reduced density matrix Ω_n expressed in terms of two reduced density matrices Ω_i, Ω_j, of lower order, $i+j=n$, and of an infinite number of factors describing the relevant correlations between Ω_i and Ω_j.

The present investigation is restricted to normal systems including a single type of atoms (bosons or fermions). These particles are described by non-relativistic wave operator $\psi(\boldsymbol{x})$ satisfying the appropriate commutation or anti-commutation relations. The n-th reduced density matrix depending on n non local pairs of space points (in the case of a system of fermions, the spin coordinates are supposed included in the parameters $\boldsymbol{x}$) and on time t, is defined as the trace

$$\begin{aligned}\Omega_n(\boldsymbol{x}_1' \ldots \boldsymbol{x}_n'; \boldsymbol{x}_1'' \ldots \boldsymbol{x}_n'') &= \mathrm{Sp}\{\psi^+(\boldsymbol{x}_1'') \cdot \ldots \cdot \psi^+(\boldsymbol{x}_n'')\, \psi(\boldsymbol{x}_n') \cdot \ldots \cdot \psi(\boldsymbol{x}_1')\, \Omega(t)\} \\ &= \Omega_n^*(\boldsymbol{x}_1'' \ldots \boldsymbol{x}_n''; \boldsymbol{x}_1' \ldots \boldsymbol{x}_n')\end{aligned} \tag{1.1}$$

including the density matrix $\Omega(t)$, which satisfies the equation of motion

$$i\hbar\, \partial_t\, \Omega(t) = [H, \Omega(t)] \tag{1.2}$$

where H is the Hamiltonian operator of the system. Assuming an H whose potential energy is expressed only in terms of an arbitrary two-body interaction potential $V(|\mathbf{x}-\mathbf{y}|)$, depending on relative distance only, the Eq. (1.2) leads to the hierarchical set of exact equations of motion (m = atomic mass),

$$\begin{aligned} L_n\Omega_n \equiv & \Big\{ i\hbar\,\partial_t + \frac{\hbar^2}{2m}\sum_{k=1}^{n} \partial^2_{\mathbf{x}'_k} - \partial^2_{\mathbf{x}''_k}) \\ & - \frac{1}{2}\sum_{i,k=1}^{n} [V(|\mathbf{x}'_i - \mathbf{x}'_k|) - V(|\mathbf{x}''_i - \mathbf{x}''_k|)]\Big\}\,\Omega_n(\mathbf{x}'_1 \ldots; \ldots \mathbf{x}''_n) \\ = & W_n(\mathbf{x}'_1 \ldots; \ldots \mathbf{x}''_n) \end{aligned} \tag{1.3}$$

where

$$\begin{aligned} W_n(\mathbf{x}'_1 \ldots; \ldots \mathbf{x}''_n) = \int d^3\mathbf{x}_{n+1} \sum_{k=1}^{n} [V(|\mathbf{x}'_k - \mathbf{x}_{n+1}|) - V(|\mathbf{x}''_k - \mathbf{x}_{n+1}|)] \\ \cdot\, \Omega_{n+1}(\mathbf{x}'_1 \ldots \mathbf{x}'_n\,\mathbf{x}_{n+1};\ \mathbf{x}''_1 \ldots \mathbf{x}''_n\,\mathbf{x}_{n+1}). \end{aligned} \tag{1.4}$$

W_n describes the influence of the rest of the system, it involves differences between the average potential energies at the n pairs of points of Ω_n with respect to their surrounding. In the following the Eq. (1.3) will be written symbolically as $L_n\Omega_n = W_n$.

From the definition (1.1) of Ω_n and the commutation relations satisfied by the $\psi(\mathbf{x})$, follow the normalisation condition

$$\int d^3\mathbf{x}_{n+1}\,\Omega_{n+1}(\mathbf{x}'_1 \ldots \mathbf{x}'_n\,\mathbf{x}_{n+1};\ \mathbf{x}''_1 \ldots \mathbf{x}''_n\,\mathbf{x}_{n+1}) = (N-n)\,\Omega_n(\mathbf{x}'_1 \ldots \mathbf{x}'_n;\ \mathbf{x}''_1 \ldots \mathbf{x}''_n), \tag{1.5}$$

and the various symmetry properties obeyed by the reduced density matrices. In the particular case of the second reduced density matrix these properties are

$$\Omega_2(\mathbf{x}'\,\mathbf{y}';\ \mathbf{x}''\,\mathbf{y}'') = \Omega_2(\mathbf{y}'\,\mathbf{x}';\ \mathbf{y}''\,\mathbf{x}'') = \Omega_2^*(\mathbf{x}''\,\mathbf{y}'';\ \mathbf{x}'\,\mathbf{y}') = \pm\,\Omega_2(\mathbf{y}'\,\mathbf{x}';\ \mathbf{x}''\,\mathbf{y}'') \tag{1.6}$$

where the $\pm$ refers to bosons (+) or fermions (−).

For the later purpose of describing translation invariant parts of the second reduced density matrix, we shall introduce yet another type of coordinates consisting of one absolute coordinate $\boldsymbol{\rho}$, and three relative coordinates $\mathbf{r}$, $\boldsymbol{\mu}$, $\boldsymbol{\nu}$ defined as (the corresponding Jacobian is equal to unity),

$$\begin{aligned} \boldsymbol{\rho} &= \tfrac{1}{4}\,(\mathbf{x}' + \mathbf{x}'' + \mathbf{y}' + \mathbf{y}'') & \mathbf{x}' &= \left(\boldsymbol{\rho} + \tfrac{\mathbf{r}}{2}\right) + \tfrac{1}{2}\left(\tfrac{\boldsymbol{\mu}}{2} + \boldsymbol{\nu}\right) \\ \mathbf{r} &= \tfrac{1}{2}\,(\mathbf{x}' + \mathbf{x}'' - \mathbf{y}' - \mathbf{y}'') & \mathbf{x}'' &= \left(\boldsymbol{\rho} + \tfrac{\mathbf{r}}{2}\right) - \tfrac{1}{2}\left(\tfrac{\boldsymbol{\mu}}{2} + \boldsymbol{\nu}\right) \\ \boldsymbol{\mu} &= \mathbf{x}' - \mathbf{x}'' + \mathbf{y}' - \mathbf{y}'' & \mathbf{y}' &= \left(\boldsymbol{\rho} - \tfrac{\mathbf{r}}{2}\right) + \tfrac{1}{2}\left(\tfrac{\boldsymbol{\mu}}{2} - \boldsymbol{\nu}\right) \\ \boldsymbol{\nu} &= \tfrac{1}{2}\,(\mathbf{x}' - \mathbf{x}'' - \mathbf{y}' + \mathbf{y}'') & \mathbf{y}'' &= \left(\boldsymbol{\rho} - \tfrac{\mathbf{r}}{2}\right) - \tfrac{1}{2}\left(\tfrac{\boldsymbol{\mu}}{2} - \boldsymbol{\nu}\right) \end{aligned} \tag{1.7}$$

$\boldsymbol{\rho}$ and $\mathbf{r}$ are the new local coordinates; $\boldsymbol{\mu}$ and $\boldsymbol{\nu}$ are two non local coordinates because they vanish at the local limit defined by $\mathbf{x}' = \mathbf{x}''$, $\mathbf{y}' = \mathbf{y}''$. Both types of coordinates will be used simultaneously according to their relevance to the different parts of expressions. The following properties related to these new coordinates will be of importance in the Section 2. From Eq. (1.7), it follows that

$$\begin{aligned}
\partial_{\boldsymbol{\rho}} &= \partial_{\boldsymbol{x}'} + \partial_{\boldsymbol{x}''} + \partial_{\boldsymbol{y}'} + \partial_{\boldsymbol{y}''} \\
\partial_{\boldsymbol{r}} &= \tfrac{1}{2}(\partial_{\boldsymbol{x}'} + \partial_{\boldsymbol{x}''} - \partial_{\boldsymbol{y}'} - \partial_{\boldsymbol{y}''}) \\
\partial_{\boldsymbol{\mu}} &= \tfrac{1}{4}(\partial_{\boldsymbol{x}'} - \partial_{\boldsymbol{x}''} + \partial_{\boldsymbol{y}'} - \partial_{\boldsymbol{y}''}) \\
\partial_{\boldsymbol{v}} &= \tfrac{1}{2}(\partial_{\boldsymbol{x}'} - \partial_{\boldsymbol{x}''} - \partial_{\boldsymbol{y}'} + \partial_{\boldsymbol{y}''})
\end{aligned} \tag{1.8}$$

and the kinetic energy terms of L_2 take the form

$$\frac{\hbar^2}{2m}\left[\partial^2_{\boldsymbol{x}'} - \partial^2_{\boldsymbol{x}''} + \partial^2_{\boldsymbol{y}'} - \partial^2_{\boldsymbol{y}''}\right] = \frac{\hbar^2}{m}\left[\partial_{\boldsymbol{\rho}} \cdot \partial_{\boldsymbol{\mu}} + \partial_{\boldsymbol{r}} \cdot \partial_{\boldsymbol{v}}\right]. \tag{1.9}$$

The scalar products of derivative on the r.h.s. of Eq. (1.9) associate a definite non local coordinate to each local one. In the case of L_1, one has $\frac{1}{2}(\partial^2_{\boldsymbol{x}'} - \partial^2_{\boldsymbol{x}''}) = \partial_{\boldsymbol{x}} \cdot \partial_{\boldsymbol{\xi}}$, where $\boldsymbol{x} = \frac{1}{2}(\boldsymbol{x}' + \boldsymbol{x}'')$, and $\boldsymbol{\xi} = \boldsymbol{x}' - \boldsymbol{x}''$.

Written in terms of $(\boldsymbol{\rho}, \boldsymbol{r}, \boldsymbol{\mu}, \boldsymbol{v})$ the symmetry properties (1.6) obeyed by Ω_2 take the form

$$\Omega_2(\boldsymbol{\rho}\,\boldsymbol{r}\,\boldsymbol{\mu}\,\boldsymbol{v}) = \Omega_2(\boldsymbol{\rho}, -\boldsymbol{r}, \boldsymbol{\mu}, -\boldsymbol{v}) = \Omega_2^*(\boldsymbol{\rho}, \boldsymbol{r}, -\boldsymbol{\mu}, -\boldsymbol{v}) = \pm\,\Omega_2(\boldsymbol{\rho}, -\boldsymbol{v}, \boldsymbol{\mu}, -\boldsymbol{r}), \tag{1.10}$$

where the description of the relevant permutation of the spin variables is omitted and the same symbol Ω_2 is used in Eqs. (1.6) and (1.10) for the sake of simplifying the notation.

Various scalar products of space derivatives will form parts of the series expression of Ω_2. Such scalar products have to be invariant with respect to the transformations implied in the Eqs. (1.6) and (1.10), viz,

1. Exchange $\boldsymbol{x} \leftrightarrow \boldsymbol{y}$, and change of sign of $\boldsymbol{r}$ and $\boldsymbol{v}$;
2. Simultaneous exchanges $\boldsymbol{x}' \leftrightarrow \boldsymbol{x}''$ and $\boldsymbol{y}' \leftrightarrow \boldsymbol{y}''$, complex conjugation and change of sign of $\boldsymbol{\mu}$ and $\boldsymbol{v}$;
3. Exchanges $\boldsymbol{x}' \leftrightarrow \boldsymbol{y}'$ (or $\boldsymbol{x}'' \leftrightarrow \boldsymbol{y}''$), and $\boldsymbol{r} \leftrightarrow -\boldsymbol{v}$.

The possible scalar products of space derivatives which satisfy such invariance requirements are restricted to

$$i\,\partial_{\boldsymbol{g}} \cdot \partial_{\boldsymbol{\Gamma}} = \{i\,\partial_{\boldsymbol{\rho}} \cdot \partial_{\boldsymbol{\mu}};\ i\,\partial_{\boldsymbol{r}} \cdot \partial_{\boldsymbol{v}};\ \partial^2_{\boldsymbol{\rho}};\ \partial^2_{\boldsymbol{\mu}}\} \tag{1.11}$$

where $(i\,\partial_{\boldsymbol{g}} \cdot \partial_{\boldsymbol{\Gamma}})$ is a common name for the scalar products of the r.h.s. of Eq. (1.11). [The linear combination $(\partial^2_{\boldsymbol{r}} + \partial^2_{\boldsymbol{v}})$ is not relevant as we shall find in the following section.] It will also be convenient to introduce the notation (with a lower case γ)

$$\partial_{\boldsymbol{g}} \cdot \partial_{\boldsymbol{\gamma}} = \{\partial_{\boldsymbol{\rho}} \cdot \partial_{\boldsymbol{\mu}};\ \partial_{\boldsymbol{r}} \cdot \partial_{\boldsymbol{v}}\} \tag{1.12}$$

when referring to those scalar products which occur in the kinetic energy part (1.9) of L_2.

2. A Hypothetical Infinite Series Expression of the Second Reduced Density Matrix of Normal System

We shall assume that Ω_2 can be expressed in terms of two Ω_1's and of translation invariant factors describing the relevant correlations between the two Ω_1's, i.e. of such correlations depending only on the relative vectors $\boldsymbol{r}, \boldsymbol{\mu}, \boldsymbol{v}$ [cf. Eqs. (1.7)] between the two non local pairs of points and on time t. The assumption of translation invariance is supposed to require an infinite series expression of Ω_2. Whether these hypotheses are or are not compatible with $L_2\Omega_2=W_2$, will not be investigated here. They should be related to the problem of diatomic molecular formation. However we shall prove that the form of the various terms of Ω_2 is then determined by the symmetry properties (1.6) of Ω_2, and by the requirement of unambiguous formal description (i) of translation invariance of the correlation factors and (ii) of the transfer of the action of the rest of the system (as described by W_2) on such factors.

Let us assume that Ω_2 factorizes, viz,

$$\Omega_2^{as}(\boldsymbol{x}'\boldsymbol{y}';\boldsymbol{x}''\boldsymbol{y}'')=\Omega_1(\boldsymbol{x}';\boldsymbol{x}'')\,\Omega_1(\boldsymbol{y}';\boldsymbol{y}'')\pm\Omega_1(\boldsymbol{y}';\boldsymbol{x}'')\,\Omega_1(\boldsymbol{x}';\boldsymbol{y}'') \tag{2.1}$$

when the two non local pairs of points are in any one of the following asymptotic configurations

$$\{\boldsymbol{x}',\boldsymbol{x}''\}\rightleftarrows\{\boldsymbol{y}',\boldsymbol{y}''\},\quad\text{or}\quad\{\boldsymbol{y}',\boldsymbol{x}''\}\rightleftarrows\{\boldsymbol{x}',\boldsymbol{y}''\} \tag{2.2}$$

where the symbol $\rightleftarrows$ indicates that all the space points in one set of brackets are far apart from those in the other brackets, viz, by more than a distance of the order of the range of the two-body interaction potential V. The relative distance between the points enclosed in any set of brackets of the relations (2.2) is arbitrary. Because of the short range property of Ω_1, the r.h.s. of Eq. (2.1) consists of only one non vanishing term in each case of asymptotic configuration (2.2). The Eq. (2.1) satisfies the symmetry properties (1.6) of Ω_2; the $\pm$ sign occurs according to the relevant type of statistics and in the case of a system of fermions the spin coordinates are supposed included in the parameters $\boldsymbol{x}$ and $\boldsymbol{y}$.

Ω_2 may involve a translation invariant, i.e. $\boldsymbol{\rho}$-independent short range correlation term which is called $\mathfrak{K}_2(\boldsymbol{r}\boldsymbol{\mu}\boldsymbol{v},t)$ in Eq. (2.3) below. We shall regard Ω_1 as the $\boldsymbol{\rho}$-dependent subunits or building blocks of the remaining part of Ω_2, which we assume to be of the form

$$\begin{aligned}\Omega_2(\boldsymbol{x}'\boldsymbol{y}';\boldsymbol{x}''\boldsymbol{y}'')=\mathfrak{K}_2(\boldsymbol{r}\boldsymbol{\mu}\boldsymbol{v},t)+\Omega_1(\boldsymbol{x}';\boldsymbol{x}'')\,\Omega_1(\boldsymbol{y}';\boldsymbol{y}'')\,K_2(\boldsymbol{r}\boldsymbol{\mu}\boldsymbol{v},t)\\ \pm\Omega_1(\boldsymbol{y}';\boldsymbol{x}'')\,\Omega_1(\boldsymbol{x}';\boldsymbol{y}'')\,K_2(-\boldsymbol{v},\boldsymbol{\mu},-\boldsymbol{r},t)+\dots\end{aligned} \tag{2.3}$$

where K_2 is supposed to include the $\boldsymbol{\rho}$-independent short range correlations which are not described by the factorized form (2.1) in subunits Ω_1's. $\mathfrak{K}_2$ takes the asymptotic value zero when any one point is far apart from the three other points; from Eq. (2.1) K_2 takes the value 1 when one non local pair of points is far apart from the other pair. For instance

$$K_2(\boldsymbol{y}'\boldsymbol{x}';\boldsymbol{x}''\boldsymbol{y}'')\to 1 \tag{2.4}$$

when the points are in the asymptotic configuration described on the r.h.s. of the relations (2.2). $\mathfrak{K}_2$ is assumed to satisfy the symmetry properties (1.10) of Ω_2. As these properties have to result also from the relevant choice of $\pm$ sign occurring in the Eq. (2.1) and from the corresponding properties obeyed by the subunits Ω_1's, it follows quite generally that the symmetry properties of K_2 are restricted to

$$K_2(\boldsymbol{r}\boldsymbol{\mu}\boldsymbol{v},t)=K_2(-\boldsymbol{r},\boldsymbol{\mu},-\boldsymbol{v},t)=K_2^*(\boldsymbol{r},-\boldsymbol{\mu},-\boldsymbol{v},t). \tag{2.5}$$

However, the most general case requires

$$K_2(\boldsymbol{r}\boldsymbol{\mu}\boldsymbol{v},t) \neq \pm K_2(-\boldsymbol{v},\boldsymbol{\mu},-\boldsymbol{r},t). \tag{2.6}$$

The general $\boldsymbol{\rho}$-independence of $\mathfrak{K}_2$ and K_2 must follow from their respective eqs. of motion which we write

$$\mathfrak{L}_2\mathfrak{K}_2(\boldsymbol{r}\boldsymbol{\mu}\boldsymbol{v},t) = w_{\mathfrak{K}_2}(\boldsymbol{r}\boldsymbol{\mu}\boldsymbol{v},t) \tag{2.7}$$

$$\mathfrak{L}_2 K_2(\boldsymbol{r}\boldsymbol{\mu}\boldsymbol{v},t) = w_{K_2}(\boldsymbol{r}\boldsymbol{\mu}\boldsymbol{v},t). \tag{2.8}$$

As $\mathfrak{K}_2$ and K_2 must have a good deal of properties in common with Ω_2, one guesses that $\mathfrak{L}_2$ is derived from L_2 simply by dropping the scalar product of associated derivatives $\boldsymbol{\partial}_{\boldsymbol{\rho}} \cdot \boldsymbol{\partial}_{\boldsymbol{\mu}}$, in Eq. (1.9), because of the $\boldsymbol{\rho}$-independence of $\mathfrak{K}_2$ and K_2. Furthermore one has to remember that each two-body interaction potential V occurring in any $L_n\Omega_n = W_n$, induces e.g. a two-point screening due to the hard core of V in each term of Ω_n which depend on the relevant two single-dashed or two double-dashed coordinates. As $\mathfrak{K}_2$ and K_2 describe correlation between two physical objects, it follows that $\mathfrak{L}_2$ must indeed have the same V-terms as L_2. It follows also that the other subunits of the $\boldsymbol{\rho}$-dependent part of Ω_2 are necessarily Ω_1's, because these Ω_1's do not describe correlations between physical objects and thus do not require the V-terms of $L_2\Omega_2$ already used in Eqs. (2.7) and (2.8). The Section 3 will offer more opportunities to use the V-terms as heuristic tool to arrive at the form of the evolution * operator of the equation of motion of a particular subunit of Ω_n and to find the possible subdivisions of Ω_n in subunits. One concludes e.g. that Eq. (2.8) has the translation invariant form

$$\begin{aligned}\mathfrak{L}_2 K_2 \equiv \Big\{ i\hbar\,\partial_t + \frac{\hbar^2}{m}\,\boldsymbol{\partial}_{\boldsymbol{r}} \cdot \boldsymbol{\partial}_{\boldsymbol{v}} - [V(|\boldsymbol{r}+\boldsymbol{v}|) - V(|\boldsymbol{r}-\boldsymbol{v}|)]\Big\}& \\ \cdot K_2(\boldsymbol{r}\boldsymbol{\mu}\boldsymbol{v},t) = w_{K_2}(\boldsymbol{r}\boldsymbol{\mu}\boldsymbol{v},t).&\end{aligned} \tag{2.8'}$$

One of our tasks will be to dissociate formally the Eqs. (2.7) and (2.8) from $L_2\Omega_2 = W_2$, In general the latter eq. is not translation invariant; a requirement of translation invariance of $L_2\Omega_2 = W_2$, would lead to Eq. (2.7) only.

From the assumed $\boldsymbol{\rho}$-independence property of K_2, one expects the terms of Ω_2 which are explicitly written in Eq. (2.3) to require the presence of additional terms indicated by dots in Eq. (2.3). These additional terms will again be expressed in terms of two submits Ω_1's and of translation invariant correlation factors. The problem of accommodating $\mathfrak{L}_2\mathfrak{K}_2$ and $\mathfrak{L}_2 K_2$ in $L_2\Omega_2$ is assumed to require an infinite number of such additional terms of Ω_2.

It is obvious that a prerequisite to the condition of unambiguous formal description (i) of the $\boldsymbol{\rho}$-independence of $\mathfrak{K}_2$ and K_2 and (ii) of the transfer of the action of the rest of the system on such correlations, is to get rid of the following mixed terms of $L_2\Omega_2$; each such mixed term resulting from the kinetic energy part (1.9) of L_2, when associated derivatives act on two different subunits Ω_1 and

* The word "evolution" of a particular subunit of Ω_n is introduced to name the result of the action on that subunit of the operator L of its eq. of motion, i.e. of all the operators of that eq. except the integral term describing the potential energy with the surrounding. The term evolution usually refers only to the action of the operator $i\hbar\partial_t$, hence the present use of the term differs from the usual practice.

K_2 of a particular term of Ω_2. There is apparently not many ways to achieve it formally, the only manner is to cancel these mixed terms by means of suitably chosen subsequent terms of Ω_2. Let us see it in detail. The action of the operator L_2 on those terms of Ω_2 which are written explicitly in the r.h.s. of Eq. (2.3) leads to

$$\begin{aligned} L_2\Omega_2(\boldsymbol{x}'\boldsymbol{y}';\boldsymbol{x}''\boldsymbol{y}'') &= \mathfrak{L}_2\mathfrak{K}_2(\boldsymbol{r}\boldsymbol{\mu}\boldsymbol{v},t) \\ &+ [W_1(\boldsymbol{x}';\boldsymbol{x}'')\,\Omega_1(\boldsymbol{y}';\boldsymbol{y}'') + \Omega_1(\boldsymbol{x}';\boldsymbol{x}'')\,W_1(\boldsymbol{y}';\boldsymbol{y}'')]\,K_2(\boldsymbol{r}\boldsymbol{\mu}\boldsymbol{v},t) \pm \text{Exch} \\ &+ \Omega_1(\boldsymbol{x}';\boldsymbol{x}'')\,\Omega_1(\boldsymbol{y}';\boldsymbol{y}'')\,\mathfrak{L}_2K_2(\boldsymbol{r}\boldsymbol{\mu}\boldsymbol{v},t) \pm \text{Exch} \\ &+ \frac{\hbar^2}{m}\sum_{\boldsymbol{g},\boldsymbol{\gamma}}\{\boldsymbol{\partial}_{\boldsymbol{g}}[\Omega_1(\boldsymbol{x}';\boldsymbol{x}'')\,\Omega_1(\boldsymbol{y}';\boldsymbol{y}'')]\}\,\boldsymbol{\partial}_{\boldsymbol{\gamma}}K_2(\boldsymbol{r}\boldsymbol{\mu}\boldsymbol{v},t) \pm \text{Exch} + \cdots. \end{aligned} \tag{2.9}$$

The abreviation "Exch" in Eq. (2.9) and in all future relations refers to those terms obtained by exchanging $\boldsymbol{x}' \leftrightarrow \boldsymbol{y}'$ and $\boldsymbol{r} \leftrightarrow -\boldsymbol{v}$, in the relevant part of the expression standing before. The terms of the form $(W_1\Omega_1K_2)$ and $(\Omega_1\Omega_1\mathfrak{L}_2K_2)$ on the r.h.s. of Eq. (2.9) are diagonal* terms resulting from the concentrated action of the relevant evolution operator on one subunit Ω_1 or K_2. The last group of terms in Eq. (2.9) are the above defined mixed terms generated by the kinetic energy part of L_2. The notation $\Sigma\,\boldsymbol{\partial}_{\boldsymbol{g}}\cdot\boldsymbol{\partial}_{\boldsymbol{\gamma}}$, cf. Eq. (1.12), refers to the two scalar products of derivatives on the r.h.s. of Eq. (1.9) and to all possible combinations of the implied single derivatives acting on $(\Omega_1\Omega_1)$ and on K_2. Because of the particular dependence in space coordinates of these scalar products and of each subunit Ω_1, no real mixed term (whose presence would require the actual solving of $L_1\Omega_1 = W_1$) occurs in $L_2\Omega_2$ with a product of associated derivatives acting partly on one Ω_1 and partly on the other Ω_1; this property is easily verified by referring to the original coordinates $(\boldsymbol{x}', \boldsymbol{x}'', \boldsymbol{y}', \boldsymbol{y}'')$.

The cancellation of the above mixed terms of $L_2\Omega_2$ requires a well-defined form of the subsequent terms of the series Ω_2. Indeed such a cancellation can be obtained in a purely formal way only if the subsequent terms of Ω_2 are of the same form (called III) as the unwanted mixed terms, viz.

$$\text{III} = \{\boldsymbol{\partial}_{\boldsymbol{g}}[\Omega_1(\boldsymbol{x}';\boldsymbol{x}'')\,\Omega_1(\boldsymbol{y}';\boldsymbol{y}'')]\}\,i\,\boldsymbol{\partial}_{\boldsymbol{\Gamma}}\,K_2^{g/\Gamma}(\boldsymbol{r}\boldsymbol{\mu}\boldsymbol{v},t) \tag{2.10}$$

including new translation invariant short range correlations $K_2^{g/\Gamma}$ (the short range character of $K_2^{g/\Gamma}$ is guaranteed by the derivative $\boldsymbol{\partial}_{\boldsymbol{\Gamma}}$, which is itself required in a scalar contribution to Ω_2) and $\boldsymbol{\partial}_{\boldsymbol{g}}$ being chosen identical to the corresponding part of the mixed term one tries to get rid of. The notation $(i\,\boldsymbol{\partial}_{\boldsymbol{g}}\cdot\boldsymbol{\partial}_{\boldsymbol{\Gamma}})$ (with a capital $\boldsymbol{\Gamma}$) refers to any one scalar product (1.11) because such products have of course to comply with the symmetry property requirements (1.6) of Ω_2. Taking III into account, it follows that $L_2\Omega_2$ involves two terms with a particular common factor $(\boldsymbol{\partial}_{\boldsymbol{g}}\Omega_1\Omega_1)$, viz. the unwanted mixed term and the evolution term of type $(\boldsymbol{\partial}_{\boldsymbol{g}}\Omega_1\Omega_1)(\mathfrak{L}_2\,i\,\boldsymbol{\partial}_{\boldsymbol{\Gamma}}\,K_2^{g/\Gamma})$, resulting from III. The unwanted mixed term disappears by mutual cancellation of these two terms if the correlations K_2 and $K_2^{g/\Gamma}$ are assumed to solve the corresponding homogeneous differential eq. obtained by equat-

* Taking the permission of referring to L_2 as if it was a matrix and treating the various subunits of Ω_2 as the independent components of a general vector, it follows that the elimination procedure of the mixed terms defined above bears some similarity with the process of diagonalization of a matrix by referring to a suitable orthogonal frame of reference.

ing the two terms in question [cf. the first Eq. (2.15)]. Thus we have achieved two results in one step: we have simplified the form of $L_2\Omega_2$ by removing an unwanted term and have obtained a relationship between two subsequent translation invariant correlation factors of Ω_2. Such a relationship will obviously be required in describing the transfer of the action of the rest of the system upon these correlations in an unequivocal manner.

At that point let us refer to the remark below the expression (1.11) which implies that the linear combination $(\partial_r^2 + \partial_v^2)$ is not suited for the cancellation procedure of the mixed terms of $L_2\Omega_2$. Indeed this procedure is possible only if the relevant scalar products $(i\boldsymbol{\partial}_g \cdot \boldsymbol{\partial}_\Gamma)$ factorize in terms of the components of the products $(\boldsymbol{\partial}_g \cdot \boldsymbol{\partial}_\gamma)$ which occur on the r.h.s. of Eq. (1.9). Furthermore the $\boldsymbol{\rho}$-independence of K_2 and $K_2^{g/\Gamma}$, is easily seen to restrict the scalar products of derivatives required in the general form III above, to those $(i\boldsymbol{\partial}_g \cdot \boldsymbol{\partial}_\gamma)$ occurring in the relation (1.12).

Taking the last remark into account the action of L_2 on the terms III of Ω_2 leads to the two following types of terms, viz. the diagonal terms

$$\{\boldsymbol{\partial}_g [W_1(\boldsymbol{x}'; \boldsymbol{x}'')\, \Omega_1(\boldsymbol{y}'; \boldsymbol{y}'') + \Omega_1(\boldsymbol{x}'; \boldsymbol{x}'')\, W_1(\boldsymbol{y}'; \boldsymbol{y}'')]\}\, i\boldsymbol{\partial}_\gamma\, K_2^{g/\gamma}(\boldsymbol{r}\boldsymbol{\mu}\boldsymbol{v}, t), \tag{2.11}$$

and the subsequent mixed terms

$$\frac{\hbar^2}{m} \sum_{\substack{k,l=1,2,3\\ g',\gamma'}} \{\partial_{g'_k}\, \partial_{g_l} [\Omega_1(\boldsymbol{x}'; \boldsymbol{x}'')\, \Omega_1(\boldsymbol{y}'; \boldsymbol{y}'')]\}\, i\,\partial_{\gamma'_k}\, i\,\partial_{\gamma_l}\, K_2^{g/\gamma}(\boldsymbol{r}\boldsymbol{\mu}\boldsymbol{v}, t) \qquad (2.12)^*$$

which are to be removed in a manner similar to what was done before, that is by requiring about the form of the next terms of Ω_2 and about their mutual relationship.

The step by step cancellation of the mixed terms of $L_2\Omega_2$ by means of suited subsequent terms of the series Ω_2 leads straightforwardly to the hypothetical infinite series expression [see the footnote related to Eq. (2.12)]

$$\begin{aligned}\Omega_2(\boldsymbol{x}'\boldsymbol{y}'; \boldsymbol{x}''\boldsymbol{y}'') &= \mathfrak{K}_2(\boldsymbol{r}\boldsymbol{\mu}\boldsymbol{v}, t) + \Omega_1(\boldsymbol{x}'; \boldsymbol{x}'')\, \Omega_1(\boldsymbol{y}'; \boldsymbol{y}'')\, K_2(\boldsymbol{r}\boldsymbol{\mu}\boldsymbol{v}, t) \pm \text{Exch}\\ &+ d^2 \sum_{g,\gamma} \{\boldsymbol{\partial}_g [\Omega_1(\boldsymbol{x}'; \boldsymbol{x}'')\, \Omega_1(\boldsymbol{y}'; \boldsymbol{y}'')]\}\, i\boldsymbol{\partial}_\gamma\, K_2^{g/\gamma}(\boldsymbol{r}\boldsymbol{\mu}\boldsymbol{v}, t) \pm \text{Exch}\\ &+ d^4 \sum_{g\gamma,g'\gamma'} \{\boldsymbol{\partial}_{g'}\, \boldsymbol{\partial}_g [\Omega_1(\boldsymbol{x}'; \boldsymbol{x}'')\, \Omega_1(\boldsymbol{y}'; \boldsymbol{y}'')]\}\, i\boldsymbol{\partial}_{\gamma'}\, i\boldsymbol{\partial}_\gamma\, K_2^{g'g/\gamma'\gamma}(\boldsymbol{r}\boldsymbol{\mu}\boldsymbol{v}, t) \pm \text{Exch} + \dots\end{aligned} \tag{2.13}$$

d is an atomic length introduced for dimensional reason. $(\sum \boldsymbol{\partial}_g \cdot \boldsymbol{\partial}_\gamma)$ and $(\sum \boldsymbol{\partial}_{g'} \cdot \boldsymbol{\partial}_{\gamma'})$ are identical but independent sums including the two scalar products (1.12) and all the possible conbinations of the implied single derivatives acting on $(\Omega_1\Omega_1)$ and on $K_2^{\cdot|\cdot}$.

The symbol $K_2^{\cdot|\cdot}$ refers to any one of the infinite set of translation invariant correlation factors which appears with superscripts added to K_2. Each $K_2^{\cdot|\cdot}$ obeys the symmetry properties (2.5) and the inequality (2.6). Their associa-

* For the sake of simplifying the notation we shall drop the Cartesian indices involved in the scalar products of derivatives with respect to associate coordinates; we shall assume implicitely that two derivatives with respect to the variables $\boldsymbol{g}$ and $\boldsymbol{\gamma}$ occurring with the same number of dashes in all following similar expressions are components of a scalar product.

tion with derivatives in Eq. (2.13) makes sure that these $K_2^{\cdot|\cdot}$ are of short range character, it implies that

$$K_2^{\cdot|\cdot}(\boldsymbol{x}'\boldsymbol{y}';\boldsymbol{x}''\boldsymbol{y}'')\to 0 \tag{2.14}$$

if any point is far apart from the three other points. The original correlation K_2 includes a long range component according to Eq. (2.4) and a short range component as well.

Let us remark that the abreviation "Exch" defined immediately after the Eq. (2.9) implies that the variables of the derivatives are also to be exchanged accordingly. However the superscripts of the correlation factors $K_2^{\cdot|\cdot}$ must be left unchanged!

According to the previous description, the cancellation procedure of the so-called mixed terms of $L_2\Omega_2$, leads to the following infinite set of simultaneous differential equations

$$\begin{aligned}&\frac{\hbar^2}{m}\,\partial_\gamma K_2(\boldsymbol{r}\boldsymbol{\mu}\boldsymbol{v},t)+d^2\,\mathfrak{L}_2\,i\,\partial_\gamma K_2^{g/\gamma}(\boldsymbol{r}\boldsymbol{\mu}\boldsymbol{v},t)=0\\&\frac{\hbar^2}{m}\,\partial_{\gamma'_k}\,i\,\partial_{\gamma_l}K_2^{g/\gamma}(\boldsymbol{r}\boldsymbol{\mu}\boldsymbol{v},t)+d^2\,\mathfrak{L}_2\,i\,\partial_{\gamma'_k}\,i\,\partial_{\gamma_l}K_2^{g'g/\gamma'\gamma}(\boldsymbol{r}\boldsymbol{\mu}\boldsymbol{v},t)=0\,.\\&\dots\end{aligned}\tag{2.15}$$

The part of $L_2\Omega_2$ which consists of the exchange terms leads to a similar set of eqs.; they can be obtained from Eqs. (2.15) by the relevant exchange of parameters, viz. $\boldsymbol{x}'\leftrightarrow\boldsymbol{y}'$ and $\boldsymbol{r}\leftrightarrow-\boldsymbol{v}$. According to the relation (1.12) the set of possible variables implied in $\boldsymbol{g}$ and γ are

$$(\boldsymbol{g},\gamma):(\boldsymbol{\rho},\boldsymbol{\mu});\ (\boldsymbol{r},\boldsymbol{v});\ (\boldsymbol{v},\boldsymbol{r}) \tag{2.16}$$

$(\boldsymbol{g}',\gamma'),\dots,$ take the same values independently. It follows that the derivatives ∂_γ, $\partial_{\gamma'}$, ... in Eq. (2.15) do not generally commute with the potential energy terms of $\mathfrak{L}_2$ [cf. Eq. (2.8')]. This prevents systematic simplifications of the Eqs. (2.15). These eqs. relate the various translation invariant factors K_2, $K_2^{\cdot|\cdot}$, in an unambiguous way, such a property is required if Ω_2 is to be a hypothetical solution of $L_2\Omega_2=W_2$.

The cancellation procedure of the mixed terms of $L_2\Omega_2$, leads to the following remaining diagonal terms, viz.

$$\begin{aligned}L_2\Omega_2(\boldsymbol{x}'\boldsymbol{y}';\boldsymbol{x}''\boldsymbol{y}'')&=\mathfrak{L}_2\mathfrak{K}_2(\boldsymbol{r}\boldsymbol{\mu}\boldsymbol{v},t)\\&+\{W_1(\boldsymbol{x}';\boldsymbol{x}'')\,\Omega_1(\boldsymbol{y}';\boldsymbol{y}'')+\Omega_1(\boldsymbol{x}';\boldsymbol{x}'')\,W_1(\boldsymbol{y}';\boldsymbol{y}'')\}\,K_2(\boldsymbol{r}\boldsymbol{\mu}\boldsymbol{v},t)\pm\text{Exch}\\&+\Omega_1(\boldsymbol{x}';\boldsymbol{x}'')\,\Omega_1(\boldsymbol{y}';\boldsymbol{y}'')\,\mathfrak{L}_2K_2(\boldsymbol{r}\boldsymbol{\mu}\boldsymbol{v},t)\pm\text{Exch}\\&+C(\boldsymbol{x}'\boldsymbol{y}';\boldsymbol{x}''\boldsymbol{y}''),\end{aligned}\tag{2.17}$$

where the first four terms on the r.h.s. of Eq. (2.17) already appeared in Eq. (2.9) and C is the infinite series [see the footnote related to Eq. (2.12)]

$$\begin{aligned}C(\boldsymbol{x}'\boldsymbol{y}';\boldsymbol{x}''\boldsymbol{y}'')=d^2&\sum_{\boldsymbol{g},\gamma}\{\partial_{\boldsymbol{g}}[W_1(\boldsymbol{x}';\boldsymbol{x}'')\,\Omega_1(\boldsymbol{y}';\boldsymbol{y}'')+\Omega_1(\boldsymbol{x}';\boldsymbol{x}'')\,W_1(\boldsymbol{y}';\boldsymbol{y}'')]\}\\&\cdot i\,\partial_\gamma K_2^{g/\gamma}(\boldsymbol{r}\boldsymbol{\mu}\boldsymbol{v},t)\pm\text{Exch}+d^4\sum_{\boldsymbol{g}\gamma,\boldsymbol{g}'\gamma'}\\&\cdot\{\partial_{\boldsymbol{g}'}\partial_{\boldsymbol{g}}[W_1(\boldsymbol{x}';\boldsymbol{x}'')\,\Omega_1(\boldsymbol{y}';\boldsymbol{y}'')+\Omega_1(\boldsymbol{x}';\boldsymbol{x}'')\,W_1(\boldsymbol{y}';\boldsymbol{y}'')]\}\\&\cdot i\,\partial_{\gamma'}\,i\,\partial_\gamma K_2^{g'g/\gamma'\gamma}(\boldsymbol{r}\boldsymbol{\mu}\boldsymbol{v},t)\pm\text{Exch}+\dots\,.\end{aligned}\tag{2.18}$$

K_2 (and W_2 as well) must reflect a good deal of the information stored in Ω_2; this is particularly true if the subunits Ω_1's do not depend much on space and time. Then one would expect the various $K_2^{\cdot|\cdot}$ to be small corrections and the hypothetical series Ω_2 (2.13) and C (2.18) to converge well.

The formal description of the $\boldsymbol{\rho}$-independence of all the factors K_2, $K_2^{\cdot|\cdot}$, springs obviously from the following restrictive condition on W_2 which is deduced from Eq. (2.17)

$$
\begin{aligned}
&W_2(\boldsymbol{x}'\boldsymbol{y}';\boldsymbol{x}''\boldsymbol{y}'')\\
&\qquad\equiv\int d^3\boldsymbol{z}\{V(|\boldsymbol{x}'-\boldsymbol{z}|)-V(|\boldsymbol{x}''-\boldsymbol{z}|)+V(|\boldsymbol{y}'-\boldsymbol{z}|)-V(|\boldsymbol{y}''-\boldsymbol{z}|)\}\\
&\qquad\quad\cdot\Omega_3(\boldsymbol{x}'\boldsymbol{y}'\boldsymbol{z};\boldsymbol{x}''\boldsymbol{y}''\boldsymbol{z})\\
&\qquad= w_{\mathfrak{R}_2}(\boldsymbol{r}\boldsymbol{\mu}\boldsymbol{v},t)+\{W_1(\boldsymbol{x}';\boldsymbol{x}'')\,\Omega_1(\boldsymbol{y}';\boldsymbol{y}'')+\Omega_1(\boldsymbol{x}';\boldsymbol{x}'')\,W_1(\boldsymbol{y}';\boldsymbol{y}'')\}\\
&\qquad\quad\cdot K_2(\boldsymbol{r}\boldsymbol{\mu}\boldsymbol{v},t)\pm\mathrm{Exch}+\Omega_1(\boldsymbol{x}';\boldsymbol{x}'')\,\Omega_1(\boldsymbol{y}';\boldsymbol{y}'')\,w_{K_2}(\boldsymbol{r}\boldsymbol{\mu}\boldsymbol{v},t)\\
&\qquad\quad\pm\mathrm{Exch}+C(\boldsymbol{x}'\boldsymbol{y}';\boldsymbol{x}''\boldsymbol{y}'')
\end{aligned}
\tag{2.19}
$$

because it leads to the Eqs. (2.7) and (2.8) and to the corresponding eqs. with the relevant exchange of parameters. The Eq. (2.8′) and the infinite set of differential Eqs. (2.15), which the correlation factors have to satisfy, are compatible with the assumption of a hypothetical series Ω_2 with $\boldsymbol{\rho}$-independent K_2, $K_2^{\cdot|\cdot}$, because the V-factors of $\mathfrak{L}_2$ are also $\boldsymbol{\rho}$-independent.

In the present paper we shall neither investigate whether the assumption of a hypothetical series Ω_2 (2.13) with translation invariant correlation factors is compatible with $L_2\Omega_2=W_2$, nor analyse the consequential restrictive condition on Ω_3 corresponding to the assumption (2.19). A rigorous analysis would require solving the N body problem in detail. Such a hypothesis including "diatomic correlation" should be related to molecular formation.

The basic ideas operating in the present section are the following. We assume a hypothetical series Ω_2 expressed in terms of two Ω_1's and of translation invariant correlation factors. We found that the form of the terms of such a series is determined by the cancellation procedure of the so-called mixed terms of $L_2\Omega_2$. This procedure is required to allow for the formal description of the translation invariance of the correlation factors. It has also to link the various correlation factors of Ω_2; this is obviously needed in an unambiguous description by Eq. (2.19) of the action of the rest of the system on these correlations. These requirements lead to the type of procedure introduced above. There is a characteristic difference between the set of Eqs. (2.15) which negociate the link between the various K_2, $K_2^{\cdot|\cdot}$, and the Eq. (2.19). If one may use an expression of statistical mechanics, one would say that Eq. (2.15) confront the various K_2, $K_2^{\cdot|\cdot}$, from a microscopic point of view, whereas the global Eq. (2.19) operates on a coarse grained level, because of the presence of local averages of correlation factors included in the W_1's and W_2. By itself the Eq. (2.19) is of course not sufficient to correlate the microscopic structure of the various K_2, $K_2^{\cdot|\cdot}$. The step by step cancellation of the so-called mixed terms of $L_2\Omega_2$ by requiring an adequate form of the subsequent terms of Ω_2 including new correlations $K_2^{\cdot|\cdot}$, induces the linear arrangement of these K_2, $K_2^{\cdot|\cdot}$, and the following sense in the transfer of the action of W_2 (playing here the role of an external condition) on all correlations K_2, $K_2^{\cdot|\cdot}$

$$(W_2, K_2, K_2^{\cdot|\cdot}) \xrightarrow{(2.19)} (w_{\mathfrak{K}_2}, w_{K_2}) \xrightarrow{(2.8)} K_2 \xrightarrow{(2.15)} K_2^{g/\gamma} \xrightarrow{(2.15)} K_2^{g' g/\gamma'\gamma} \longrightarrow \dots \tag{2.20}$$

The unambiguous transfer of the action of the rest of the system on the various K_2, $K_2^{\cdot|\cdot}$, follows from that ordering. The step leading to the translation invariant functions $w_{\mathfrak{K}_2}$ and w_{K_2} is clearly unequivocal because w_{K_2} is multiplied with two Ω_1-factors in Eq. (2.19); hence the two relevant terms have different transformation properties with respect to translations. $w_{\mathfrak{K}_2}$ and w_{K_2} are assumed to be deduced from Eq. (2.19) after calculating the integral form of W_2; with an approximate method of calculating these functions, the assumption (2.19) should be approximately valid by taking as many terms of the series C as it is required by its convergence property.

Let us add a few comments related to particular terms of Eq. (2.19). In principle the term $w_{\mathfrak{K}_2}$ in Eq. (2.19) could be easily obtained from the exact integral form of W_2 including Ω_3. Indeed a translation invariant part of Ω_3 induces a corresponding term in W_2, which in turn requires the presence of $\mathfrak{K}_2$ in Ω_2. The two terms of the type $(W_1 \Omega_1 K_2)$ in Eq. (2.19) are expected in such a diagonal form of W_2. Indeed in the asymptotic configurations (2.2) of the two non local pairs of points, each term of the series C vanishes as well as $w_{\mathfrak{K}_2}$ and w_{K_2} because of the short range character of the correlation factors involved. It follows from Eqs. (2.19) and (2.4)

$$W_2^{as}(\boldsymbol{x}'\boldsymbol{y}'; \boldsymbol{x}''\boldsymbol{y}'') = W_1(\boldsymbol{x}'; \boldsymbol{x}'')\, \Omega_1(\boldsymbol{y}'; \boldsymbol{y}'') + \Omega_1(\boldsymbol{x}'; \boldsymbol{x}'')\, W_1(\boldsymbol{y}'; \boldsymbol{y}'') \pm \text{Exch} \tag{2.21}$$

as it has to be the case if Ω_2 factorizes in the asymptotic configurations (2.2). Then the need for other terms on the r.h.s. of Eq. (2.19) becomes obvious by considering the local limit $\boldsymbol{x}' = \boldsymbol{x}''$, $\boldsymbol{y}' = \boldsymbol{y}''$, of that eq., where

$$W_1(\boldsymbol{x}; \boldsymbol{x}) = 0 = W_2(\boldsymbol{x}\boldsymbol{y}; \boldsymbol{x}\boldsymbol{y}), \quad \text{but} \quad W_1(\boldsymbol{x}; \boldsymbol{y}) \neq 0.$$

Supposing cases where the series C gives only a small contribution to Eq. (2.19), the essential term related to Eq. (2.21) is the non vanishing purely imaginary exchange term

$$\pm \{W_1(\boldsymbol{y}; \boldsymbol{x})\, \Omega_1(\boldsymbol{x}; \boldsymbol{y}) + \Omega_1(\boldsymbol{y}; \boldsymbol{x})\, W_1(\boldsymbol{x}; \boldsymbol{y})\}\, K_2(0, 0, -\boldsymbol{r}, t) \tag{2.22}$$

which clearly requires the presence in Eq. (2.19) of other $\boldsymbol{\rho}$-dependent terms of a different type, if $\boldsymbol{x}$ and $\boldsymbol{y}$ are sufficiently close together.

The present derivation of the hypothetical series expression (2.13) of Ω_2 must be completed by adding the normalisation condition

$$\int d^3\boldsymbol{y}\, \Omega_2(\boldsymbol{x}'\boldsymbol{y}; \boldsymbol{x}''\boldsymbol{y}) = (N-1)\, \Omega_1(\boldsymbol{x}'; \boldsymbol{x}''). \tag{2.23}$$

The set of Eq. (2.15) (and the corresponding eqs. obtained by the relevant exchanges of parameters), combined with the Eqs. (2.19) and (2.23), are a system of simultaneous integro-differential eqs. for the determination of the various correlation factors K_2, $K_2^{\cdot|\cdot}$, of the series Ω_2. They are equivalent to $L_2\Omega_2 = W_2$, in the case of the hypothetical form (2.13) of Ω_2.

We shall end this deduction by raising the following question. There is no point in assuming an unrestricted correlation factor K_2 in Eq. (2.3), because it would merely be a substitute to Ω_2 and no series expression would be needed. Could one replace the series (2.13) of Ω_2 with its translation invariant correlations by another similar series with correlation factors characterised by a requirement other than translation invariance? The formal description of this other condition would again require the use of the equations of motion of these other correlation factors and hence would again lead to the previous procedure of cancelling the so-called mixed terms of $L_2\Omega_2$, and to a series of a similar form as before. However as the new kind of correlations are $\boldsymbol{\rho}$-dependent, the expression (2.13) would have to be generalized by writing $(i\,\partial_{\boldsymbol{g}} \cdot \partial_{\boldsymbol{r}})$, [cf. Eq. (1.11)] instead of $(i\,\partial_{\boldsymbol{g}} \cdot \partial_{\boldsymbol{\gamma}})$ in Eq. (2.13). This would lead to more correlation factors $\bar{K}_2^{\cdot|\cdot}$ and obviously to some ambiguity in the problem of linking these correlations. Thus additional restrictive conditions would be required. It follows that the above formalism seems to be suited only to the case of translation invariant correlation factors K_2, $K_2^{\cdot|\cdot}$.

FRÖHLICH [1] showed that the force term of the Navier-Stokes eq. essentially follows from the symmetry properties (1.6) of Ω_2. Let us discuss the consequence on this force of the more specific series expression (2.13) of Ω_2. A characteristic property of this hypothetical series is its space dependence which is completely determined in terms of Ω_1. It follows that the space dependence of the Navier-Stokes force is also completely determined in the same way. In the following we shall refer to FRÖHLICH's original derivation of this force from the general expression of the force density [cf. Eqs. (3.7) and (3.10) of Ref. 1, where the local limit of the second reduced density matrix, viz. $\Omega_2(\boldsymbol{x}\,\boldsymbol{y};\,\boldsymbol{x}\,\boldsymbol{y})$ is called $P(\boldsymbol{x},\,\boldsymbol{y})$]; the present discussion will be limited to what is particular to the hypothetical series (2.13) of Ω_2.

Useful information about the local value of the translation invariant correlation factors $K_2^{\cdot|\cdot}$ is derived from their symmetry properties, viz.

$$K_2^{\cdot|\cdot}(\boldsymbol{r}\,\boldsymbol{\mu}\,\boldsymbol{\nu},\,t) = K_2^{\cdot|\cdot}(-\boldsymbol{r},\,\boldsymbol{\mu},\,-\boldsymbol{\nu},\,t) = K_2^{\cdot|\cdot\,*}(\boldsymbol{r},\,-\boldsymbol{\mu},\,-\boldsymbol{\nu},\,t). \tag{2.5'}$$

For instance $\partial_{\boldsymbol{\mu}} K_2^{\varrho|\mu}|_{\boldsymbol{\mu}=\boldsymbol{\nu}=0} = 0$, because this derivative is a pure imaginary even function of $\boldsymbol{r}$; from similar considerations one finds

$$\partial_{\boldsymbol{\nu}} K_2^{r|\nu}|_{\boldsymbol{\mu}=\boldsymbol{\nu}=0} = \boldsymbol{r} \cdot i\Gamma_2(r,\,t)\,; \qquad \partial_{\boldsymbol{r}} K_2^{\nu|r}|_{\boldsymbol{\mu}=\boldsymbol{\nu}=0} = \boldsymbol{r} \cdot \Gamma_3(r,\,t)$$

where Γ_2 and Γ_3 are real scalar functions of the magnitude of $\boldsymbol{r}$ and of t.

Approximating the near local development of Ω_1 in terms (i) of the local density field $\sigma_1(\boldsymbol{x})$, (ii) of the internal kinetic energy density $T^0_{\mathrm{kl}}(\boldsymbol{x})$, [assumed rotationally symmetrical $T^0_{\mathrm{kl}}(\boldsymbol{x}) = \delta_{\mathrm{kl}}\,\frac{1}{3}\,T^0(\boldsymbol{x})$], and (iii) of the velocity field $\boldsymbol{v}(\boldsymbol{x})$ only, viz. $(\boldsymbol{x} = \frac{1}{2}[\boldsymbol{x}' + \boldsymbol{x}''],\ \boldsymbol{\xi} = \boldsymbol{x}' - \boldsymbol{x}'')$

$$\Omega_1(\boldsymbol{x}';\,\boldsymbol{x}'') \equiv \sigma_1(\boldsymbol{x},\,\boldsymbol{\xi}) \cdot \exp\,[i\chi_1(\boldsymbol{x},\,\boldsymbol{\xi})] \tag{2.24}$$

where

$$\sigma_1(\boldsymbol{x},\,\boldsymbol{\xi}) \simeq \sigma_1(\boldsymbol{x}) - \frac{m}{\hbar^2}\,\xi_{\mathrm{k}}\,\xi_{\mathrm{l}}\,T^0_{\mathrm{kl}}(\boldsymbol{x}) \tag{2.25}$$

$$\chi_1(\boldsymbol{x},\,\boldsymbol{\xi}) \simeq \frac{m}{\hbar}\,\xi_{\mathrm{k}}\,v_{\mathrm{k}}(\boldsymbol{x})\,; \tag{2.26}$$

[this approximation affects the exchange terms of the series (2.13)] and neglecting the following space derivatives

$$\partial^m\,\sigma_1^2(\boldsymbol{x}),\ m \geqq 2;\ \partial^n\,\boldsymbol{v}(\boldsymbol{x}),\ n \geqq 3;\ [\partial^p\,\sigma_1^2(\boldsymbol{x}) \cdot \partial^q\,\boldsymbol{v}(\boldsymbol{x})],\ p \text{ and } q \geqq 1; \tag{2.27}$$

($T^0(\boldsymbol{x})$ is treated on the same footing as $\sigma_1(\boldsymbol{x})$); one finds that all the terms written explicitly in the series expression (2.13) lead to the Navier Stokes force term

$$-\text{grad}\, p(\boldsymbol{x}) - \eta_1(\boldsymbol{x})\ \text{curl curl}\ \boldsymbol{v}(\boldsymbol{x}) + \eta_2(\boldsymbol{x})\ \text{grad div}\ \boldsymbol{v}(\boldsymbol{x}) \tag{2.28}$$

where the pressure is

$$p(\boldsymbol{x}) = \tfrac{2}{3}\, T^0(\boldsymbol{x}) + a_p\, \sigma_1^2(\boldsymbol{x}) + b_p\, \sigma_1(\boldsymbol{x})\ T^0(\boldsymbol{x}) + c_p\ T^{0\,2}(\boldsymbol{x}) + \dots \tag{2.29}$$

and the viscosity coefficients are ($i=1, 2$)

$$\eta_i(\boldsymbol{x}) = a_{\eta_i}\, \sigma_1^2(\boldsymbol{x}) + b_{\eta_i}\, \sigma_1(\boldsymbol{x})\ T^0(\boldsymbol{x}) + c_{\eta_i}\ T^{0\,2}(\boldsymbol{x}) + \dots. \tag{2.30}$$

The first term on the r.h.s. of Eq. (2.29) originates from the l.h.s. of $L_1\Omega_1 = W_1$; it is the usual pressure term in the case of an ideal gas, thus $T^0(\boldsymbol{x})$ is then expected to be linear in the density and in the absolute temperature. All the other terms in the Eqs. (2.29) and (2.30) arise from $W_1(\Omega_2)$, i.e. from the action of the rest of the system on Ω_1. The factors σ_1^2, $(\sigma_1 T^0)$, $T^{0\,2}$, and the space derivatives in Eq. (2.28) are obtained (i) from the products $(\Omega_1\Omega_1)$ occurring in the series (2.13) by using the near local approximation (2.24)—(2.26), and (ii) from an eventual development of σ_1, T^0, $\boldsymbol{v}$, about $\boldsymbol{x}$, in the case where these macroscopic fields depend on $\boldsymbol{y}$ or $\boldsymbol{\rho} = \frac{1}{2}(\boldsymbol{x}+\boldsymbol{y})$. $\mathfrak{R}_2$ does not contribute to the Navier-Stokes force. The coefficients $a(t)$, $b(t)$, $c(t)$, ... (with suitable units) result from space integration over $\boldsymbol{r} = \boldsymbol{x} - \boldsymbol{y}$, of expressions including the local value of K_2 or $\partial K_2^{\cdot|\cdot}$, times $[\partial_{\boldsymbol{r}} V(r)]$ and some power r^n; [cf. Eqs. (4.14)–(4.16) of Ref. 1]. Under normal conditions the correlations K_2, $K_2^{\cdot|\cdot}$, and the coefficients a, b, c, ... are not expected to depend on the information included in Ω_1, such as the density σ_1.

The term $(-a_p\sigma_1^2)$ in Eq. (2.29) is the van der Waals cohesion forces term; it arises from the original term $[\Omega_1\Omega_1 K_2 + \text{Exch}]$ of Ω_2 (2.13) only and as such it is expected to be an important part of the action of the rest of the system on Ω_1. Its σ_1^2-dependence is a direct consequence of the construction of the hypothetical series (2.13) of Ω_2 in terms of two subunits Ω_1's. From Eqs. (2.24) and (2.25) it follows that $W_1(\Omega_2)$ also includes contributions to the pressure which are proportional to $(\sigma_1 T^0)$ and $T^{0\,2}$ [cf. Eq. (2.29)]; such terms lead to a generalised form of the van der Waals equation with cohesion forces depending on σ_1 and T^0. The internal kinetic energy density T^0 is relatively important in the case of gases, it is of smaller importance in the case of ordinary liquids; thus the occurrence of both $\sigma_1(\boldsymbol{x})$ and $T^0(\boldsymbol{x})$ in the hypothetical results (2.29) and (2.30) is interesting. The main question is whether T^0 can be measured! At the approximation defined by (2.27) the next contribution to the pressure arises from the $\boldsymbol{\xi}^3$-term (energy flux) of χ_1. The appearance of the van der Waals cohesion forces term indicates that the series (2.13) of Ω_2 also includes physically unstable processes.

The first term of Ω_2 (2.13) which leads to the standard viscous force terms is $[(i\,\partial_{\boldsymbol{v}}\,\Omega_1\Omega_1)\ \partial_{\boldsymbol{r}} K_2^{v|r}]$; it is directly related to $\Omega_1\Omega_1 K_2$ through Eq. (2.15). Many more subsequent terms of Ω_2 which include at least one $\partial_{\boldsymbol{v}}$ operating on $(\Omega_1\Omega_1)$ lead also to the viscous force terms [$\partial_{\boldsymbol{v}}$ is the only available derivative leading to a current $(\sigma_1\,\boldsymbol{v})$]. Thus if the series Ω_2 (2.13) were rapidly converging the main part of the viscous force terms would also be related to a single term of Ω_2 only, viz. $[(i\,\partial_{\boldsymbol{v}}\,\Omega_1\Omega_1)\ \partial_{\boldsymbol{r}}\, K_2^{v|r}]$. Such a possible correspondence between the van der Waals cohesion forces and the viscous force terms on the one hand and the first two terms of Ω_2 on the other, seems to support the present hypothetical result of Ω_2.

In the context of the kinetic theory of gases the standard description of the viscosity in terms of transfer of momentum and of mean free path leads to a viscosity coefficient which is independent of the density σ_1. Such a property does not appear in the result (2.30), (where all the terms written explicitly on the r.h.s. of that equation are proportional to σ_1^2 in this case) no doubt because the relevant term, viz. the momentum flux, in the near local expansion of Ω_1 has not been considered in the approximation (2.24). The real problem is the determination of the relative magnitude of the coefficient $a, b, c, \ldots$. This will be discussed in another publication.

Let us emphasize that the series (2.13) of Ω_2 does not lead to Galilean variant force terms such as expression proportional to $\boldsymbol{v}(\boldsymbol{x})$, [which would arise out of a term of Ω_2 including $(\partial_\mu \Omega_1 \Omega_1)$ inappropriate to the hypothetical series (2.13)] or including a single space derivative of $[\sigma_1^2(\boldsymbol{x}) \cdot v^2(\boldsymbol{x})]$ or of $[\sigma_1(\boldsymbol{x}) \cdot T^0(\boldsymbol{x}) \cdot v^2(\boldsymbol{x})]$ (cf. Hyland and Rowlands [2]).

3. Comments about the Derivation of Hypothetical Infinite Series Expressions of Ω_n

The previous procedure of derivation of a hypothetical series of Ω_2 can obviously be applied to any Ω_n. The generalization is straightforward; in this section we shall discuss only few points of interest. One assumes again without proof that Ω_n can be expressed in terms of two reduced density matrices of lower order and of an infinite number of translation invariant correlation factors. Such a hypothesis including polyatomic correlations is expected to be related to the corresponding molecular formation.

First let us assume that Ω_n factorizes, viz.

$$\Omega_n^{as}(\boldsymbol{x}_1' \ldots \boldsymbol{x}_n'; \boldsymbol{x}_1'' \ldots \boldsymbol{x}_n'') = \Omega_{n-1}(\boldsymbol{x}_1' \ldots \boldsymbol{x}_{n-1}'; \boldsymbol{x}_1'' \ldots \boldsymbol{x}_{n-1}'')\, \Omega_1(\boldsymbol{x}_n'; \boldsymbol{x}_n'') \pm \cdots \quad (3.1)$$

when the n pairs of points are in the asymptotic configuration

$$\{\boldsymbol{x}_1' \ldots \boldsymbol{x}_{n-1}', \boldsymbol{x}_1'' \ldots \boldsymbol{x}_{n-1}''\} \rightleftarrows \{\boldsymbol{x}_n', \boldsymbol{x}_n''\}. \quad (3.2)$$

The symmetry properties of Ω_n require n^2 terms of the type $(\Omega_{n-1}\Omega_1)$ on the r.h.s. of Eq. (3.1). Ω_{n-1} and Ω_1 are the $\boldsymbol{\rho}$-dependent subunits of Ω_n which is obtained by a straightforward generalization of Eq. (2.3)

$$\begin{aligned}\Omega_n(\boldsymbol{x}_1' \ldots \boldsymbol{x}_n'; \boldsymbol{x}_1'' \ldots \boldsymbol{x}_n'') &= \mathfrak{K}_n(\boldsymbol{r}_1 \ldots \boldsymbol{r}_{n-1}, \boldsymbol{\mu} \boldsymbol{v}_1 \ldots \boldsymbol{v}_{n-1}, t)\\ &+ \sum_{i,k=1}^{n} (\pm)\, \Omega_{n-1}(\boldsymbol{x}_1' \ldots \boldsymbol{x}_{i-1}' \boldsymbol{x}_n' \boldsymbol{x}_{i+1}' \ldots \boldsymbol{x}_{n-1}'; \boldsymbol{x}_1'' \ldots \boldsymbol{x}_{k-1}'' \boldsymbol{x}_n'' \boldsymbol{x}_{k+1}'' \ldots \boldsymbol{x}_{n-1}'') \quad (3.3)\\ &\cdot \Omega_1(\boldsymbol{x}_i'; \boldsymbol{x}_k'')\, K_n(\boldsymbol{x}_1' \ldots \boldsymbol{x}_{i-1}' \boldsymbol{x}_n' \boldsymbol{x}_{i+1}' \ldots \boldsymbol{x}_{n-1}' \boldsymbol{x}_i'; \boldsymbol{x}_1'' \ldots \boldsymbol{x}_{k-1}'' \boldsymbol{x}_n'' \boldsymbol{x}_{k+1}'' \ldots \boldsymbol{x}_{n-1}'' \boldsymbol{x}_k'', t) + \cdots\end{aligned}$$

where $\mathfrak{K}_n$ is a translation invariant short range correlation term which depend on the relative coordinates $\boldsymbol{\mu}, \boldsymbol{r}_i, \boldsymbol{v}_i$; $(i=1, \ldots, n-1)$, and on time t. [Use is made of an obvious generalization of the previous case with 2 pairs of points, $\boldsymbol{\mu}$ is the non local coordinate associated with the absolute coordinate $\boldsymbol{\rho}$, $2n\boldsymbol{\rho} = \sum_{i=1}^{n} (\boldsymbol{x}_i' + \boldsymbol{x}_i'')$, cf. Eq. (1.7).] K_n is supposed to include the other $\boldsymbol{\rho}$-independent short range correlations which are not described by the factorized form (3.1). For the sake of simplifying the notation the coordinates $\boldsymbol{x}_i'$, $\boldsymbol{x}_i''$, are used as the variables of K_n. $\mathfrak{K}_n$ takes the asymptotic value zero when at least one point is far apart from the $(2n-1)$ other points; each K_n takes the asymptotic value 1 when its first $(n-1)$ pairs of points are far apart from the remaining pair. The symmetry properties obeyed by K_n are

$$\begin{aligned} & K_n(\boldsymbol{x}_1' \ldots \boldsymbol{x}_i' \ldots \boldsymbol{x}_j' \ldots \boldsymbol{x}_n'; \boldsymbol{x}_1'' \ldots \boldsymbol{x}_i'' \ldots \boldsymbol{x}_j'' \ldots \boldsymbol{x}_n'', t) \\ & = K_n(\boldsymbol{x}_1' \ldots \boldsymbol{x}_j' \ldots \boldsymbol{x}_i' \ldots \boldsymbol{x}_n'; \boldsymbol{x}_1'' \ldots \boldsymbol{x}_j'' \ldots \boldsymbol{x}_i'' \ldots \boldsymbol{x}_n'', t), \quad \text{for} \quad (i,j)=(1,\ldots,n) \\ & = K_n^*(\boldsymbol{x}_1'' \ldots \boldsymbol{x}_i'' \ldots \boldsymbol{x}_j'' \ldots \boldsymbol{x}_n''; \boldsymbol{x}_1' \ldots \boldsymbol{x}_i' \ldots \boldsymbol{x}_j' \ldots \boldsymbol{x}_n', t), \quad \text{for} \quad (i,j)=(1,\ldots,n) \\ & = K_n(\boldsymbol{x}_1' \ldots \boldsymbol{x}_j' \ldots \boldsymbol{x}_i' \ldots \boldsymbol{x}_n'; \boldsymbol{x}_1'' \ldots \boldsymbol{x}_i'' \ldots \boldsymbol{x}_j'' \ldots \boldsymbol{x}_n'', t), \quad \text{for} \quad (i,j)=(1,\ldots,n-1). \end{aligned} \tag{3.4}$$

However

$$\begin{aligned} & K_n(\boldsymbol{x}_1' \ldots \boldsymbol{x}_i' \ldots \boldsymbol{x}_{n-1}'\, \boldsymbol{x}_n'; \boldsymbol{x}_1'' \ldots \boldsymbol{x}_i'' \ldots \boldsymbol{x}_n'', t) \\ & \quad \neq \pm K_n(\boldsymbol{x}_1' \ldots \boldsymbol{x}_n' \ldots \boldsymbol{x}_{n-1}'\, \boldsymbol{x}_i'; \boldsymbol{x}_1'' \ldots \boldsymbol{x}_i'' \ldots \boldsymbol{x}_n'', t), \quad \text{for} \quad i=(1,\ldots,n-1). \end{aligned} \tag{3.5}$$

The general $\boldsymbol{\rho}$-independence of $\mathfrak{K}_n$ and K_n must follow from their respective equations of motion, viz.

$$\mathfrak{L}_n \mathfrak{K}_n(\boldsymbol{r}_1 \ldots \boldsymbol{r}_{n-1}, \boldsymbol{\mu}\boldsymbol{v}_1 \ldots \boldsymbol{v}_{n-1}, t) = w_{\mathfrak{K}_n}(\boldsymbol{r}_1 \ldots \boldsymbol{r}_{n-1}, \boldsymbol{\mu}\boldsymbol{v}_1 \ldots \boldsymbol{v}_{n-1}, t) \tag{3.6}$$

$$\overline{\mathfrak{L}}_n K_n(\boldsymbol{r}_1 \ldots \boldsymbol{r}_{n-1}, \boldsymbol{\mu}\boldsymbol{v}_1 \ldots \boldsymbol{v}_{n-1}, t) = w_{K_n}(\boldsymbol{r}_1 \ldots \boldsymbol{r}_{n-1}, \boldsymbol{\mu}\boldsymbol{v}_1 \ldots \boldsymbol{v}_{n-1}, t). \tag{3.7}$$

$\mathfrak{L}_n$ and $\overline{\mathfrak{L}}_n$ include the same differential operators, they are obtained from L_n by dropping the kinetic energy term $(\hbar^2 m^{-1}\, \boldsymbol{\partial}_{\boldsymbol{\rho}} \cdot \boldsymbol{\partial}_{\boldsymbol{\mu}})$ of L_n. For any subdivision of the n pairs of coordinates into a group of $(n-1)$ pairs and a remaining pair, L_n can of course always be expressed in the form

$$L_n = i\hbar\, \partial_t + \frac{\hbar^2}{m}\left[\boldsymbol{\partial}_{\boldsymbol{\rho}} \cdot \boldsymbol{\partial}_{\boldsymbol{\mu}} + \sum_{i=1}^{n-1} \boldsymbol{\partial}_{\boldsymbol{r}_i} \cdot \boldsymbol{\partial}_{\boldsymbol{v}_i}\right] - \mathfrak{V}_{\Omega_{n-1}} - \mathfrak{V}_{K_n} \tag{3.8}$$

where $\mathfrak{V}_{\Omega_{n-1}}$ and $\mathfrak{V}_{K_n}$ involve all the relevant two-body interaction potentials V which matter in the description of the evolution of the subunits Ω_{n-1} and K_n of Ω_n. For instance in the case of the particular term $(\Omega_{n-1} \cdot \Omega_1 \cdot K_n)$ of Eq. (3.3) with the same distribution of parameters as in the term written explicitly on the r.h.s. of Eq. (3.1), one finds that the potential energy part of L_n is subdivided into the following groups,

$$\begin{aligned} \mathfrak{V}_{\Omega_{n-1}} &= \sum_{\substack{i,k=1 \\ i<k}}^{n-1} \left[V(|\boldsymbol{x}_i' - \boldsymbol{x}_k'|) - V(|\boldsymbol{x}_i'' - \boldsymbol{x}_k''|)\right] \\ \mathfrak{V}_{K_n} &= \sum_{k=1}^{n} \left[V(|\boldsymbol{x}_k' - \boldsymbol{x}_n'|) - V(|\boldsymbol{x}_k'' - \boldsymbol{x}_n''|)\right]. \end{aligned} \tag{3.9}$$

Then $\mathfrak{V}_{K_n}$ is the potential energy part of $\overline{\mathfrak{L}}_n$, whereas $\mathfrak{L}_n$ includes all the V-terms of L_n. The Eqs. (3.6) and (3.7) have to be completed by similar eqs. obtained by permutation of variables.

The unambiguous formal description of the translation invariance of the correlation factors of Ω_n and of the transfer of the action of the rest of the system on such factors requires the formal dissociation of Eqs. (3.6) and (3.7) from $L_n \Omega_n = W_n$, and the elimination of the so-called mixed term of $L_n \Omega_n$. This leads to a well defined hypothetical series expression of Ω_n with a form similar to the series Ω_2 (2.13).

An arbitrariness inherent in the series expression (3.3) appears in the assumption made about the choice of factorization of Ω_n in subunits. It is obvious that the previous choice in terms of Ω_{n-1} and Ω_1 is not the only possible one. Using the V-factors of L_n as heuristic tool, one finds easily that there are $\left[\frac{n}{2}\right]$ possible types of factorisation of Ω_n in suitable subunits which are constructed from the products $(\Omega_{n-k} \cdot \Omega_k)$, $k = \left(1, \ldots, \left[\frac{n}{2}\right]\right)$. (We use the notation $[x]$ = greatest integer contained in x.) Each choice leads to a hypothetical unambiguously defined series expression of Ω_n of the particular type in question.

Acknowledgements

We would like to express our thanks to Professor H. Fröhlich for useful remarks.

References

1. Fröhlich, H.: Physica **37**, 215 (1967).
2. Hyland, G. J., Rowlands, G.: Phys. Letters **33** A, 197 (1970).

Towards a Generalized Ginzburg-Landau Equation

G. J. HYLAND and G. ROWLANDS

University of Warwick, Department of Physics, Coventry/England

Abstract

By making reasonable assumptions on the space-time dependence of the first reduced density matrix in superconductors, a time-dependent generalization of the Ginzburg-Landau equation is obtained which is valid at all temperatures, provided only that the scale of the space-time variation of the macroscopic wave-function be sufficiently slow in comparison to that of the phonon-induced electron-electron interaction—a condition equivalent only to assuming the existence of local equilibrium.

The structure of the macroscopic wave equation is found to be wave-like over the whole temperature range, in accord with phenomenological considerations but in contradiction to the results of certain analyses based on microscopic theory.

1. Introduction

"Much effort is at present spent on attempts of improving and generalizing the Ginzburg-Landau equations for superconductors, by going into great details of the microscopic theory. Such efforts, I think, are out of place ..." [1].

In the years following FRÖHLICH's epochal contribution to the microscopic theory of superconductivity, which sprang from his introduction of field theoretical methods into solid state physics, it became increasingly clear that the connection between micro and macro physics was by no means as straightforward as was implied by the *tour de force* approaches to the quantum mechanical many-body problem that were quickly developed after 1950 and that new concepts were required.

Already by 1961 the seeds of one such new concept were sown in a conjecture that the difficulties attendant on current quantum hydrodynamics, associated with the necessity for atomistic considerations, could be "... overcome by the introduction of non-local features into hydrodynamics" [2]. Elaborated somewhat further in the BETHE Festschrift in 1966 [3] this concept of non-locality, finally came to fruition the following year with the realization that the required non-local features were contained in a natural way in certain complex, time-dependent, macroscopic fields known as the *reduced density matrices*, whose exact equations of motion formed a hierarchy leading gradually from macro to microscopic descriptions of the system. As illustrative of the power of this formalism FRÖHLICH derived [4], in a non-truncative way and with the minimum of mathematical effort, the Navier-Stokes' equations of classical hydrodynamics. The derivation revealed these equations to be a consequence of a very basic symmetry property of the system—in particular, that of the two-particle correlation density (a quantity closely related to the second reduced density matrix) under particle interchange. The procedure required only the existence of a hydrodynamical equilibrium state of the fluid and involved no obscure mathematical approximations—so typical of conventional many-body techniques—whose physical implications evade interpretation. The penalty paid for the simplicity of FRÖHLICH's approach—and its extension by the authors [5] requiring the existence only of a state of *local* equilibrium—is that it yields only the structure of the macroscopic equations and *not* the magnitudes of the parameters (e.g. transport coefficients) which appear therein; it is only at this stage, in connection with the calculation of these quantities should consideration of specific material properties enter. "Such treatments must, of course, make use of specific detailed methods used in many-body problems. Such methods invariably are approximations and, therefore, allow for considerable flexibility. They should be devised such that the general (structural) equations remain satisfied automatically, an important rule that is frequently ignored" [6, adapted].

Perhaps of more interest is the possible application of this philosophy to the derivation of macroscopic equations describing the flow properties of superfluids and superconductors. It has long been suggested, since the pioneering work of F. LONDON [7], that the peculiar flow properties of such systems are connected with the existence therein of macroscopic wave-functions and (to quote LONDON) an associated "long-range order of the average momentum". Crucial to the realization of this programme was YANG's empirical observation [8] that this long-range order is an asymptotic property (in configuration space) of an appropriate reduced density matrix. On entering the superfluid (superconductive) phase this reduced density matrix assumes a dominating eigenvalue whose associated eigenfunction is the macro wave-function contemplated by LONDON; the dominating eigenvalue being the density, in momentum space, of the condensed part of the system.

Utilizing the asymptotic properties of the reduced density matrices in the presence of this "off-diagonal-long-range-order" (O.D.L.R.O.], FRÖHLICH showed how the macro wave equations satisfied by these macro wave-functions could be derived in a virtually *exact* way [1, 9, 6]. They have the form of integro-differential equations and involve a coupling between the condensed and uncondensed parts of the system. In the case of superfluid boson systems (liquid ^{4}He II), in which O.D.L.R.O. occurs in the first reduced (or one-particle) density matrix, the macro wave-function is a complex, time-dependent field function of one spatial variable, whilst for superfluid fermion systems (superconductors) wherein O.D.L.R.O. can first occur in the second reduced (or two-particle) density matrix—the Pauli principle preventing its earlier occurrence—the macro wave-function is a similar function of two spatial variables; in the light of the microscopic theory this two-point function finds interpretation as the macro wave-functon of the Schafroth condensate of electron (Cooper) pairs.

At this stage it must be pointed out that the reduced density matrix describing the uncondensed part of the system—which through the FRÖHLICH integro-differential equations is coupled to the condensed part—itself satisfies a hierarchial system of equations. From these equations the authors have recently derived [10], in a non-truncative way, a closed set of hydrodynamic equations closely resembling those of the thermodynamic two-fluid model of superfluid ^{4}He II. A first step towards the derivation of similar equations for superconductors must be the establishment of a (generalized) Ginzburg-Landau equation. For FRÖHLICH has already pointed out [11] the close connection between this equation and those of inviscid hydrodynamics. The validity of these much used equations, first derived phenomenologically in 1950 [12], is restricted to equilibrium situations in the immediate vicinity of the superconducting transition temperature, and of late there have been many attempts [13] based exclusively on GORKOV's [14] formulation of the microscopic theory, towards a generally valid, time-dependent generalization.

However, as stressed repeatedly by FRÖHLICH ... "such efforts are out of place" [1], and that a better way to proceed is in two distinct stages: firstly obtain the structure of the macroscopic equations and *then* calculate the magnitudes and temperature dependences, for example, of parameters occuring therein. For in this way those approximations necessary for the calculation of the latter are clearly separated from those (usually of a very general and unrestrictive nature) necessary to obtain the structure.

In view of FRÖHLICH's fundamental contribution to the microscopic theory of superconductivity and his recent work on the connection between micro and macro physics, it is, perhaps, particularly fitting that this contribution to "Cooperative Phenomena"—a publication in his honour—should concern itself with a persuance of the above considerations with respect to the derivation of one of the best known macroscopic equations of superconductivity, itself being one of the most striking examples of all cooperative phenomena, leading, as it does, to the establishment of a thermodynamic phase exhibiting quantum mechanical effects on a macroscopic scale.

2. Equations of Motion

Following YANG [8] it is assumed that the onset of the superconducting phase is characterized by the appearance of O.D.L.R.O. in the second reduced density matrix Ω_2, which accordingly then factorizes in the following way (wherein the variables x and y are understood to include spin as well as space coordinates)

$$\Omega_2(\underline{x}', \underline{y}'; \underline{x}'', \underline{y}'' |t) = \Phi(\underline{x}', \underline{y}' |t)\, \Phi^*(\underline{x}'', \underline{y}'' |t) + \Lambda_2(\underline{x}', \underline{y}'; \underline{x}'', \underline{y}'' |t). \tag{1}$$

This introduces the two-point macro wave-function Φ, and Λ_2, the second reduced density matrix describing normal correlations between electrons, which is negligible unless pairs of coordinates are close together.

In terms of fermion wave operators (ψ), Ω_2 is defined by

$$\Omega_2(\underline{x}', \underline{y}'; \underline{x}'', \underline{y}'' |t) = \mathrm{Tr}\,\left(\psi^+(\underline{x}'', t)\, \psi^+(\underline{y}'', t)\, \psi(\underline{y}', t)\, \psi(\underline{x}', t)\, \Omega(t)\right) \tag{2}$$

where $\Omega(t)$ is the von Neumann density matrix, satisfying

$$i\hbar\, \partial_t \Omega = [H, \Omega]_- \tag{3}$$

H is the electron hamiltonian operator defined by

$$H = \frac{\hbar^2}{2m} \int \partial\psi^+(\underline{x})\, \partial\psi(\underline{x})\, d^3\underline{x} + \frac{1}{2} \oint\!\!\oint V(|\underline{x}-\underline{y}|)\, \psi^+(\underline{x})\psi^+(\underline{y})\psi(\underline{y})\psi(\underline{x})\, d^3\underline{x}\, d^3\underline{y} \quad (4)$$

where V is an instantaneous approximation to the FRÖHLICH's phonon-induced electron-electron interaction.

The required macro wave equation for Φ follows from the exact equation of motion for Ω_2, readily derivable using (2) (3) and (4), by exploiting the asymptotic behaviour of (1) and the finite range of V. The result, first given by FRÖHLICH [9] and later via a slightly different route by TAYLOR [15], is

$$i\hbar\, \partial_t\, \Phi(\underline{x}, \underline{y}) + \left\{ \frac{\hbar^2}{2m} (\partial^2_{\underline{x}} + \partial^2_{\underline{y}}) - V(|\underline{x}-\underline{y}|) \right\} \Phi(\underline{x}, \underline{y}) = S(\underline{x}, \underline{y}) \quad (5)$$

where

$$\begin{aligned} S(\underline{x}, \underline{y}) = \oint d^3\underline{z}\, [V(|\underline{z}-\underline{x}|) + V(|\underline{z}-\underline{y}|)]\, [\Omega_1(\underline{z}; \underline{z})\, \Phi(\underline{x}, \underline{y}) - \Omega_1(\underline{x}; \underline{z})\, \Phi(\underline{z}, \underline{y}) \\ + \Omega_1(\underline{y}; \underline{z})\, \Phi(\underline{z}, \underline{x})]. \end{aligned} \quad (6)$$

It is seen that this equation couples Φ to Ω_1, the first reduced density matrix [whose local and near local limits are, respectively, the total electron density and total electron current density], and this arises because the equation of motion for Ω_2 involves, in the usual hierarchial way, Ω_3, the third reduced density matrix. Following TAYLOR [15], Ω_3 can be written

$$\Omega_3 \approx \Omega_1 \Omega_2 + \Lambda_3 \quad (7)$$

where, like Λ_2, Λ_3 vanishes if any two of its space coordinates are sufficiently far from the others.

Ω_1 is further related to Ω_2 through the integration condition

$$(N-1)\, \Omega_1(\underline{x}'; \underline{x}'') = \oint \Omega_2(\underline{x}', \underline{y}; \underline{x}'', \underline{y})\, d^3\underline{y} \quad (8)$$

where $N (= \oint \Omega_1(\underline{x}; \underline{x})\, d^3\underline{x})$ is the total mean number of electrons.

Utilizing the definition of $\Omega_1(\underline{x}'; \underline{x}'') = \mathrm{Tr}\, (\psi^+(\underline{x}'')\, \psi(\underline{x}')\, \Omega)$, together with (3) and (4), a corresponding equation of motion for Ω_1 is obtained in terms of Ω_2; in the presence of O.D.L.R.O. in Ω_2, as defined by (1), the equation takes the form

$$i\hbar\, \partial_t\, \Omega_1(\underline{x}'; \underline{x}'') + \frac{\hbar^2}{2m} (\partial^2_{\underline{x}'} - \partial^2_{\underline{x}''})\, \Omega_1(\underline{x}'; \underline{x}'') = W_1(\underline{x}', \underline{x}'') \quad (9)$$

where

$$\begin{aligned} W_1(\underline{x}', \underline{x}'') = \oint \{V(|\underline{x}'-\underline{y}|) - V(|\underline{x}''-\underline{y}|)\} \\ \cdot [\Phi(\underline{x}', \underline{y})\, \Phi^*(\underline{x}'', \underline{y}) + \Lambda_2(\underline{x}', \underline{y}; \underline{x}'', \underline{y})]\, d^3\underline{y}. \end{aligned} \quad (10)$$

Thus the coupled pair of Eqs. (5) and (9) do not form a *closed* set unless some assumption is made on the form of Λ_2. To date the most widely used assumption (c.f. GORKOV [14]) is the Hartree-Fock approximation

$$\Lambda_2^{\mathrm{H.F.}}(\underline{x}', \underline{y}; \underline{x}'', \underline{y}) = \Omega_1(\underline{x}'; \underline{x}'')\, \Omega_1(\underline{y}; \underline{y}) - \Omega_1(\underline{x}'; \underline{y})\, \Omega_1(\underline{y}; \underline{x}''). \quad (11)$$

Within the framework of the present approach no such approximation is necessary, however.

For the purpose of the following it will be assumed that the pair condensate is in a singlet state, whence $\Phi(\underline{x}, \underline{y})$ is a symmetric function† of its spatial coordinates. In this case, the spin summations are such that the spin indices which occur in Ω_1 on the R.H.S. of (6) are always equal; accordingly, the equation of motion for $\Omega_1^{\uparrow\uparrow}$ (or $\Omega_1^{\downarrow\downarrow}$) only is required. Assuming now that

$$\Omega_1^{\uparrow\uparrow}(\underline{x}'; \underline{x}'') = \Omega_1^{\downarrow\downarrow}(\underline{x}'; \underline{x}'') = \tfrac{1}{2}\Omega_1(\underline{x}'; \underline{x}'') \tag{12}$$

where $\Omega_1(\underline{x}; \underline{x}) = \varrho(\underline{x})$, the total electron density at $\underline{x}$ (irrespective of spin), the spin functions can be removed yielding for $S(\underline{x}, \underline{y})$ and $W_1(\underline{x}', \underline{x}'')$, respectively, the following forms (where now $\underline{x}$, $\underline{y}$ and $\underline{z}$ denote space coordinates only)

$$\begin{aligned} S(\underline{x}, \underline{y}) = \tfrac{1}{2}\int \{V(|\underline{x}-\underline{z}|) + V(|\underline{y}-\underline{z}|)\} \\ \cdot [2\varrho(\underline{z})\,\Phi(\underline{x}, \underline{y}) - \Omega_1(\underline{x}; \underline{z})\,\Phi(\underline{z}, \underline{y}) - \Omega_1(\underline{y}; \underline{z})\,\Phi(\underline{x}, \underline{z})]\, d^3\underline{z} \end{aligned} \tag{13}$$

and

$$\begin{aligned} W_1(\underline{x}', \underline{x}'') = \tfrac{1}{2}\int \{V(|\underline{x}'-\underline{y}|) - V(|\underline{x}''-\underline{y}|)\} \\ \cdot [4\Phi(\underline{x}', \underline{y})\,\Phi^*(\underline{x}'', \underline{y}) + 2\varrho(\underline{y})\,\Omega_1(\underline{x}'; \underline{x}'') - \Omega_1(\underline{x}'; \underline{y})\,\Omega_1(\underline{y}; \underline{x}'') \\ + 2\lambda_2(\underline{x}', \underline{y}; \underline{x}'', \underline{y})]\, d^3\underline{y} \end{aligned} \tag{14}$$

where $\lambda_2 = \Lambda_2 - \Lambda_2^{\text{H.F.}}$.

3. Towards the Ginzburg-Landau Equations

The macro wave equation [Eq. (5) with (13)] presented in the preceding section couples the two-point macro wave-function Φ to the first reduced density matrix Ω_1 of the electron system, whose own equation of motion [(9) with (14)] involves, in turn, a coupling to Φ; as stressed by FRÖHLICH on numerous occasions [1, 6, 9], these coupled equations must form the basis for the derivation of the Ginzburg-Landau equation and any generalization thereof.

The Ginzburg-Landau equations are, however, differential equations involving but one field quantity, the one-point macro wave-function $\Psi(\underline{x})$. To proceed, therefore, it is clearly necessary first to define Ψ in terms of Φ, and secondly to treat the non-local (integral) terms in the wave equation which couple Φ (and eventually Ψ) to Ω_1; the non-linearity of the equation of motion for Ω_1 precludes, however, its solution (except in the trivial case of translational invariance, with $\lambda_2 = 0$, whence it reduces to $0=0$ [15]). A somewhat analogous situation was encountered in FRÖHLICH's derivation of the Navier-Stokes' equation in terms of the reduced density matrices [4]. In that case, the equation of motion for Ω_1, from which follows, in the near-local limit, the equation of motion of the mass current density, involves the second reduced density matrix, Ω_2. The important point which emerged from that work was that solution for Ω_2 was *not required* and that progress could be made simply by exploiting the microscopic symmetry properties of Ω_2 and requiring the existence of a state of hydrodynamic equilibrium.

† From the definition of Ω_2 given by (2) it follows that $\Omega_2(\underline{x}', \underline{y}'; \underline{x}'', \underline{y}'') = -\Omega_2(\underline{y}', \underline{x}'; \underline{x}'', \underline{y}'')$, implying that Φ is a totally antisymmetric function of its space and spin coordinates.

The present work was stimulated by the belief that treatment of the wave equation [Eq. (5) with (13)] for superconductors in a similar way should lead to the Ginzburg-Landau equations and their extensions; this is indeed found to be the case.

It is instructive at this stage to refer to a closely allied problem—that of the derivation of hydrodynamic equations for superfluid boson systems (e.g. ^{4}He II). In this case, FRÖHLICH has shown [1, 6, 9] that the imposition of O.D.L.R.O. on Ω_1 leads to two coupled, non-linear, integro-differential equations for the one-point macro wave-function $\phi(\underline{x})$ and for a non-local quantity $\Lambda_1(\underline{x}'; \underline{x}'')$, characterizing respectively, the condensed and uncondensed (in momentum space) parts of the system. As in the superconducting case, however, the coupled equations are not closed, since the equation of motion for Λ_1 involves, in a hierarchial way, a quantity Λ_2, closely related to Ω_2, describing two-particle correlations in the uncondensed part of the system. Progress in this case proved possible [5] by making a CHAPMAN-ENSKOG-type expansion of ϕ, Λ_1 and Λ_2 in terms of the gradients of their associated macroscopic intensive variables about a suitably defined state of local equilibrium; examination of the system of non-local, coupled equations revealed this state to be one of constant mass density in which there existed a constant relative velocity between the velocity fields $\underline{v}_c$ and $\underline{v}_d$ associated respectively with ϕ and Λ_1. In this way the system of equations was closed (it never being necessary to solve for Λ_2) and simultaneously made local. The resulting set of differential, hydrodynamic equations were not, however, those of the thermodynamic two-fluid model since the latter are based on the concepts of a superfluid and normal mass densities (ϱ_s and ϱ_n) and their associated velocity fields, $\underline{v}_s$ and $\underline{v}_n$. Now whilst $\underline{v}_s = \underline{v}_c$, $\varrho_s \neq |\phi|^2$; for at the absolute zero of temperature ϱ_s becomes equal to the total mass density of the system, whilst recent theoretical and experimental work [16] reveals that $|\phi|^2$ is here equal to only about 10% of the total density.

It is in terms of these super/normal variables, however, that the Ginzburg-Pitaevskii equations [17] (and their time-dependent extensions, given later by PITAEVSKII [18] and KHALATNIKOV [19])—the analogue for superfluidity of the Ginzburg-Landau equations—are formulated, i.e. $|\psi|^2 = \varrho_s$; thus $\phi \neq \psi$, where ψ is the macro wave-function introduced by PITAEVSKII and KHALATNIKOV. Now whilst the connection between condensate/depletion (noncondensate) and super/normal variables has been elucidated [10] at the level of the associated hydrodynamic equations, the same has not yet proved possible at the $\phi - \psi$ level.

The situation in superconductivity is, in this respect, somewhat simpler; for at the absolute zero of temperature all the electrons occupy identical Cooper pair states, i.e. there is no depletion in the ^{4}He II sense, but only in the sense implied by the integration condition (8) together with (1), namely that $\int \Phi(\underline{x}', \underline{y})\, \Phi^*(\underline{x}'', \underline{y})\, d^3\underline{y}$ is of order N, whilst $\int \Lambda_2(\underline{x}', \underline{y}; \underline{x}'', \underline{y})\, d^3\underline{y}$ is of order N^2, N being the total mean number of electrons. This discrepancy in order is simply a reflection of the relative dominance of bound ($\Phi\Phi^*$) versus unbound (Λ_2) pair correlations, the latter dominating because of the great spatial separation (of the order of the Pippard coherence length $\sim 10^{-4}$ cms) between the partners of individual Cooper pairs, in consequence of which many other electrons each (at the absolute zero of temperature) belonging to different bound pairs can penetrate and participate in normal, two-electron correlations expressed through

Λ_2. Thus the existence of a dominant (incoherent) part of Ω_2 is in no way inconsistent with a total condensation into pair states.

To proceed then for superconductivity in a way analogous to that developed by the authors for superfluids we do not attempt to solve for Ω_1, but rather close the macro wave equation by making an expansion of Ω_1, in terms of gradients of its associated macroscopic intensive variables about a state of local superconducting equilibrium. This we now proceed to consider together with the definition of Ψ in terms of Φ.

4. Local Equilibrium and the Macro Wave-Function Ψ

Before discussing the form of Ω_1 appropriate to local equilibrium in a superconductor, it is necessary to introduce the Ginzburg-Landau wave-function Ψ in terms of the two-point macro wave-function Φ.

To this end an uncoupling between centre-of-gravity and internal motions of the condensed electron pairs is assumed possible*, and following FRÖHLICH (e.g. [6]) we write**

$$\Phi(\underline{x}, \underline{y}) = u(\underline{r})\, \Psi(\underline{R}) \tag{15}$$

where $\underline{r} = \underline{y} - \underline{x}$ and $\underline{R} = \frac{\underline{x} + \underline{y}}{2}$. This identifies the Ginzburg-Landau wave-function with the centre-of-gravity wave-function of the Schafroth pair condensate. Unlike the internal wave-function $u(\underline{r})$ which is significant only over the dimensions of a Cooper pair (of order the Pippard coherence distance $\approx 10^{-4}$ cms), $\Psi(\underline{R})$ spreads throughout the entire volume of the whole system; the appearance of O.D.L.R.O. in Ω_2 is thus equivalent to the existence of long-range phase correlations in the $\Psi(\underline{R})$.

Clearly $\Psi(\underline{R})$ should be normalized to the pair centroid density of the Schafroth condensate.

Although O.D.L.R.O. does not (because of the Pauli principle) occur intrinsically in Ω_1, Ω_1 is related, via the integration condition (8) to Ω_2. Accordingly, the existence of a macro wave-function Ψ in Ω_2 must be reflected in Ω_1, wherein it constitutes an intensive variable, *additional* to those necessary for a characterization of Ω_1 appropriate to the normal (non-superconducting) state—e.g. total density σ_n, current density $\underline{J}_n$, internal kinetic energy density, etc.

Thus

$$\Omega_1^s(\underline{x}'; \underline{x}''|t) = \Omega_1^s(|\underline{r}|, \sigma_n(\underline{R}), \underline{J}_n(\underline{R}), \ldots; \Psi(\underline{R})\, |t) \tag{16}$$

where Ω_1^s denotes the first reduced density matrix appropriate to the superconductive state. In this case, the local limit ($\underline{x}' = \underline{x}''$ or $r = 0$) of Ω_1^s defines the total electron density at $\underline{x}$, which, of course, at finite temperatures, receives contributions from both paired and unpaired electrons.

Discussion of the local equilibrium form of Ω_1^s is greatly facilitated by the assumption that the normal intensive variables of Ω_1 relax *instantaneously* to

* The validity of this approximation becomes doubtful in the case of films of a thickness comparable to the internal size (coherence length) of the electron pairs.

** The symmetry of Φ with respect to spatial coordinates assumed in § 2 on the basis of the assumption of a singlet condensate implies u is an *even* function of $\underline{r}$; further u can be chosen to be real.

their absolute (constant) time-independent equilibrium values: $\sigma_n(\underline{R})=\sigma_n^0$, $\underline{J}_n(\underline{R})=0$ etc., Ω_1 ceasing to be *explicitly* time-dependent. The validity of this assumption requires the characteristic time for interaction between the heat bath (lattice vibrations) and the thermal excitations of the system, i.e. the normal fluid, to be short compared with the scale of the time variation of $\Psi(\underline{R})$. (There is of course no analogous simplifying assumption in the case of superfluid ^{4}He II owing to the absence of a crystal lattice; in this case the normal fluid constitutes its own heat bath.)

Accordingly, the only space ($\underline{R}$) and time dependence of Ω_1 is through $\Psi(\underline{R}, t)$; the existence of a non-trivial *local* equilibrium thus relies on a space-time dependent Ψ, Ω_1^s retaining its absolute equilibrium form in so far as its dependence on the normal intensive variables is concerned. An anticipated consequence of this infinitely fast relaxation of the normal processes is the absence of any dissipation in the eventual Ψ equation.

Finally suppressing any supercurrent in local equilibrium (for this current is derived from spatial gradients of Ψ, which in the spirit of the expansion method arise only upon departure from local equilibrium) we write

$$\Omega_1^s(\underline{x}'; \underline{x}'')=\Omega_1^{\text{loc}}(|\underline{r}|, |\Psi(\underline{R}, t)|^2; \ldots)[1+\delta\Omega_1^s] \tag{17}$$

where ... denotes normal intensive variables which retain their constant (space and time independent) values appropriate to absolute equilibrium†.

$\delta\Omega_1^s$ embodies departures from the local equilibrium described by Ω_1^{loc} and involves spatial gradients of Ψ and Ψ^* (treated as independent variables); these gradients introduce, amongst other things, the supercurrent density, $\underline{J}_s$, defined by

$$\underline{J}_s(\underline{R})=\frac{\hbar}{4im}\{\Psi^*(\underline{R})\,\underline{\nabla}\,\Psi(\underline{R})-\Psi(\underline{R})\,\underline{\nabla}\,\Psi^*(\underline{R})\}. \tag{18}$$

The general structure of $\delta\Omega_1^s$ is dictated solely by the microscopic symmetry of $\Omega_1^s(\underline{x}'; \underline{x}'')$ under coordinate interchange

$$\Omega_1^s(\underline{x}'; \underline{x}'')=\{\Omega_1^s(\underline{x}''; \underline{x}')\}^* \tag{19}$$

and by the requirement that Ω_1^s be invariant under phase transformations of the first kind; thus Ψ, Ψ^* and their gradients must occur always in pairs.

Finally, to obtain equations of maximum significance we take $\underline{J}_s$ (itself a gradient of the phase of Ψ) to be of the same order as the gradient of the amplitude of Ψ. This effective expansion in $\underline{J}_s$ (and its derivatives) is well motivated by the fact that the range of convergence of the $\underline{J}_s$ expansion far exceeds that appropriate to the critical supercurrent above which superconductivity is destroyed. With this ordering the coefficients which appear in the eventual Ginzburg-Landau equation, though functions of $|\Psi(\underline{R})|^2$, are *not* functions of $\underline{J}_s$; this is somewhat analogous to the ordering adopted in the superfluid case [5] where the final coefficients which appear in the hydrodynamic equations are functions of the condensate density but not of the relative velocity between the condensate and depletion.

† The presence of $|\Psi|^2$ in Ω_1^{loc} reflects the absence of any supercurrent; Ψ is real here.

5. The Generalized Ginzburg-Landau Equation

To proceed we rewrite (5) in terms of $u(\underline{r})$ and $\Psi(\underline{R})$ using (15). The required equation for $\Psi(\underline{R})$ follows upon multiplying by $u(\underline{r})$ and integrating over $\underline{r}$; in this way we obtain, assuming that on the time scale of Ψ, u is time-independent

$$i\hbar\, \partial_t \Psi(\underline{R}) + \frac{\hbar^2}{8m} \nabla^2 \Psi(\underline{R}) - V_0 \Psi(\underline{R}) = \bar{S}(\underline{R}) \tag{20}$$

where

$$\begin{aligned}\bar{S}(\underline{R}) \int u^2(|\underline{r}|)\, d^3\underline{r} = {} & 2\Psi(\underline{R}) \int V(|\underline{z}|)\, u^2(|\underline{r}|)\, \varrho\left(\underline{R}+\underline{z}+\frac{\underline{r}}{2}\right) d^3\underline{r}\, d^3\underline{z} \\ & - \int V(|\underline{z}|)\, u(|\underline{z}|)\, u(|\underline{r}|)\, \Omega_1\left(\underline{r}+\underline{z}, \underline{R}+\frac{\underline{z}}{2}\right) \\ & \cdot \Psi\left(\underline{R}+\frac{\underline{r}}{2}+\frac{\underline{z}}{2}\right) d^3\underline{r}\, d^3\underline{z} - \int V(|\underline{z}|)\, u(|\underline{r}|)\, u(|\underline{r}+\underline{z}|) \\ & \cdot \Omega_1\left(\underline{z}, \underline{R}+\frac{\underline{r}}{2}+\frac{\underline{z}}{2}\right) \Psi\left(\underline{R}+\frac{\underline{z}}{2}\right) d^3\underline{r}\, d^3\underline{z},\end{aligned} \tag{21}$$

where

$$V_0 \int u^2(|\underline{r}|)\, d^3\underline{r} = \int \left[V(|\underline{r}|)\, u^2(|\underline{r}|) + \frac{\hbar^2}{2m}\left(\frac{\partial u}{\partial r}\right)^2\right] d^3\underline{r} \tag{22}$$

and, for example $\Omega_1(\underline{x}; \underline{y})$ is rewritten as $\Omega_1(\underline{r}, \underline{R})$.

To obtain from (20) with (21) and (22) a local (differential) equation for Ψ it is only necessary to assume that Ψ is slowly varying in comparison to u and V; in practice, this means slow in comparison to u, since the range of V is surely less than 10^{-6} cms, whilst u is significant over the Pippard coherence length ($\sim 10^{-4}$ cms).

Developing the integrands under these assumptions, and using the form of Ω_1^s discussed in § 4, we obtain a local expression for $\bar{S}$, whence (20) can be written

$$\begin{aligned} & i\hbar\, \partial_t \Psi(\underline{R}) + \frac{\hbar^2}{8m} \nabla^2 \Psi(\underline{R}) \\ & \quad = [A + \{B \nabla^2 \varrho_s + C \operatorname{div} \underline{J}_s + D \underline{J}_s^2 + E \underline{J}_s \cdot \underline{\nabla} \varrho_s + F(\underline{\nabla} \varrho_s)^2\}] \Psi(\underline{R}) \\ & \quad\quad + G \nabla^2 \Psi(\underline{R}) \end{aligned} \tag{23}$$

with

$$\frac{\varrho_s}{2} \equiv |\Psi|^2 \tag{24}$$

where the unknown coefficients A, B, C etc. are all expressible in terms of the coefficients which arise in the gradient expansion of $\delta\Omega_1^s$; certain inter-relations between these latter coefficients may well arise from the requirement that the Ω_1^s ansatz be consistent with the equation of motion for Ω_1^s, (9). Further from the structure of $\delta\Omega_1^s$ it can be concluded that A, D and G are all real; in the case where V is taken to be a delta function, B and F are also real. [All coefficients are, remember, functions of $|\Psi(\underline{R})|^2$].

Clearly the term $G\nabla^2\Psi$ can be absorbed into the second term on the L.H.S. of (23) leading to a renormalization of the electron mass, m.

It is to be stressed that the form of Eq. (23) follows from an expansion in the gradients of Ψ and *not* in the magnitude of Ψ; accordingly, the validity of this structure is *not* restricted to temperatures close to the superconducting transition temperature. If, however, one does expand in Ψ then the usual non-linear term in the Ginzburg-Landau equation derives from the expansion of $A(\underline{R})$.

The relative ease with which the generalized, time-dependent Ginzburg-Landau Eq. (23) was obtained illustrates the advantage of the present method, whereas the plethora of undetermined coefficients which arise is a characteristic limitation; as stressed earlier, it is only at this stage, in connection with the evaluation of these coefficients, need one resort to the details of the microscopic theory. As illustrative of the procedure to be adopted one could, for example, following GORKOV take the Hartree-Fock approximation to Λ_2 [Eq. (11)], assume V to be a delta-function, and actually solve the linearized Ω_1^s equation. In this way (c.f. [13, 14]) explicit expressions for A, B, C etc. can be obtained.

It is gratifying to note that if, in the present analysis, the assumption is made that V is a delta-function, but not that $\Lambda_2 = \Lambda_2^{\mathrm{H.F.}}$ (unnecessary since *we* do not attempt to solve the Ω_1^s equation), then our generalized Ginzburg-Landau equation (23) reduces to the form given by WERTHAMER [20], starting from microscopic theory.

Inclusion of an external electromagnetic field in the usual gauge invariant way is straightforward provided only that the spatial variation of the potentials be slow in comparison with the range of $u(r)$ (a condition which could well restrict the analysis to type II superconductors) and under the assumption that V, the phonon-induced electron-electron interaction be instantaneous. Again no difficulties are anticipated so long as the time scale of V be short in comparison to that of Ψ, in which case the only modification will be a replacement of the coefficients A, B, C etc. by their time-averaged equivalents.

6. Comparison with other Works

The results of the foregoing section thus fully vindicate FRÖHLICH's philosophy towards the derivation of macroscopic equations from microphysics; indeed, the mathematical simplicity and transparency of the underlying assumptions are unrivalled in existing treatments of this problem. As mentioned in § 1, such treatments do, however, go further in that they derive the desired equations along with the magnitudes—and, for example, the temperature dependence—of the parameters appearing therein in one fell swoop; unfortunately, however, there are usually two particular drawbacks to such "all-in-one" approaches: either it becomes impossible to unravel structural and parametrical implications of certain approximation—or, if it is possible to do so, the domain of validity of the equation is dictated by approximations used in the derivation of parameters; in the latter case one does not know, therefore, whether the general structure of the equation persists outside this domain.

Existing attempts to derive the Ginzburg-Landau equation from the microscopic theory of superconductivity provide good examples of the difficulties into which one is led by the conventional approach; in particular, the attempts by ABRAHAMS and TSUNETO [21] and by PIKE [22] to derive time-dependent generalizations warrant mention.

ABRAHAMS and TSUNETO, by extending GORKOV's formulation of the microscopic theory to non-equilibrium situations, find—provided variations in intensive variables are sufficiently slow—that at the absolute zero of temperature Ψ satisfies a wave-like equation; for non-zero temperatures, however, a diffusion-like equation is obtained.

The analysis of PIKE, on the other hand, is, in consequence of certain approximations used, restricted to temperatures close to the transition temperature where he obtains a wave-like equation for Ψ, in agreement with the phenomenological considerations of KULIK [23].

Whether or not the Ψ equation should, at non-zero temperatures, be wave or diffusion-like is basically a structural question about which the present work, being itself primarily a structural theory, can make definitive statements; the general "master" equation for $\Phi(\underline{x}, \underline{y})$ given by FRÖHLICH is valid at all temperatures, whilst the approximations used in this essay to derive the Ψ equation are again valid for all temperatures.

The present analysis clearly yields a wave-like Ψ equation at all temperatures, as is readily seen by differentiating (23) with respect to time, eliminating the first time derivative and retaining, consistent to the order worked to above, only second order space derivatives.

Basically, the wave-like structure derives from our assumption that the only time dependence of Ω_1^s is through that of Ψ, an assumption which is physically equivalent to assuming that the "normal" aspects of the system relax instantaneously to their absolute (constant) equilibrium values.

In fine it is our pleasant task to thank Professor H. FRÖHLICH, F.R.S. for initiating our work in the theory of superfluids and superconductivity, for cultivating the attitude expressed in this essay and for his continuing interest and encouragement.

References

1. FRÖHLICH, H.: J. Phys. Soc. Japan **26** (Suppl.), 189 (1969).
2. FRÖHLICH, H.: Rept. Progr. Phys. **24**, 1 (1961).
3. FRÖHLICH, H.: Contribution to "Perspectives in Modern Physics", Ed. R. E. MARSHAK, p. 539–352. New York: Interscience 1966.
4. FRÖHLICH, H.: Physica **37**, 215 (1967).
5. HYLAND, G. J., ROWLANDS, G.: Physica **54**, 542 (1971).
6. FRÖHLICH, H.: Phys. Kondens. Materie **9**, 350 (1969).
7. LONDON, F.: Superfluids, Vols. 1 and 2. New York: Dover Publications 1961.
8. YANG, C. N.: Rev. Mod. Phys. **34**, 694 (1962).
9. FRÖHLICH, H.: Contribution to "Problems of Theoretical Physics", p. 373–378. Moscow: Nauka 1969.
10. HYLAND, G. J., ROWLANDS, G.: J. Low Temp. Phys. **7**, 271 (1972).
11. FRÖHLICH, H.: Proc. Phys. Soc. (London) **87**, 330 (1966).
12. GINZBURG, V. L., LANDAU, L. D.: Zh. Eksperim. i Theor. Fiz. **20**, 1064 (1950).
13. WERTHAMER, N. R.: Contribution to "Superconductivity", Ed. R. D. PARKS, Vol. 1, p. 321–370. New York: Marcel Dekker Inc. 1969.
14. GORKOV, L. P.: Soviet Phys. JETP **7**, 505 (1958).
15. TAYLOR, A. W. B.: J. Phys. C **3**, 52 (1970).
16. CUMMINGS, F. W., HYLAND, G. J., ROWLANDS, G.: Phys. Kondens. Materie **12**, 90 (1970).
17. GINZBURG, V. L., PITAEVSKII, L. P.: Soviet Phys. JETP **7**, 858 (1958).
18. PITAEVSKII, L. P.: Soviet Phys. JETP **8**, 282 (1959).
19. KHALATNIKOV, I. M.: Soviet Phys. JETP **30**, 268 (1970).
20. WERTHAMER, N. R.: Phys. Rev. **132**, 663 (1963).
21. ABRAHAMS, E., TSUNETO, T.: Phys. Rev. **152**, 416 (1966).
22. PIKE, E. R.: Proc. Adv. Summer Study Instit. McGill Univ., Montreal, "Superconductivity", Ed. P. R. WALLACE, Vol. II, p. 691–717. New York: Gordon & Breach 1969.
23. KULIK, I. O.: Soviet Phys. JETP **23**, 1077 (1966).

Bose Condensation and Superfluid Hydrodynamics

K. WEISS and H. HAUG

Philips Research Laboratories, Eindhoven-Netherlands

Abstract

The four basic conservation laws for a superfluid which is characterized by ODLRO are derived without any further assumptions. All relevant physical quantities of the superfluid hydrodynamics are thereby expressed in terms of the first two reduced density matrices. For a diluted, weakly interacting Bose system, for which a solution of the many-body problem exists, the velocity and temperature dependences of all quantities are calculated explicitly. For the general case, in which no such solution is available, the symmetry of the reduced density matrices is exploited to show that the structure of all relevant physical quantities is compatible with LANDAU's nonlinear two-fluid model. A relation between the densities of the condensate and of the supercomponent is obtained and discussed.

1. Introduction

In 1938 F. LONDON [1] suggested that the transition of liquid helium-4 from its normal to its superfluid phase is related to the inset of Bose-Einstein condensation. At that time it was known that in the case of an ideal Bose gas below a certain critical temperature a finite fraction of all particles gathers in the lowest momentum state. Since, however, such a condensation in the momentum space for a system of interacting Bose particles was not yet understood, it was not possible to derive directly from LONDON's concept the flow properties of helium in the superfluid phase. In 1941 LANDAU [2] developed the phenomenological theory of the hydrodynamic behaviour of superfluid helium in terms of the two-fluid model. The essential feature of this theory as compared with the hydrodynamics of a normal fluid is that it introduces an additional independent hydrodynamic variable—the velocity of the supercomponent. It took another 15 years before the interrelation between LONDON's microscopic and LANDAU's phenomenological concepts became firmly established through the work of BOGOLUBOV [3] and of HOHENBERG and MARTIN [4]. But even after these pioneering contributions the situation was still far from transparent and many questions remained open. One puzzling problem was for instance: What is the relation between the fraction of particles in the condensate and the fraction of particles that forms the supercomponent in the two-fluid model?

In 1967 FRÖHLICH [5] showed for a normal fluid how the structure of the Navier-Stokes equation follows very simply from the equation of motion of the first reduced density matrix by only exploiting the symmetries of the density matrices. For superfluids the reduced density matrices differ from those of normal fluids in their asymptotic behaviour for widely separated spatial coordinates [6]. This concept of off-diagonal long-range order (ODLRO) of the reduced density matrices was YANG's [7] fruitful generalization of the Bose-Einstein condensation for an ideal gas to interacting systems. It was again FRÖHLICH [8] who first studied the information that can be gained about the dynamics of superfluids from the general symmetries of the reduced density matrices combined with their properties of ODLRO. FRÖHLICH's investigations initiated wider studies to determine the connection between the microscopic equations and the structure of the phenomenological equations of superfluids, as can be seen from various other contributions in the present volume. In the present paper a coherent review is given of the work of the present authors [9], and some as yet unpublished derivations and new results are included.

The second chapter presents the derivation of the conservation laws for the mass density $m\varrho$, and the densities of the momentum $\vec{p}$ and the energy E of an interacting Bose system with an arbitrary two-particle potential. This enables the three densities and their related fluxes—i.e. $\vec{p}$, the stress tensor $\boldsymbol{\Pi}$ and the energy flux $\vec{Q}$—to be expressed in terms of the first two reduced density matrices. These three conservation laws form the basis of the hydrodynamics of a normal fluid. For a superfluid an additional hydrodynamic variable is needed which, as already mentioned, is the velocity of the supercomponent $\vec{v}_s$. Microscopically the superfluid is characterized by ODLRO in the first reduced density matrix, which means that a part of this matrix splits into the product of a wave function and its complex conjugate. This wave function is known as the order parameter [10] or the macroscopic wave function of the superfluid. The connection between the microscopic and the macroscopic description is obtained by identifying $\vec{v}_s$ with the gradient of the phase of the order parameter. To obtain the equation for $\vec{v}_s$, in Chapter 3, the equations of motion for the order parameter and its conjugate complex are derived. These equations can be transformed into a pair of equations for the phase and the amplitude of the order parameter. They have the form of canonical equations. ANDERSON [11] used a pair of canonical equations to discuss the Josephson effect in He-II. The Anderson equations certainly hold for a system in a coherent state [12], but it has not yet been possible to derive such canonical equations for arbitrarily interacting Bose systems, inspite of recent claims [13]. To proceed further, it is assumed—not proved—that only the gradient of the phase and not the amplitude (i.e. not the density of the particles in the condensate ϱ_c) is an independent hydrodynamic variable, for which a slowly varying spatial variation is linked with a slow temporal change. As mentioned above, the equation of motion for the gradient of the phase is now identified with that for $\vec{v}_s$. In this way an expression for the flux which is related to $\vec{v}_s$, namely the chemical potential μ, is obtained again in terms of the first two reduced density matrices. To elaborate further the quantities $\varrho, \vec{p}, E, \boldsymbol{\Pi}, \vec{Q}$ and μ, which appear in the four fundamental equations of motion for a superfluid, assumptions have to be made about the second reduced density matrix. The resulting expressions of the microscopic calculations can then systematically be compared with the corresponding phenomenological expressions of the two-fluid model. A connection is thus found between the density of the supercomponent ϱ_s and that of the condensate fraction ϱ_c.

In Chapter 4 the procedure is elucidated and checked by using the reduced density matrices in the form which they have in the Bogolubov approximation [14] for a dilute, weakly interacting Bose system. In this model the temperature and velocity dependences of the reduced density matrices are known, so that all relevant physical quantities can be evaluated explicitly. It will be demonstrated that the resulting microscopic expressions are compatible with LANDAU's non-linear two-fluid model. The procedure automatically gives a relation between ϱ_s and ϱ_c, which depends on the temperature and the relative velocity between the super and the normal component. The results are checked to obey the correct thermodynamic relations. In the limit of low temperatures, the decrease of ϱ_c and ϱ_s is calculated to be proportional to T^2 and T^4, respectively. These results are in agreement with those obtained by FERRELL et al. [15] for He-II from a consideration of fluctuations. This agreement is to be expected, because the results depend only on the existence of a phonon part in the excitation spectrum. Following FRÖHLICH's program, in Chapter 5 an ansatz for the second reduced density matrix R_2 is made which contains, besides the terms expressible in R_1, an unfactorized part. It is again demonstrated that this model is compatible with LANDAU's nonlinear two-fluid equations and that the resulting ϱ_s, ϱ_c relation gives a finite depletion of the condensate at zero temperature.

2. Conservation Laws

It will now be shown that the conservation laws for the particle density ϱ, the momentum density $\vec{p}$ and the energy density E hold rigorously for an interacting Bose system with an arbitrary two-particle potential, and not only in the hydrodynamic limit,where the variation of the macroscopic variables are small over the range of the two-particle potential [16]. The three basic conservation laws are

$$\partial_t \varrho + \partial_i p_i / m = 0, \tag{2.1}$$

$$\partial_t p_i + \partial_j \Pi_{ij} = 0, \tag{2.2}$$

$$\partial_t E + \partial_i Q_i = 0. \tag{2.3}$$

Here ∂_t and ∂_i stand for $\partial / \partial t$ and $\partial / \partial x_i$. In the following the notation $\partial_i' = \partial / \partial x_i'$, and $D_i = \partial_i - \partial_i'$, or $\partial_i^z = \partial / \partial z_i$, etc., together with the summation convention, will be used.

The first and second reduced density matrices are defined by

$$\begin{aligned} R_1(\vec{x}; \vec{x}') &= \operatorname{tr} \hat{\varrho}\, \hat{\psi}^+(\vec{x}')\, \hat{\psi}(\vec{x}), \\ R_2(\vec{x}, \vec{y}; \vec{x}', \vec{y}') &= \operatorname{tr} \hat{\varrho}\, \hat{\psi}^+(\vec{x}')\, \hat{\psi}^+(\vec{y}')\, \hat{\psi}(\vec{y})\, \hat{\psi}(\vec{x}), \end{aligned} \tag{2.4}$$

where $\hat{\psi}^+$ and $\hat{\psi}$ are the Bose field operators obeying the commutation relations $[\hat{\psi}(\vec{x}), \hat{\psi}^+(\vec{y})]_- = \delta^{(3)}(\vec{x} - \vec{y})$. $\hat{\varrho}$ is the statistical operator of the system. Taking the commutators of the field operators with the Hamiltonian

$$\mathscr{H} = \int d^3x \hat{H}(\vec{x}) \quad \text{with} \quad \hat{H}(\vec{x}) = \hbar^2/(2m)(\partial_i\hat{\psi}^+)(\partial_i\hat{\psi}) + \tfrac{1}{2}\int d^3y\, V(|\vec{x}-\vec{y}|)\,\hat{\psi}^+(\vec{x})\,\hat{\psi}^+(\vec{y})\,\hat{\psi}(\vec{y})\,\hat{\psi}(\vec{x}), \tag{2.5}$$

one can calculate the time derivative of R_1 and R_2 and obtains

$$i\hbar\,\partial_t R_1(\vec{x};\vec{x}') + \hbar^2/(2m)(\partial_j + \partial_j')\,D_j\,R_1(\vec{x};\vec{x}') = \int d^3y\,[V(|\vec{x}-\vec{y}|) - V(|\vec{x}'-\vec{y}|)]\,R_2(\vec{x},\vec{y};\vec{x}',\vec{y}), \tag{2.6}$$

where $V(z)$ is the isotropic two-particle potential. For R_2 only the time derivative of the diagonal part is needed below

$$i\hbar\,\partial_t R_2(\vec{x},\vec{y};\vec{x},\vec{y}) = -[\partial_j J_j(\vec{x},\vec{y}) + \partial_j^y J_j(\vec{y},\vec{x})], \tag{2.7}$$

where

$$J_j(\vec{x},\vec{y}) = -i\hbar/(2m)\lim_{\vec{x}'\to\vec{x}} D_j\,R_2(\vec{x},\vec{y};\vec{x}',\vec{y}).$$

Note that the diagonal part of R_2 does not couple to R_3. The quantities ϱ, $\vec{p}$ and E are given by

$$\varrho(\vec{x}) = \lim_{\vec{x}'\to\vec{x}} R_1(\vec{x};\vec{x}'), \tag{2.8}$$

$$p_j(\vec{x}) = \hbar/(2i)\lim_{\vec{x}'\to\vec{x}} D_j\,R_1(\vec{x};\vec{x}'), \tag{2.9}$$

$$E(\vec{x}) = \hbar^2/(2m)\lim_{\vec{x}'\to\vec{x}}\partial_j\,\partial_j'\,R_1(\vec{x};\vec{x}') + \tfrac{1}{2}\int d^3y\,V(|\vec{x}-\vec{y}|)\,R_2(\vec{x},\vec{y};\ldots). \tag{2.10}$$

If the second pair of arguments in R_2 is not written explicitly, it is understood to be same as the first pair. The diagonal part of Eq. (2.6) immediately yields Eq. (2.1). The time derivative of the momentum density (2.9) together with Eq. (2.6) yields

$$\partial_t p_j + \partial_l[-\hbar^2/(4m)\lim_{\vec{x}'\to\vec{x}} D_l\,D_j\,R_1(\vec{x};\vec{x}')] + \int d^3y\,R_2(\vec{x},\vec{y};\ldots)\,\partial_j V(|\vec{x}-\vec{y}|) = 0. \tag{2.11}$$

While the kinetic term already has the form required by Eq. (2.2), the potential term is not yet a divergence of a tensor and needs further treatment. Noting that

$$\partial_j V(|\vec{x}-\vec{y}|) = (z_j/z)\,\partial V(z)/\partial z,$$

where $\vec{z} = \vec{x}-\vec{y}$, one gets

$$-\tfrac{1}{2}\int d^3z\, z_j/z\,[R_2(\vec{x},\vec{x}+\vec{z};\ldots) - R_2(\vec{x},\vec{x}-\vec{z};\ldots)]\,\partial V/\partial z,$$

where $-z_j$ has been substituted for z_j in the first term. Because R_2 is symmetric in the first and last two arguments, respectively, the above formula can be written as the result of an integration

$$-\tfrac{1}{2}\int d^3z\,(z_j/z)\,(\partial V/\partial z)\int_0^z d\varrho\;\partial R_2(\vec{x}+(\varrho-z)\,\vec{z}/z,\vec{x}+\varrho\vec{z}/z;\ldots)/\partial\varrho.$$

The derivative with respect to ϱ can be changed into a derivative with respect to $\vec{x}$, i.e. $\partial R_2/\partial\varrho = (z_i/z)\,\partial R_2/\partial x_i$. The derivative with respect to x_i can be pulled in front of all integrals, so that this term too is now in the form required by Eq. (2.2). The total resulting stress tensor is

$$\Pi_{ij} = -\frac{\hbar^2}{4m} \lim_{\vec{x}' \to \vec{x}} D_i D_j R_1(\vec{x}; \vec{x}') - \frac{1}{2} \int d^3z \frac{\partial V}{\partial z} \cdot \frac{z_i z_j}{z^2} \int_0^z R_2\left(\vec{x} + (\varrho - z)\frac{\vec{z}}{z}, \vec{x} + \varrho \frac{\vec{z}}{z}; \ldots\right) d\varrho. \tag{2.12}$$

For the short range potential, i.e. whose range is small compared with a characteristic hydrodynamic length, the second term of (2.12) reduces to

$$-\frac{1}{2} \int d^3z \frac{\partial V}{\partial z} \frac{z_i z_j}{z} R_2(\vec{x}, \vec{x} + \vec{z}; \ldots). \tag{2.13}$$

This is the form given by MARTIN and SCHWINGER [16].

The time derivative of the energy density (2.10) together with Eqs. (2.6) and (2.7) yields

$$\partial_t E + \partial_j \left[\frac{i\hbar^3}{16m^2} \lim_{\vec{x}' \to \vec{x}} D_j D_l^2 R_1(\vec{x}; \vec{x}') + \frac{1}{2} \int d^3y\, V(z)\, J_j(\vec{x}, \vec{y})\right] + \frac{1}{2} \int d^3y\, V(z)\, \partial_j^y [J_j(\vec{x}, \vec{y}) + J_j(\vec{y}, \vec{x})] = 0. \tag{2.14}$$

The last term of Eq. (2.14) needs again further treatment. Integrating by parts yields

$$-\tfrac{1}{2} \int d^3y \cdot \partial V / \partial y_j [J_j(\vec{x}, \vec{y}) + J_j(\vec{y}, \vec{x})].$$

Now the problem is the same as that of transforming the last term of Eq. (2.11). The energy flux then takes the form

$$Q_j = \frac{i\hbar^3}{16m^2} \lim_{\vec{x}' \to \vec{x}} D_j D_l^2 R_1(\vec{x}; \vec{x}') + \frac{1}{2} \int d^3y\, V(z)\, J_j(\vec{x}, \vec{y}) - \frac{1}{4} \int d^3z \frac{z_l z_j}{z^2} \frac{\partial V}{\partial z} \int_0^z d\varrho \cdot \left[J_l\left(\vec{x} + (\varrho - z)\frac{\vec{z}}{z}, \vec{x} + \varrho\frac{\vec{z}}{z}\right) + J_l\left(\vec{x} + \varrho\frac{\vec{z}}{z}, \vec{x} + (\varrho - z)\frac{\vec{z}}{z}\right)\right]. \tag{2.15}$$

In the hydrodynamic limit the last term in Eq. (2.15) reduces to

$$-\tfrac{1}{4} \int d^3z\, z_j [J_l(\vec{x}, \vec{x} + \vec{z}) + J_l(\vec{x} + \vec{z}, \vec{x})]\, \partial_l^z V(z). \tag{2.16}$$

Eqs. (2.8–2.10), (2.12) and (2.15) determine all quantities that appear in the conservation laws (2.1–2.3) in terms of the first density matrices.

3. Order Parameter and Superfluid Velocity

The three conservation laws for $\varrho, \vec{p}$ and E derived in the previous chapter form the basis for the hydrodynamics of a normal fluid. For a superfluid, a further hydrodynamic variable has to be introduced, namely the velocity of the super-component. This velocity is microscopically given as the gradient of the phase of the order parameter. The order parameter or macroscopic wave function [10] characterizes the degree of extra order in the superfluid as compared with a normal fluid. This extra order is expressed as ODLRO in the first reduced density matrix [6, 7]:

$$R_1(\vec{x}; \vec{x}') = \Phi^*(\vec{x}')\, \Phi(\vec{x}) + \tilde{R}_1(\vec{x}; \vec{x}'). \tag{3.1}$$

Here

$$\Phi(\vec{x}) = \langle \hat{\psi}(\vec{x}) \rangle = \operatorname{tr} \hat{\varrho}\, \hat{\psi}(\vec{x}) \tag{3.2}$$

is the order parameter which vanishes for a normal fluid.

The equation of motion for Φ can be obtained from that for the field operator $\hat{\psi}$:

$$i\hbar\, \partial_t \hat{\psi}(\vec{x}) = [\hat{\psi}(\vec{x}), \hat{\mathscr{H}}[\hat{\psi}, \hat{\psi}^+]]_- = \frac{\delta \hat{\mathscr{H}}[\hat{\psi}, \hat{\psi}^+]}{\delta \hat{\psi}^+(\vec{x})}. \tag{3.3}$$

Here $\hat{\mathscr{H}}[\hat{\psi}, \hat{\psi}^+]$ is the total Hamiltonian (2.5) which is a functional of $\hat{\psi}$ and $\hat{\psi}^+$. $\delta\hat{\mathscr{H}}/\delta\hat{\psi}^+(\vec{x})$ is the functional derivative of $\hat{\mathscr{H}}$ with respect to $\hat{\psi}^+(\vec{x})$. The second equality in (3.3) is a generalization of the well-known operator identity $[\hat{b}, \hat{f}(\hat{b}, \hat{b}^+)]_- = \partial \hat{f}(\hat{b}, \hat{b}^+)/\partial \hat{b}^+$, where $\hat{b}$ and $\hat{b}^+$ obey the commutation relation $[\hat{b}, \hat{b}^+]_- = 1$.

Taking the expectation value of (3.3), the equation of motion for the order parameter is

$$i\hbar\, \partial_t\, \Phi(\vec{x}) = \left\langle \frac{\delta \hat{\mathscr{H}}[\hat{\psi}, \hat{\psi}^+]}{\delta \hat{\psi}^+(\vec{x})} \right\rangle. \tag{3.4}$$

This equation can be transformed into

$$i\hbar\, \partial_t\, \Phi(\vec{x}) = \delta\mathscr{H}[\Phi, \Phi^*]/\delta\Phi^*(\vec{x}), \tag{3.5}$$

where $\mathscr{H} = \int d^3x E$ is the expectation value of the Hamiltonian. This expectation value $\mathscr{H}[\Phi, \Phi^*]$ is a functional of the order parameter and its conjugate complex. To prove the validity of Eq. (3.5), one decomposes the field operator into

$$\hat{\psi}(\vec{x}) = \Phi(\vec{x}) + \hat{\tilde{\psi}}(\vec{x}), \quad \text{with} \quad \langle \hat{\tilde{\psi}}(\vec{x}) \rangle = 0. \tag{3.6}$$

For the kinetic energy term the identity of (3.4) and (3.5) is obvious, if one applies the derivative only with respect to the explicitly appearing Φ^*; i.e. the term $\widetilde{R}_1$ is not involved in the operation $\delta/\delta\Phi^*$, although a many-body solution would show that it depends implicitly on Φ^*. Inserting (3.6) into the second reduced density matrix R_2 which appears in the expectation value of the potential energy, one sees again that the operation (3.5) yields the same result as (3.4), as long as all terms formed by $\hat{\tilde{\psi}}$ and $\hat{\tilde{\psi}}^+$ are not differentiated. With these restrictive derivatives Eq. (3.5) and its conjugate complex have the form of a pair of canonical equations. It can easily be checked that Eq. (3.5) agrees with FRÖHLICH's order parameter equation [8], if his ansatz for R_2 is inserted. FRÖHLICH obtained the order parameter equation by considering Eq. (2.6) in the nonlocal limit $|\vec{x}' - \vec{x}| \to \infty$. Writing the order parameter in terms of a phase θ and the density of the condensate fraction ϱ_c

$$\Phi(\vec{x}) = \sqrt{\varrho_c(\vec{x})}\, \exp i\,\theta(\vec{x}), \tag{3.7}$$

one gets from Eq. (3.5) and its conjugate complex a pair of equations for ϱ_c and θ:

$$\hbar\, \partial_t\, \theta(\vec{x}) = -\delta\mathscr{H}/\delta\varrho_c(\vec{x}), \tag{3.8}$$

$$\hbar\, \partial_t\, \varrho_c(\vec{x}) = \delta\mathscr{H}/\delta\theta(\vec{x}). \tag{3.9}$$

It must be stressed again that the functional derivatives in Eqs. (3.8–3.9) have to be taken only with respect to ϱ_c and θ which appear explicitly in $\mathscr{H}$. Due to these restrictions, Eqs. (3.8–3.9) cannot literally be considered as canonical equations. A pair of canonical equations for the phase θ and the *total* density ϱ has been used by ANDERSON [11] to discuss the Josephson effect in He-II. These canonical equations for θ and ϱ hold for a system in a coherent state [12], where $\varrho_c = \varrho$. However, for an interacting Bose-system no rigorous proof for the Anderson equations exists [12, 13].

The connection with the two-fluid model is achieved by introducing the extra hydrodynamical variable $\vec{v}_s$ as the gradient of θ

$$v_{s\,i}(\vec{x}) = (\hbar/m)\, \partial_i \theta(\vec{x}). \tag{3.10}$$

Eq. (3.8) then gives an equation of motion for the irrotational velocity field $\vec{v}_s$ which again has the structure of a conservation law

$$\partial_t v_{s\,i}(\vec{x}) + \partial_i\left(\tfrac{1}{2} v_s^2(\vec{x}) + \delta \mathscr{H}_{\text{pot}}/\delta \varrho_c(\vec{x})\right) = 0, \tag{3.11}$$

where $\frac{1}{2}v_s^2$ is the contribution from the kinetic energy $\mathscr{H}_{\text{kin}}$ with $\mathscr{H} = \mathscr{H}_{\text{kin}} + \mathscr{H}_{\text{pot}}$.

The two-fluid model has no room for a further hydrodynamic variable. Therefore, in order to get the two-fluid model from the quantum mechanical equations of motion, one has to assume that Eq. (3.9) is no independent hydrodynamic equation. In local equilibrium the condensate density is then completely determined through the local independent hydrodynamic variables. To prove the assumption, that ϱ_c is no independent hydrodynamic variable, one would have to show that strong restoring forces drive ϱ_c with a high frequency $\omega(\vec{k})$ towards its local equilibrium value. The extra mode corresponding to Eq. (3.9) is then a nonhydrodynamic mode, i.e. $\lim_{k\to 0} \omega(\vec{k}) \neq 0$. This behaviour of the condensate density was suggested by HALPERIN and HOHENBERG [17] by analogy with the ferromagnetic problem. Eqs. (2.1–2.3) and an equation for $\vec{v}_s$ of the form (3.11) are the basis of the two-fluid model. In this model the flux corresponding to $\vec{v}_s$ is, besides the kinetic term $\frac{1}{2}v_s^2$, the chemical potential μ, so that one obtains from a comparison

$$m\mu(\vec{x}) = \delta \mathscr{H}_{\text{pot}}/\delta \varrho_c(\vec{x}). \tag{3.12}$$

A heuristic connection with the thermodynamic definition of the chemical potential $m\mu = (\delta \mathscr{H}_{\text{pot}}/\delta \varrho)_s$, where s is the entropy per unit mass, is obtained by noticing that the functional derivative $\delta/\delta \varrho_c$ does not operate on terms formed by $\hat{\tilde{\psi}}$ and $\hat{\tilde{\psi}}^+$. Specifically, the depletion ϱ_d and the entropy s (the condensate carries no entropy) are not involved in the differentiation. This suggests the equality

$$\delta \mathscr{H}_{\text{pot}}/\delta \varrho_c = (\delta \mathscr{H}_{\text{pot}}/\delta \varrho)_s.$$

The relations [18] between the various quantities of the phenomenological two-fluid model are especially simple in the coordinate system K^0, in which the supercomponent is at rest:

$$\vec{v}_s^{\,0} = 0, \quad \vec{v}_n^{\,0} = \vec{v}_n - \vec{v}_s, \tag{3.13}$$

where $\vec{v}_n$ is the velocity of the normal component. Here one has

$$\varrho = \varrho_s + \varrho_n, \tag{3.14}$$

$$p_i^0 = m \varrho_n v_{ni}^0, \tag{3.15}$$

$$\Pi_{ij}^0 = P \delta_{ij} + m \varrho_n v_{ni}^0 v_{nj}^0, \tag{3.16}$$

$$Q_i^0 = v_{ni}^0 (E^0 + P - m \varrho_s \mu), \tag{3.17}$$

where P is the pressure.

Finally the entropy s per unit mass is given by

$$m \varrho s T = E^0 + P - m \varrho \mu - m \varrho_n v_n^{0^2}. \tag{3.18}$$

The various quantities which appear in the four conservation laws and which have been determined in the last two chapters in terms of R_1 and R_2 can now be calculated in certain approximations. The resulting expressions have to be compared with the phenomenological two-fluid model (3.14–3.17).

4. Bogolubov Approximation

Before pursuing FRÖHLICH's programm [5, 8], to evaluate the various expressions by using essentially only the symmetries and ODLRO for R_1 and R_2, an evaluation of the previously derived results will be carried out in the Bogolubov approximation [14]. Though the Bogolubov approximation which is valid for a dilute, weakly interacting Bose system is no quantitative description of He-II, it has often served [14, 19, 20] as a useful model for various aspects of superfluidity. Everything can be calculated explicitly in this approximation. According to BOGOLUBOV, the field operator in a homogeneous situation takes the form

$$\hat{\psi} = \Phi + \hat{\tilde{\psi}} = e^{i\theta(\vec{x})} \left[\varrho_c^{\frac{1}{2}} + \frac{1}{V^{\frac{1}{2}}} \sum{}' e^{i\vec{k}\vec{x}} \frac{\hat{\xi}_{\vec{k}} + L_k \hat{\xi}^+_{-\vec{k}}}{\sqrt{1 - L_k^2}} \right], \tag{4.1}$$

where V is the volume. In the following quantum mechanical calculations the condensate density ϱ_c has to be considered as a given parameter. The prime on the summation symbol indicates that the lowest momentum value $\vec{k} = 0$ has to be omitted. The function L_k is given by

$$\begin{aligned} L_k &= (E_k - T_k - \Delta)/\Delta; \qquad E_k = (T_k^2 + 2T_k \Delta)^{\frac{1}{2}}; \\ T_k &= \hbar^2 k^2/(2m) \quad \text{and} \quad \Delta = V_0 \varrho_c. \end{aligned} \tag{4.2}$$

E_k is the energy spectrum of the elementary exitations. For simplicity, the potential is assumed to be $V(\vec{x}) = V_0 \delta^{(3)}(\vec{x})$. Within the Bogolubov approximation, where $(\varrho - \varrho_c)/\varrho \ll 1$, one can write with the same degree of accuracy $\Delta = V_0 \varrho$. The Hamiltonian is diagonal in the $\hat{\xi}_{\vec{k}}$ operators

$$\hat{\mathscr{H}} = \mathscr{H}_0 + \Sigma' E_k \hat{\xi}^+_{\vec{k}} \hat{\xi}_{\vec{k}},$$

where

$$\mathscr{H}_0 = \tfrac{1}{2} (V_0 \varrho^2 + \Delta \Sigma' L_k). \tag{4.3}$$

In thermal equilibrium the number of elementary excitations is given by

$$\langle \hat{\xi}^+_{\vec{k}} \hat{\xi}_{\vec{k}} \rangle = n_{\vec{k}}(\vec{v}^0_n, T) = \{\exp[(E_k - \vec{v}^0_n \hbar \vec{k})/k_B T - 1]\}^{-1}. \tag{4.4}$$

In the coordinate system K^0 the reduced density matrices are

$$R^0_1(\vec{x}; \vec{x}') = \varrho_c + \tilde{R}^0_1(\vec{x}; \vec{x}'),$$

where

$$\tilde{R}^0_1(\vec{x}; \vec{x}') = \langle \hat{\tilde{\psi}}^+(\vec{x}') \hat{\tilde{\psi}}(\vec{x}) \rangle \exp[+i(\theta(\vec{x}') - \theta(\vec{x}))], \tag{4.5}$$

and

$$\begin{aligned} R^0_2(\vec{x}, \vec{y}; \vec{x}', \vec{y}') = \varrho_c^2 + \varrho_c [&\tilde{R}^0_1(\vec{x}; \vec{x}') + \tilde{R}^0_1(\vec{y}; \vec{y}') + \tilde{R}^0_1(\vec{x}; \vec{y}') \\ &+ \tilde{R}^0_1(\vec{y}; \vec{x}') + \tilde{R}^0_a(\vec{x}, \vec{y}) + \tilde{R}^{0*}_a(\vec{x}', \vec{y}')]. \end{aligned} \tag{4.6}$$

Terms containing more than two $\hat{\tilde{\psi}}$ operators have been neglected in (4.6), which is consistent with the Bogolubov approximation. The anomalous functions are defined as

$$\tilde{R}^0_a(\vec{x}, \vec{y}) = \langle \hat{\tilde{\psi}}(\vec{x}) \hat{\tilde{\psi}}(\vec{y}) \rangle \exp[-i(\theta(\vec{x}) + \theta(\vec{y}))]. \tag{4.7}$$

With Eqs. (4.1–4.7) one can calculate ϱ, $\vec{p}^0$, and E^0 according to (2.8–2.10), as well as Π^0, $\vec{Q}^0$ and μ with (2.12–2.13), (2.15–2.16) and (3.12). Note that the two formulae (2.13) and (2.16) which are only valid in the hydrodynamic limit become exact for the δ-function potential. All six quantities will be calculated to second nonvanishing order in the relative velocity $\vec{v}^0_n$. From (4.5) the total density is obtained as

$$\varrho = \varrho_c + \frac{1}{V} \sum{}' \frac{L_k^2}{1 - L_k^2} + \frac{1}{V} \sum{}' \frac{n_{\vec{k}}(\vec{v}^0_n, T) + n_{-\vec{k}}(\vec{v}^0_n, T) L_k^2}{1 - L_k^2}. \tag{4.8}$$

An elementary integration gives for the first sum $(1/3\pi^2)(m\Delta/\hbar^2)^{3/2}$. This is for $T = 0$ the density of the depletion ϱ_d, which is defined as $\varrho_d = \varrho - \varrho_c$. At finite temperatures, the second term also contributes to ϱ_d. Expanding the population factor for small relative velocities

$$n_{\vec{k}}(\vec{v}^0_n, T) = n_k(0, T) - \vec{v}^0_n \cdot \hbar \vec{k} \frac{\partial n_k(0, T)}{\partial E_k} + \frac{\hbar^2}{2} (\vec{v}^0_n \cdot \vec{k})^2 \frac{\partial^2 n_k(0, T)}{\partial E_k^2} + \dots, \tag{4.9}$$

one gets in second order of $\vec{v}^0_n$

$$\varrho = \varrho_c + \frac{1}{3\pi^2} \left(\frac{m\Delta}{\hbar^2}\right)^{3/2} + \frac{1}{V} \sum{}' \frac{T_k + \Delta}{E_k} \left[n_k(0, T) + \frac{m v_n^{0^2}}{3} T_k \frac{\partial^2 n_k(0, T)}{\partial E_k^2}\right]. \tag{4.10}$$

For the momentum density one gets according to (2.9)

$$p^0_i = \frac{1}{V} \sum{}' n_{\vec{k}}(\vec{v}^0_n, T) \hbar k_i. \tag{4.11}$$

With the expansion (4.9), one gets up to third order in $\vec{v}^0_n$

$$p^0_i = m v^0_{ni} \left[-\frac{2}{3V} \sum{}' T_k \left(\frac{\partial n_k(0, T)}{\partial E_k} + \frac{m v_n^{0^2}}{5} T_k \frac{\partial^3 n_k(0, T)}{\partial E_k^3}\right)\right]. \tag{4.12}$$

Comparing (4.12) with the two-fluid form (3.15) of $\vec{p}^0 = m\varrho_n \vec{v}^0_n$, one sees that the expression in the square brackets is equal to ϱ_n, the density of the normal component. The total density ϱ (4.10) in the two-fluid model is equal to $\varrho_s + \varrho_n$. Taking ϱ_n from (4.12), one gets a relation between ϱ_s and ϱ_c:

$$\varrho_s = \varrho_c + \frac{1}{3\pi^2}\left(\frac{m\Delta}{\hbar^2}\right)^{3/2} + \frac{1}{V}\sum{}' \left\{ n_k(0,T)\frac{T_k+\Delta}{E_k} + \frac{2}{3}\frac{\partial n_k(0,T)}{\partial E_k} T_k + \frac{m v_n^{0^2}}{3} T_k \right.$$
$$\left. \cdot \left[\frac{\partial^2 n_k(0,T)}{\partial E_k^2}\frac{T_k+\Delta}{E_k} + \frac{2}{5}\frac{\partial^3 n_k(0,T)}{\partial E_k^3} T_k\right]\right\}. \tag{4.13}$$

Omitting for simplicity the velocity-dependent terms in (4.13), one can reformulate the relation by using the formula

$$\partial n_k/\partial E_k = m/(\hbar^2 k)\,[E_k/(T_k+\Delta)]\,\partial n_k/\partial k.$$

An integration by parts yields

$$\varrho_s = \varrho_c + \frac{1}{3\pi^2}\left(\frac{m\Delta}{\hbar^2}\right)^{3/2} + \frac{\Delta^2}{3V}\sum{}' n_k(0,T)\frac{T_k+3\Delta}{E_k(T_k+\Delta)^2}. \tag{4.14}$$

Eq. (4.14) has also been obtained in Ref. [21] (apart from a missing factor of $\frac{1}{2}$ in the last sum) as an illustration of a construction of ϱ_s in terms of a response kernel.

After the evaluation of the particle density and the momentum density, the energy density E^0 will now be calculated again up to second order in the relative velocity $\vec{v}_n^0$. From Eq. (2.10) one gets

$$E^0 = \frac{1}{2}V_0\varrho^2 + \frac{1}{V}\sum{}'\left[\frac{1}{2}(E_k - T_k - \Delta) + n_k(0,T)E_k + \frac{m v_n^{0^2}}{3}\frac{\partial^2 n_k(0,T)}{\partial E_k^2} T_k E_k\right]. \tag{4.15}$$

To calculate the chemical potential according to (3.12) as a functional derivative of the potential energy with respect to ϱ_c, one cannot use Eq. (4.15), because in this explicit form it is no longer possible to keep the terms formed with $\hat{\tilde{\psi}}$ and $\hat{\tilde{\psi}}^+$ constant. Going back to (2.10), together with (4.5–4.7), one has to differentiate only with respect to the explicitly appearing ϱ_c and gets

$$m\mu = V_0\left[\varrho + \frac{1}{V}\sum{}'\left(\frac{1}{2}\frac{T_k - E_k}{E_k} + n_k(0,T)\frac{T_k}{E_k} + \frac{m v_n^{0^2}}{3}\frac{T_k^2}{E_k}\frac{\partial^2 n_k(0,T)}{\partial E_k^2}\right)\right]. \tag{4.16}$$

As a next step, the part of the stress tensor which contributes only to the diagonal $i=j$ will be evaluated. According to (3.16) this part is equal to the pressure P. An integration by parts shows that (2.13) gives, besides the term $E_{\text{pot}}\delta_{ij}$, a term

$$\tfrac{1}{2}V_0 \int d^3z\, z_i\, \partial_j^z R_2(\vec{x}, \vec{x}+\vec{z}; \ldots)\, \delta^{(3)}(z),$$

which contributes for $i=j$, because R_2 contains in the Bogolubov approximation a singularity for $\vec{z}\to 0$ (see appendix of Ref. [9e]). The final result is

$$P = E^0 - \frac{1}{2V}\sum{}'\left[(E_k - T_k - \Delta)(E_k - T_k) + \frac{2}{3}T_k(T_k+\Delta)\,n_k(0,T) \right.$$
$$\left. - \frac{m v_n^{0^2}}{5}\frac{\partial^2 n_k(0,T)}{\partial E_k}\cdot T_k^2(T_k+\Delta)\right]E_k^{-1}. \tag{4.17}$$

The off-diagonal part of the stress tensor $i \neq j$ which according to (3.16) is equal to $m\varrho_n v_{ni}^0 v_{nj}^0$ yields again a determination of ϱ_n which is identical with the result obtained above.

As the last quantity, the energy flux is calculated according to (2.15–2.16). Here again, one has to go up to third order in $\vec{v}_n^0$ and obtains

$$Q_i^0 = -v_{ni}^0 \frac{2}{3V} \sum' E_k^2 \left(\frac{\partial n_k(0,T)}{\partial E_k} + \frac{m v_n^{0^2}}{5} \frac{\partial^3 n_k(0,T)}{\partial E_k^3} T_k \right). \tag{4.18}$$

Using the calculated expressions for ϱ_s, E^0, P, μ, and $\vec{Q}^0$ of Eqs. (4.13) and (4.15–4.18), one checks that the two-fluid relation (3.17) $Q_i^0 = v_{ni}^0 (E^0 + P - m\varrho_s \mu)$ is valid up to the considered order in the relative velocity. From Eq. (3.18) one gets for the entropy

$$\begin{aligned} m\varrho s T = \frac{1}{V} \sum' \Big\{ & n_k(0,T) \left[E_k + \frac{2}{3} \frac{T_k(T_k + \Delta)}{E_k} \right] \\ & + \frac{m v_n^{0^2}}{3} \frac{\partial^2 n_k(0,T)}{\partial E_k^2} \left[E_k T_k - \frac{2}{5} T_k^2 \frac{T_k + \Delta}{E_k} \right] \Big\}. \end{aligned} \tag{4.19}$$

A brief discussion of the thermodynamics of the results obtained will now be given. The results (4.10–4.19) are in such a form that it is natural to use the particle density ϱ, the temperature T, and the relative velocity $\vec{v}_n^0$ as the independent thermodynamic variables. Introducing the free energy density

$$F^0(\varrho, T, \vec{v}_n^0) = E^0 - T m \varrho s - \vec{v}_n^0 \vec{p}^0, \tag{4.20}$$

one gets with the aid of the thermodynamic identity for the energy $dE^0 = m\mu\, d\varrho + m T\, d(\varrho s) + \vec{v}_n^0 d\vec{p}^0$ the corresponding identity for the free energy density F^0:

$$dF^0 = m\mu\, d\varrho - m\varrho s\, dT - \vec{p}^0 d\vec{v}_n^0. \tag{4.21}$$

Eq. (4.21) shows that

$$\varrho (\partial F^0 / \partial \varrho)_{T, \vec{v}_n^0} = m\varrho\mu, \tag{4.22}$$

$$T (\partial F^0 / \partial T)_{\varrho, \vec{v}_n^0} = -m\varrho s T, \tag{4.23}$$

$$v_{nj}^0 (\partial F^0 / \partial v_{nj}^0)_{T, \varrho} = -v_{nj}^0 p_j^0. \tag{4.24}$$

Using the thermodynamic relation (3.18), one finds for the free energy density

$$F^0 = -P + m\varrho\mu. \tag{4.25}$$

One can check that the calculated quantities fulfill (4.22–4.24).

The results which have been obtained so far as an illustration for the general results of Chapters 2 and 3 can be checked in the model. The energy flux e.g. can be calculated [18] with the formula

$$Q_i^0 = \mu p_i^0 + \sum' u_i(\vec{k}) E_k n_{\vec{k}},$$

where $u_i = \partial E_k / \partial \hbar k_i$ is the group velocity of the thermal excitations. The entropy is given by $m\varrho s = k_B \sum' \{(n_{\vec{k}} + 1) \ln (n_{\vec{k}} + 1) - n_{\vec{k}} \ln n_{\vec{k}}\}$ and is, as the momentum density (4.11), a function of $n_{\vec{k}}$ only. The chemical potential is according to (4.22)

$$m\mu = (\partial F^0 / \partial \varrho)_{T, \vec{v}_n^0} = \left(\partial E^0 (n_{\vec{k}}, \varrho) / \partial \varrho \right)_{n_{\vec{k}}},$$

where with (4.3)

$$E^0(n_{\vec{k}}, \varrho) = (1/2V) V_0 \varrho^2 + (1/V) \sum' (\tfrac{1}{2} L_k \Delta + n_{\vec{k}} E_k).$$

The resulting expressions for $\vec{Q}^0$, $m\varrho s$, and $m\mu$ are in agreement with formulae (4.18), (4.19) and (4.16), respectively.

To end this chapter, an evaluation of the low temperature behaviour of ϱ_c and ϱ_s will be given. The asymptotic temperature dependence for $T \to 0$ of the Bogolubov model is the same as that of real He-II, because it depends only on the existence of a linear phonon part in the excitation spectrum. For simplicity, the case $\vec{v}_n^0 = 0$ will be considered only. Because only small k-values are involved for $T \to 0$, the excitation spectrum E_k is

$$E_k = k_B T x \{1 + (x\tau/2)^2 + \cdots\}, \tag{4.26}$$

where $x = \hbar k c / k_B T$ is the phonon energy normalized with the thermal energy and $\tau = k_B T / \Delta$ is the thermal energy normalized with the interaction energy $\Delta = mc^2$, where c is the sound velocity. The Bose population factor reduces to

$$n_k(0, T) = (e^x - 1)^{-1} + \tfrac{1}{4} x^3 \tau^2 \,\partial (e^x - 1)^{-1} / \partial x + \cdots. \tag{4.27}$$

With these expansions (4.10) can now be calculated and gives

$$\varrho_c = \varrho - \varrho_0 (1/3\,\pi^2 + \tau^2/12 - \pi^2 \tau^4/40), \tag{4.28}$$

with $\varrho_0 = (mc/\hbar)^3$.

ϱ_0 can be seen as a density determined by the zero point motion. The mean square amplitude of a harmonic oscillator is $a^2 = \hbar^2/(4me)$. Taking for the zero point energy e the interaction energy per particle $\Delta = mc^2$, one gets $\varrho_0 \sim a^{-3} \sim (mc/\hbar)^3$. One sees from (4.28) that the decrease of ϱ_c begins with a T^2 term for a fixed value of ϱ. This result agrees with the results of FERRELL et al. [15], where the same temperature dependence of ϱ_c is obtained with the aid of fluctuation arguments. The calculation of the asymptotic temperature dependence of (4.14) yields

$$\varrho_s = \varrho_c + \varrho_0 (1/3\,\pi^2 + \tau^2/12 - 5\,\pi^2 \tau^4/72). \tag{4.29}$$

Inserting ϱ_c from (4.28) shows that ϱ_s decreases proportional to T^4 or that the increase of the normal fluid density is given by

$$\varrho_n = \frac{2\pi^2}{45} \varrho_0 \tau^4, \tag{4.30}$$

which is again in agreement with the general consideration of Ref. [15]. The functions E^0, P, $m\varrho\mu$, $\vec{Q}^0$, and $Tm\varrho s$ all vary as T^4 in the low temperature limit.

5. Ansatz for the Second Reduced Density Matrix

In this chapter the general case will be treated in which a many-body solution is not available. It will be demonstrated that the symmetries of the reduced density matrices already guarantee that the relevant physical quantities have a structure which is compatible with the nonlinear two-fluid hydrodynamics [9f]. For this purpose the depletion part of the first reduced density matrix [see (3.1)] is written as [8]

$$\tilde{R}_1^0(\vec{x}; \vec{x}') = \sigma(\vec{x}, \vec{x}') \,\exp i\beta^0(\vec{x}, \vec{x}'), \tag{5.1}$$

where the symmetry of R_1 implies

$$\sigma(\vec{x},\vec{x}') = \sigma(\vec{x}',\vec{x}) \quad \text{and} \quad \beta^0(\vec{x},\vec{x}') = -\beta^0(\vec{x}',\vec{x}). \tag{5.2}$$

The formulae for ϱ, $\vec{p}$, E, $\boldsymbol{\Pi}$ and $\vec{Q}$ in Chapter 2 show that in the terms which are independent of the potential the following quantities appear:

the depletion density

$$\varrho_d(\vec{x}) = \lim_{\vec{x}'\to\vec{x}} \sigma(\vec{x},\vec{x}') \tag{5.3 a}$$

the particle flux

$$\varrho_d v^0_{di}(\vec{x}) = (\hbar/2m)\,\varrho_d \lim_{\vec{x}'\to\vec{x}} D_i \beta^0(\vec{x},\vec{x}') \tag{5.3 b}$$

the internal kinetic energy tensor

$$T_{ij}(\vec{x}) = -(\hbar^2/8m) \lim_{\vec{x}'\to\vec{x}} D_i D_j \sigma(\vec{x},\vec{x}') \tag{5.3 c}$$

and the internal energy flux tensor

$$\vec{Q}^0_{ijk}(\vec{x}) = -(\hbar^3/16m^2)\,\varrho_d \lim_{\vec{x}'\to\vec{x}} D_i D_j D_k \beta^0(\vec{x},\vec{x}'). \tag{5.3 d}$$

Because of the symmetries of σ and β^0, the zeroth and second derivative of σ and the first and third derivative of β^0 appear. It must be stressed that the quantities (5.3) are purely formal and are no measurable quantities, at least in hydrodynamic experiments. The depletion density, however, can be determined with different experiments, e.g. with neutron scattering [22].

The idea is now to evaluate the expressions in E^0, $\boldsymbol{\Pi}^0$, $\vec{Q}^0$ and μ which depend on the potential in terms of these four depletion quantities and of the condensate density ϱ_c. The resulting expressions for ϱ, $\vec{p}^0$, $\boldsymbol{\Pi}^0$ and $\vec{Q}^0$ can then be systematically compared with the corresponding two-fluid quantities which are given at the end of Chapter 3. This comparison expresses four of the five normal quantities in terms of the physical quantities of the two-fluid model. One scalar quantity remains undetermined, but symmetry arguments will be used to obtain information about its qualitative behaviour.

The density matrix R_2 is written in the form [8]

$$\begin{aligned} R^0_2(\vec{x},\vec{y};\vec{x}',\vec{y}) = {} & R^0_1(\vec{x};\vec{y})\,R^0_1(\vec{y};\vec{x}') + R^0_1(\vec{x};\vec{x}')\,R^0_1(\vec{y};\vec{y}) \\ & - |\Phi^0(\vec{y})|^2\,\Phi^{*0}(\vec{x}')\,\Phi^0(\vec{x}) + \widetilde{R}^0_2(\vec{x},\vec{y};\vec{x}',\vec{y}), \end{aligned} \tag{5.4}$$

where the parts which simply factorize in R^0_1 are written out explicitly. The unfactorized part $\widetilde{R}^0_2$ no longer possesses any ODLRO. This function consists in the Bogolubov approximation of the anomalous functions $\varrho_s(\widetilde{R}^0_a + \widetilde{R}^{0}_a{}^*)$ [see (4.6)]. In the general case also $\widetilde{R}^0_a\widetilde{R}^0_a{}^*$ and the "three-leg" anomalous functions of the type $\varrho_c^{\frac{1}{2}}\langle\hat{\tilde{\psi}}^+\hat{\tilde{\psi}}\hat{\tilde{\psi}}\rangle$ should be included. For the time being, those types of functions will be omitted; at the end of this chapter comments will be made concerning their contributions. The remaining part of $\widetilde{R}^0_2$ is the term $\langle\hat{\tilde{\psi}}^+\hat{\tilde{\psi}}^+\hat{\tilde{\psi}}\hat{\tilde{\psi}}\rangle$, which does not depend explicitly on ϱ_c. It will be assumed that this quantity is of the form [23]

$$\widetilde{R}^0_2(\vec{x},\vec{y};\vec{x}',\vec{y}) = \Lambda^0(|\vec{x}-\vec{y}|,\,|\vec{x}'-\vec{y}|)\,\exp i\beta^0(\vec{x},\vec{x}'), \tag{5.5}$$

where Λ^0 is a real equilibrium function. The form (5.4), together with (5.5), is compatible with all symmetries and the normalization condition which R_2 has to fulfill [8]. The two-particle potential is assumed to have an already appropriately screened hard-core. The integrals in (2.10), (2.12) and (2.15) are now evaluated approximately by making a Taylor expansion of the reduced density matrices around the origin of the potential [8] which is assumed to be spherically symmetric. The derivatives of the reduced density matrices can all be expressed in terms of the four depletion variables $\varrho_d, \vec{v}_d^0, T, \boldsymbol{Q}^0$, and of the condensate density ϱ_c [9b, d].

Note that in this approximation the formulae (2.13) and (2.16) for the hydrodynamic limit are again equal to the exact ones (2.12) and (2.15). The interaction is then characterized by the zeroth and second momentum of the potential w_0 and w_2, respectively

$$w_0 = \int d^3z\, V(z), \qquad w_2 = \tfrac{1}{3}\int d^3z\, z^2\, V(z). \tag{5.6}$$

It is convenient to introduce the dimensionless interaction constant

$$\varkappa = w_2\, m\, \varrho/\hbar^2. \tag{5.7}$$

The second moment w_2 and $\varkappa$ are negative due to the attractive part of the screened potential. The following results are obtained:

$$\varrho = \varrho_c + \varrho_d \tag{5.8}$$

$$p_i = m\,\varrho_d\, v_{di}^0, \tag{5.9}$$

$$E^0 = K_{ii}^0 + I_{ii}^0 + w_0(\varrho^2 - \tfrac{1}{2}\varrho_c^2) + F, \tag{5.10}$$

$$\Pi_{ij}^0 = 2(K_{ij}^0 + I_{ij}^0) + \delta_{ij}(I_{ll}^0 + w_0(\varrho^2 - \tfrac{1}{2}\varrho_c^2) + F + G), \tag{5.11}$$

$$\begin{aligned} Q_i^0 = {} & (1-2\varkappa)\, Q_{ill}^0 + m\mu\,\varrho_d\, v_{di}^0 + \tfrac{1}{2} m\,\varrho_d\, v_d^{0^2} v_{di}^0 \left(1 + \varkappa(\varrho_d - 2\varrho_c)/\varrho\right) \\ & + \left(1 - \varkappa(\varrho_d + 2\varrho_c)/\varrho\right)(v_{di}^0 T_{ll} + 2 v_{dl}^0 T_{li}) \\ & + v_{di}^0(2F + G + \varrho_c\,\varrho_d\, w_0), \end{aligned} \tag{5.12}$$

$$\mu = -(\varkappa/\varrho)\, K_{ii}^0 + 2 w_0(\varrho - \tfrac{1}{2}\varrho_c); \tag{5.13}$$

where

$$K_{ij}^0 = \tfrac{1}{2} m\,\varrho_d\, v_{di}^0\, v_{dj}^0 + T_{ij}, \tag{5.14a}$$

$$I_{ij}^0 = -(\varkappa/\varrho)(\tfrac{1}{2} m\,\varrho_c\,\varrho_d\, v_{di}^0\, v_{dj}^0 + \varrho T_{ij}), \tag{5.14b}$$

$$F = \tfrac{1}{2}\int d^3z\, \Lambda^0(z, z)\, V(z), \tag{5.14c}$$

$$G = \tfrac{1}{2}\int d^3z\, z_i\left(\partial_i^z \Lambda^0(z, z)\right) V(z). \tag{5.14d}$$

These expressions will now systematically be compared with the corresponding two-fluid expressions (3.14)–(3.17). Equating the densities and momentum densities yields

$$v_{di}^0 = v_{ni}^0\, \varrho_n/\varrho_d. \tag{5.15}$$

The off-diagonal elements of the stress-tensor Π_{ij}^0 determine the off-diagonal elements of the internal kinetic energy tensor, which is conveniently written in the form

$$T_{ij} = T_0\,\delta_{ij} + T_{ij}', \tag{5.16}$$

where the part with a diagonal singularity has been written out explicitly. T'_{ij} is then obtained as

$$T'_{ij} = a\, v^0_{ni}\, v^0_{nj}, \qquad a = \tfrac{1}{2} m\, \varrho_n \big(1 - (1 - \varkappa\, \varrho_c/\varrho)\, \varrho_n/\varrho_d\big)/(1-\varkappa). \tag{5.17}$$

It can be seen that the quantity a vanishes at the limiting temperatures $T=0$ (because $\varrho_n=0$) and $T=T_\lambda$ (because $\varrho_n=\varrho_d=\varrho$), where only one component is present in each case. In such a situation, where only one component exists, one always can transform to a coordinate system in which the fluid is at rest, so that, without viscous terms, no vector is left to construct the off-diagonal part T'_{ij} of the internal kinetic energy tensor. $T'_{ij} \neq 0$ is therefore typical of a two-fluid situation. In Ref. [23], T'_{ij} was assumed to be zero because of which the microscopic expressions in Ref. [23] were not compatible with the nonlinear two-fluid hydrodynamics. The comparison of the diagonal parts of Π^0_{ij} relates T_0 with the pressure

$$P = 2T_0\big(1 - (5/2)\varkappa\big) + F + G + w_0\big(\varrho^2 - \tfrac{1}{2}\varrho_c^2\big) - \varkappa\, v_n^{0^2}\big(a + \tfrac{1}{2} m\, \varrho_n^2\, \varrho_c/(\varrho\, \varrho_d)\big)$$

which enables T_0 to be determined in the form

$$T_0 = b + c\, v_n^{0^2}. \tag{5.18}$$

From the microscopically introduced, formal quantities still a density, say ϱ_d, and the internal energy flux Q^0_{ill} are not determined. But one knows that Q^0_{ill} has to be of the form

$$Q^0_{ill} = q\, v^0_{di}, \tag{5.19}$$

where q again has, as the quantity a, to vanish at $T=0$ and T_λ. Equating finally the energy flux (5.12) with the two-fluid expression (3.17) $Q^0_i = v^0_{ni}(E^0 + P - m\varrho_s \mu)$ and using (5.10), (5.11) and (5.13), one gets a relation between ϱ_n and ϱ_d which, with the aid of (3.18), can be put into the form

$$\varrho_n = \varrho_d \frac{m\varrho s T + m\varrho_n v_n^{0^2}}{m\varrho s T + m\varrho_n v_n^{0^2} + d}, \tag{5.20}$$

where d is given by

$$d = (1-2\varkappa)\, q - 5\varkappa\, T_0\, \varrho_c/\varrho + m\, \varrho_n\, v_n^{0^2}\left(1 - \frac{\varrho_n}{\varrho_d} + \frac{1}{2}\left(4\frac{\varrho_n}{\varrho_d} - 3\right)\varkappa\, \varrho_c/\varrho\right). \tag{5.21}$$

The quantity q in d remains undetermined apart from its known limiting behaviour.

At T_λ d vanishes, so that (5.20) gives $\varrho_n = \varrho_d (=\varrho)$. At zero temperature d is positive and finite, so that $\varrho_n = 0$ for $\varrho_d \neq 0$, i.e. there is a finite depletion density at zero temperature. This fact is due to the unfactorized part $\widetilde{R}^0_2$ in the second reduced density matrix [9e]. At finite temperatures $T < T_\lambda$ it can be seen that $d > 0$, as long as

$$\big(5T_0 - \tfrac{1}{2} m\, \varrho_n\, v_n^{0^2}\big)\, \varkappa\, \varrho_c/\varrho \leqq q(1-2\varkappa),$$

so that $\varrho_n < \varrho_d$, as it should be. In the Bogolubov model, where T_0 and q can be calculated explicitly, one finds $5T_0 > \frac{1}{2} m\varrho_n v_n^{0^2}$, which suggests that the above condition is automatically fulfilled.

All results obtained in this chapter are correct up to second order in the relative velocity $\vec{v}_n^0$. The results show that the ansatz (5.4) for R_2 gives expressions, which are fully compatible with LANDAU's nonlinear hydrodynamics. Additional inclusion of anomalous functions, which have been discussed in connection with (5.4), does not change the structure of the result obtained but, of course, introduces new unknown functions in (5.21). Specifically they do not change the above discussed conclusions for $T-0$ and T_λ, but they do change the value of the depletion at $T=0$. A relation between the densities of the condensate (depletion) and that of the super (normal) component, which holds for He-II, is of importance, because there are attemps to measure the condensate density [22]. But numerical predictions can, at least without additional work, not be obtained from the present result.

Acknowledgement

We thank Prof. D. POLDER for his critical comments on the manuscript.

References

1. LONDON, F.: Nature **141**, 643 (1938); — Phys. Rev. **54**, 947 (1938).
2. LANDAU, L.: J. Phys. USSR **5**, 71 (1941).
3. BOGOLUBOV's work is extensively described in: GALASIEWICZ, Z. M.: Superconductivity and Quantum Fluids. Oxford: Pergamon press 1970.
4. HOHENBERG, P. C., MARTIN, P. C.: Ann. Phys. (N.Y.) **34**, 291 (1956).
5. FRÖHLICH, H.: Physica **37**, 215 (1968).
6. PENROSE, O., ONSAGER, L.: Phys. Rev. **104**, 576 (1956).
7. YANG, C. N.: Rev. Mod. Phys. **34**, 694 (1962).
8. FRÖHLICH, H.: Phys. Kondens. Materie **9**, 350 (1969).
9. HAUG, H., WEISS, K.:
   a) Phys. Letters **33** A, 263 (1970).
   b) Phys. Rev. **3** A, 717 (1971).
   c) Lettre al Nuovo Cimento **2**, 887 (1971).
   d) Physica **59**, 29 (1972).
   e) Phys. Kondens. Materie **14**, 324 (1972).
   f) Phys. Letters **40** A, 19 (1972).
10. Unfortunately, besides the above described order parameter, there is another not always clearly distinguished order parameter in the literature of He-II whose amplitude is given by the square root of the density of the supercomponent. JOSEPHSON, B. D.: Phys. Letters **32** A, 349 (1965).
11. ANDERSON, P. W.: Rev. Mod. Phys. **38**, 288 (1966).
12. CARRUTHERS, P., NIETO, M. M.: Rev. Mod. Phys. **40**, 411 (1968).
13. BISWAS, A. C.: Phys. Letters **30** A, 296 (1969). The derivation of the canonical equations for the action-angle variables and their subsequent use and interpretation appear to the present authors to be erroneous.
14. BOGOLUBOV, N.: J. Phys. USSR **11**, 23 (1947). Reprinted in PINES, D.: The Many Body Problem. New York: Benjamin 1962.
15. FERRELL, R. A., MENYHARD, N., SCHMIDT, H., SCHWABL, F., SZÉPFALUSY, P.: Ann. Phys. (N.Y.) **47**, 565 (1968).
16. In an often quoted paper [MARTIN, P. C., SCHWINGER, J.: Phys. Rev. **115**, 1342 (1959)] the conservation law of the momentum density is said to be valid only in the hydrodynamic region. In PUFF, R. D., GILLIS, N. S.: Ann. Phys. (N.Y.) **46**, 346 (1968) this error was recognized. The results of PUFF and GILLIS for the operator conservation laws are in agreement with ours. For the comparison note that in Eq. (2.4) of PUFF and GILLIS

the asymmetric part of the operator corresponding to $\vec{J}(\vec{x}, \vec{y})$ does not contribute to the double integral.
17. HALPERIN, B. J., HOHENBERG, P. C.: Phys. Rev. **188**, 898 (1969).
18. KHALATNIKOV, I. M.: An Introduction to the Theory of Superfluidity. New York: Benjamin 1965.
19. GLASSGOLD, A. E., KAUFMAN, A. N., WATSON, K. M.: Phys. Rev. **120**, 660 (1960).
20. FETTER, A. L.: Ann. Phys. (N.Y.) **70**, 67 (1972).
21. DE PASQUALE, F., TABET, E.: Ann. Phys. (N.Y.) **51**, 223 (1969).
22. See e.g. HARLING, O. K.: Phys. Rev. A**3**, 1073 (1971).
23. HYLAND, G. J., ROWLANDS, G.: J. Low Temp. Phys. **7**, 271 (1972).

Chapter V. Phase Transitions

Phase transitions of multiple nature are right in the centre of contemporary investigation of the collective behaviour of many-body systems. There are many other contributions in this book which could with some justification also have been included in this chapter. Thus, many kinds and facets of phase transitions are to be found in the other chapters, and the reader should be aware of them. In this chapter we have included only the four contributions which are directly and completely confined to phase transitions. The first (SEWELL) utilizes the algebraic formulation of statistical mechanics for a sharp characterization of thermodynamic phases and transitions between them in the thermodynamic limit; precise solutions for the structural properties of the equilibrium states of certain tractable models are also presented. The second paper (MATSUDA and HIWATARI) reviews the features of solid-liquid phase transitions within the frame of the "ideal three-phase model", the "one-species model" and the "two-species model". The third paper (PARANJAPE and HUGHES) relates to an original idea of Mott that there is a sharp transition of specific semiconductors from a non-conducting to a conducting state. FRÖHLICH's theory of dielectrics is employed to calculate the critical magnetic field for this transition. In the final paper (DUYNEVELDT, SOETEMAN and GORTER) relaxation measurements near the magnetic phase transition of a specific antiferromagnetic material are presented. A single relaxation time is found and its sharp maximum as a function of temperature or external magnetic field is shown to be ascribable to the anomalies of the specific heat in the transition region.

The Description of Thermodynamical Phases in Statistical Mechanics

G. L. SEWELL

Queen Mary College, Department of Physics, London/England

1. Introduction

As has often been emphasised by FRÖHLICH, a perequisite of any treatment of a many-body problem is the specification of the macroscopic variables in terms of which the problem may be correctly posed (cf. [1]). In this article we shall discuss an approach to statistical mechanics, which is explicitly designed to specify the states of large assemblies of particles in both macroscopic and microscopic terms and thereby to pose the relevant many-body problems in the correct macroscopic terms. Such an approach lends itself naturally to the theory of phase transitions, since different phases are usually describable in terms of different sets of macroscopic variables.

Let us first recall that it is often conceptually advantageous to formulate the statistical mechanics of large assemblies of particles in the so-called thermodynamical limit. This is the limit in which the volume of the system concerned tends to infinity, while the intensive variables remain finite. The formulation of statistical mechanics in this limit has the merit of exposing certain intrinsic properties of many-particle systems that would otherwise be masked by boundary or other finite-size effects. For example, it is only in this limit that a thermodynamical function may exhibit a discontinuity or singularity, thereby revealing in sharp mathematical form the existence of a phase transition [2].

Until relatively recently, the use of the thermodynamical limit was confined to the formulation of the thermodynamical functions (free energy density, etc.) rather than of the states of many-body systems. During the past few years, however, an algebraic generalisation of quantum (and classical) mechanics to infinite systems has led to a new approach to statistical mechanics which is ideally suited to the rigorous study of general structural properties (e.g. symmetry and stability) of the various states of such systems (cf. [3, 4]). The object of this article will be to provide a relatively simple and slightly novel formulation of this algebraic approach to statistical mechanics, surveying some results to which it has led in the theory of thermodynamical phases and phase transitions, and speculating on further developments to which it might lead.

In order to give a preliminary idea of the nature of the algebraic approach, we first recall that the observables of a quantal system of a finite number of particles have an algebraic structure generated by the canonical commutation relations. According to a theorem of VON NEUMANN [5], there is only one irreducible Hilbert space representation of the observables, and consequently one may formulate the properties of the system in terms of this representation. Von Neumann's theorem does not apply, however, to systems with an infinity of degrees of freedom. In general, the observables of such a system admit an infinity of mutually inequivalent representations, each representation carrying with it a family of states of the system. Indeed, it turns out that the state of an infinite system requires a macroscopic (i.e., global) as well as a microscopic (i.e., local) specification, the macroscopic parameters serving to determine a family of microscopically differing states carried by the same representation of the observables. The formulation of macroscopically different states in terms of mutually inequivalent representations opens up interesting possibilities for the theory of phase transitions. For example, one might conjecture that different thermodynamical phases of a system might correspond to representations with qualitatively different mathematical properties. In fact, as we shall see, this conjecture is not only feasible on general grounds but is realised in some exactly soluble models.

The following Sections will be concerned with the mathematical formulation of general properties of thermodynamical phases and of phase transitions for systems with an infinity of degrees of freedom. In Section 2, we shall outline the formulation of observables, states and dynamics first of finite then of infinite systems. In Section 3, we shall employ a form of the fluctuation-dissipation theorem to define thermal equilibrium states in a way that is sufficiently general to encompass infinite as well as finite systems. In Section 4, we prescribe a specification of those equilibrium states (of infinite systems) that correspond to pure thermodynamical phases. We then show how this specification can lead to a theory of symmetry breakdown in pure phases, and of symmetry changes in phase transitions. In Section 5, we consider the problem of the existence of metastable states and argue that, in principle, there are two distinct types of metastability. In Section 6, we summarise our conclusions.

Throughout the article, we aim to emphasize the physical interpretation, rather than the mathematical intricacies, of the algebraic formalism. Thus the results of relevant theorems will merely be stated, with appropriate references, but not proved.

2. The Mathematical Model

In general, a model of a physical system Σ consists of a mathematical structure containing a description of its observables, states and dynamics. Its observables and states correspond to sets $\mathcal{O}$, $\mathcal{S}$, respectively, such that for $A \in \mathcal{O}$ and $\phi \in \mathcal{S}$, there is a well-defined real number $\langle \phi; A \rangle$ which is taken to represent the (mean) value of A for the state ϕ. The dynamics of the system corresponds to a law governing the time-dependence of the states of Σ. Thus, the model is specified by the mathematical structures of $\mathcal{O}$, $\mathcal{S}$ and of its dynamical law. We shall now sketch the forms of these structures first for finite then for infinite quantum mechanical systems. In both cases, $\mathcal{O}$ will be taken to consist of bounded observables, in the sense that, if $A \in \mathcal{O}$, then $|\langle \phi; A \rangle|$ is uniformly bounded as ϕ runs through $\mathscr{S}$. The assumption of bounded observables is introduced for technical simplicity: in fact our main conclusions can be generalised to unbounded observables (cf. [6]).

a) Finite Systems

The observables $\mathcal{O}$ and the states $\mathscr{S}$ of a quantal system, of a finite number of degrees of freedom, correspond to the self-adjoint bounded operators and the density matrices in an appropriate separable Hilbert space $\mathscr{H}$ (cf. VON NEUMANN [7]). Specifically, if $\phi \in \mathscr{S}$, then there exists a unique density matrix ϱ in $\mathscr{H}$ such that $\langle\phi; A\rangle = \mathrm{Tr}\,(\varrho A)$, for all observables A. In particular, ϕ is a pure state if ϱ is the projection operator for some vector f in $\mathscr{H}$, in which case $\langle\phi; A\rangle \equiv (f, Af)$. The dynamics of Σ are governed by the Hamiltonian H, a self-adjoint operator* in $\mathscr{H}$, whose explicit form is determined by the masses, spins, interactions, etc. of the particles of the system. Time-translations in Σ correspond to the one-parameter group $\{\tau(t)\}$ of transformations of $\mathscr{S}$ given by:

$$\langle\tau(t)\,\phi; A\rangle \equiv \langle\phi; e^{iHt} A e^{-iHt}\rangle. \tag{2.1}$$

b) Infinite Systems

The mathematical model for an infinite system Σ is constructed so that its observables for any bounded spatial region Λ correspond to those of a finite system Σ_Λ, of the same species of particles, confined to Λ. We now sketch the construction of the model for cases were Σ consists of an assembly of particles of a single species (bosons or fermions) in a Euclidean space X, whose position vectors will be denoted by x. One may similarly construct models of the Heisenberg or Ising type, where X is a lattice. For a detailed description of the construction, both for continuous and lattice systems, see for example Refs. [3, 4].

We formulate the model of Σ in terms of a Hilbert space $\mathscr{H}_F$ (the Fock space), uniquely defined by the following properties:

(i) There exist quantised field operators $\psi(x)$, $\psi^*(x)$ in $\mathscr{H}_F$ which satisfy the canonical commutation or anticommutation rules, according to whether Σ consists of bosons or fermions:

$$[\psi(x), \psi^*(x')]_\pm = \delta(x-x'); \qquad [\psi(x), \psi(x')]_\pm = 0.$$

Strictly speaking**, these relations should be expressed in terms of "smeared fields" $\psi(f) = \int \psi(x)\, f(x)\, dx$ and $\psi^*(f) = \int \psi^*(x)\, f(x)\, dx$, corresponding to real square-integrable functions f of x:

$$[\psi(f), \psi^*(g)]_\pm = \int f(x)\, g(x)\, dx; \qquad [\psi(f), \psi(g)]_\pm = 0.$$

(ii) There exists a vector Ω_F (the vacuum vector) in $\mathscr{H}_F$, such that $\psi(f)\,\Omega_F \equiv 0$.

(iii) The space $\mathscr{H}_F$ is generated by application to Ω_F of all polynomials in the $\psi^*(f)$'s.

We represent space-translations, gauge transformations and Galilean transformations by unitary groups $\{U_F(x)\}$, $\{G_F(\alpha)\}$, and $\{\Gamma_F(v)\}$, respectively, in $\mathscr{H}_F$, which

* H is in general an unbounded operator, and thus does not belong to $\mathcal{O}$.

** The point is that $\psi(x)$, $\psi^*(x)$ are distributions, in the sense of L. SCHWARTZ, and unlike the smeared fields $\psi(f)$, $\psi^*(f)$, they are not bona fide operators in $\mathscr{H}_F$.

are uniquely defined by the following equations

$$U_F(x)\,\Omega_F \equiv \Omega_F; \qquad U_F(x)\,\psi^*(y)\,U_F^{-1}(x) \equiv \psi^*(x+y) \tag{2.2}$$

$$G_F(\alpha)\,\Omega_F \equiv \Omega_F; \qquad G_F(\alpha)\,\psi^*(x)\,G_F^{-1}(\alpha) \equiv \psi^*(x)\,e^{-i\alpha} \tag{2.3}$$

and

$$\Gamma_F(v)\,\Omega_F \equiv \Omega_F; \qquad \Gamma_F(v)\,\psi^*(x)\,\Gamma_F^{-1}(v) \equiv \psi^*(x)\,e^{-ivx}. \tag{2.3}$$

In order to represent the local properties of Σ, we associate each bounded open region Λ of X with the subspace $\mathscr{H}_\Lambda$ of $\mathscr{H}_F$ generated by polynomials in those $\psi^*(f)$'s for which f vanishes outside some closed subregion of Λ. We then define a finite localised system Σ_Λ whose observables $\mathcal{O}_\Lambda$, states $\mathscr{S}_\Lambda$ and Hamiltonian H_Λ are specified as follows. $\mathcal{O}_\Lambda$ is taken to be the set of bounded self-adjoint gauge-invariant* operators A $\left(=G_F(\alpha)\,A\cdot G_F^{-1}(\alpha)\right)$ in $\mathscr{H}_\Lambda$: this set corresponds precisely to the bounded functions of coordinates and moments of particles confined to Λ, as formulated in first quantisation. It follows from our definition that, if $\Lambda' \supset \Lambda$, then $\mathcal{O}_{\Lambda'} \supset \mathcal{O}_\Lambda$. The states $\mathscr{S}_\Lambda$ of Σ_Λ are taken to correspond to the linear functionals ϕ_Λ of $\mathcal{O}_\Lambda$ induced by the density matrices ϱ_Λ in $\mathscr{H}_\Lambda$ $\left(\langle\phi_\Lambda;\,A\rangle \equiv \mathrm{Tr}_{\mathscr{H}_\Lambda}(\varrho_\Lambda A)\right)$. The Hamiltonian H_Λ can be constructed in the usual way in terms of the masses, interactions etc. of the particles of the system. In the case of a system with two-body interactions and appropriate boundary conditions, H_Λ is given by the right-hand side of Eq. (2.7) below, with integrations restricted to Λ.

The infinite system Σ is defined in terms of the finite systems $\{\Sigma_\Lambda\}$ as follows. The set $\mathcal{O}$ of observables** of Σ is taken to be the union of all the sets $\mathcal{O}_\Lambda$ of local observables: thus, every observable of Σ corresponds to an observable of some finite system. The states*** $\mathscr{S}$ of Σ are taken to correspond to the set of linear functionals ϕ of $\mathcal{O}$ such that, for each bounded region Λ, the restriction of ϕ to $\mathcal{O}_\Lambda$ is a state of Σ_Λ; i.e. $\exists$ a density matrix ϱ_Λ in $\mathscr{H}_\Lambda$ such that $\langle\phi;\,A\rangle = \mathrm{Tr}_{\mathscr{H}_\Lambda}(\varrho_\Lambda A)$ for all A in $\mathcal{O}_\Lambda$.

It can be inferred from our definitions that, if $A \in \mathcal{O}$, then $U_F(x)\,A\,U_F^{-1}(x)$ and $\Gamma_F(v)\,A\,\Gamma_F^{-1}(v)$ also belong to $\mathcal{O}$, and that $\{U_F(x)\}$, $\{\Gamma_F(v)\}$ induce groups of transformations $\{\sigma(x)\}$, $\{\gamma(v)\}$ of $\mathscr{S}$, defined by the formulae

$$\langle\sigma(x)\,\phi;\,A\rangle \equiv \langle\phi;\,U_F(x)\,A\,U_F^{-1}(x)\rangle \tag{2.5}$$

and

$$\langle\gamma(v)\,\phi;\,A\rangle \equiv \langle\phi;\,\Gamma_F(v)\,A\,\Gamma_F^{-1}(v)\rangle. \tag{2.6}$$

Thus space translations and Galilean transformations correspond to groups $\{\sigma(x)\}$, $\{\gamma(v)\}$ of transformations of $\mathscr{S}$.

It should be noted that, although the restriction to $\mathcal{O}_\Lambda$ of any state ϕ of $\mathscr{S}$ is induced by a density matrix ϱ_Λ in $\mathscr{H}_\Lambda$, ϕ itself is not necessarily induced by a density matrix ϱ in $\mathscr{H}_F$, in the sense that $\langle\phi;\,A\rangle = \mathrm{Tr}\,(\varrho A)$ for all observables A.

---

* To the best of my knowledge, the restriction of gauge-invariance has not previously been imposed on the observables in algebraic statistical mechanics, though it has been imposed in field theory [8].

** Thus we take $\mathcal{O}$ to correspond to what are usually termed the local observables, whereas it is more usual to take $\mathcal{O}$ to be a larger set comprising the so-called quasi-local observables.

*** States defined in this way are usually referred to as "locally normal".

In fact (cf. Ref. [9]) the only states thus induced by density matrices are the ones for which the number of particles in the infinitely extended system is finite. We shall denote the set of such states by $\mathscr{S}_F$. Although $\mathscr{S}_F$ does not exhaust $\mathscr{S}$, it has the important property* that any state $\phi\,(\in\mathscr{S})$ is the limit of some sequence $\{\phi_n\}$ in $\mathscr{S}_F$, in the sense that $\lim_{n\to\infty}\langle\phi_n;A\rangle=\langle\phi;A\rangle$ for all A in $\mathcal{O}$. We note here that, if $\phi\,(\in\mathscr{S}_F)$ is induced by a density matrix ϱ', then in view of the gauge invariance of the observables, $\langle\phi;A\rangle=\mathrm{Tr}\,(\varrho A)$, where ϱ is the gauge-invariant density matrix $\frac{1}{2\pi}\int_0^{2\pi} d\alpha\, G_F(\alpha)\,\varrho'\,G_F^{-1}(\alpha)$. Further it can be shown** that for each $\phi\in\mathscr{S}_F$, there is only one gauge-invariant density matrix ϱ such that

$$\langle\phi;A\rangle\equiv\mathrm{Tr}\,(\varrho A).$$

In order to formulate the dynamics of Σ, we first introduce the Hamiltonian operator H_F in $\mathscr{H}_F$, governing time-translations in $\mathscr{S}_F$. For a system interacting via two-body forces, H_F takes the usual form:

$$H_F=\tfrac{1}{2}\int\nabla\psi^*(x)\cdot\nabla\psi(x)\,dx+\int dx\,dy\,V(x-y)\,\psi^*(x)\,\psi^*(y)\,\psi(y)\,\psi(x). \quad (2.7)$$

Let $\phi\,(\in\mathscr{S}_F)$ be the state of Σ induced by the gauge-invariant density matrix ϱ. We define $\tau(t)\,\phi$, the time-translate of ϕ, to be the state induced by the gauge-invariant density matrix $e^{-iH_Ft}\,\varrho\, e^{iH_Ft}$. Thus, time-translations in $\mathscr{S}_F$ correspond to the group $\{\tau(t)\}$ of transformations of that set.

We note here that one cannot define time-translations in the whole of $\mathscr{S}$ by direct analogy with our definition of $\sigma(x)$ [Eq. (2.5)], since if $A\in\mathcal{O}$, then in general $e^{iH_Ft}Ae^{-iH_Ft}$ is not in $\mathcal{O}$. Thus, we need some dynamical assumption which serves to extend $\tau(t)$ from $\mathscr{S}_F$ to $\mathscr{S}$. For this purpose we introduce the following postulate (P). If $A\in\mathcal{O}$ and t is real, then for each $\varepsilon>0$ there exists an observable $A_{t,\varepsilon}$ such that $|\langle\phi;A_{t,\varepsilon}\rangle-\langle\tau(t)\,\phi;A\rangle|<\varepsilon$ for all ϕ in $\mathscr{S}_F$. This signifies that, even though the operator $e^{iH_Ft}Ae^{-iH_Ft}$ might not belong to $\mathcal{O}$, one may nevertheless find some observable which approximates to that operator with any desired accuracy. We point out here that the postulate (P) is chosen largely for simplicity; in fact, it is known to be sometimes though not universally valid***. Other formulations of time-translations have been made on the basis of slightly different assumptions, known to be valid for a wide class of systems, and all lead to the same physical results (cf. [10, 11 and 12]).

Having introduced the postulate (P), we now utilise the above-mentioned fact that any state ϕ of Σ is the limit of some $\mathscr{S}_F$-class sequence $\{\phi_n\}$. It may readily be inferred from (P) that the convergence of ϕ_n, in the sense defined above,

* This sequence $\{\phi_n\}$ could, for example, be constructed as follows. Let $\{A_n\}$ be an increasing sequence of bounded regions whose union covers the space X. Let E_n be the projection operator for the subspace $\mathscr{H}_{A_n}$ of $\mathscr{H}_F$, and let ϕ_n be the state of Σ given by $\langle\phi_n;A\rangle\equiv\langle\phi;E_nAE_n\rangle$. Then it follows from our definitions that $\phi_n\in\mathscr{S}_F$ and that $\langle\phi;A\rangle=\lim_{n\to\infty}\langle\phi_n;A\rangle$ for all A in $\mathcal{O}$.

** This follows from the fact that all gauge invariant self-adjoint operators in $\mathscr{H}_F$ are (ultra-weak) limit points of sequences of elements of $\mathcal{O}$.

*** For example one can easily show that it is valid for free fermions, but not for free bosons. Also, it has been shown to be valid for a large class of lattice systems [13].

ensures that $\lim_{n\to\infty} \tau(t)\, \phi_n$ exists, and further that this limit is independent of the sequence chosen to approach ϕ. We therefore extend $\tau(t)$ unambiguously for $\mathscr{S}_F$ to $\mathscr{S}$ by defining $\tau(t)\,\phi = \lim_{n\to\infty} \tau(t)\, \phi_n$. It follows easily from our definitions that $\{\tau(t)\}$ is a one-parameter group.

It follows from our definitions that the model of Σ, as represented by $\mathcal{O}$, $\mathscr{S}$ and τ, possesses the following algebraic properties:

(P1) If $A, B \in \mathcal{O}$, then $A \pm B$, $[A, B]_+$ and $i[A, B]_-$ also belong to $\mathcal{O}$.

(P2) If $A \in \mathcal{O}$ and λ is a real number, then $\lambda A \in \mathcal{O}$.

(P3) The elements ϕ of $\mathscr{S}$ are real-valued linear functionals of the observables such that $\langle \phi; I \rangle = 1$ and $\langle \phi; A^2 \rangle \geqq 0$ for all A in $\mathcal{O}$.

(P4) $\mathscr{S}$ is equipped with transformation groups $\{\tau(t)\}$, $\{\sigma(x)\}$, $\{\gamma(v)\}$, corresponding to time translations, space translations and Galilean transformations, respectively. It may be inferred from our definitions of these groups that

$$\sigma(x)\, \tau(t) \equiv \tau(t)\, \sigma(x)$$

and

$$\gamma(v)\, \tau(t)\, \gamma(-v) \equiv \tau(t)\, \sigma(vt).$$

The composition laws governing the algebraic operations in (P1)–(P4) are determined by the Fock space formulation of the model, e.g. for given A, B in $\mathcal{O}$, the value of $[A, B]_+$ is determined by the algebra of the canonical commutation (or anticommutation) rules for the field operators ψ, ψ^* in $\mathscr{H}_F$. However, since these various composition laws are given, we may abstract the algebraic structure of Σ from its Fock space origins. Thus, we shall henceforth regard the model as the algebraic structure generated by (P1–4), with its given composition laws—without further reference to $\mathscr{H}_F$.

It may be inferred from standard algebraic theorems* that this model for Σ also has the following properties:

(P5) Each state ϕ of Σ induces a representation π_ϕ of $\mathcal{O}$ in a Hilbert space $\mathscr{H}_\phi$, such that

a) π_ϕ preserves the algebraic structure of $\mathcal{O}$

$$\left(\text{i.e. } \pi_\phi(A)^* \equiv \pi_\phi(A),\ \pi_\phi(A+B) \equiv \pi_\phi(A) + \pi_\phi(B), \text{ etc.}\right)$$

b) $\mathscr{H}_\phi$ contains a vector Ω_ϕ such that $\langle \phi; A \rangle \equiv \left(\Omega_\phi, \pi_\phi(A)\, \Omega_\phi\right)$; and

c) $\mathscr{H}_\phi$ is generated by application to Ω_ϕ of all operators of the form $\pi_\phi(A) + i\,\pi_\phi(B)$, with $A, B \in \mathcal{O}$.

Further, there is only one representation of $\mathcal{O}$, up to unitary equivalence, which satisfies a)–c). We shall henceforth denote by $\mathscr{S}_\phi$ the family of states ω of Σ induced by density matrices ϱ in $\mathscr{H}_\phi$:— $\langle \omega; A \rangle \equiv \mathrm{Tr}\left(\varrho\, \pi_\phi(A)\right)$. Physically, $\mathscr{S}_\phi$ corresponds to the set of states generated from ϕ by local perturbations. Mathematically, $\mathscr{S}_\phi$ contains all those states of Σ that induce representations of $\mathcal{O}$ that are unitarily equivalent to π_ϕ.

* Specifically, (P5) follows from the GELFAND-NAIMARK-SEGAL construction (described in Refs. [3, 4]) as applied to the C^*-algebra generated by $\mathcal{O}$, with $\mathscr{H}_F$-operator norm; while (P6) and (P7) follow from the same construction, supplemented by theorems due to KADISON ([14], Theorem 3.4) and SEGAL [15].

(P6) If $\phi(\varepsilon\mathscr{S})$ is invariant under $\{\sigma(x)\}$, then space-translations in $\mathscr{S}_\phi$ correspond to a unitary group $\{U_\phi(x)\}$ of operators in $\mathscr{H}_\phi$ such that $U_\phi(x)\,\Omega_\phi \equiv \Omega_\phi$, and if $\omega(\varepsilon\,\mathscr{S}_\phi)$ is induced by the density matrix ϱ in $\mathscr{H}_\phi$, then $\sigma(x)\,\omega$ is the state induced by $U_\phi^{-1}(x)\,\varrho\,U_\phi(x)$. The infinitessimal generator P_ϕ of $\{U_\phi(x)\}$ is then the momentum operator for the states $\mathscr{S}_\phi$.

Likewise, if $\phi(\varepsilon\mathscr{S})$ is invariant under $\{\tau(t)\}$, then time-translations in $\mathscr{S}_\phi$ correspond to a unitary group $\{T_\phi(t)\}$ of operators in $\mathscr{H}_\phi$ such that $T_\phi(t)\,\Omega_\phi \equiv \Omega_\phi$ and, if $\omega(\varepsilon\,\mathscr{S}_\phi)$ is induced by the density matrix ϱ in $\mathscr{H}_\phi$, then $\tau(t)\,\omega$ is the state induced by $T_\phi^{-1}(t)\,\varrho\,T_\phi(t)$. The infinitessimal generator H_ϕ of $\{T_\phi(t)\}$ is the Hamiltonian operator for $\mathscr{S}_\phi$.

It is important to note that H_ϕ, P_ϕ are representation-dependent quantities. Thus H_ϕ may have properties which are completely different from those of the Fock Hamiltonian H_F, which served merely to determine the dynamical group $\{\tau(t)\}$.

(P7) If ϕ is invariant under both $\{\sigma(x)\}$ and $\{\tau(t)\}$, then its Galilean transform $\gamma(v)\,\phi$ is also time-translationally invariant.. Further, the transformation $\phi \to \gamma(v)\,\phi$ induces a unitary transformation* $V(v)\colon \mathscr{H}_\phi \to \mathscr{H}_{\gamma(v)\phi}$ such that

$$V^{-1}(v)\,H_{\gamma(v)\phi}\,V(v) = H_\phi + v\cdot P_\phi;$$

and therefore the spectra of $H_{\gamma(v)\phi}$ and $H_\phi + v\cdot P_\phi$ are identical

Finally, we introduce the concept of a symmetry group of Σ as follows. Let G be some group such that, for each g in G, there exists a transformation $\alpha^*(g)$ of $\mathscr{O}$ such that:

(i) α^* preserves the group structure of G, i.e. $\alpha^*(g_1)\,\alpha^*(g_2) = \alpha^*(g_1 g_2)$, and $\alpha^*(I) = I$; and

(ii) for each g in G, the action of $\alpha^*(g)$ on $\mathscr{O}$ preserves the algebraic structure of that set, i.e.

$$\alpha^*(g)\,(\lambda_1 A_1 + \lambda_2 A_2) \equiv \lambda_1 \alpha^*(g)\,A_1 + \lambda_2\,\alpha^*(g)\,A_2;$$

$$\alpha^*(g)\,[A, B]_+ \equiv [\alpha^*(g)\,A, \alpha^*(g)\,B]_+,$$

and

$$\alpha^*(g)\,i\,[A, B]_- \equiv i\,[\alpha^*(g)\,A, \alpha^*(g)\,B]_-.$$

Let $\{\alpha(g)\}$ be the corresponding group of transformations of $\mathscr{S}$ defined by the formula

$$\langle \alpha(g)\,\phi; A\rangle \equiv \langle \phi; \alpha^*(g)\,A\rangle.$$

Then we term G a symmetry group of Σ if $\{\alpha(g)\}$ commutes with $\{\tau(t)\}$.

3. The Equilibrium States

In order to specify the thermal equilibrium states of an infinite system, one needs some generalised equilibrium conditions which, when applied to a finite system, yield the Gibbs canonical states. It has been proposed that the appropriate conditions are those of Kubo, Martin and Schwinger (KMS) (cf. Refs. [16, 17]), a proposal we support. In the present Section, we shall first state these conditions and then give our reasons for regarding them as thermal equilibrium conditions.

* In fact $V(v)$ is uniquely defined by the equation $V(v)\,\pi_\phi(A)\,\Omega_\phi \equiv \pi_{\gamma(v)\phi}(A_v)\,\Omega_{\gamma(v)\phi}$, with $A_v \equiv \Gamma_F^{-1}(v)\,A\,\Gamma_F(v)$.

Let ϕ be a time-translationally invariant state of Σ, and let $\{T_\phi(t)\}$ be the unitary group governing time-translations in $\mathscr{H}_\phi$, as described in (P6) above. Then for each pair of elements A, B in $\mathcal{O}$, we define functions $F_{AB}(t)$, $G_{AB}(t)$ by the formulae

$$F_{AB}(t)=(\Omega_\phi, \pi_\phi(A) T_\phi(t)\, \pi_\phi(B)\, \Omega_\phi)\,; \qquad G_{AB}(t)=(\Omega_\phi, \pi_\phi(B)\, T_\phi(-t)\, \pi_\phi(A)\, \Omega_\phi).$$

We denote by $\hat{F}_{AB}$, $\hat{G}_{AB}$ the Fourier transforms* of F_{AB}, G_{AB}, respectively (i.e. $F_{AB}(\omega)=\int dt e^{-i\omega t} F_{AB}(t)$, etc.). Then, for temperature $(k\beta)^{-1}$, the KMS conditions on ϕ are that:

(i) F_{AB}, G_{AB} are continuous functions of t, and

(ii) $\hat{F}_{AB}(\omega)=\hat{G}_{AB}(\omega)\, e^{\beta\omega}$, for all observables A, B.

We now list our reasons for supporting the postulate that the equilibrium states of Σ are those which satisfy these conditions.

1. In the case of a finite system, the KMS conditions uniquely specify the Gibbs equilibrium state: $\phi(A)\equiv\text{Tr}\,(Ae^{-\beta H})/\text{Tr}\,(e^{-\beta H})$ (cf. Ref. [4], p. 189).

2. The infinite volume limit (appropriately defined) of a sequence of finite volume Gibbs states satisfies the KMS conditions [10].

3. For a class of lattice systems, it has been shown that the states which satisfy the KMS conditions are precisely those corresponding to a natural generalisation to infinite systems of the Gibbs probability distribution [18].

4. In the case of some exactly soluble models, it has been shown that the set of states satisfying the KMS conditions is precisely the set of stable and metastable equilibrium states, as obtained by traditional methods in the thermodynamical limit [19].

5. The KMS conditions are equivalent to the fluctuation-dissipation theorem, up to one or two technicalities [20]. It has been argued by CALLEN and WELTON [21] that the states which satisfy this theorem are those for which Σ behaves as a thermal reservoir in the sense that, when suitably coupled to a finite system, it drives it into thermal equilibrium at temperature $(k\beta)^{-1}$. In fact one may demonstrate the equivalence between the KMS conditions and this reservoir property, within the framework of the present formalism, by means of a generalisation of the techniques employed in Ref. [20].

In view of the arguments (1–5), we shall henceforth identify the thermal equilibrium states with those that satisfy the KMS conditions. We note that these conditions reduce, at absolute zero, to the requirement that the Hamiltonian H_ϕ for $\mathscr{S}_\phi$ is a positive operator, i.e. $(f, H_\phi f)\geqq 0$ for all f in $\mathscr{H}_\phi$ (cf. [22]). This simply means that all excitation energies from ϕ, as generated by local perturbations, are positive.

Finally we note that, whereas a finite system can have only one equilibrium state at a given temperature [cf. (1) above], an infinite system can have a plurality of such states (cf. Ref. [4], Ch.2). As will be shown in the following sections, this opens up the possibility of describing the properties of metastable states and states with broken symmetries within the framework of the algebraic formalism for infinite systems.

* Here F, G, $\hat{F}$, $\hat{G}$ should be considered to be distributions.

4. Pure Thermodynamical Phases

Let $\mathscr{K}_\beta$ denote the set of all equilibrium (i.e. KMS) states of Σ for temperature $(k\beta)^{-1}$. Then it is a simple matter to show that $\mathscr{K}_\beta$ corresponds mathematically to a convex compact set, from which it follows by standard theorems that:

(i) $\mathscr{K}_\beta$ contains a subset $\mathscr{E}_\beta$, whose elements ϕ cannot be decomposed into positive linear combinations of other $\mathscr{K}_\beta$-class states [i.e. combinations $\lambda\phi_1+(1-\lambda)\,\phi_2$ with $\phi_1, \phi_2 \in \mathscr{K}_\beta$ and $0<\lambda<1$]; and

(ii) All other $\mathscr{K}_\beta$-class states can be decomposed as positive sums or integrals of $\mathscr{E}_\beta$-class states.

Consequently, it is natural to identify $\mathscr{E}_\beta$ with the pure thermodynamical phases and the other $\mathscr{K}_\beta$-class states with mixtures of such phases at temperature $(k\beta)^{-1}$ (cf. Ref. [16]). This identification is supported by the fact that $\mathscr{E}_\beta$ consists precisely of those $\mathscr{K}_\beta$-class states which possess spatial clustering properties of a type generally associated with pure phases [16].

We shall therefore take $\mathscr{E}_\beta$ to correspond to the equilibrium states of Σ in pure thermodynamical phases at temperature $(k\beta)^{-1}$. The total set of states in the pure phase corresponding to $\phi\,(\in\mathscr{E}_\beta)$ is then $\mathscr{S}_\phi$. In general, it is to be anticipated that the elements of $\mathscr{E}_\beta$ may be labelled by the values of a finite distinguishing set of parameters ξ, corresponding to the thermodynamical variables required to specify a pure phase at that temperature (e.g., density, magnetic polarisation, etc.). This means that the mathematical structure of $\mathscr{K}_\beta$ determines the complete set of thermodynamical variables ξ required to specify a pure phase at temperature $(k\beta)^{-1}$.

Suppose now that G is a symmetry group of represented by transformations $\{\alpha(g)\,|\,g\in G\}$ of $\mathscr{S}$, as described at the end of Section 2. It follows from our definitions that, if $\phi\in\mathscr{E}_\beta$, then $\alpha(g)\,\phi$ is also in $\mathscr{E}_\beta$. This implies that $\mathscr{E}_\beta$ is invariant under $\alpha(g)$, i.e. that $\mathscr{E}_\beta\equiv\{\alpha(g)\,\phi\,|\,\phi\in\mathscr{E}_\beta\}$. It does not imply, however, that the elements of $\mathscr{E}_\beta$ are individually invariant under $\alpha(g)$, unless $\mathscr{E}_\beta$ consists of one state, as in the case of a finite system. Hence, in general, we have the possibility that $\mathscr{E}_\beta$ may contain states which are not invariant with respect to some symmetry group G of Σ. In such cases we say that the system undergoes a symmetry breakdown in the states concerned.

This phenomenon of symmetry breakdown is of course known to occur in many states of real systems, e.g., the crystalline phases of systems with translationally invariant forces, the magnetically polarised phases of systems with rotationally invariant forces, etc. Consequently, it is of real physical significance that the present algebraic formalism opens up the possibility of formulating this phenomenon in a clean mathematical manner. Moreover, this possibility has been realised in certain solvable models which have been shown to exhibit symmetry breakdown associated with magnetic polarisation below a certain transition temperature [23, 24]*. On the other hand, there are as yet no solvable models exhibiting the breakdown of Euclidean symmetry in favour of some crys-

* In fact, the symmetry breakdown as defined in Ref. [24] is not identical with that considered here. However, it can be inferred from Ref. [25] (Theorem 3.2.1 and subsequent remark) and Ref. [18] the results of [24] imply symmetry breakdown, in the present sense, below the transition temperature.

tallographic symmetry. The construction of such a model would surely constitute a significant advance in the theory of the solid state.

The phenomenon of symmetry breakdown is known empirically to be linked, at least sometimes, with that of phase transitions. Indeed one might consider a phase transition to be characterised in a number of ways; e.g., a change in symmetry, a discontinuity or singularity in a thermodynamical function, a divergence of some suitable defined correlation length (in the case of critical phenomena) etc. Presumably a general theory of phase transitions should inter-relate these various characteristics in some way. At present, no such theory exists. A limited advance in the direction of such a theory has been made by M. MARINARO and the author [24], who prove that the phase transitions in a certain class of Ising spin systems exhibit all these characteristics. However, it would be erroneous to assume that all critical phenomena are similarly characterised. For, if one accepts our thesis that the observables $\mathcal{O}$ should be taken to be gauge invariant, then it is easy to show, by an adaptation of the analysis of ARAKI and WOODS [26], that an ideal bose gas exhibits no symmetry breakdown* either above or below its thermodynamical transition temperature. The same thing can presumably be said of real superfluids. On the other hand the ideal bose gas (and presumably a real superfluid) does exhibit off-diagonal long range order, as defined by YANG [27], below the condensation point.

5. Metastable States

A state of a system is regarded as metastable, from the empirical standpoint, if it simulates an equilibrium state over suitably long observational times, even though it does not correspond to an absolute minimum of free energy for the given thermodynamical conditions. Familiar examples of such states are provided by the persistent current-carrying states of a superfluid and the states of a ferromagnet whose polarisation opposes that of some suitable weak applied field.

From the theoretical standpoint, it is clear that a metastable state of a real finite system should be characterised by a lifetime (suitably defined) which, although long by some relevant observational standards, is nevertheless finite. The question then arises as to whether this lifetime depends on some extensive variable of the system in such a way that mathematically it tends to infinity with this variable. If so, one has the possibility that, in the infinite volume limit, the state is stable with respect to local perturbations and satisfies the KMS conditions. Moreover, this possibility is known to be mathematically feasible, since it is realised in certain exactly soluble models [19, 28].

Thus one can conceive of two different types of metastable states, as described within the framework of the above formalism. These are (a) the states which satisfy the KMS conditions but which do not correspond to absolute minima of the free energy density (appropriately defined); and (b) states which do not satisfy the KMS conditions but which have very long lifetimes by some appropriate observational standards. We shall now discuss the two examples of metastable states cited above

* On the other hand, if one does not enforce the restriction that the observables are gauge invariant, then the ideal bose gas exhibiting a gauge symmetry breakdown below but not above the transition temperature [26].

(the superfluid and the ferromagnet), with the view to determining whether they are metastable in the sense (a) or (b).

Example 1. Superfluidity. Let us examine the possibility that a system Σ may exhibit superfluidity in the sense that is possesses current-carrying states which satisfy the KMS conditions: such states would be metastable in the sense (a). For simplicity we restrict our considerations to the states of Σ at the absolute zero of temperature.

We first suppose that Σ possesses a translationally invariant equilibrium state ϕ, which does not carry a current, at absolute zero. The fact that ϕ is an equilibrium state at that temperature is equivalent to the fact that the Hamiltonian H_ϕ is a positive operator in $\mathscr{H}_\phi$, as pointed out in Section 3. Further, since ϕ is current-free, its Galilean transform $\gamma(v)\ \phi$ carries a current if $v \neq 0$. We now ask whether this current-carrying state may satisfy the equilibrium condition that $H_{\gamma(v)\phi}$ is a positive operator in $\mathscr{H}_{\gamma(v)\phi}$. In fact, it follows directly from property (P 7), formulated at the end of Section 2, that $\gamma(v)\ \phi$ fulfills this condition if and only if $H_\phi + v \cdot P_\phi$ is a positive operator in $\mathscr{H}_\phi$. Hence we see that a sufficient condition for superfluidity of Σ at absolute zero is that $(H_\phi + v \cdot P_\phi)$ is a positive operator for some non-zero range of values for the velocity v. This condition corresponds to that of Landau [29].

Example 2. Ferromagnet in Opposing Field. We now consider the possible existence of metastable states of a ferromagnet polarised in a direction opposite to that of an applied magnetic field. For definiteness, we base our discussion on the theory of an Ising model of spins on a simple cubic lattice. In a usual way the spins are kinematically constrained to take values ± 1 in the Oz direction. It is assumed that the interactions are confined to nearest neighbouring pairs, the potential energy of such a pair with spins σ and σ' being $-J\sigma\sigma'$, where J is a positive constant. This interaction favours alignment of the spins and, in the absence of any external field, leads to ferromagnetically ordered states at temperatures T below some critical value T_c [30].

If a magnetic field H, parallel to Oz, is applied to the system, then it sill still exhibit ordered equilibrium states with polarisation along Oz, for $T < T_c$. One might imagine that the system might also have equilibrium states in which the polarisation were opposed to the applied field: for it might be argued that, if the external field were weaker than the internal one generated by the magnetic ordering, then it would cost energy to reverse a spin in such a state. However, it has been rigorously shown that no states of the model exist in which the equilibrium conditions are satisfied and the applied field is opposed to the polarisation of the system [31]*. Thus the system does not exhibit metastable states of type (a), in which the polarisation opposes the applied field.

We now give a argument indicating that the system might nevertheless exhibit (b)-type metastable states with polarisation opposing the applied field. For this purpose we start by considering the situation at absolute zero. Suppose then that the system were in the ordered state with all spins antiparallel to Oz. Then the energy required to reverse a spin would be $6J - H$, which would be

* The equilibrium conditions of Ref. [31] have been shown by Brascamp [18] to be equivalent to the KMS conditions.

positive if $6J > H$. Likewise, it would cost energy to reverse two or three spins, if J/H is large enough. However, the situation changes if one takes account of the possibility of reversing a suitably large number of spins. For consider a cube within the lattice, whose sides coincide with principal axes of the lattice and are of length equal to n times the lattice spacing. Then the internal energy required to reverse the spins in this cube is $6n^2J$, while the resultant energy gain from the external field is n^3H. Thus, for n sufficiently large, one achieves a net reduction in energy by reversing the spins in the cube, which confirms that the "wrongly polarised" state is not stable. However, in order to reverse such a block of n^3 spins, one would need to overcome an energy barrier corresponding to the maximum (w.r.t. n) of $6n^2J - n^3H$. Thus the energy barrier would be $16J^3/H^2$. It is reasonable to suppose that one would require to overcome an energy barrier E of the same order of magnitude at a finite temperature $T(>T_c)$ in order to align a block of spins with the external field. This means that the lifetime of the "wrongly polarised" state would be proportional to $\exp(E/kT)$, apart from other presumably milder factors. For reasonable values of $J(\sim kT_c)$, H and T, it is easy to see that $E/kT\left(\sim \frac{16J^3}{kT\cdot H^2}\right)$ could be very large by comparison with unity. Thus one can envisage that the lifetime of the "wrongly polarised" state might be astronomically long, in which case it would be a metastable state of type (b).

6. Conclusion

The algebraic formulation of statistical mechanics in the thermodynamical limit provides a natural framework for the sharp characterisation of thermodynamical phases and of phase transitions. In this formulation, an equilibrium state ϕ of a system Σ in a pure phase at temperature $(k\beta)^{-1}$ corresponds to an element of the set $\mathscr{E}_\beta$, defined in Section 4, and is designated by the values of a set of thermodynamical variables ξ. The pure phase associated with ϕ then comprises the set $\mathscr{S}_\phi$ of states generated by local perturbations of ϕ. Different phases of the system are then mutually inaccessible by such perturbations, this inaccessibility being represented by the inequivalence of the corresponding representations of the observables.

The formalism has been used both to classify the various phases of matter in terms of symmetry, stability and other characteristics; and also to solve precisely the structural properties of the equilibrium states of certain tractable models (cf. Refs. [19, 23, 24]).

For each phase $\mathscr{S}_\phi$ of a system, the dynamical group $\{\tau(t)\}$ induces a Hamiltonian H_ϕ governing time-translations in $\mathscr{S}_\phi$. The phase-dependence of the properties of H_ϕ opens up some interesting possibilities which have not yet been greatly explored. For example, an indication of how certain general features of H_ϕ may control the qualitative properties of the phase is given by our discussion of superfluidity in Section 5. Presumably one could go much further and construct a theory of superfluid hydrodynamics on the basis of some very general assumptions concerning the Hamiltonian for the phase. Similarly, one ought to be able to base theories of other phases on some species properties of the relevant Hamiltonians.

References

1. FRÖHLICH, H.: Theoretical Physics and Biology, p. 13–22, Ed. M. MAUROIS. Amsterdam: North-Holland Publishing Company 1969.
2. YANG, C. N., LEE, T. D.: Phys. Rev. **87**, 404 (1952).
3. RUELLE, D.: Statistical Mechanics. New York: W. A. Benjamin, Inc. 1969.
4. EMCH, G. G.: Algebraic Methods in Statistical Mechanics and Quantum Field Theory. New York: J. Wiley & Sons, Inc. 1972.
5. VON NEUMANN, J.: Math. Ann. **104**, 570 (1931).
6. SEWELL, G. L.: J. Math. Phys. **11**, 1868 (1970).
7. VON NEUMANN, J.: Mathematical Foundations of Quantum Mechanics. Princeton: Princeton University Press 1955.
8. DOPLICHER, S., HAAG, R., ROBERTS, J. E.: Commun. Math. Phys. **13**, 1 (1969).
9. DELL'ANTONIO, G. F., DOPLICHER, S., RUELLE, D.: Commun. Math. Phys. **2**, 223 (1966).
10. HAAG, R., HUGENHOLTZ, N. M., WINNINK, M.: Commun. Math. Phys. **5**, 215 (1967).
11. DUBIN, D. A., SEWELL, G. L.: J. Math. Phys. **11**, 2990 (1970).
12. SIRUGUE, M., WINNINK, M.: Preprint.
13. ROBINSON, D. W.: Commun. Math. Phys. **6**, 151 (1967); **7**, 337 (1968).
14. KADISON, R. V.: Topology **3**, Suppl. 2, 177 (1965).
15. SEGAL, I., DUKE, E.: Math. J. **18**, 221 (1951).
16. EMCH, G. G., KNOPS, H. J. F., VERBOVEN, E.: J. Math. Phys. **11**, 1655 (1970).
17. LANFORD, O. E.: Cargese Lectures in Physics, Vol. 4, p. 113–136, Ed. D. KASTLER. New York: Gordon and Breach 1970.
18. BRASCAMP, H. J.: Commun. Math. Phys. **18**, 82 (1970).
19. EMCH, G. G., KNOPS, H. J. F.: J. Math. Phys. **11**, 3008 (1970).
20. PRESUTTI, E., SCACCIATELLI, E., SEWELL, G. L., WANDERLINGH, F.: J. Math. Phys. (in Press).
21. CALLEN, H. B., WELTON, T. A.: Phys. Rev. **83**, 34 (1951).
22. SIRUGUE, M., TESTARD, D.: Preprint.
23. THIRRING, W.: Commun. Math. Phys. **7**, 181 (1968).
24. MARINARO, M., SEWELL, G. L.: Commun. Math. Phys. **24**, 310 (1972).
25. RUELLE, D.: Cargese Lectures in Physics, Vol. 4, p. 169–194, Ed. D. KASTLER. New York: Gordon and Breach 1970.
26. ARAKI, H., WOODS, E. J.: J. Math. Phys. **4**, 637 (1963).
27. YANG, C. N.: Rev. Mod. Phys. **34**, 694 (1962).
28. EMCH, G. G.: J. Math. Phys. **8**, 13 (1967).
29. LANDAU, L. D.: J. Phys. USSR **5**, 71 (1941).
30. DOBRUSHIN, R. L.: J. Funct. Anal. Appl. **2**, 44 (1968).
31. LANFORD, O. E., RUELLE, D.: Commun. Math. Phys. **13**, 194 (1969).

The Effect of Interatomic Potential on the Feature of Solid-Liquid Phase Transition

HIROTSUGU MATSUDA* and YASUAKI HIWATARI**

Kyoto University, Research Institute for Fundamental Physics, Kyoto/Japan

1. Introduction

Most substances in nature are in the solid state at low temperatures, and the solid undergoes a solid-liquid phase transition at a certain temperature under given pressure. The features of solid-liquid phase transitions of various substances should be a reflection of the nature of the interatomic potential, and the purpose of this article is to give a brief review of some recent work carried out with the aim of elucidating the effect which the interatomic potential has on the features of solid-liquid phase transition.

Among various substances, there is one group whose molecules can be treated as spherical. The liquid state of such substances is called simple liquid, and the intensive study of simple liquids has been carried out both theoretically and experimentally [1–3]. Molecular liquids, such as liquefied inert gases or liquefied vapour of spherical molecules, and liquids of fused metals are typical examples of simple liquids. The intermolecular potential of these simple liquids may be assumed to be a sum of effective pair potential in a first approximation. In this article we confine ourselves within the frame of such a pair theory. The effective pair potential of real simple liquids is characterized by a short-range strong repulsion and a longer-range attraction. Although we have virtually no rigorous theory at present as to what conditions on the pair potential are sufficient for the existence of solid-liquid phase transition, numerical studies of ALDER and WAINWRIGHT [4], and WOOD and JACOBSON [5] have made it almost certain that even an assembly of hard spheres may crystallize and that the most important factor for the solid-liquid transition is the existence of a strong repulsive potential rather than an attractive potential. Moreover, the hard-sphere model has been found to represent the equation of state [6], the liquid structure factor [7], and the diffusion constant [8] of real simple liquids fairly well, if the diameter of the hard sphere is suitably chosen as a function of density and temperature

* Present Address: Kyushu University, Department of Biology, Faculty of Science, Fukuoka, Japan.

** On leave of absence from Institute for Spectroscopic Study of Matter, Faculty of Science, Kanazawa University, Kanazawa.

[9]. Thus, the hard-sphere model has been advocated as a kind of ideal liquid [10].

The hard-sphere system has the following simplicity: if one knows a physical quantity such as the equation of state, diffusion constant and viscosity for all densities at one temperature, then one can easily obtain the corresponding quantities at any temperature by scaling. However, apart from its conceptual simplicity, this physical simplicity is not a privilege that only the hard-sphere system can enjoy. As shown in § 2, we have this scaling property in all the classical systems of atoms with a pair potential [11],

$$\phi_I(r) = \varepsilon(\sigma/r)^n - \alpha\gamma^3 \exp(-\gamma r), \qquad [\varepsilon > 0, \sigma > 0, n > 3, \alpha \geqq 0, \gamma \to 0+], \tag{1.1}$$

which includes the hard-sphere system as a special case; $n \to \infty$, $\alpha = 0$. Here $\gamma \to 0+$ means that we let positive γ tend to zero after taking a thermodynamic limit.

When $\alpha = 0$, $n < \infty$, this model is called the soft-sphere model, and the equation of state of the soft-sphere model has been obtained by computer experiment using the Monte Carlo method [12]. From analyses on the basis of the Monte Carlo studies of HOOVER, GRAY and JOHNSON, the above model with the pair potential (1.1) turns out to be capable of exhibiting three phases of matter, that is, gas, liquid and solid phases, in a certain temperature range when α is a positive constant. Thus, we call it the ideal three-phase model [13]. In § 2 we summarize some important properties of this model.

Now, among substances forming simple liquids, metals have melting phenomena which are systematically different from those of molecular crystals, as shown in Table 1–3: **1.** The melting temperature, which is close to the triple-point temperature T_t under a normal pressure, measured in the unit of cohesive energy is lower in metals. **2.** The fractional volume and entropy change upon melting is smaller. **3.** In alkali and alkaline-earth metals the fractional increase of melting temperature per fractional volume change under pressure is smaller than in molecular crystals. In addition, the critical temperature T_c of metals measured in the unit of cohesive energy is larger in metals; this fact together with **1.** results in the larger liquid range for metals.

Studying the thermodynamic property of the ideal three-phase model, HIWATARI and MATSUDA [13] pointed out that the above systematically different properties of metals and molecular crystals can be roughly accounted for in the frame of the ideal three-phase model by assuming that metals have smaller values of n than molecular crystals.

The interatomic potential of inert gases is usually given by the Lennard-Jones potential,

$$\phi_{LJ}(r) = 4\varepsilon\{(\sigma/r)^{12} - 2(\sigma/r)^6\}, \tag{1.2}$$

which is by no means close to the potential (1.1) in the limit $\gamma \to 0+$. The effective pair potential of metals calculated on the basis of the pseudopotential for electrons has a more complicated structure such as a long-range oscillation [14]. Moreover, the effective pair potential of metals depends on density.

However, a detail of the potential may not be so important for determining thermodynamic quantities of the substance. Indeed, SCHIFF [15] has shown on the basis of his computer studies that the shape of the long-range part of the

Table 1. Experimental value[a] of the melting temperature and critical temperature measured in the unit of cohesive energy, and liquid range.

Element	$kT_t/\|u_0\|$	$kT_c/\|u_0\|$	T_c/T_t	Element	$kT_t/\|u_0\|$	$kT_c/\|u_0\|$	T_c/T_t
Ne	0.0810	0.147	1.82	Zn	0.0443	0.220	4.95
A	0.0820	0.148	1.80	Cd	0.0440	0.220	5.00
Kr	0.0816	0.148	1.81	Hg	0.0302	0.226	7.50
Xe	0.0810	0.146	1.80	Al	0.0241	0.221	9.16
Li	0.0236	0.168	7.1	Ga	0.0093	0.234	25.2
Na	0.0283	0.196	6.9	In	0.0149	0.231	15.6
K	0.0308	0.204	6.6	Tl	0.0264	0.253	9.52
Rb	0.0306	0.206	6.7	Ge	0.0270	0.188	6.95
Cs	0.0312	0.213	6.81	Sn	0.0139	0.240	17.2
Cu	0.0333	0.218	6.56	Pb	0.0254	0.228	9.00
Ag	0.0359	0.218	6.06	Mn	0.0450	—	—
Au	0.0302	0.214	7.09	Fe	0.0362	0.135	3.73
Mg	0.0519	0.216	4.17	Co	0.0345	—	—
Ca	0.0525	0.216	4.12	Ni	0.0335	0.119	3.48
Sr	0.0528	—	—				

[a] Experimental data are taken from the table listed in the paper of G. L. POLLACK [Rev. Mod. Phys. **36**, 748 (1964]) for inert gases, and of K. A. GSCHNEIDER [Solid State Physics vol. **16**, 412 (1964)] for melting temperature and cohesive energy of metals. Critical temperatures for metals except Cs and Hg are estimated ones and they are listed in the paper of D. A. YOUNG and B. J. ALDER [Phys. Rev. **3**A, 364 (1971)].

Table 2. Fractional volume and entropy change on melting[a]

Element	$\Delta V_f/V_{ts}$ (%)	S_f (e.u.)	Element	$\Delta V_f/V_{ts}$ (%)	S_f (e.u.)
Ne	15.3	3.26	Au	5.1	2.29
Ar	14.4	3.35	Mg	3.05	2.25
Kr	15.1	3.36	Ca		
Xe	15.1	3.40	Sr		
Li	1.65	1.53	Zn	4.2	2.48
Na	2.5	1.70	Cd	4.7	2.57
K	2.55	1.70	Hg	3.7	2.37
Rb	2.5	1.68	Al	6.0	2.70
Cs	2.6	1.65	Ga	−3.2	4.42
Cu	4.51	2.29	In	2.7	1.82
Ag	3.30	2.22			

[a] From A. R. UBBELOHDE: Melting and Crystal Structure, Clarendon Press, Oxford, 1965.

pair potential $\phi(r)$ is rather irrelevant to the liquid structure factor and velocity autocorrelation function, whereas the shape of the short-range repulsive part of $\phi(r)$ has some effect on them. Moreover, since the density of the solid phase is not so different from that of the liquid phase at the melting point, it may be possible to assume, as a first approximation, the same pair potential in both phases of real substances. Then it might be possible to approximate the thermodynamic properties of the real substance by our model with parameters n, $C \equiv \varepsilon\sigma^n$ and α determined from the knowledge of the solid phase. The result of this analysis will be given in § 3.

Table 3. Value of n

Element	n [a]	n [b]	Element	n [a]	n [b]
Ne	15.9	—	Zn	12.7	13.5
Ar	20.0	16.8	Cd	16.2	13.8
Kr	17.8	15	Hg	18.5	—
Xe	22.5	15	Al	6.74	17.7
Li	2.83	1.6	Ga	7.45	—
Na	4.33	4.8	In	8.08	15.4
K	4.50	4.8	Tl	10.2	13.1
Rb	5.90	4.8	Ge	8.51	—
Cs	5.04	4.8	Sn(w)	8.80	9.6
Cu	8.27	—	Pb	12.0	—
Ag	10.8	—	Mn	4.70	—
Au	14.5	—	Fe	8.60	8.4
Mg	10.0	—	Co	8.65	—
Ca	6.78	6.2	Ni	8.60	—
Sr	7.19	5.7			

[a] One as deduced from $3Bv_0/|u_0|$.
[b] One as deduced from $-3(d \ln T_m/d \ln v_s)$.

The analyses of § 2 and § 3 strongly suggest that it is a softer repulsive potential which primarily determines the different features of the melting behavior of metals and molecular crystals. However, the volume dependence of melting temperatures of noble metals and trivalent metals under pressure cannot adequately be fitted by this model, which seems to distinguish these metals from alkali and alkaline-earth metals.

Although the ideal three-phase model represents features of alkali metals under relatively low pressures, it does not have the melting curve maximum which Cs, Rb and K exhibit under high pressure [16–18] (see Fig. 1). In order to understand the occurrence of the melting curve maximum in terms of interatomic potential, Rapoport [19] proposed a two-species model. In this model each atom is assumed to have two states with different atomic diameters. At low pressures most atoms are in the state with a larger atomic diameter. As the pressure increases, there may exist a pressure range in which some of the atoms assume the state with a smaller diameter in the liquid state, whereas most atoms in the solid are still in the state with a larger diameter because of the periodic structure of the crystal. Thus, as the pressure increases, at a certain pressure the density of the liquid catches up with that of the solid in equilibrium, at which point the melting curve maximum occurs according to the Clausius-Clapeyron relation of thermodynamics. On the other hand, Matsuda [20] has pointed out that a pair potential $\phi(r)$ with a substantial deviation from that of the ideal three-phase model, such as having an inflection point, would be sufficient to give the melting curve maximum; this kind of model may be called a one-species model. Recently, Yoshida and Okamoto [21] studied the one-species model, and Kuramoto and Furukawa [22] studied the two-species model in detail. Their results will be reviewed in § 4.

The solid-liquid transition is not restricted to the melting of crystalline solids. There are non-crystalline solids, which exhibit a transition called glass transition

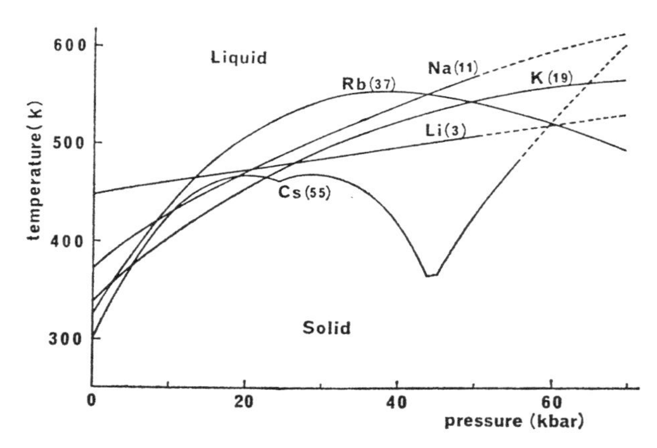

Fig. 1. Melting point curves and phase diagrams of alkali metals. This figure is taken from Ref. [18]

when the temperature is raised. Experimentally, it is difficult to make an amorphous solid from a simple liquid; a pure simple liquid crystallizes without forming an amorphous solid by the present experimental technique. However, if it could be cooled or compressed with sufficient rapidity, then even a simple liquid might become an amorphous solid. On the other hand, it might equally well be that the amorphous solid of the simple liquid is in principle nonexistent.

In order to elucidate this point and also to investigate the nature of the glass transition, HIWATARI, OGAWA, OGITA, MATSUDA and UEDA [23] are carrying out a molecular dynamical study of the amorphous solid of the soft-core model. In the computer simulation, rapid cooling and rapid compression are rather easy, and the state is obtained in which almost no diffusive motion of atoms takes place, which may be called the amorphous solid. When the temperature is raised, the diffusive motion of each atom sets in in a cooperative manner, which may be associated with the glass transition. An outline of this study will be given in § 5.

So far we have discussed the solid-liquid transition on the basis of classical mechanics. This may be justifiable in view of the fact that in most solids, except those of the lighter elements, the melting temperature is higher than the Debye temperature. In quantum solids, because of HEISENBERG's uncertainty principle, the ground-state eigenfunction provides for the uncertainty in the position of atoms in order to reduce the kinetic energy. Since the effect of the kinetic energy on the atomic structure becomes more important as the mass of an atom decreases, if one could gradually decrease the mass of each atom of a crystal at 0 °K under a fixed pressure, the crystal eventually would melt, becoming a superfluid with off-diagonal long-range order (ODLRO) [24]. Before the melting, whether the crystal provides for the uncertainty of the atomic positions simply by increasing its lattice constant of whether it produces lattice vacancies will depend on the type of the interatomic potential. ANDREEV and LIFSHITZ [25] pointed out the possibility of superfluidity in solids in which the Bose condensation of lattice vacancies plays an essential role. This corresponds to the existence of ODLRO in the crystal. MATSUDA and TSUNETO [26] discussed the conditions on the interatomic potential and the mass of an atom for the existence of ODLRO in the crystal, which may be called supersolid. They pointed out that if the condition is fulfilled, the normal crystal at 0 °K will make a phase transition to a supersolid and then to a superfluid as a pressure is changed. The result of this study will be given in § 6.

2. Properties of the Ideal Three-phase Model [11, 13]

Consider a set of $N(\gg 1)$ classical particles with a pair potential

$$\phi(r) = C/r^n - \alpha\gamma^3 \exp(-\gamma r), \tag{2.1}$$

where r is an interparticle distance, $C(>0)$, $\alpha(\geqq 0)$ and $n(>3)$ are the parameters characterizing the dynamical structure of the system, which is confined in a cubic box with volume V. The second term represents the Kac potential [27], where a positive constant γ is to be made to tend to zero after taking the thermodynamic limit.

First, consider the case for $\alpha = 0$ [28]. The model is then called soft-sphere model. As was already pointed out and as is shown in more detail in the Appendix, if one measures length by the unit

$$l = (V/N)^{1/3} \equiv v^{1/3}, \tag{2.2}$$

and time by the unit

$$\tau = l(m/kT)^{1/2}, \tag{2.3}$$

then a set of representative points in the $6N$-dimensional phase space with a canonical distribution makes an identical time development for any system with the same value of n for the same value of the reduced atomic volume

$$v^* = (kT/C)^{3/n} v. \tag{2.4}$$

Here m is a mass of a particle, T is a temperature and k is the Boltzmann constant. Therefore, the analysis by scaling is possible for the physical quantities which can be represented by the above set of representative points.

As a consequence, the soft-sphere model has the following properties:

I. The Lindemann hypothesis [29], which states that the ratio of the root-mean-squared displacement to the nearest-neighbor distance is a constant along the melting line, exactly holds for the system with the same value of n. According to the Monte Carlo studies, this ratio is slightly dependent on n and falls in the interval between 0.13 and 0.15.

II. The pressure P can be written as

$$P = (kT)(kT/C)^{\frac{3}{n}} P_0^{(n)}(v^*), \tag{2.5}$$

where $P_0^{(n)}(v^*)$ is a function of only v^* with n as a parameter. The function $P_0^{(n)}(v^*)$ is schematically shown in Fig. 2, where the reduced freezing and melting volumes $v_\pm^{(n)}$ and the reduced melting pressure $\overline{P_0^{(n)}}$ are the constants depending only on n.

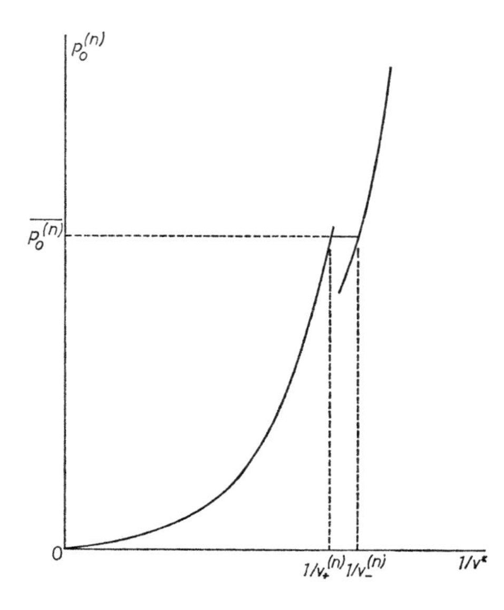

Fig. 2.
Schematic drawing of $P_0^{(n)}(v^*)$ as a function of $1/v^*$. The $v_+^{(n)}$ and $v_-^{(n)}$ are the reduced freezing and melting volumes, respectively. $\overline{P_0^{(n)}}$ is the reduced melting pressure

III. Therefore, the atomic volumes $v_\pm$ of the fluid and solid phase at the melting point $T = T_m$ are given by

$$v_\pm = (C/k T_m)^{\frac{3}{n}} v_\pm^{(n)}. \tag{2.6}$$

IV. According to the linear response theory, the self-diffusion constant D and viscosity coefficient η can be expressed as a time integral of the canonical average of the suitable correlation functions, so that they can be written as

$$D = \frac{C^{\frac{1}{n}}}{\sqrt{m}} (k T)^{\frac{1}{2} - \frac{1}{n}} D_n(v^*), = \sqrt[3]{v} \left(\frac{k T}{m}\right)^{\frac{1}{2}} D_n^*(v^*) \tag{2.7}$$

$$\eta = \frac{\sqrt{m}}{C^{2/n}} (k T)^{\frac{1}{2} + \frac{2}{n}} \eta_n(v^*), = \sqrt{k T m}\, v^{-\frac{2}{3}} \eta_n^*(v^*), \tag{2.8}$$

where $D_n(v^*)$, $D_n^*(v^*)$, $\eta_n(v^*)$ and $\eta_n^*(v^*)$ are the functions only of v^* with n as a parameter. Since at the melting point the reduced volumes v^* of the fluid and solid phases are constants $v_\pm^{(n)}$, Eqs. (2.7) and (2.8) give the temperature dependence of D and η along the melting line. Using (2.6) in (2.8), we obtain the viscosity coefficient η in the liquid phase along the melting line $T = T_m$ in the form

$$\eta = \frac{\sqrt{m}\sqrt{T_m}}{v_+^{\frac{2}{3}}} \eta_n(v_+^{(n)}), \tag{2.9}$$

which coincides with the Andrade formula [30].

Thus, the case $\alpha = 0$ is very simple but the model cannot have the liquid and gaseous phases as two distinct phases. In order to include this, we proceed to the model with $\alpha > 0$. By virtue of the so-called Van der Waals theory [27, 31], the pressure is now given by

$$P = (k T)(k T/C)^{\frac{3}{n}} \overline{P_A^{(n)}}(v^*), \tag{2.10}$$

where

$$P_A^{(n)}(v^*) = P_0^{(n)}(v^*) - A/v^{*2}, \tag{2.11}$$

$$A = 4\pi\alpha C^{-\frac{3}{n}} (kT)^{-1+\frac{3}{n}}, \tag{2.12}$$

where $\overline{P_A^{(n)}}(v^*)$ is the corrected value of $P_A^{(n)}(v^*)$ by MAXWELL's equal-area rule when $P_A^{(n)}(v^*)$ has a region such that $dP_A^{(n)}(v^*)/dv^* > 0$.

Referring to the result of Monte Carlo studies for $\alpha = 0$, our model with $\alpha > 0$ is found to exhibit the equation of state with phase transitions between solid, liquid and gas in a certain temperature range [13]. Let A_t and A_c be the value of A at the triple point and critical point. Let v_c^* be the value of v^* at the critical point and v_{ts}^*, v_{tl}^* and v_{tg}^* be the value of v^* at the triple point in the solid, liquid and gaseous phases, respectively. They are constants dependent on the value of n. From (2.12) we may write

$$kT_\gamma = \left(4\pi\alpha/A_\gamma C^{\frac{3}{n}}\right)^{\frac{n}{n-3}} \qquad (\gamma = t, c). \tag{2.13}$$

From (2.4) and (2.13) we have

$$\begin{aligned} v_c &= \left(\frac{A_c c}{4\pi\alpha}\right)^{\frac{3}{n-3}} v_c^* \\ v_{t\gamma} &= \left(\frac{A_t C}{4\pi\alpha}\right)^{\frac{3}{n-3}} v_{t\gamma}^* \qquad (\gamma = s, l, g). \end{aligned} \tag{2.14}$$

On the other hand, the potential energy per atom in the crystal is given by

$$\begin{aligned} u &= \tfrac{1}{2} C \sum_j R_{ij}^{-n} - 4\pi\alpha/v \\ &= \tfrac{1}{2} C C_n d^{-n} - 4\pi\alpha l_i d^{-3}, \end{aligned} \tag{2.15}$$

where d is a nearest-neighbor distance and R_{ij} is a distance between lattice point i and j, l_i is a positive constant dependent on the lattice structure, and C_n is a positive constant dependent on n and on the lattice structure. A minimization of (2.15) yields for the atomic volume v_0 and the potential energy per atom u_0 at 0 °K:

$$v_0 = v_{no}(C/\alpha)^{\frac{3}{n-3}} \tag{2.16}$$

$$u_0 = u_{no}\left(\alpha^{\frac{n}{3}}/C\right)^{\frac{3}{n-3}} \tag{2.17}$$

where v_{no} and u_{no} are the constants dependent on n and the crystal structure. Therefore, the temperature and atomic volumes at the critical and triple points measured in the unit of u_0 and $v_0 = d^3/l_i$ are constants dependent only on n; the estimated values of these constants are shown in Figs. 3, 4 and 5.

From (2.10) and (2.11) we can estimate the effect of α on the atomic volumes v_s and v_l of the solid and liquid phases at the melting point. The analysis on the basis of Monte Carlo studies indicates that the effect is rather small; the deviation of v_s and v_l from $v_\mp$ turns out to be less than 6% above the triple-point temperature [13]. Primarily, when we take T and v as independent variables char-

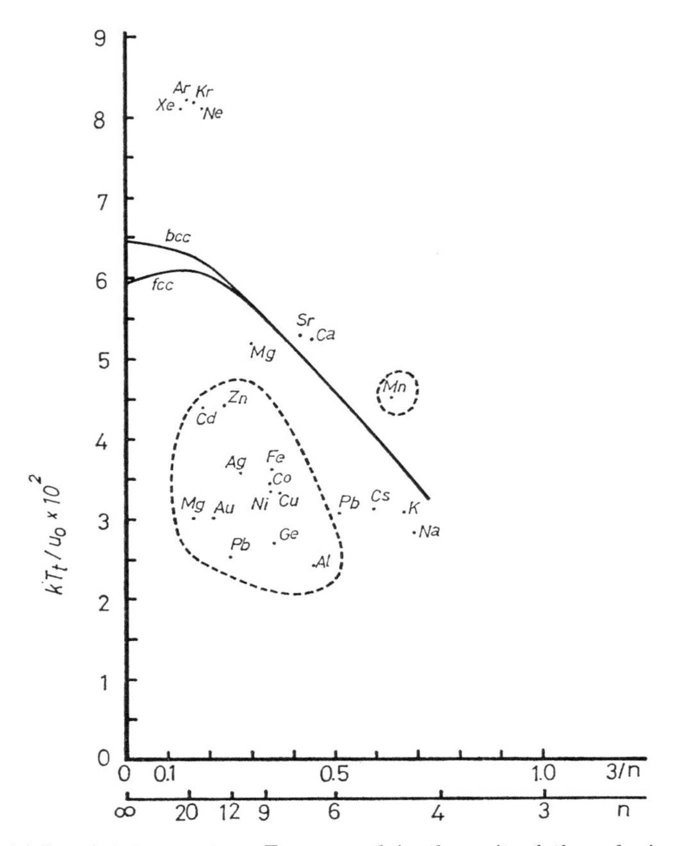

Fig. 3. The triple-point temperature T_t measured in the unit of the cohesive energy $|u_0|$ versus the value of n, as determined from (3.2). Here k is the Boltzmann constant. The solid lines are theoretical ones corresponding to the lattice structures (fcc, bcc) of the solid at 0 °K. Metals other than alkali and alkaline-earth are encircled by a broken line

acterizing the state of the system, then for the uniform phase the physical quantity which can be defined with no reference to the surface or volume change is not influenced by the presence of the Kac potential. The smallness of the effect of α on v^* at the melting point suggests that the properties I., III., and IV. approximately hold for our model with $\alpha > 0$.

Taking the effect of α on the melting pressure as a perturbation, the reduced melting pressure becomes

$$P_{mA}^{(n)} = \overline{P}_0^{(n)} - A/v_-^{(n)}\, v_+^{(n)}. \tag{2.18}$$

Therefore, using (2.10), (2.16) and (2.17) we can express the relation between the melting pressure P_m and the melting temperature T_m as a linear relation such that

$$Y = C_{no} - C_{n1} X, \tag{2.19}$$

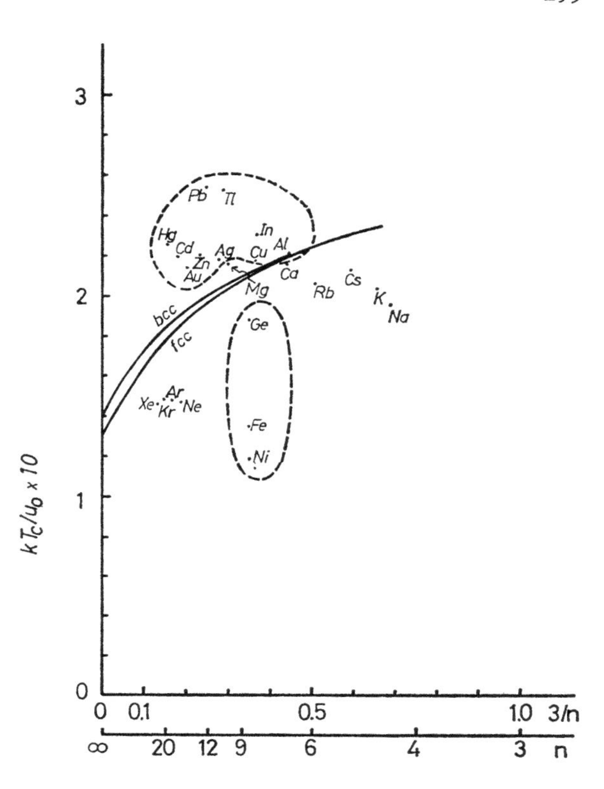

Fig. 4.
The critical-point temperature T_c measured in the unit of the cohesive energy $|u_0|$ versus the value of n, as determined from (3.2). Here k is the Boltzmann constant. The solid lines are theoretical ones corresponding to the lattice structures (fcc, bcc) of the solid at 0 °K. Metals other than alkali and alkaline-earth are encircled by a broken line

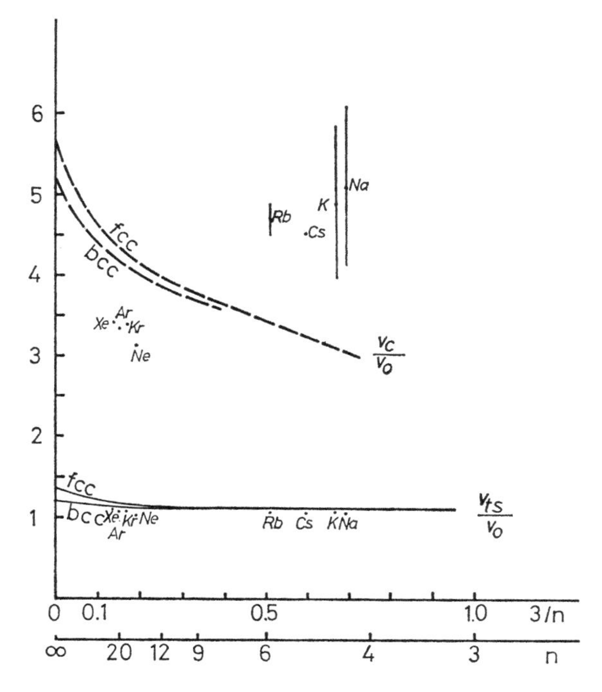

Fig. 5.
The atomic volumes at the triple point (v_{ts}) and the critical point (v_c) measured in the unit of the atomic volume at 0 °K (v_0) versus the value of n, as determined from (3.2). Each vertical bar shows an uncertainty of estimated value of v_c. The solid and broken lines are theoretical ones corresponding to the lattice structure (fcc, bcc) of the solid at 0 °K

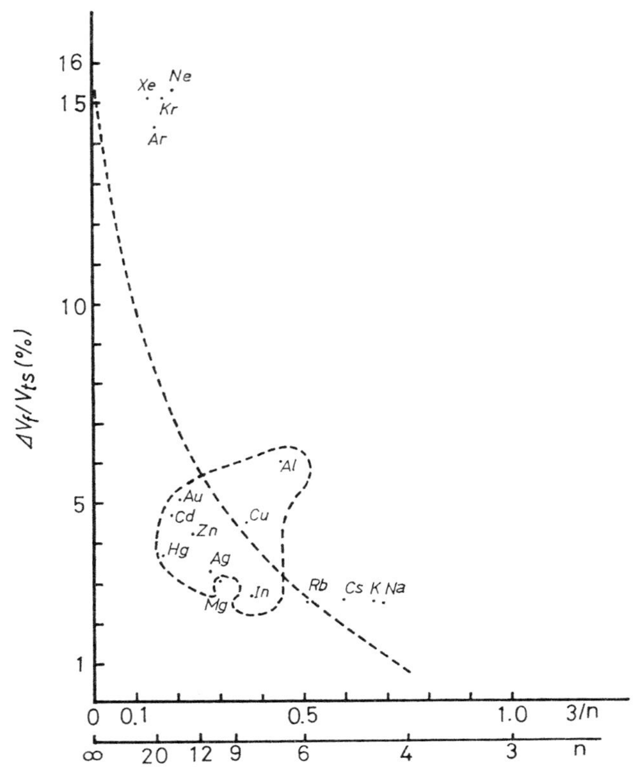

Fig. 6. The fractional volume change on melting $\Delta V_f/V_{ts}$ versus the value of n, as determined from (3.2). The dotted line is the theoretical one. Metals other than alkali and alkaline-earth are encircled by a broken line

where

$$X \equiv (kT_m/u_0)^{-1+\frac{3}{n}}, \quad Y \equiv (P_m v_0/kT_m)\,(kT_m/u_0)^{-\frac{3}{n}} \tag{2.20}$$

$$C_{no} \equiv \overline{P}_0^{(n)}\, v_{no}\, u_{no}{}^{\frac{3}{n}}, \quad C_{n1} \equiv \left(\frac{4\pi}{v_-^{(n)}\, v_+^{(n)}}\right) \overline{P}_0^{(n)}\, v_{no}\, u_{no}{}^{-1+\frac{6}{n}}. \tag{2.21}$$

Finally, the estimated values of the fractional volume change on melting are drawn as a function of n in Fig. 6. In view of the above simple scaling properties and capability of exhibiting three phases of matter, we call our model with $\alpha > 0$ the ideal three-phase model.

3. Ideal Three-phase Model and Real Substances in Nature

As shown by a curve in Figs. 3–7 and the relation (2.6), the difference between the thermodynamical properties of our models with smaller n and those with larger n is qualitatively in parallel with the difference between metals and molecular crystals, as pointed out in § 1. Therefore, we are tempted to examine how far this parallelism can be carried over and how far our model can be useful as a reference system for classifying simple liquids.

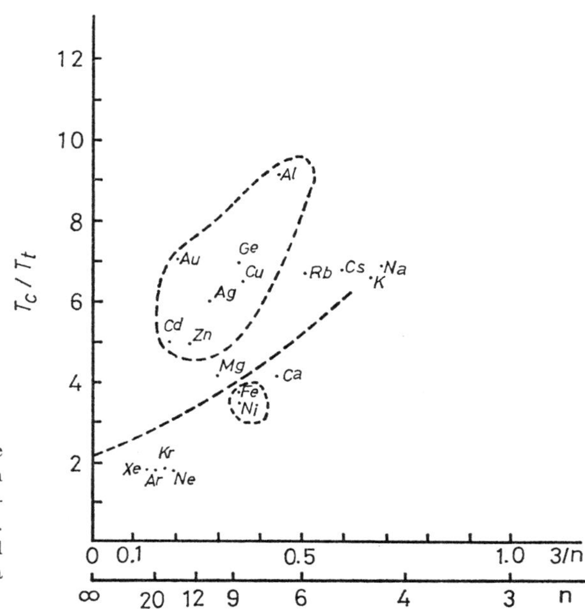

Fig. 7.
The liquid range T_c/T_t versus the value of n, as determined from (3.2). The accuracy of the theoretical line (dotted) is not so high. Metals other than alkali and alkaline-earth are encircled by a broken line

Previously HIWARARI and MATSUDA [13] assigned the value of n to real substances by comparing the relation (2.6) with the empirical relation, the so-called Kraut-Kennedy plot [32], between T_m and $v_s \cong v_-$, obtaining n values of about 15 for solidified inert gases and about 4 or 5 for alkali metals. Then they compared various thermodynamic quantities of our model with those of solidified inert gases and alkali metals, concluding that the difference between these substances can consistently be accounted for by this model, at least as a first approximation. However, the practical use of Eq. (2.6) for assigning the value of n to real substances is limited by the empirical difficulty of obtaining a reliable melting volume v_- under pressure. Moreover, one may question the validity of assuming the same pair potentials under high pressures.

On the other hand, under low pressures the density of the solid at the melting point is not so different from that at 0 °K. Then it will be of significance to inquire how the characteristic properties at 0 °K are related to those at higher temperatures. Thus, here we determine n by the experimental data at very low temperatures. At 0 °K we obtain from (2.15)

$$n = 3\,\frac{B_0 - 2P}{P - u/v}\,, \tag{3.1}$$

where $B_0 = -v\,(\partial P/\partial v)_{T=0}$ is the bulk modulus at 0 °K.

At zero pressure, we have

$$n = -3\,B v_0/u_0. \tag{3.2}$$

We tabulate in Table 3 the values of n determined from (3.2) as well as from (2.6) for various substances [33]. Eq. (3.2) shows that n is a measure of the hardness of the solid at 0 °K in the sense that it is proportional to the fractional

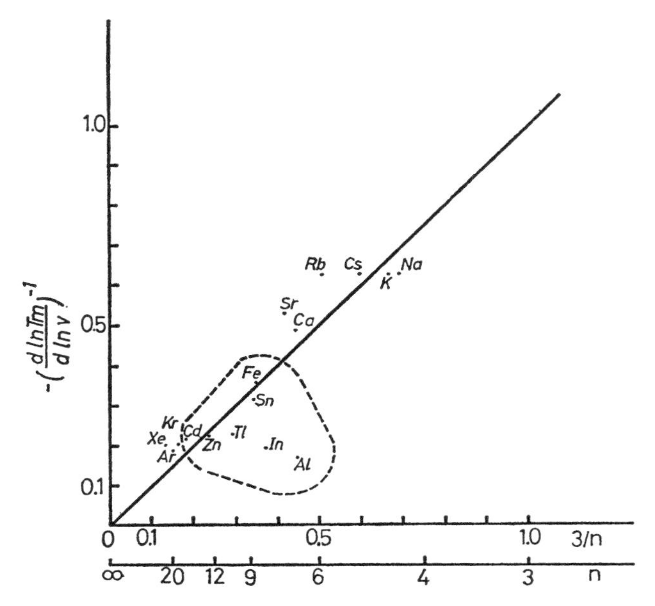

Fig. 8. The value of $-(d \ln T_m / d \ln v_s)^{-1}$ against $3/n$ determined from (3.2)

increase of energy per unit fractional volume change. In Fig. 8 we plot the experimental value of $-\left(\frac{d \ln T_m}{d \ln v_s}\right)^{-1}$ for various substances against $3/n$ determined from (3.2). For most substances the former is approximately $3/n$ as predicted by our model.

Another quantity which can be a measure of the hardness of the solid is the Grüneisen constant [34] as defined by

$$\gamma = -d(\log \Theta)/d(\log v), \tag{3.3}$$

where Θ is the Debye temperature. For the ideal three-phase model we have a simple relation

$$n = 6\gamma - 2. \tag{3.4}$$

As shown in Fig. 9, almost all substances approximately satisfy (3.4).

In Figs. 3–7 we plot experimental data of elements against the values of n determined from (3.2). From these plots we find that for inert gases, alkalis and alkaline-earth metals the plots are in accord with the n-dependence of the ideal three-phase model except for v_c/v_0 [35]*. On the other hand, most of the other metals show considerable deviations from the behavior of the ideal three-phase model; the plots for these metals are encircled by broken lines. As shown in Fig. 3, some multivalent metals have larger n while their values for $kT_t/|u_0|$ are

* The discord in v_c/v_0 may be due to the difficulty of simulating the real system up to the critical density with our model. At the same time we note that the analysis of the computer experiment for obtaining the critical point may include more errors than that for the triple point.

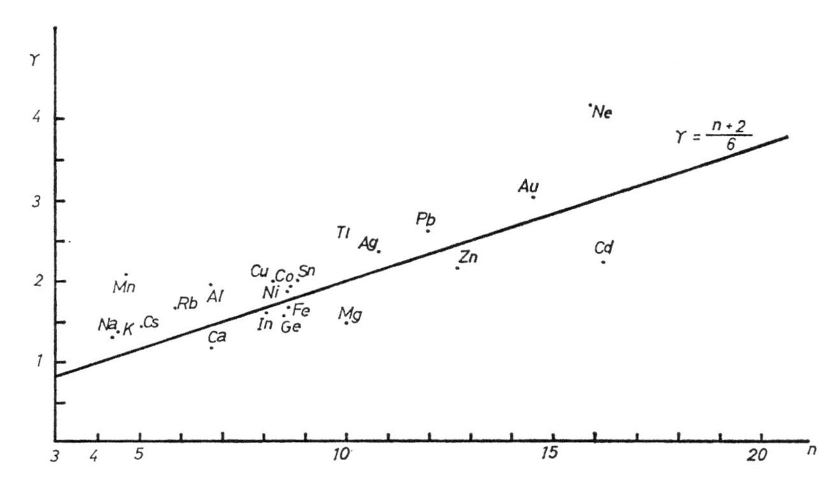

Fig. 9. The Grüneisen constant versus the value of n, as determined from (3.2). The straight line is the theoretical one $(\gamma = (n + 2)/6)$

as small as those of alkalis having smaller n. Why we should have such discrepancies may be an interesting problem to be answered on the basis of the electron theory of metals.

4. Melting Curve Maximum at High Pressure

In the laboratory, the melting temperature T_m of most substances increases with pressure, but some loose-packed solids like Ge, Ga, Sb, Bi and ice show a melting temperature depression under pressure [36]. This depression of T_m is considered to be related to the loose-packed structure of the solid, and does not necessarily indicate a singular feature of the pair potential. However, there exists a solid like Cs which has a melting curve maximum even in the close-packed fcc phase [17]. In addition to this anomaly, the Cs solid in fcc phase has an isostructural phase transition with a density change of about 10% at about 42 Kbar. Klement and Jayaraman, and Kawai and Inokuchi [18] have surveyed experimental data for the pressure dependence of melting temperatures, and the latter authors conjectured that all the substance may have a melting curve maximum.

Since Rapoport suggested [19] the two-species model to account for the melting anomaly, Matsuda [20] has proposed what we may call a one-species model and pointed out, using the Lennard-Jones and Devonshire (LJD) cell theory of melting [37], that a pair potential $\phi(r)$ which has an inflection point may show an anomaly like the fcc solid of Cs. Estimating the parameters characterizing the pair potential from the experimental equation of state of the solid Cs, he conjectured that the solid in the premelting region may have a negative thermal expansion coefficient, and there the bulk modulus will be smaller than in the liquid phase. It must be noted, however, that this conjecture is based on the LJD theory and on the assumption of the existence of significant premelting phenomena the validity of which is still an open question.

A little later, the one-species model was studied in detail by YOSHIDA and OKAMOTO [21], who proposed a model with a pair potential of the form.

$$\psi(r) = \varepsilon \exp(-r/r_0) \tag{4.1}$$

which does not have an inflection point. On the basis of the LJD theory and the Bragg-Williams approximation, they determined the parameters ε and r_0 in (4.1) in such a way that the maximum melting point (T_M, P_M) obtained experimentally in Rb and Cs coincides with that predicted by the present theory. This shows that the calculated melting curves are in fairly good agreement with the observed values near the maximum melting point. According to their theory, however, the ratio of the atomic volumes at the maximum of the melting temperature v_M to that under normal pressure v_0 is 0.149 for Rb and 0.182 for Cs in the fcc crystal, which is too small compared to the experimental values of about 0.6*. They argued that the discrepancy is due to the Bragg-Williams approximation, ignoring the short-range order and the LJD theory itself. At the melting curve maximum their theory predicts that about 40% of the normal lattice sites will be vacant, which is an unbelievably large percentage. Although experimental verification for or against this is highly desirable, we may surmise that it indicates the substantial failure of the LJD theory in which melting is treated by analogy with the order-disorder phenomena of alloys.

On the other hand, KURAMOTO and FURUKAWA [22] studied a kind of two-species model. Each atom is assumed to have an internal degree of freedom corresponding to an electronic transition such as $6s \to 5d$ in Cs [40]. Let the internal degree of freedom or the "species" of the j-th atom be represented by the variable s_j which takes the value 1 or -1. They assumed that the potential energy of the total system is given by

$$U = \varepsilon \sum_{i<j}\sum \left[\left\{\left(\frac{\sigma_A + \sigma_B}{2}\right) + \left(\frac{\sigma_A - \sigma_B}{2}\right)(s_i + s_j)\right\} / r_{ij}\right]^n - \frac{\Delta}{2}\sum_j (s_j - 1), \tag{4.2}$$

where ε, σ_A, σ_B, n and Δ are the positive constants. Using the cell model [41] and the Bragg-Williams approximation as to the configurations of both species of atoms they determined the equilibrium concentration of both species in such a way that the free energy is minimum.

In order to find the melting curve, they made use of the "generalized Lindemann law" proposed by ROSS [42] which states that the free volume divided by an atomic volume is constant along the melting curve. Then they explicitly showed that, both for $n = \infty$ and 4, the model exhibits a solid-solid phase transition at sufficiently low temperatures so long as $\sigma_B/\sigma_A < 1$. As σ_B/σ_A decreases, the critical point of the solid-solid transition approaches to the melting curve, T_m versus P_m, which comes to have a maximum and minimum.

* Very recently, YOSHIDA and KAMAKURA [38] improved this point by assuming the pair potential to have the form

$$\phi(r) = A\left[\exp\left\{a\left(1 - \frac{r}{r_0}\right) - b\left(1 - \frac{r}{r_0}\right)^l \log\frac{r}{r_0}\right\} - (r/r_0)^{-d} \exp\left\{c\left(1 - \frac{r}{r_0}\right)\right\}\right]$$

(A, a, b, c, d, r_0 and l are positive constants)

and using the expandable lattice model developed by MORI, OKAMOTO and ISA [39].

Although all the above theories are based on approximations the validity of which might be questioned, especially in the vicinity of the anomalous melting region, these studies indicate that a softening of the pair potential as interatomic distance decreases will be essential for the melting anomaly. What is not yet clear is which kind of model is closer to the anomaly of real substances, the one-species or the two-species model. The two models are expected to show considerable differences in entropy because of the difference in the internal degrees of freedom, so that the question could be experimentally tested. Theoretically, it will be an interesting problem to investigate whether and how we can derive an effective pair potential from the electron theoretical point of view which was phenomenologically introduced.

5. Molecular Dynamical Study of the Amorphous State of the Soft-sphere Model [23]

As an initial state of the molecular dynamical study of the soft-sphere model with $n=12$, HIWATARI et al. took $N(=32)$ atoms situated at the lattice points of the fcc lattice in the cubic box with periodic boundaries. The initial velocity to each atom is determined according to the Maxwell distribution at a certain temperature, and a computer is made to simulate the motion of these atoms by solving the Newtonian equation of motion [43].

We define an instantaneous temperature $\tilde{T}$ and pressure $\tilde{P}$ by

$$k\tilde{T} = \frac{2}{3N} \cdot \frac{m}{2} \sum_{j=1}^{N} \boldsymbol{v}_j^2 \tag{5.1}$$

$$\frac{\tilde{P} v}{k\tilde{T}} = 1 + \frac{nC}{3k\tilde{T}} \sum_{i<j} r_{ij}^{-n}, \tag{5.2}$$

where $\boldsymbol{v}_j$ is the velocity of the j-th atom and r_{ij} is the distance between the i-th and j-th atoms. Eq. (5.2) is a direct consequence of the virial theorem. In the computer simulation, when $\tilde{T}$ and $\tilde{P}$ begin to fluctuate around some constant values and suitable time averages of $\tilde{T}$ and $\tilde{P}$ become stationary, we assume that an equilibrium state is reached. We take the time average of $\tilde{P}$ as the equilibrium pressure corresponding to the temperature equal to the time average of $\tilde{T}$.

Thus, varying the initial temperature defining the Maxwell distribution, we obtain the equation of state relating v, T and P. Because of the scaling property given in § 2, both Pv/kT and $P_0^{(n)} = (P/kT)(C/kT)^{3/n}$ should be the universal function of $\varrho^* = v^{*-1} = v^{-1}(C/kT)^{3/n}$. We plot Pv/kT against ϱ^* in Fig. 10. Our results with $N=32$ are in excellent agreement with the Monte Carlo results of HOOVER et al. [12] with $N=32$ and 500.

When the initial temperature of the simulation is high enough, the fcc structure becomes unstable and the system soon passes into the fluid state. Now we suddenly change the velocities of all the atoms in the fluid phase to zero, leaving the positions of the atoms as they were. Taking this as a new initial state, we let the computer do the simulation as before. Since the configuration of atoms is usually not at the minimum of the potential energy, the atoms gradually gain non-zero velocities. Repeating the above rapid cooling process and applying the same

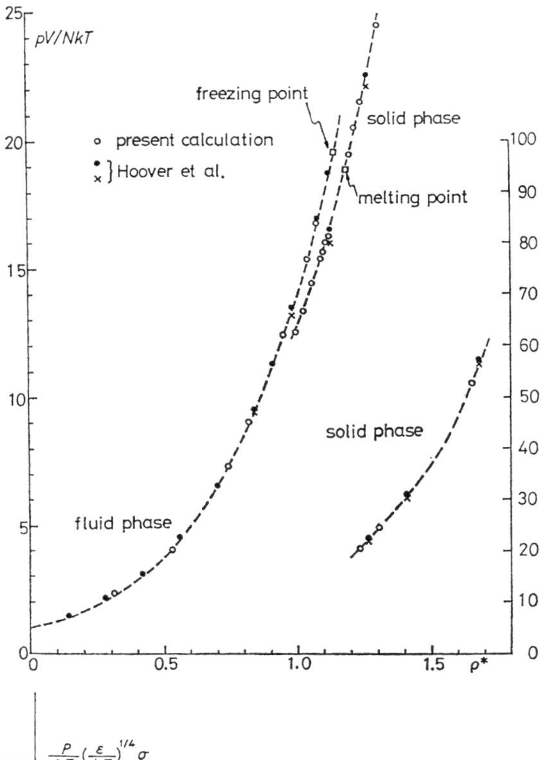

Fig. 10.
The equation of state (compressibility factor) of inverse-12 system; ○ result of molecular dynamics ($N = 32$) by HIWATARI et al. [23] and •, × results of Monte Carlo studies ($N = 500$, 32) by HOOVER et al. [12]

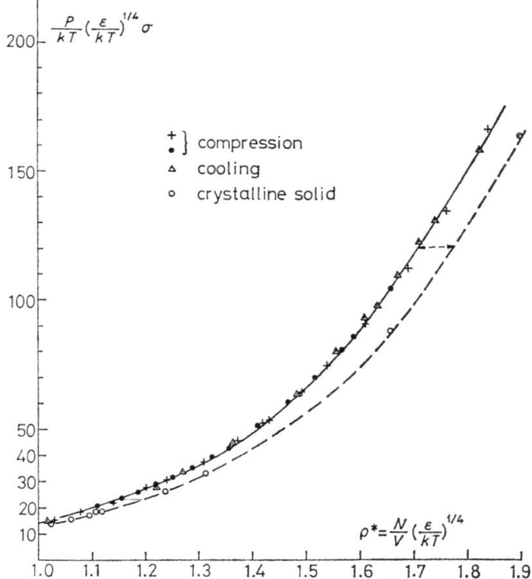

Fig. 11.
The reduced pressure $P_0^{(12)}$ versus ϱ^*; •, + results obtained using the compression method, ▵ result obtained using the cooling method, ○ that of crystalline solid. The solid horizontal line shows the liquid-solid phase transition. The dotted horizontal line shows the volume difference between supercooled liquid (or amorphous solid) and crystalline solid, which is nearly constant over a wide range of density

procedure as before, we obtained the equation of state of the amorphous state with ϱ^* which is larger than the freezing density $\varrho^* \cong 1.15$. A similar amorphous state was also obtained by rapid compression instead rapid cooling. In Fig. 11 we plot $(P/kT)(C/kT)^{1/4} = P_0^{(12)}(v^*)$ against $\varrho^* = v^{*-1}$. In this figure the amorphous

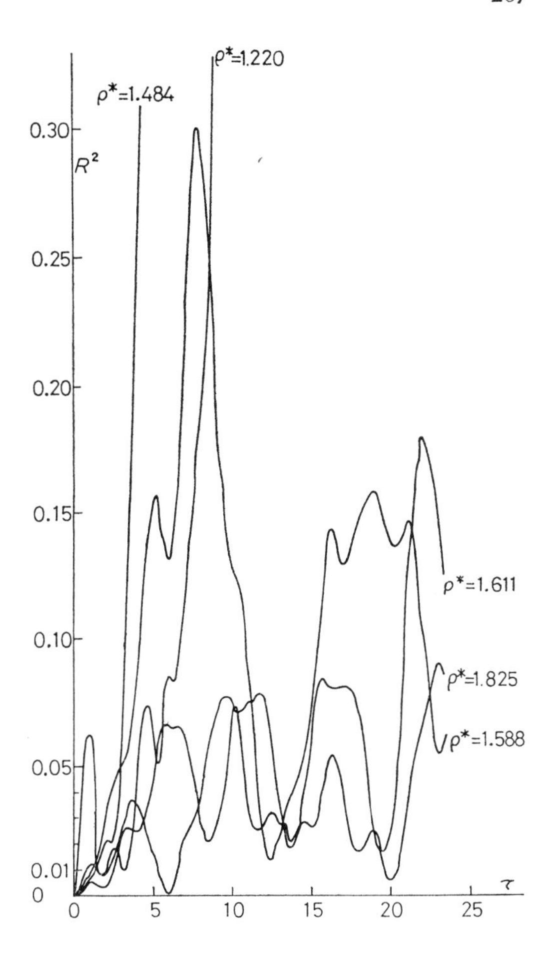

Fig. 12.
The squared displacement R^2 of an arbitrarily chosen atom from the initial position against time for various ϱ^*. The length and time are measured in the unit of $(V/2N)^{\frac{1}{3}}$ and $\tau = (m\sigma^2/12\varepsilon)^{\frac{1}{2}}$, respectively

state forms a curve which smoothly continues to the fluid branch. From the plot we obtain an approximate relation between the density of the crystalline phase ϱ_s and that of the fluid or amorphous phase ϱ_f:

$$\frac{\varrho_s - \varrho_f}{\varrho_s} = \frac{\varrho_s^* - \varrho_f^*}{\varrho_s^*} \simeq \frac{0.06}{\varrho_s^*} \tag{5.3}$$

where ϱ_s^* and ϱ_f^* are the reduced quantities corresponding to ϱ_s and ϱ_f, respectively.

In Fig. 12 we plot the squared displacement of an arbitrarily chosen atom from the initial position against time for various ϱ^*. In Fig. 13 we show the above quantity averaged over all the atoms. From these figures we may roughly say that for $\varrho^* \lesssim 1.5$ the atomic motion is still of the diffusion-type as in a liquid, while for $\varrho^* \gtrsim 1.5$ it is mainly of the vibration-type as in a crystal. If we tentatively assume that the reduced density $\varrho^* = 1.5$ corresponds to glass transition, then from (2.5), (2.4) and Fig. 11 we obtain the ratio of the melting temperature

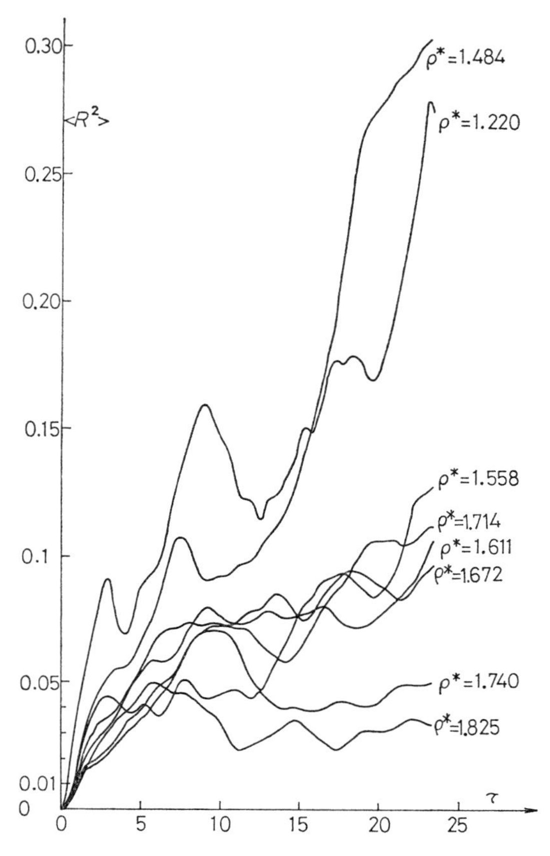

Fig. 13.
The squared-displacement averaged over all the atoms from their initial positions against time. The units of length and time are the same as those used in Fig. 12

T_m to the glass transition temperature T_g under the constant pressure:

$$T_g/T_m = (P_m^*/P_g^*)^{\frac{n}{n+3}} \cong (20/70)^{\frac{4}{5}} \cong 0.4. \tag{5.4}$$

This value is somewhat less than 2/3, which is the approximate empirical value for many amorphous solids [44].

Thus, even a soft-sphere model, which is a model for simple liquids, may assume an amorphous state in which the atomic motion is more like that in a crystal than in a liquid. However, because of the limited computer time and small N of the above study, it is still a matter needing further scrutiny whether the above change in the type of atomic motion could be associated with a possible glass transition of the simple liquid represented by our model.

6. Possibility of Off-diagonal Long-range Order in Crystals [26]

Generally, a state of matter can be classified according as the presence or absence of diagonal and off-diagonal long-range order. Let ϱ be the normalized density operator of the system and let $a(x)$ and $a^\dagger(x)$ be the annihilation and creation operators of a particle at the point x in the second-quantization scheme. Then, the coordinate representation of the one-particle reduced density operator ϱ_1 is given by

$$\langle x' | \varrho_1 | x \rangle = \mathrm{Tr}\,[\varrho\, a(x')\, a^\dagger(x)]. \tag{6.1}$$

The one-particle distribution function $n(x)$ is given by the diagonal element of (6.1):

$$n(x) = \langle x | \varrho_1 | x \rangle. \tag{6.2}$$

If and only if there exists a set of independent vectors a_α such that

$$n(x + a_\alpha) = n(x) \qquad (\alpha = 1, 2, 3) \tag{6.3}$$

and the length of not all of such vectors can be chosen arbitrarily small, then the state of the system is said to have a diagonal long-range order (DLRO). Therefore, the crystal has DLRO while the fluid (liquid and gas) has no DLRO. On the other hand, if and only of $\langle x' | \varrho_1 | x \rangle \nrightarrow 0$ as $|x' - x| \to \infty$, then the state of the system is said to have off-diagonal long-range order (ODLRO) with respect to a one-particle reduced density operator. Although it is possible to define ODLRO with respect to two-particle and many-particle reduced density operators, here we do not consider such higher-order ODLRO [24].

Penrose and Onsager [45] argued that the superfluid He^4 has ODLRO, whereas the crystalline He^4 has no ODLRO at 0 °K. They based their argument on the characteristics of the ground-state eigenfunction of the system, where they assumed that a crystal at 0 °K is a "perfect crystal" in the sense that the value of the ground-state eigenfunctions of N bosons $\psi(x_1 \ldots x_N)$ is very small unless every particle is near each lattice site. However, as stated in § 1, there must be some conditions for the interatomic potential in order that a crystal at 0 °K can be a perfect crystal. For the sake of simplicity, we invoke the quantum lattice model (QLM) for the assembly of bosons with hard core which was introduced by Matsubara and Matsuda [46].

According to QLM, we take the second-quantization scheme and consider the following Hamiltonian for n bosons with hard core in the lattice space consisting of N lattice points:

$$\mathfrak{H}_n = \sum_{i<j} v_{ij} n_i n_j + \tfrac{1}{2} \sum_{i<j} u_{ij} (a_i^\dagger a_j + a_i a_j^\dagger) + A\, n. \tag{6.4}$$

Here, $a_i^\dagger$ and a_i are the creation and annihilation operators of a boson at the i-th lattice point. As bosons, we assume the commutation relations:

$$[a_i^\dagger, a_j^\dagger]_- = [a_i, a_j]_- = [a_i, a_j^\dagger]_- = 0. \qquad (i \neq j) \tag{6.5}$$

On the other hand, in order to exclude the multiple occupation of atoms at each lattice point due to the atomic core, we impose for $i = j$ the following commutation relation of fermion type:

$$[a_i^\dagger, a_i^\dagger]_+ = [a_i, a_i]_+ = 0, \qquad [a_i, a_i^\dagger]_+ = 1. \tag{6.6}$$

From (6.6), $n_i = a_i^\dagger a_i$ has the value either 0 or 1. $v_{ij} (= v_{ji})$ is a real constant representing a pair potential between the atoms occupying the i-th and j-th lattice points, and $u_{ij} (= u_{ji})$ is a real constant representing the quantum mechanical transfer of atoms between the i-th and j-th lattice points. A is a constant including the effect of quantum-mechanical zero-point energy and n is the total number of atoms confined in the lattice space:

$$n = \sum_i n_i. \tag{6.7}$$

We may determine the values of u_{ij} and A in such a way that the kinetic energy of the system is given by

$$K = \frac{\hbar^2}{2m} \sum_k{}' k^2 a_k^\dagger a_k, \tag{6.8}$$

where $\sum_k'$ denotes the sum over k in the first Brillouin zone of the simple cubic lattice and

$$a_k = N^{-\frac{1}{2}} \sum_j \exp(i k R_j) a_j. \tag{6.9}$$

Here R_j is a position vector of the j-th lattice point of the s.c. lattice. Then they are given by

$$A = (\hbar \pi)^2/(2 m d^2), \tag{6.10}$$

$$u_{ij} = u(R_i - R_j) \tag{6.11}$$

$$u(R) = \begin{cases} \pm 2\hbar^2/m|R|^2 & \text{(for } R \text{ lying on the } x, y \text{ or } z \text{ axis. The upper sign is for } \mathfrak{A} \text{ sublattice and the lower sign is for } \mathfrak{B} \text{ sublattice.)} \\ 0 & \text{(otherwise)} \end{cases}$$

Here, the whole s.c. lattice is subdivided into two equivalent sublattices $\mathfrak{A}$ and $\mathfrak{B}$; and $\mathfrak{A}$ includes the origin of the s.c. lattice.

Now the operators satisfying (6.5) and (6.6) are isomorphic to a set of those of N spins each of which is localized at each lattice point and has the magnitude of $\frac{1}{2}$ as follows:

$$\begin{aligned} a_j &\leftrightarrow S_j^x + i S_j^y \\ a_j^\dagger &\leftrightarrow S_j^x - i S_j^y \end{aligned} \tag{6.12}$$

$$n_j = a_j^\dagger a_j \leftrightarrow \tfrac{1}{2} - S_j^z. \tag{6.13}$$

Therefore, a down spin, $S_j^z = -\frac{1}{2}$, corresponds to a particle and an up-spin, $S_j^z = \frac{1}{2}$, corresponds to a hole occupying the j-th lattice point.

Then, it can be shown that the grand canonical ensemble of the QLM is equivalent to the canonical ensemble of the corresponding spin system with the Hamiltonian

$$\tilde{\mathfrak{H}} = \sum_{i<j} [v_{ij} S_i^z S_j^z + u_{ij}(S_i^x S_j^x + S_i^y S_j^y)] - HM, \tag{6.14}$$

where

$$M = \sum_j S_j^z = \frac{N}{2} - n, \tag{6.15}$$

and an external magnetic field H is to be determined such that

$$\langle M \rangle \equiv T_r[M e^{-\tilde{\mathfrak{H}}/kT}]/T_r[e^{-\tilde{\mathfrak{H}}/kT}] = \frac{N}{2} - n. \tag{6.16}$$

The pressure of the QLM is given by

$$p = \frac{N}{V} \int_H^\infty \left(\frac{1}{2} - \frac{\langle M \rangle}{N}\right) dH. \tag{6.17}$$

Table 4. The state of matter classified by the presence or absence of DLRO and ODLRO

		ODLRO	
		No	Yes
DLRO	No	↑↑ Ferro Para Normal Fluid (Gas, Liquid)	↖ ↗ Spin Flop Superfluid
	Yes	↑↓ Antiferro Normal Solid	↖ ↘ Intermediate State Supersolid

Now, the DLRO of the QLM corresponds to the state of the spin system in which the whole lattice consists of sublattices which are not mutually equivalent with respect to the value of $\langle S_j^z\rangle$. The ODLRO of the QLM corresponds to the state such that $\langle S_i^x S_j^x + S_i^y S_j^y\rangle \not\to 0$ as $|R_i - R_j| \to \infty$, since we have

$$T_r[\varrho\, a_i^\dagger a_j] \leftrightarrow \langle S_i^x S_j^x + S_i^y S_j^y\rangle. \tag{6.18}$$

Then we can classify the state of the spin system having at most two sublattices, as in Table 4. By choosing smaller lattice constants for the QLM, we could have a state with more than two sublattices which is closer to the state in real continuous space. However, since the two-sublattice model is just enough to discuss the possibility of ODLRO in crystals, or the coexistence of ODLRO and DLRO, we consider what state of the two-sublattice model gives the lowest energy for various densities, or equivalently for various H. Using the molecular field approximation for the sake of simplicity, Matsuda and Tsuneto [26] gave the state diagram at 0 °K as shown in Fig. 14a, b. From the figure and from Eq. (6.14) we find that the ODLRO in crystals which might be called supersolid can exist only under pressure as an intermediate state between the normal solid and superfluid when the pressure is varied. In this respect solid He^4, which can be solid only under pressure, could be a candidate for the substance becoming supersolid just before it melts to a superfluid at 0 °K.

However, from Fig. 14 we find that in order to have ODLRO in crystals we must have the inequalities:

(i)
$$8C \equiv \sum_{j:j\in\mathfrak{A},\, i\in\mathfrak{A}} v_{ij} + \Big|\sum_{j:j\in\mathfrak{A},\, i\in\mathfrak{A}} u_{ij}\Big| > 0, \tag{6.19}$$

and

(ii)
$$8(B-D) \equiv \sum_{j:j\in\mathfrak{B},\, i\in\mathfrak{A}} v_{ij} - \Big|\sum_{j:j\in\mathfrak{B},\, i\in\mathfrak{A}} u_{ij}\Big| > 0. \tag{6.20}$$

Since in the crystal a majority of atoms occupy one of the two sublattices, the condition (i) states that the interatomic potential between atoms in the crystal must not be so negative. Indeed, a molecular field approximation gives the expression for C:

$$C = (2V_{ns} + K_{ns})/n, \tag{6.21}$$

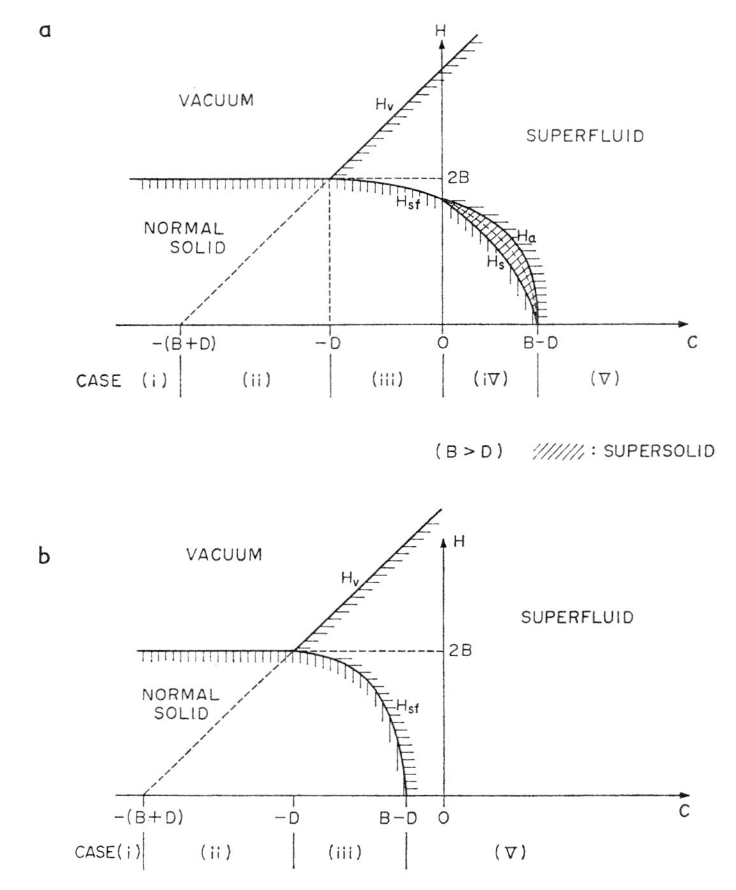

Fig. 14a and b. Phase diagrams of the quantum lattice model at 0 °K obtained by the molecular field approximation. a The case for $B > D$. b The case for $B \leqq D$

$$H_v = 2(B + C + D),$$

$$H_{sf} = 2\sqrt{B^2 - (C + D)^2},$$

$$H_s = 2\sqrt{(B - C)^2 - D^2},$$

$$H_a = H_s(B + C + D)/(B - C + D)$$

where V_{ns} and K_{ns} are the potential and kinetic energy of the normal solid at 0 °K. In solid He^4, C is estimated to be negative, so that in view of (6.19) solid He^4 is unlikely to be a supersolid. On the other hand, conditions (6.19), (6.20) and (6.8) indicate that a system with essentially repulsive pair potential like the soft-sphere model may have ODLRO in crystals.

Of course, since our discussion is based on the QLM and molecular field approximation, the result must be taken with considerable reservation. Even so, we think it interesting to observe how the presence or absence of ODLRO in crystals depends on the feature of interatomic potential, so that it is not necessarily out of the question from a physical point of view.

Acknowledgement

The authors feel it a great honour to dedicate this paper to Professor H. FRÖHLICH. They are indebted to the Japan Institute for Promotion of Sciences for financial support.

Appendix

Proof of the scaling property of the soft-sphere model.

Let the coordinates and momenta of N particles be denoted by the $3N$-dimensional vectors x and p. Let us define the dimensionless coordinates and momenta x^* and p^* by

$$x = l\,x^* \qquad (l = v^{\frac{1}{3}}) \tag{A.1}$$

$$p = (m\,k\,T)^{\frac{1}{2}}\,p^*. \tag{A.2}$$

Then we have

$$\exp\{-p^2/2m\,k\,T\} = \exp\{-p^{*2}/2\}, \tag{A.3}$$

so that the canonical distribution of p^* is given by a universal function. On the other hand, the total potential energy of the soft-sphere model can be written as

$$V(x) = C \sum_{i<j}\sum r_{ij}^{-n} = C\,l^{-n}\,V^*(x^*), \tag{A.4}$$

where

$$V^*(x^*) = \sum_{i<j}\sum r_{ij}^{*\,-n} \tag{A.5}$$

and r_{ij}^* is a dimensionless distance between the i-th and j-th particles measured by the unit of l. Then we have

$$\exp\{-V(x)/k\,T\} = \exp\{-V^*(x^*)/v^{*\frac{n}{3}}\}, \tag{A.6}$$

where v^* is given by (2.4). Considering the system to be confined in a cubic box of side length $V^{\frac{1}{3}} = N^{\frac{1}{3}} l$, the canonical distribution of x^* is given by the universal function (A.6) with n and v^* as parameters.

Now the Newtonian equations of motion

$$m\frac{dx}{dt} = p \tag{A.7}$$

$$\frac{dp}{dt} = -\frac{\partial V(x)}{\partial x} \tag{A.8}$$

are easily shown to reduce to the universal equations

$$dx^*/dt^* = p^* \tag{A.9}$$

$$dp^*/dt^* = -\frac{1}{v^{*\,n/3}}\,\frac{\partial V^*(x^*)}{\partial x^*} \tag{A.10}$$

by using (A.1), (A.2), (A.4) and defining t^* by

$$t = l(m/kT)^{\frac{1}{2}} t^*. \tag{A.11}$$

Thus our proposition in § 2 is proved.

References

1. HIRSCHFELDER, J. O., CURTISS, C. F., BIRD, R. B.: Molecular Theory of Gases and Liquids. New York: John Willey & Sons 1954.
2. FRISCH, H. L., SALSBURG, Z. W., Edited: Simple Dense Fluids. New York and London: Academic Press 1968.
3. TEMPERLEY, H. N. V., ROWLINSON, J. S., RUSHBROOKE, G. S., Edited: Physics of Simple Liquids. Amsterdam: North-Holland 1968.
4. ALDER, B. J., WAINRIGHT, T. E.: J. Chem. Phys. **27**, 1208 (1957); **33**, 1439 (1960).
5. WOOD, W. W., JACOBSON, J. D.: J. Chem. Phys. **27**, 1207 (1957).
   WOOD, W. W., PARKER, F. R., JACOBSON, J. D.: Nuovo Cimento, Suppl. [10] **9**, 133 (1958). See also Chap. 5 of Ref. [3].
6. LONGUET-HIGGINS, H. C., WIDOM, B.: Mol. Phys. **8**, 549 (1964).
   VERLET, L.: Phys. Rev. **165**, 201 (1968).
7. ASHCROFT, N. W., LEKNER, J.: Phys. Rev. **145**, 83 (1966).
8. LONGUET-HIGGINS, H. C., POPLE, J. A.: J. Chem. Phys. **25**, 884 (1956).
9. Relationship between the hard sphere fluids and fluids with realistic repulsive pair potentials is discussed, based on the perturbation theory of a generalized cluster expansion by ANDERSON, H. C., CHANDLER, D., and WEEKS, J. D.: J. Chem. Phys. **56**, 3812 (1972) and the references cited therein.
10. ALDER, B. J., HOOVER, W. G.: Chap. 4 of Ref. [3].
11. MATSUDA, H.: Progr. Theoret. Phys. (Kyoto) **42**, 140 (1969).
12. HOOVER, W. G., ROSS, M., JOHNSON, K. W., HENDERSON, D., BARKER, J. A., BROWN, B. C.: J. Chem. Phys. **52**, 4931 (1970).
    HANSEN, J. P.: Phys. Rev. **2**, A221 (1970).
    HOOVER, W. G., GRAY, S. G., JOHNSON, K. W.: J. Chem. Phys. **55**, 1128 (1971).
13. HIWATARI, Y., MATSUDA, H.: Progr. Theoret. Phys. (Kyoto) **47**, 741 (1972).
14. HARRISON, W.: Pseudopotentials in the Theory of Metals. New York: W. A. Benjamin Inc. 1966.
    STROUD, D., ASHCROFT, N. W.: Phys. Rev. B**5**, 371 (1972).
    HASEGAWA, M., WATABE, M.: J. Phys. Soc. Japan **32**, 14 (1972).
15. SHIFF, D.: Phys. Rev. **186**, 151 (1969).
16. BUNDY, F. P.: Phys. Rev. **115**, 274 (1959).
    BUNDY, F. P., STRONG, H. M.: Solid State Physics, Edit. by SEITZ, F., and TURNBULL, P., Vol. 13, p. 81. New York and London: Academic Press 1962.
17. NEWTON, R. C., JAYARAMAN, A., KENNEDY, G. C.: J. Geophys. Res. **67**, 2559 (1962).
    KENNEDY, G. C., JAYARAMAN, A., NEWTON, R. C.: Phys. Rev. **126**, 1363 (1962).
18. KAWAI, N., INOKUTI, Y.: Japan J. Appl. Phys. **7**, 989 (1968).
    KLEMENT, W., Jr., JAYARAMAN, A.: Progress in Solid State Chemistry, p. 289. Oxford and New York: Pergamon 1966.
19. RAPOPORT, E.: Phys. Rev. Letters **19**, 345 (1967); — J. Chem. Phys. **46**, 2891 (1967); **48**, 1433 (1968).
20. MATSUDA, H.: Progr. Theoret. Phys. (Kyoto) **42**, 414 (1969); — Busseikenkyu [in Japanese] **13**, F–32 (1970).
21. YOSHIDA, T., OKAMOTO, H.: Progr. Theoret. Phys. (Kyoto) **45**, 663 (1971).
22. KURAMOTO, Y., FURUKAWA, H.: Progr. Theoret. Phys. (Kyoto) **47**, 1069 (1972).
23. HIWATARI, Y., OGAWA, T., OGITA, N., MATSUDA, H., UEDA, A.: Busseikenkyu [in Japanese] **18** (1972) (to be published).
24. YANG, C. N.: Rev. Mod. Phys. **34**, 694 (1962).
25. ANDREEV, A. F., LIFSHITZ, I. M.: Zh. Eksperim. i Teor. Fiz. **56**, 2057 (1969); — Soviet Phys. JETP **29**, 1107 (1969).

26. MATSUDA, H., TSUNETO, T.: Progr. Theoret. Phys. (Kyoto), Suppl. No. 46, p. 411 (1970).
27. KAC, M.: Phys. Fluids **2**, 8 (1959).
KAC, M., UHLENBECK, G. E., HEMMER, P. C.: J. Math. Phys. **7**, 216 (1963).
28. HOOVER, W. G., ROSS, M.: Contemp. Phys. (GB) **12**, No. 4, 339 (1971).
29. LINDEMANN, F. A.: Phys. Z. **11**, 609 (1910).
30. ANDRADE, E. N. DA C.: Phil. Mag. **17**, 497, 698 (1934).
31. LEBOWITZ, J. L., PENROSE, O.: J. Math. Phys. **7**, 98 (1966).
32. KRAUT, E. A., KENNEDY, G. C.: Phys. Rev. Letters **16**, 608 (1966);—Phys. Rev. **151**, 668 (1966).
33. Experimental data of the bulk modulus and the volume are taken from following references; for inert gases: POLLACK, G. L.: Rev. Mod. Phys. **36**, 748 (1964); for metals: Ref. [29]. Only for inert gases $|u_0|$ is corrected for zero-point energy.
34. GSCHNEIDER, K. A., Jr.: Solid State Physics, Edit. by SEITZ, F., and TURNBULL, D., Vol. 16, p. 412 (1964).
35. v_c is taken from YOUNG, D. A., and ALDER, B. J. [Phys. Rev. A**3**, 364 (1971)]. Notice that except for rare gases data used here are estimated ones by GROOSE and his colleagues. The atomic volume at 0 °K (v_0) is taken from Ref. [33] for rare gases and that of alkali metals were calculated from the lattice constants (5 °K) as listed by PEARSON, W. B. (Handbook of Lattice Spacing and Structure of Metals. New York: Pergamon Press 1958).
36. UBBELOHDE, A. R.: Melting and Crystal Structure. Oxford: Clarendon Press 1965.
37. LENNARD-JONES, J. E., DEVONSHIRE, A. F.: Proc. Roy. Soc. (London), Ser. A**169**, 317 (1939); A**170**, 464 (1939).
38. YOSHIDA, T., KAMAKURA, S.: Progr. Theoret. Phys. (Kyoto) **47**, 1069 (1972).
39. MORI, H., OKAMOTO, H., ISA, S.: Progr. Theoret. Phys. (Kyoto) **47**, 1087 (1972).
40. STERNHEIMER, R.: Phys. Rev. **78**, 235 (1950).
41. LENNARD-JONES, J. E., DEVONSHIRE, A. F.: Proc. Roy. Soc. (London), Ser. A**163**, 53 (1937).
42. ROSS, M.: Phys. Rev. **184**, 233 (1969).
43. The method of numerical integration which we employ is the same as that of RAHMAN, A. [Phys. Rev. A**136**, 405 (1964)].
44. SAKKA, S., MACKENZIE, J. D.: J. Non-Crys. Solid **6**, 145 (1971).
45. PENROSE, O.: Phil. Mag. **42**, 1373 (1951).
PENROSE, O., ONSAGER, L.: Phys. Rev. **104**, 576 (1956).
46. MATSUBARA, T., MATSUDA, H.: Progr. Theoret. Phys. (Kyoto) **16**, 569 (1956).

The Effect of a Magnetic Field on Impurity Conduction in Semiconductors

V. V. PARANJAPE and D. G. HUGHES

Lakehead University, Department of Physics, Thunder Bay, Ontario, Canada

Abstract

A dielectric approach is used to calculate the critical magnetic field at which a transition from a conducting to a non-conducting state occurs in semiconductors in a strong magnetic field.

1. Introduction

MOTT [1] has pointed out that if the spacing between the atoms, initially far apart, in an assembly is decreased a sharp transition from a non-conducting to a conducting state will occur at some critical value of the spacing; i.e., of the number density of the atoms. He showed, in the same paper, how this idea can be applied to impurity conduction in semiconductors if the impurities are regarded as hydrogen-like centers embedded in a dielectric medium. YAFET, KEYES and ADAMS [2] have discussed the effect of a magnetic field, B, on a hydrogen atom; they point out that if the field is strong enough the charge distribution shrinks in all dimensions. Following MOTT, clearly then if the impurity concentration in a semiconductor is large enough for the conduction to take place, we can restore the material to a non-conducting state by applying a sufficiently large magnetic field. It is this critical value, B_c, of the magnetic field which we estimate in the present paper. FENTON and HAERING [3] were the first to estimate B_c taking into account the screening of the impurity by conduction electrons. DURKAN and MARCH [4] considered, in addition to screening, the effect on the binding energy of the first excited state of the impurity, with particular reference to the overlap of the impurity wave functions. In all this work, the approach is essentially microscopic.

FROOD [5] has pointed out that dielectric theory provides an alternative approach. On this basis, the assumption is that conduction occurs when the permittivity of a dielectric, consisting of a crystal plus its impurities, becomes very large.

We adopt this dielectric approach, taking for our model hydrogen-like impurities in a semiconductor crystal providing a background of permittivity, ε_0. We assume that $T=0$ °K, so that there are no electrons in the conduction band

and the impurities are all in the ground state (i.e., they have no intrinsic electric dipole moment). We then carry out a variational calculation and obtain the average polarizability, $\bar{\alpha}$, of the impurity in a magnetic field, B. The permittivity, ε_s, of the crystal and impurities together is then given by the Clausius-Mosotti formula [6]

$$\frac{\varepsilon_s - \varepsilon_0}{\varepsilon_s + 2\varepsilon_0} = \frac{4\pi}{3\varepsilon_0} N \bar{\alpha} \tag{1.1}$$

where N is the impurity concentration. According to Frood, conduction occurs when $\varepsilon_s \to \infty$, giving the criterion, from (1.1), to be

$$\frac{4\pi}{3\varepsilon_0} N \bar{\alpha} \to 1. \tag{1.2}$$

We shall see in the following sections that $\bar{\alpha}$ decreases with B. Thus the critical value of B necessary to bring about a transition from the conducting to the non-conducting state can be obtained from Eq. (1.2).

2. Calculations

In this section, we shall first calculate the polarizability of a hydrogen atom embedded in a medium of dielectric constant, ε_0 and in the presence of a uniform magnetic field, B. We may write the Hamiltonian for such a system and introduce additional perturbation due to a vanishing small electric field F on the system. The effect of the electric field on the energy of the electron would give us the polarizability. Trial wave function for the electron with suitable adjustable parameters is chosen. The energy may then be calculated and minimized with respect to the parameters. The expression for the minimum energy contains a term, E_F, which is proportional to F^2. The polarizability, α, is then obtained using the well known relationship

$$E_F = -\frac{\alpha}{2} F^2. \tag{2.1}$$

Employing the standard notation, the Hamiltonians in cylindrical polar coordinates, for the two cases in which the electric field is parallel and perpendicular to the magnetic field, are, respectively,

$$H_{\parallel} = \frac{1}{2m}\left(\frac{\hbar}{i}\underset{\sim}{\nabla} + \frac{e}{c}\underset{\sim}{A}\right)^2 - \frac{e^2}{(r^2+z^2)^{\frac{1}{2}}} + Fez \tag{2.2}$$

$$H_{\perp} = \frac{1}{2m}\left(\frac{\hbar}{i}\underset{\sim}{\nabla} + \frac{e}{c}\underset{\sim}{A}\right)^2 - \frac{e^2}{(r^2+z^2)^{\frac{1}{2}}} + Fer\cos\theta \tag{2.3}$$

in which m is the effective mass of the electron, e is the magnitude of the effective charge [$=$actual charge$\times(\varepsilon_0)^{-\frac{1}{2}}$] and A is the vector potential with components [0, $(Br/2)$, 0]. We then choose normalized trial wave functions for each case

$$\Phi_{\parallel} = \frac{1}{a\sqrt{\pi b(1+\beta^2 b^2)}} \exp\left[-\left(\frac{r^2}{a^2} + \frac{z^2}{b^2}\right)^{\frac{1}{2}}\right](1+\beta z) \tag{2.4}$$

$$\Phi_{\perp} = \frac{1}{a\sqrt{\pi b(1+\beta^2 a^2)}} \exp\left[-\left(\frac{r^2}{a^2} + \frac{z^2}{b^2}\right)^{\frac{1}{2}}\right](1+\beta\cos\theta) \tag{2.5}$$

in which a, b and β are parameters to be varied. For each case we calculate the energy, E, given by

$$E = \int \Phi^* H \Phi \, d\tau. \tag{2.6}$$

Substituting Eqs. (2.2), (2.4) and (2.3), (2.5) in Eq. (2.6), we get

$$E_{\parallel} = \frac{1}{(1+\beta^2 b^2)} (X + 2 F e b^2 \beta + Y \beta^2) \tag{2.7}$$

and

$$E_{\perp} = \frac{1}{(1+\beta^2 a^2)} (X + 2 F e a^2 \beta + Z \beta^2) \tag{2.8}$$

in which, introducing the eccentricity, γ, defined by

$$\gamma = 1 - \frac{a^2}{b^2} \tag{2.9}$$

he quantities X, Y, Z are given by

$$X = \frac{\hbar^2}{2m}\left(\frac{2}{3a^2} + \frac{1}{3b^2}\right) + \frac{e^2 B^2}{4mc^2} a^2 + \frac{e}{2b} \gamma^{-\frac{1}{2}} \ln\left(\frac{1-\gamma^{\frac{1}{2}}}{1+\gamma^{\frac{1}{2}}}\right) \tag{2.10}$$

$$\begin{aligned} Y = {} & \frac{\hbar^2}{2m}\left(\frac{2b^2}{5a^2} + \frac{3}{5}\right) + \left(\frac{e^2 B^2}{4mc^2}\right)\left(\frac{3a^2 b^2}{2}\right) - \left(\frac{e^2\, 3a^2}{8b}\right) \\ & \cdot \left[\gamma^{-\frac{3}{2}} \ln\left(\frac{1-\gamma^{\frac{1}{2}}}{1+\gamma^{\frac{1}{2}}}\right) + \frac{2}{\gamma(1-\gamma)}\right] \end{aligned} \tag{2.11}$$

$$\begin{aligned} Z = {} & \frac{\hbar^2}{2m}\left(\frac{a^2}{5b^2} + \frac{4}{5}\right) + \frac{e^2 B^2}{4mc^2} \cdot 3a^4 + \frac{e^2}{2b} \cdot \frac{3a^2}{8} \\ & \cdot \left[(\gamma^{-\frac{1}{2}} + \gamma^{-\frac{3}{2}}) \ln\left(\frac{1-\gamma^{\frac{1}{2}}}{1+\gamma^{\frac{1}{2}}}\right) + \frac{2}{\gamma}\right]. \end{aligned} \tag{2.12}$$

Minimizing E with respect to β, a and γ (in place of b) and treating β as a small quantity, we get from Eq. (2.1) the expressions for the polarizabilities in the two cases in the following forms

$$\alpha_{\parallel} = \frac{2e^2 b^4}{Y - b^2 X} \tag{2.13}$$

$$\alpha_{\perp} = \frac{2e^2 a^4}{Z - a^2 X} \tag{2.14}$$

in which the values of a and γ (and hence b) to be used are the solutions of the two equations

$$\frac{\hbar^2}{2m}\left(1 - \frac{\gamma}{3}\right)\left(-\frac{2}{a^3}\right) + \frac{e^2 B^2}{2mc^2} a - \frac{e^2}{2a^2} \frac{(1-\gamma)^{\frac{1}{2}}}{\gamma^{\frac{1}{2}}} \ln\left(\frac{1-\gamma^{\frac{1}{2}}}{1+\gamma^{\frac{1}{2}}}\right) = 0 \tag{2.15}$$

$$\frac{\hbar^2}{2m} \cdot \frac{1}{3a^2} + \frac{e^2}{4a} \cdot \frac{1}{\gamma^{\frac{3}{2}}(1-\gamma)^{\frac{1}{2}}} \ln\left(\frac{1-\gamma^{\frac{1}{2}}}{1+\gamma^{\frac{1}{2}}}\right) + \frac{e^2}{2a} \cdot \frac{1}{\gamma(1-\gamma)^{\frac{1}{2}}} = 0. \tag{2.16}$$

We may now define the average polarizability, $\bar{\alpha}$, by

$$\bar{\alpha} = \frac{\alpha_{\parallel} + 2\alpha_{\perp}}{3}. \tag{2.17}$$

Eqs. (2.15) and (2.16) are solved giving a and γ for different values of the magnetic field, B. The values so obtained are then used in Eqs. (2.13), (2.14) and (2.17) to

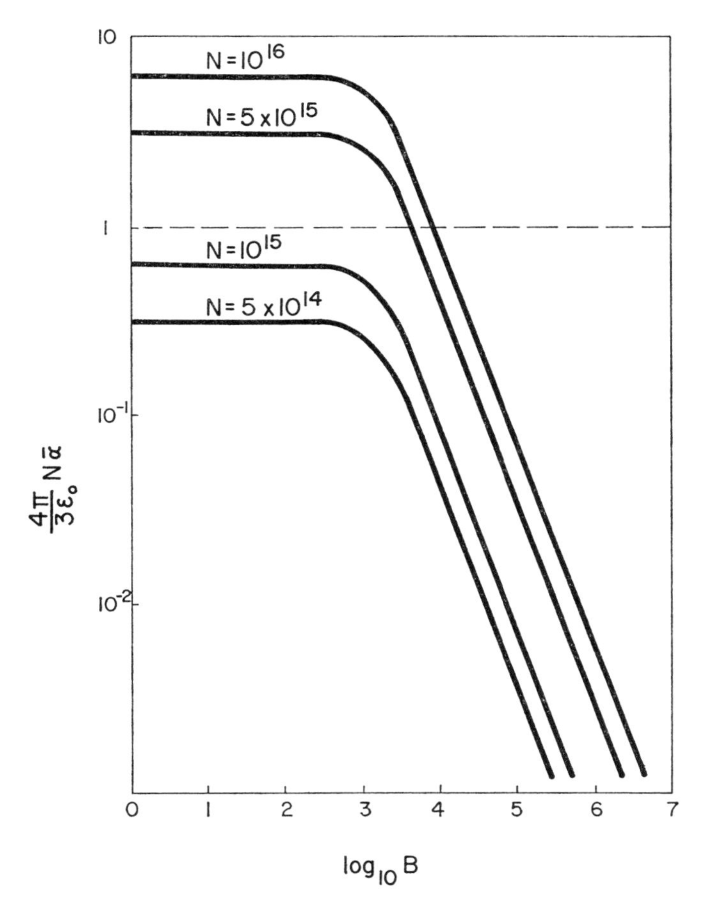

Fig. 1. $4\pi N\bar{\alpha}/3\varepsilon_0$ as a function of B for *InSb*. B is in gauss, N in cm^{-3}. The broken line marks the boundary between regions in which impurity conduction does (above the line) and does not occur

determine $\bar{\alpha}$ as a function of B. The calculations were done for indium antimonide taking $\varepsilon_0=16$ and $m=0.01\times$ free electron mass. In Fig. 1, which relates to *InSb*, $4\pi N\bar{\alpha}/3\varepsilon_0$ is shown as a function of B.

3. Discussion

Our figure has been drawn with our dielectric criterion for conduction [Eq. (1.2)] in mind. In the region above the horizontal dashed line, the system is conducting; below this line, it is non-conducting. We see that, for an impurity concentration of, for example, $10^{16}\,cm^{-3}$, the system is non-conducting for magnetic fields greater than 10000 gauss.

The figure also yields information about the dependence on N of the critical field, B_c, at which the transition takes place. We note first that if B is sufficiently large the curves of Fig. 1 all have the same slope, s (a negative number), which can be calculated. For a fixed value of N

$$\bar{\alpha} \propto B^s. \tag{3.1}$$

The critical values of $\bar{\alpha}$ and B are therefore related by

$$\bar{\alpha}_c \propto B_c^s \tag{3.2}$$

Now α_c is related to N by our conduction criterion, namely

$$\frac{4\pi}{3\varepsilon_0} N \bar{\alpha}_c = 1$$

therefore

$$\bar{\alpha}_c \propto N^{-1}. \tag{3.3}$$

Hence, from (3.2),

$$B_c^s \propto N^{-1} \tag{3.4}$$

and therefore

$$B_c \propto N^{-1/s}. \tag{3.5}$$

From Fig. 1 (or, more readily, from the computed values of $\bar{\alpha}$ as a function of B), we can determine s. We obtain $s = -(10/9)$ so that

$$B_c \propto N^{0.90}. \tag{3.6}$$

This is close to the relationship observed by Beckman, Hanamura and Neuringer [7]

$$B_c \propto N^{6/7} = N^{0.86}. \tag{3.7}$$

Our macroscopic model in spite of its simplicity yields results consistent with experimental observations.

References

1. Mott, N. F.: Can. J. Phys. **34**, 1356–1368 (1956).
2. Yafet, J., Keyes, R. W., Adams, E. N.: J. Phys. Chem. Solids **1**, 137–142 (1956).
3. Fenton, F. W., Haering, R. R.: Phys. Rev. **159**, 593–599 (1967).
4. Durkan, J., March, N. H.: J. Phys. C **1**, 1118–1127 (1968).
5. Frood, D. G. H.: Proc. Phys. Soc. (London) **75**, 2, 185–193 (1959).
6. Fröhlich, H.: Theory of Dielectrics, Second Ed. London: Oxford University Press 1958.
7. Beckman, O., Hanamura, E., Neuringer, L. J.: Phys. Rev. Letters **18**, 773–775 (1967).

Relaxation Measurements on Single Crystals of $MnCl_2 \cdot 4H_2O$ and $MnBr_2 \cdot 4H_2O$ Near the Magnetic Phase Transition

A. J. van Duyneveldt, J. Soeteman and C. J. Gorter

Kamerlingh Onnes Laboratorium der Rijksuniversiteit Leiden, Nederland

Summary

Magnetic relaxations have been studied on single crystals of $MnCl_2 \cdot 4H_2O$ and $MnBr_2 \cdot 4H_2O$. The susceptibility in the antiferromagnetic state indicates the existence of a relaxation process with a single time constant. The obtained relaxation times show a sharp maximum as a function of temperature or external magnetic field. These results can be ascribed to the anomalies of the specific heat at the magnetic phase transition.

1. Introduction

Magnetic relaxation phenomena have been studied on single crystals of $MnCl_2 \cdot 4H_2O$ and $MnBr_2 \cdot 4H_2O$ in the antiferromagnetic as well as in the paramagnetic state. These salts show spin-lattice relaxation times τ that allow a study of the temperature and field dependence of τ near the transition point. Lasheen et al. [1] were not able to detect the spin-lattice relaxation times of a single crystal of $MnCl_2 \cdot 4H_2O$ in the antiferromagnetic state. However, their results below T_N on powdered specimens did indicate the existence of an anomaly of the relaxation time near T_N. Recent work [2] demonstrated a peak in the τ versus T graph for a single crystal of $MnCl_2 \cdot 4H_2O$. Below T_N, these measurements were not accurate, since a distribution of relaxation times was observed, while the measuring technique with "step-fields", as was operated at that time, can only yield τ successfully if the relaxation proceeds with one time constant. With the present experimental technique long relaxation times can be detected with high accuracy. This allowed us to measure the relaxation times for single crystals of $MnCl_2 \cdot 4H_2O$ and $MnBr_2 \cdot 4H_2O$, down to 1.3 K and at external magnetic fields up to 25 kOe.

2. Experimental Details

The reported investigations have been performed with the so-called dispersion-absorption technique [3]. This method allows instantaneous measurements of the real and imaginary component of the susceptibility, if the sample is placed in a constant magnetic field on which a small oscillating field is superimposed. To describe spin-lattice relaxation, the magnetic salt is supposed to consist of two separate thermodynamical systems, the "spin system" and the "lattice". The spin system, connected with the magnetic properties of the sample, is supposed to be in internal equilibrium, characterized by a temperature T_S; the lattice is described by a temperature T_L. Equilibrium between the systems is established by spin-lattice interactions. The assumption is that these interactions cause a heat flow (dQ) from the spin system to the lattice, proportional to $(T_S - T_L)$:

$$dQ/dt = -\alpha(T_S - T_L).$$

For the frequency dependence of $\bar{\chi}(=\chi' - i\chi'')$ one can derive [4]:

$$\begin{aligned} \chi' &= \chi_{ad} + \frac{\chi_0 - \chi_{ad}}{1 + \omega^2\tau^2} \\ \chi'' &= (\chi_0 - \chi_{ad}) \frac{\omega\tau}{1 + \omega^2\tau^2} \end{aligned} \tag{1}$$

where $\tau = c_H/\alpha$ is the spin-lattice relaxation time (c_H is the specific heat of the spin system at constant field). The static susceptibility χ_0 is the value of $\bar{\chi}$ at $H=0$ and $\omega \ll \tau^{-1}$, the adiabatic susceptibility χ_{ad} is measured at frequencies $\omega \gg \tau^{-1}$. The Eqs. (1) express the Debye shape of the susceptibility, as is measured for single relaxation processes in the paramagnetic state. With the present experimental facilities, susceptibility measurements may be carried out in the frequency range: 0.2 Hz to 1 MHz; thus relaxation times from 1 s to 10^{-7} s may be ascertained.

3. Results

Specific heat measurements by Friedberg et al. [5] and magnetization measurements by Henry [6] and Gijsman et. al. [7] showed the antiferromagnetic behaviour of $MnCl_2 \cdot 4H_2O$ and $MnBr_2 \cdot 4H_2O$, with the crystal c-axis as the magnetically preferred axis. Therefore all samples have been examined with the magnetic field aligned parallel to the c-axis.

a) $MnCl_2 \cdot 4H_2O$

At temperatures below 2.1 K, the observed susceptibility versus frequency curves do obey Eq. (1), thus single relaxation times are obtained. As an example, Fig. 1 shows the χ''/χ_0 versus $\log \nu$ graph at $T=1.3$ K and $H=4$ kOe. Most relaxation times have been detected from the absorption curves, since the dispersion is relatively large around T_N [2] while small variations cannot be measured accurately. The part of the susceptibility occupied in the spin-lattice relaxation process decreases if the temperature decreases towards T_N. Below T_N, the absorption increases approximately by a factor 100. This means that the detection of τ slightly above T_N is very difficult.

The obtained relaxation times for an external magnetic field of 4 kOe, are exposed as a function of temperature in Fig. 2 (symbol O). A sharp peak is found at $T=1.51$ K, a value that coincides with the transition temperature at a field of 4 kOe [2].

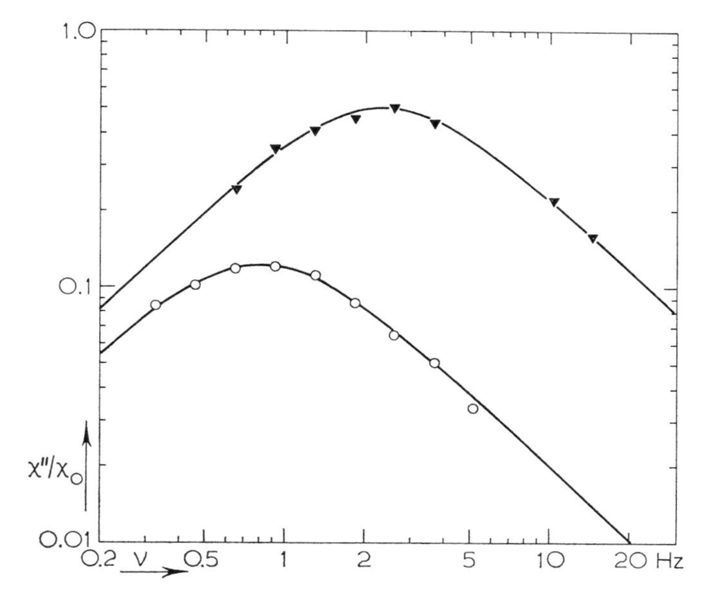

Fig. 1. Absorption versus frequency $\nu (= \omega/2\pi)$ for $MnCl_2 \cdot 4\,H_2O$ (○ $T = 1.3$ K; $H = 4$ kOe) and $MnBr_2 \cdot 4\,H_2O$ (▼ $T = 1.83$ K, $H = 8$ kOe); the drawn lines are Debye curves [Eq. (1)]

The field dependence of the relaxation has been studied at $T = 1.44$ K (Fig. 3, symbol ○). In the antiferromagnetic region the relaxation time increases if the external magnetic field increases. Above the transition field of 5.5 kOe, the relaxation is found to be roughly 10 times faster. The measurements have not been extended to fields above 10 kOe in order to avoid phonon bottleneck effects [8].

b) $MnBr_2 \cdot 4\,H_2O$

Again all susceptibility measurements obeyed Eq. (1) (example in Fig. 1), thus a description with one relaxation time is justified. The τ versus T as well as the τ versus H dependences show anomalies. The temperature dependence has been examined at 8 kOe (Fig. 2, symbol ▼). The peak is situated around $T = 1.83$ K, the transition temperature at that field [9].

The field dependence at $T = 1.98$ K (Fig. 3, symbol ▼) shows a maximum at 5.6 kOe. Both observed anomalies are less pronounced than in the chloride sample.

4. Discussion

Recent theoretical work of Barry and Harrington [10], based on statistical equilibrium theory and the thermodynamics of irreversible processes, leads to a Debye shape for the frequency dependence of the susceptibility in the antiferromagnetic region. Our present experiments clearly demonstrated this, since all susceptibilities fullfilled Eq. (1). The distribution of relaxation times, as reported in [2], has been detected after a sudden change of T_S. To gain sensitivity $T_S - T_L$ might have been to large, causing a non-exponential recovery of $\bar{\chi}$.

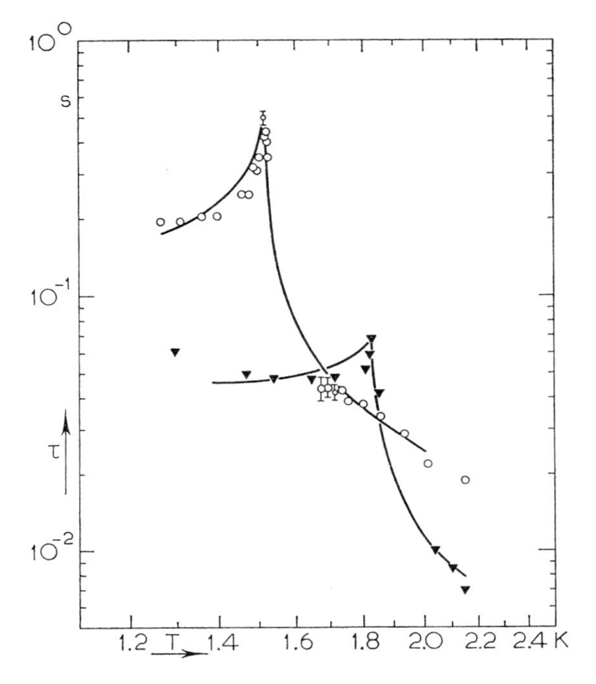

Fig. 2.
Spin-lattice relaxation times against temperature. ○ $MnCl_2 \cdot 4H_2O$; $H = 4$ kOe. ▼ $MnBr_2 \cdot 4H_2O$; $H = 8$ kOe. Drawn lines are explained in Section 4

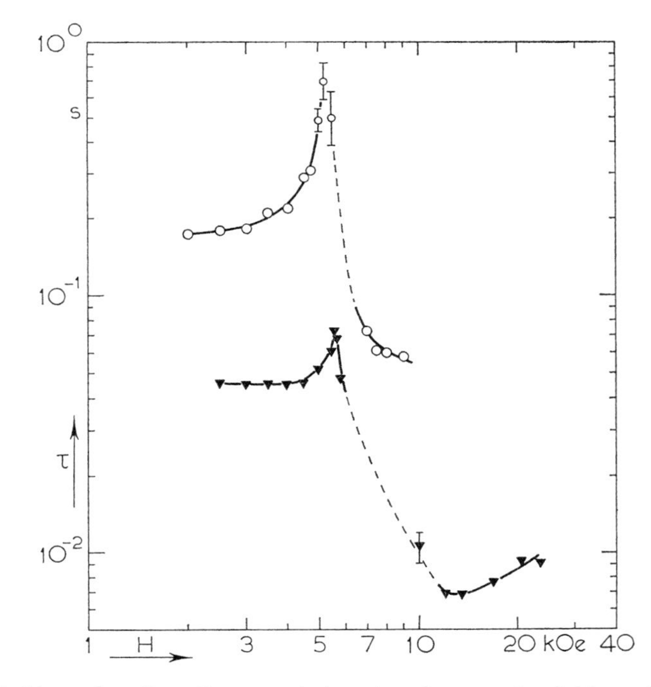

Fig. 3. Spin-lattice relaxation times against external magnetic field. ○ $MnCl_2 \cdot 4H_2O$; $T = 1.44$ K. ▼ $MnBr_2 \cdot 4H_2O$; $T = 1.98$ K. Drawn lines are for visual aid only

Barry and Harrington do predict a relaxation time with a maximum at T_N. However, detailed experimental data are required on the temperature dependence of χ' and χ'' at low as well as at high frequencies, as their theory gives results in forms amendable to experimental fitting. This study is now in progress.

At this moment it seems interesting to compare the observed relaxation times with specific heat data, as $\tau = c_H/\alpha$. The specific heats of both the chloride and the bromide salt are known from various experiments [11, 12]. If one assumes that α does not show an anomaly at T_N, one can calculate c_H/α. For $MnCl_2 \cdot 4H_2O$ this is done, taking α proportional to $T^{+0.1}$. The result is inserted in Fig. 2 and gives a remarkably good description of the observed τ versus T curve.

For $MnBr_2 \cdot 4H_2O$ a calculation based on α being proportional to $T^{+0.8}$, also gives a good connection with the experimental relaxation times.

Detailed information on c_H versus H is not known, so a comparison between τ and c_H/α cannot be made. However, the intensity of the steps in the τ versus H curves agree with c_H versus H if the value of α is not strongly dependent on external magnetic field.

From this phenomenological approach the conclusion might be that the observed anomalies in the relaxation times are closely connected with the anomalies in the specific heat. The dependences of α used in the calculations are not expected to give information about the occurring spin-lattice interactions [13]. The reported experiments will be extended to verify the theory of Barry and Harrington and also to extend the field and temperature range of the measurements, possibly leading to an insight into the spin-lattice relaxation mechanisms in these salts.

References

1. Lasheen, M. A., Broek, J. van den, Gorter, C. J.: Physics **24**, 1061 (1958). (Commun. Kamerlingh Onnes Lab., Leiden No. 312b.)
2. Soeteman, J., Duyneveldt, A. J. van, Gorter, C. J.: Physica **45**, 435 (1969). (Commun. Kamerlingh Onnes Lab., Leiden No. 375a.)
3. De Vries, A. J., Livius, J. W. M.: Appl. Sci. Res. **17**, 31 (1967). (Commun. Kamerlingh Onnes Lab., Leiden No. 349a.)
4. Casimir, H. B. G., Du Pré, F. J.: Physica **5**, 507 (1938). (Commun. Kamerlingh Onnes Lab., Leiden, Suppl. No. 85a.)
5. Friedberg, S. A., Wasscher, J. D.: Physica **19**, 1072 (1953). (Commun. Kamerlingh Onnes Lab., Leiden No. 293c.)
6. Henry, W. E.: Phys. Rev. **91**, 435 (1953); **94**, 1146 (1954).
7. Gijsman, H. M., Poulis, N. J., Handel, J. van den: Physica **25**, 954 (1959). (Commun. Kamerlingh Onnes Lab., Leiden No. 317b.)
8. Roest, J. A., Duyneveldt, A. J. van, Bilt, A. van der, Gorter, C. J.: Physica **64**, 306 (1973). (Commun. Kamerlingh Onnes Lab., Leiden No. 400a.)
9. Smidt, V. A., Friedberg, S. A.: J. Appl. Phys. **38**, 5319 (1967).
10. Barry, J. H., Harrington, D. A.: Phys. Rev. B**4**, 3068 (1971).
11. Reichert, T. A., Giauque, W. F.: J. Chem. Phys. **50**, 4205 (1969).
12. Schelling, T. H., Friedberg, S. A.: Phys. Rev. **185**, 728 (1969).
13. Duyneveldt, A. J. van, Pouw, C. L. M., Breur, W.: Physica **57**, 205 (1972). (Commun. Kamerlingh Onnes Lab., Leiden No. 387b.)

Chapter VI. Many-body Effects

There are many other newly developing fields of collective behaviour in physics. Among these are monatomic layers adsorbed on crystal surfaces. The article presented first (NAKAJIMA) treats in particular the He monolayer adsorbed on graphite. This system may be taken as a two-dimensional model provided by nature for the study of quantum liquids and quantum solids, respectively. It partially overlaps with the chapter on superfluidity and could have been placed there, too. Also in the region of nuclear physics there are attempts which owe much to the theory of superconductivity in the formulation of the collective modes of nuclei; the next paper (LE COUTEUR) pertains to this area. A rather unconventional approach to spontaneous emission is presented by Rose in our context of cooperativity. He shows that spontaneous emission can be accounted for in terms of a classical interaction between an electron and a solid and may in this way be traced back to the "hot-electron" problem. To contrast with most of the other papers in this book, we thought it amusing to include also a quasi "anti"-collective aspect in the very meritorious experiments and theoretical considerations of Jánossy about the dual nature of light. The last three tributes (HUBY, SACK, ALLCOCK) are concerned with pertinent mathematical methods and studies of more general nature.

The He Monolayer on Graphite

S. NAKAJIMA

University of Tokyo, Institute for Solid State Physics, Tokyo/Japan

1. Introduction

During the last decade a good deal of progress was made in our understanding cooperative phenomena in one and two dimensions. Experimentally, low-dimensional systems are approximately realized as extremely anisotropic magnets, very thin superconductors and so on.

Recently DASH and coworkers [1] have measured the heat capacity of the helium monolayer adsorbed on graphite at various degrees of coverage and shown that this is a new, two-dimensional quantum system. Adsorbed He atoms, when their average separation is large, move around by tunneling over the graphite surface even at very low temperatures. At higher concentrations, where the interatomic distance becomes comparable with the lattice constant of bulk solid helium, the interaction between adsorbed atoms is so important that they are localized to form a two-dimensional lattice. In fact, essentially the same quantum mechanism seems to underlie both monolayer and bulk helium. If you like, you may regard the monolayer as a two-dimensional model provided by Nature for us to understand the bulk properties of helium as well as the effects of dimensionality upon quantum liquids and solids.

The present article is an introduction to this newly developing field. But it is not a review article of the ordinary kind with a complete list of references. It seems a little premature to write such a review; there still remain many points to be cleared up by further studies before this is done.

I will first try to summarize the main features of the He monolayer which have so far been revealed by experiments. I will then point out a few theoretical problems, which emerge from the experiments and which I am particularly interested in. Finally, I will briefly describe my own attempt to understand the main features of the He^4 monolayer, at least qualitatively, by means of a simple theoretical model.

Hence the present article is rather similar to a talk at weekly seminars. I hope this style of writing is appropriate to treat the subject at the present stage of development and stimulating to those readers who are interested in cooperative phenomena in general.

2. Experimental Results

For some time, copper and argon were used as substrates, but the effects of inhomogeneity were too big to allow a consistent picture of the He monolayer to be drawn from the observed heat capacity. Experimentally, the heat capacity of the monolayer is defined as the difference between the heat capacities which an aggregate of substrate flakes show in the presence and in the absence of adsorbed helium, respectively.

Recently DASH and coworkers [1] have found that the cleavage plane of graphite is a good homogeneous substrate and shown that the heat capacity of the He monolayer adsorbed on it discloses a number of remarkable features. Before going into detail, it is convenient to introduce the parameter x_g, which is defined as the number of adsorbed He atoms per hexagonal unit cell of the graphite surface. The average distance between He atoms is then given by $r_0=(2.3\ \text{Å}/x_g^{\frac{1}{2}})$, since the carbon-carbon bond distance in the graphite plane is 1.42 Å. Note that r_0 becomes equal to the average interatomic distance in bulk liquid He^4 when x_g is around 0.4.

Now let us summarize the main features of the He monolayer on graphite as disclosed by the heat capacity experiments [1].

1° When x_g is small ($x_g \lesssim 0.1$) and the temperature T is high ($T \sim 4$ K), the heat capacity is T-independent. This T-independent value is not equal to $2k_B$ per atom as one might expect from a $2D$ (two-dimensional) vibration of adsorbed atoms, but equal to k_B, which is the value expected for a $2D$ classical gas.

2° When x_g is small and T becomes low, there appears a definite difference between the heat capacities of He^3 and He^4 monolayers. The heat capacity of the He^3 monolayer monotonously decreases with decreasing T and can be fitted by the theoretical curve of a $2D$ ideal Fermi gas if the Fermi temperature is taken as an adjustable parameter. A small shoulder observed at very low temperatures may be due to the ordering effect of nuclear spins [2]. The heat capacity of the He^4 monolayer, on the other hand, shows a peak at a higher temperature and one might be tempted to take this peak as an implication of the Bose condensation of adsorbed atoms. I will discuss this possibility later on.

From 1° and 2°, we see that for low x_g adsorbed He atoms move around by tunneling over the graphite surface and form a $2D$ quantum fluid. As x_g increases, however, the interaction between He atoms plays a more and more decisive role and eventually there appears a certain positional ordering of adsorbed atoms, as we shall see below.

3° When x_g is around 1/3 and T is around 3 K, regardless of statistics, the heat capacity exhibits a rather sharp peak which implies a positional ordering of He atoms over the graphite lattice. The simplest model [3] is to assume that the adsorbed atom is localized at each center of the hexagonal unit cell. The array of these centers may be regarded as consisting of three sublattices, each lattice point of each sublattice being surrounded by six nearest neighbors which belong to the other two sublattices. The distance between two neighboring sites, 2.45 Å, is nearly equal to the hard-core diameter of the He atom itself, and therefore a strong repulsion must appear when these two sites are occupied simultaneously.

At $x_g = 1/3$, He atoms will tend to occupy one particular sublattice, leaving the other two vacant in order to reduce the repulsive energy.

The actual situation is more complicated than this; we should take into account the quantum mechanical kinetic energy, which leads to tunneling of the adsorbed atom from site to site and is increased by the positional ordering. We should also take into account that the interaction between He atoms has not only the short-range repulsive part as represented by a hard sphere, but also has the rather long-range attractive part. As a consequence, the periodic structure which adsorbed He atoms can assume need not necessarily be commensurable with the graphite lattice.

4° When x_g is increased further beyond 1/3, the heat capacity of the He^4 monolayer as a function of T again exhibits a prominent peak. DASH and coworkers have taken this peak as an implication of the melting of $2D$ solid He^4. In fact, the melting temperature plotted against r_0^2 runs almost parallel to the melting curve of bulk hcp He^4, the difference between the two being about 2 K. Furthermore, the heat capacity at temperatures much lower than the melting point is proportional to T^2 and can be fitted by the $2D$ Debye formula. The Debye temperature plotted against r_0^2 almost coincides with the similar curve of the Debye temperature of bulk hcp He^4.

3. Some Theoretical Problems

Let us now turn to the theoretical problems which emerge from these experimental results and begin with the limit of small x_g. In this limit, we are concerned with a single He atom moving through the field of force exerted by carbon atoms in graphite. If the lattice vibration of graphite is ignored, the field of force has the same $2D$ periodicity as the graphite surface, and the energy spectrum of the He atom can be calculated by band-theoretical methods. Such a calculation has already been done [4] and the calculated bandwidth is indeed large in comparison with $k_B T$ in the temperature range we are interested in.

If the interaction with the lattice vibration is taken into account, there appears a certain modification of the spectrum and also the interaction between He atoms via graphite phonons [5]. Strictly speaking, we should also check that the interaction with the lattice vibration is not strong enough to localize the He atom.

As we have seen in § 2, the He^3 monolayer behaves like a $2D$ ideal Fermi gas. It appears that the effects of substrate potential, graphite phonons, and interaction among He atoms can all be included in the effective mass of the adsorbed atom. Anyway the problem here is a rather quantitative one and somewhat similar to the dilute mixture of He^3 in He^4.

On the other hand, the $2D$ ideal-gas model does not work in the case of He^4. Some years ago, MAY [6] gave a mathematical proof that the $2D$ ideal Bose gas has exactly the same heat capacity as the $2D$ ideal Fermi gas of the same density. This statement sounds a little surprising, but is in accord with HOHENBERG's general conclusion that the superfluid long-range order cannot exist in $2D$ Bose systems at finite temperature [7]. Experimentally, the heat capacity of the He^4 monolayer shows a peak around the degeneracy temperature $T_B = (2\pi\hbar^2 n / 2m k_B)$, where n is the number of adsorbed atoms per unit area and m is the (effective) atomic mass. To account for this observed peak, DASH and coworkers [8] adopted WIDOM's idea. WIDOM [9] showed that one- and two-dimensional ideal Bose gases have finite Bose condensation temperatures if they are placed in the field of gravity or in a rotating bucket. These external forces are macroscopic in the sense that the energy change ΔV comparable with $k_B T$ results only when the atom has travelled a macroscopic distance.

DASH et al. assumed that such a weak, but long-range field arises from substrate inhomogeneity and obtained the Bose condensation temperature as $T_B[1+\log(T_B/\Delta V)]^{-1}$. Here T_B is the degeneracy temperature defined above and ΔV is the difference of the potential energies when the atom is, say, at center and edge of the substrate respectively. The observed heat capacity peak may be accounted for by assuming ΔV which is comparable to $k_B T_B$.

WIDOM's theory was criticized later by REHR and MERMIN [10] who pointed out that the particle density becomes infinite somewhere in WIDOM's model. Furthermore, if we take account of the interaction between atoms so that the system has a finite coherence distance, the Bose condensate will no longer be so sensitive to such a terribly weak, but long-range external force. So, I take another point of view and accept HOHENBERG's conclusion; a superfluid long-range order cannot exist in our He^4 monolayer, either. But the short-range order always does exist and can lead to an anomaly of the heat capacity, as is well known in many low-dimensional systems. I will come back to this problem later.

Let us now turn to the positional ordering in the large x_g region. Some general remarks will be given first on the periodic arrangement adsorbed atoms can assume. In the presence of the substrate potential, the translational symmetry is not higher and therefore the period is not shorter than the triangualr lattice of the graphite surface. Moreover, since the lattice constant 2.45 Å is nearly equal to the hard-core diameter of the He atom, the zero-point kinetic energy will be too high if He atoms take the close packed arrangement on the graphite lattice ($x_g \simeq 1$). For $x_g < 1$, the translational symmetry can still be the same as the graphite lattice. For example, in the small x_g limit, each He atom is in a Bloch-type state, so that the average density has the same periodicity as the graphite lattice. This type of symmetry is therefore rather trivial in our problem. Let us agree to say that a long-range positional ordering exists only when the translational symmetry is lower than the graphite lattice.

The ordering we have proposed under 3° in § 2 is such an example. As we have seen, the simplest model is the tight binding approximation, in which the zeroth-order wave function of the He atom is localized around the center of the hexagonal unit cell of graphite. If we ignore the tunneling of the atom from center to center and only include the interaction between localized atoms, we obtain the classical $2D$ lattice gas defined by the Hamiltonian

$$H_{cl} = \frac{1}{2} \sum_j \sum_l J_{jl} n_j n_l - \mu \sum_j n_j. \tag{1}$$

Here n_j is the occupation number of the j-th site, J_{jl} is the interaction energy between atoms localized at j and l sites, and we have included the chemical potential μ, supposing that the system is in contact with a particle reservoir. For given μ, the atomic concentration is obtained by

$$x_g = \frac{1}{N} \sum_{j=1}^{N} \langle n_j \rangle \tag{2}$$

or vice versa, where N is the total number of adsorption centers and $\langle\ \rangle$ means the expectation value. Because of the hard-core repulsion between He atoms, the interaction J_{jj} in (1) is virtually infinite and each occupation number n_j is restricted either to 0 or to 1. As is well known, then (1) is equivalent to the $2D$ Ising model of magnets [3, 11]. Thus vacant and occupied states of the j-th site are represented respectively by eigenvalues $+1/2$ and $-1/2$ of the z-component S_{jz} of Pauli spin operators. Needless to say, this spin is a mathematical convention and has nothing to do with the physical spin, even in the case of He^3.

The positional ordering as defined above is then represented by a non-ferromagnetic ordering of spins. Mathematically, this is characterized by the condition that the Fourier transform

$$S_z(\boldsymbol{k}) = \sum_{j=1}^{N} S_{jz} \exp\left[i\boldsymbol{k} \cdot \boldsymbol{R}_j\right] \tag{3}$$

has non-vanishing expectation values $\langle S_z(\boldsymbol{k})\rangle = \langle S_z(-\boldsymbol{k})\rangle^*$ not only for $\boldsymbol{k}=0$, but also for a non-vanishing wave vector $\boldsymbol{k}$ in the first Brillouin zone of the triangular lattice whose lattice vectors are $\boldsymbol{R}_j$. In fact, the ordering proposed under 3° in § 2 is characterized by

$$\langle S_z(0)\rangle = \tfrac{1}{6} N\sigma, \qquad \langle S_z(\boldsymbol{k}_0)\rangle = -\tfrac{1}{3} N\sigma \tag{4}$$

where $\boldsymbol{k}_0$ is the wave vector which points to one of six corners of the Brillouin zone and $0 \leqq \sigma \leqq 1$. In classical theory, the ordering with $\sigma=1$ is the state of minimum energy when the Fourier transform

$$J(\boldsymbol{k}) = \sum_{j=1}^{N} J_{j0} \exp\left[i\boldsymbol{k} \cdot \boldsymbol{R}_j\right] \tag{5}$$

takes the negatively biggest value at $\boldsymbol{k}=\boldsymbol{k}_0$. Physically $\sigma=1$ means the $2D$ perfect "crystal" whose unit cell is three times as large as the unit cell of the basic graphite lattice. In quantum theory, σ can be less than unity even at $T=0$ because the occupation number can fluctuate through tunneling. This means that even at $T=0$ there are a certain finite number of "defects" which are moving through the crystal without destroying the translational symmetry—zero-point defects [12]. In the following Section I will discuss how to incorporate the tunneling in (1).

At present it is not certain whether the $2D$ solid He^4, which seems to exist for $x_g > 1/3$ and has been described under 4° in § 2, may be taken as such a crystal with zero-point defects. The answer might be negative, since experimentally there is no evidence that $x_g = 1/2$ has a particular significance. Theoretically, we usually have the particle-hole symmetry in lattice gas models, so that thermodynamic quantities are invariant when x_g is replaced by $1 - x_g$.

Another limitation of the lattice model, even after including the tunneling, is obviously that the vibrational motion of the atom is not properly taken into consideration. This point will be improved in the following way. The breadth of the zeroth-order wave function in our tight-binding approximation represents the zero-point vibration of the atom in an Einstein model. In order to obtain phonon excitations, we should at least include the first excited state at each site. This state is coupled with the ground states of neighboring sites through generalized tunneling processes. So, when one atom is excited to first excited vibrational state, this excitation is not localized on the particular site, but propagates like an exciton, and may be identified with a phonon. In order that this phonon has the vanishing energy for the vanishing wave vector, however, all matrix elements should be chosen carefully and vibrations of the substrate should also be taken into account, so that the Hamiltonian resumes its full translational invariance. The problem here is very similar to the one we are confronted with in dealing with phonons in a highly unharmonic crystal such as bulk solid helium [13]. Though it is interesting, I will not go into its detail.

4. Quantum Lattice Models

Although it has some limitations, as we have seen above, it seems worthwhile studying the tight-binding approximation in a little more detail. It will give us at least a qualitative picture of the He monolayer from a unified point of view over a wide range of x_g.

So let us generalize (1) by including the tunneling terms. The occupation number then becomes the quantum operator and may be written as $n_j = a_j^+ a_j$ in the case of the He^4 monolayer. Here a_j, a_j^+ are destruction and creation operators of the He^4 atom at the j-th site and satisfy well-known commutation rules for Bose particles. The quantum Hamiltonian is then given as

$$H = \tfrac{1}{2} \sum \sum J_{jl} n_j n_l - \tfrac{1}{2} \sum \sum K_{jl} a_j^+ a_l - \mu \sum n_j. \tag{6}$$

The second term on the right represents the tunneling. In the limit of small x_g, the first term may be ignored and the tight-binding approximation in the usual band theory is obtained. Thus the Hamiltonian is diagonalized by introducing the Fourier transform $a(\boldsymbol{k})$ of the operator a_j defined in a similar manner to (3). Each atom is in a Bloch type state, whose energy is given by $(-1/2)\, K(\boldsymbol{k})$, where $K(\boldsymbol{k})$ is the Fourier transform of K_{jl} similar to (5). We assume this energy band has a minimum at $\boldsymbol{k}=0$. If we retain K_{jl} only for nearest neighbors, it should be positive.

In the case of He^3, each operator in (6) has an extra suffix to indicate the orientation of the nuclear spin, and a, $a^\dagger$ satisfy well-known anticommutation rules for Fermi particles. The Hamiltonian in this case is obviously similar to the Hubbard Hamiltonian of electrons, which is often used in the theory of the metal-insulator transition [14]. In fact, we are also confronted with the transition between mobile and immobile states of adsorbed He atoms in the large x_g region. In our case, the interaction J_{jj} is virtually infinite, so that the first thing to do is to eliminate those states in which two atoms with opposite nuclear spins occupy one and the same site. Unfortunately there is no simple mathematical way of doing this.

In the case of He^4, as in classical theory, the self-interaction disappears from the Hamiltonian if we impose the restriction upon each occupation number that its eigenvalue shall not exceed one. Mathematically this means that we replace Bose operators by Pauli spin operators as

$$a_j \to S_{jx} + i S_{jy}, \quad n_j \to \tfrac{1}{2} - S_{jz}. \tag{7}$$

Pauli spin operators are commutative on different sites, but anticommutative on the same site. The latter property is the kinematical restriction which replaces the dynamical effect of J_{jj}.

Through (7), the Hamiltonian (6) is transcribed as

$$H = C - \eta \sum S_{jz} + \tfrac{1}{2} \sum \sum J_{jl} S_{jz} S_{lz} - \tfrac{1}{2} \sum \sum K_{jl} (S_{jx} S_{lx} + S_{jy} S_{ly}). \tag{8}$$

Here C is a constant and η is the negative chemical potential defined by

$$\eta = \tfrac{1}{2} J(0) - \mu. \tag{9}$$

As a matter of fact, the Hamiltonian (8) was proposed many years ago by MATSUBARA and MATSUDA [15] as a quantum lattice model for bulk liquid He^4. Recently the possibility of the supersolid phase of He^4 has also been discussed on the basis of this Hamiltonian [16]. In the lattice model of bulk liquid and solid, however, the basic lattice is a mere mathematical convention to simplify the continuous configuration space of the atom. In our monolayer problem, the basic lattice is defined by the substrate and the Hamiltonian (8) is almost equivalent to the tight-binding Hamiltonian (6). Thus the quantum lattice model is much more appropriate and realistic in our problem.

The great advantage of this model is that the He^4 monolayer is mathematically equivalent to a $2D$ spin system with external magnetic field and anisotropic exchange coupling, so that we can apply various methods in statistical physics of spin systems, as we see from the following examples.

A. Molecular Field Approximation

Let us begin with the molecular field approximation. If we retain J_{jl} and K_{jl} only for nearest neighbors and assume that both are positive, then this approximation predicts three possible phases, which I call I, II, III, respectively. They are characterized by three different modes of ordering of the equivalent spins S_j.

First of all the vacant substrate, on which no He atom is adsorbed, is represented by the state Φ in which all spins are parallel to the z-axis. The spin reversal means the adsorption of the atom and the totality of the states we obtain from Φ by all possible spin reversals defines Phase I. We may simply say that this phase is characterized by the ferromagnetic ordering of equivalent spins in the z-direction. Note, however, that there is no essential distinction between ferromagnetic and paramagnetic phases in the presence of magnetic field ($\eta \neq 0$). Actually the reversed spin is not localized, as we suppose in the molecular field approximation, but propagates through our system as a magnon. Thus adsorbed atoms are represented by magnons, which constitute an ideal Bose gas in the low-density limit. Phase I therefore corresponds to the region of x_g and T in which the He^4 monolayer behaves like a quantum (or classical) gas.

Phase II is characterized by the ferromagnetic ordering in a direction which makes a finite angle θ with the z-axis. So, operators $S_{j\pm} = S_{jx} \pm i S_{jy}$, or equivalently Bose operators a_j, $a_j^\dagger$, have non-vanishing expectation values and therefore our system is a superfluid. In fact, we can show that this tilted ferromagnetic ordering is equivalent to the Bose condensation of magnons and therefore of He^4 atoms. The state Ψ in which all spins are parallel to the θ-direction is related with the vacuum Φ as

$$\Psi = \prod_{j=1}^{N} \{u + v S_{j-}\} \Phi \tag{10}$$

where $u = \cos\theta/2$, $v = \sin\theta/2$. On the other hand, the state Φ_n, in which n magnons with vanishing wave vector are excited, is proportional to $[S_-(0)]^n \Phi$. Here $S_-(\boldsymbol{k})$ is the Fourier transform of S_{j-}. Expanding the product on the right of (10), we obtain

$$\Psi = \sum_{n=0}^{N} W_n^{\frac{1}{2}} \Phi_n, \qquad W_n = {}_N C_n \, u^{2(N-n)} \, v^{2n}. \tag{11}$$

Since $N \sim 10^{15}$ or so, the binomial distribution W_n has a very sharp peak at the mean value $n = N v^2$, so that Ψ is asymptotically equivalent to the state of magnon condensation.

Energetically Phase II is favorable in comparison with I when η is small, so that the gain in the transverse exchange energy exceeds the loss in the Zeeman energy caused by tilting. In fact, according to the molecular field approximation, this phase appears when η is lower than $3(K+J)$ and $k_B T$ is lower than $(3/2)K$.

However, when η is still lowered and if the nearest-neighbor repulsion J is strong enough, there appears Phase III. This is the phase which is characterized by the spin ordering (4). Thus, as was described under 3° in § 2, the graphite lattice is regarded as consisting of three sublattices, and spins are antiparallel to the z-axis on one of them and parallel on the other two. According to the molecular field theory, this phase appears when $\eta < 3J$ and $k_B T$ is lower than a critical value of the order of J.

Another interesting possibility is the supersolid phase, in which adsorbed atoms have not only the positional ordering, but also the superfluid long range order. This phase is characterized by the spin ordering in which spins are antiparallel to the z-axis on one sublattice, parallel to some θ-direction on the second sublattice, and parallel to the θ-direction on the third. So this phase is realized only when K is negative (the band minimum at the zone corner). This example shows that possible types of ordering depend upon the details of Fourier transforms $J(\boldsymbol{k})$, $K(\boldsymbol{k})$.

Anyway, for given parameters, we can even draw phase diagrams on ηT or $x_g T$ planes to show phase boundaries predicted by the molecular field approximation. In our $2D$ problem, however, the effect of fluctuation is so great that this approximation has only qualitative significance. As I have emphasized in § 3, for instance, the superfluid long-range order cannot actually exist in our $2D$ system at finite temperatures. So, if we take into consideration the fluctuation, Phase II predicted by the molecular field approximation will disappear. The phase boundary predicted by the molecular field approximation between phases I and II, for instance, merely indicates a region in which the short-range order may bring about some anomalous properties such as a broad heat-capacity peak.

B. Spin-Wave Approximation

At $T=0$, the molecular field approximation reduces to the classical theory, in which S_j are classical vectors and their ordering is determined by requiring a minimum of the total energy. The fluctuation from the ordered state may be described by the spin-wave theory. As I have already pointed out, the magnon which represents the fluctuation from the ferromagnetic ordering in the z-direction is nothing else but the adsorbed He atom in the normal gaslike phase (Phase I).

On the other hand, if we start from the tilted ferromagnetic ordering (Phase II) and apply the spin-wave theory to the fluctuation, we obtain a new sort of magnon, whose excitation energy is proportional to k in the long-wavelength limit. This magnon represents the quasiparticle in the superfluid phase. The fact that it is phononlike in the long-wavelength limit is an example of the Goldstone theorem [17]; in the superfluid phase the gauge symmetry of the Bose field is broken and this broken symmetry necessarily leads to a phonon-like excitation. The same applies also to the supersolid phase.

On the other hand, however, the average amplitude of the fluctuation associated with this Goldstone mode is divergent as $k \to 0$. In the case of $3D$ systems, the divergence is suppressed by the phase volume proportional to k^2 when we integrate over the wave vector. In the case of $2D$ systems at finite temperatures, there remains a logarithmic divergence, which implies the instability of the assumed superfluid long-range order.

If we start from the ordering (4) without the superfluid long-range order, the magnon energy takes the form const $+\alpha k^2$, so that there is no problem of instability. The magnon represents the defecton [12], which is the defect tunneling through our $2D$ crystal. It is interesting to notice that even at $T=0$ there exist finite amplitudes of spin waves, so that spins are not quite parallel to the z-axis. This zero-point precession of spins means the existence of defects even at $T=0$, as I have pointed out before.

C. Method of Green's Functions

Bogoliubov and Tyablikov [18] extended the spin-wave theory to finite temperatures by means of the method of two-time Green's functions. We start from the Green's function

$$G(k;t) = -\frac{i}{\hbar}\,\theta(t)\,\langle[S_+(\boldsymbol{k},t),\,S_-(-\boldsymbol{k},0)]\rangle \tag{12}$$

where $\theta(t)$ is the step function which changes from 0 to 1 at $t=0$. Hence

$$i\hbar\frac{d}{dt}G(\boldsymbol{k};t) = N\sigma + \theta(t)\left\langle\left[\frac{d}{dt}S_+(\boldsymbol{k},t),\,S_-(-\boldsymbol{k},0)\right]\right\rangle \tag{13}$$

where $\sigma = 2N^{-1}\langle S_z(0)\rangle$. When we insert the equation of motion into the second term on the right, we obtain higher-order Green's functions containing three spin operators. Bogoliubov and Tyablikov decoupled the chain of equations at this stage and replaced these higher-order Green's functions by approximate expressions only containing σ and $G(\boldsymbol{k};t)$. The parameter σ is determined self-consistently by means of the spectrum theorem which gives the general relation between σ and the Fourier transform $G(\boldsymbol{k},\omega)$ of $G(\boldsymbol{k};t)$ with respect to t. The same method can be applied to Phase I of our model. The magnon energy, which is defined by the pole of $G(\boldsymbol{k}, \omega)$, is given by

$$\varepsilon(\boldsymbol{k}) = \tfrac{1}{2} J(0)(1-\sigma) - \tfrac{1}{2}\sigma K(0) - \mu - \tfrac{1}{2}\sigma[K(\boldsymbol{k}) - K(0)]. \tag{14}$$

So, under the approximation $K(\boldsymbol{k}) - K(0) \cong -(\hbar^2/2m^*)\, k^2$, our model behaves like an ideal Bose gas. In fact, we can prove that in our $2D$ model the heat capacity is monotonously decreasing with decreasing temperature in spite of the fact that the effective mass m^*, being inversely proportional to σ, depends upon temperature in a complex way.

As I have emphasized before, it is physically obvious that the short-range order must be important to produce a heat-capacity peak. In fact, the reason why the Bogoliubov-Tyablikov decoupling worked in our $2D$ model is that we have finite η and therefore finite σ. In the absence of magnetic field, we are forced to go one step further beyond the Bogoliubov-Tyablikov decoupling. Then we automatically introduce two spin-correlation functions $\langle S_{jz} S_{lz}\rangle$, $\langle S_{jx} S_{lx}\rangle$ as parameters to be determined self-consistently. Recently Richards has applied this second-order decoupling theory to the one-dimensional Heisenberg antiferromagnet and successfully accounted for neutron data of $(CD_3)_4NMn\,Cl_3$ [19]. On the other hand, KONDO and YAMAJI [20] have applied the same method to the one-dimensional Heisenberg ferromagnet and obtained a broad heat-capacity peak around the Curie temperature predicted by the molecular field theory.

The same second-order decoupling theory can be applied to Phase I of our model to show the effect of transverse short-range order $\langle S_{jx} S_{lx}\rangle$. Mathematically, our problem is more complicated because we have a finite σ in addition to the two-spin correlation functions, which are anisotropic in our case. But the basic physics is the same as in the one-dimensional cases mentioned above and the only problem is how to fit experimental data quantitatively. The detail will soon be published elsewhere.

Acknowledgement

It is great pleasure for me to dedicate this article to Professor H. FRÖHLICH with my happy memories of the time which I spent as a young theorist in his Department.

My research work on the present subject has been done in the Department of Physics, University of Illinois, and I wish to thank Professor J. BARDEEN for his kind hospitality and illuminating discussions.

References

1. BRETZ, M., DASH, J. G.: Phys. Rev. Letters **26**, 963 (1971); **27**, 647 (1971).
   BRETZ, M., HUFF, G. B., DASH, J. G.: Preprint.
2. HICKERNELL, D. C., MCLEAN, E. O., VILCHES, O. E.: Phys. Rev. Letters **28**, 789 (1972).
3. CAMPBELL, C. E., SCHICK, M.: Phys. Rev. A**5**, 1919 (1972). The Bethe-Peierls approximation applied to a triangular Ising model.
4. HAGEN, D. E., NOVACO, A. D., MILFORD, F. J.: Int. Symposium on Adsorption-Desorption Phenomena. Florence: Academic Press 1971.
5. SCHICK, M., CAMPBELL, C. E.: Phys. Rev. A**2**, 1591 (1970).
6. MAY, R. M.: Phys. Rev. **135**, A1515 (1964).

7. Hohenberg, P. C.: Phys. Rev. **158**, 383 (1967).
8. Campbell, C. E., Dash, J. G., Schick, M.: Phys. Rev. Letters **26**, 966 (1971).
9. Widom, A.: Phys. Rev. **176**, 254 (1968).
10. Rehr, J. J., Mermin, N. D.: Phys. Rev. B **1**, 3160 (1970).
11. Lee, T. D., Yang, C. N.: Phys. Rev. **87**, 404, 410 (1952).
12. Andreev, A. F., Lifshitz, I. M.: Soviet Phys. JETP **29**, 1107 (1969).
13. Namaizawa, H.: Progr. Theoret. Phys. (to be published).
14. Rev. Mod. Phys. **40**, No. 4 (1968).
15. Matsubara, T., Matsuda, H.: Progr. Theoret. Phys. **16**, 569 (1956).
16. Matsuda, H., Tsuneto, T.: Progr. Theoret. Phys. **46**, 411 (1970).
Mullin, W.: Phys. Rev. Letters **26**, 611 (1971); —Phys. Rev. A **4**, 1247 (1971).
17. Anderson, P. W.: *Concepts in Solids*. N.Y., Amsterdam, Benjamin, 1963.
18. Bogoliubov, N. N., Tyablikov, S. V.: Soviet. Phys. Doklady **4**, 589 (1960).
19. Richards, P. M.: Phys. Rev. Letters **27**, 1800 (1971).
20. Kondo, J., Yamaji, K.: Progr. Theoret. Phys. **47**, 807 (1972).

Comments on Statistical Model and Collective Modes of Nuclei

K. J. Le Couteur

The Australian National University, Canberra/Australia

In a book dedicated to Professor Fröhlich it seems appropriate to write on aspects of nuclear theory which owe much to the theory of superconductivity.

There has been a gradual increase in our understanding of the range of applicability of the statistical model of nuclear reactions, and with it a better knowledge of the nuclear level distribution and its dependence on angular momentum [1]. The basic quantities of the theory are the density of states P and density of levels ϱ

$$P(U,N,Z) = \Sigma_j (2J+1)\,\varrho(U,N,Z,J) \tag{1}$$

such that $P\,du$ is the number of states and $\varrho\,dU$ is the number of levels of spin J of the nucleus N, Z in the range dU of excitation energy.

In the simplest model, the nucleus is considered made up of individual fermions moving in a common potential well with energy levels a_s for neutrons, b_s for protons and occupation numbers n_s, z_s for these levels, so that the total energy is

$$E = \Sigma_s\, a_s n_s + b_s z_s = E_0 + U \tag{2}$$

where E_0 is the energy of the ground state and

$$N = \Sigma n_s, \qquad Z = \Sigma z_s. \tag{3}$$

The evaluation of ϱ is a combinatorial problem, technically solved by calculating [2] the grand partition function ϕ which gives the Laplace transform of the level density

$$\begin{aligned}\exp -\beta\,\phi(\beta,\mu,\nu) &= \sum_{N,Z} \int dU\,\varrho(U,N,Z)\,\exp\beta(\mu N + \nu Z - U)\\ &= \text{trace}\,\exp\beta(\mu N + \nu Z - H)\end{aligned} \tag{4}$$

where H is the nuclear Hamiltonian and the parameters β, μ, ν are respectively an inverse temperature and the Fermi levels for neutrons and protons respectively. The density P is recovered from (2) by inversion of the Laplace transform

$$P(U,N,Z) = \frac{1}{(2\pi i)^3} \iiint_{\gamma - i\infty}^{\gamma + i\infty} d(\beta\mu)\,d(\beta\nu)\,d\beta\,\exp -\beta(\phi + \mu N + \nu Z - U) \tag{5}$$

and evaluation of the integral by the customary method of steepest descent leads finally to BETHE's [3] result

$$P(U,N,Z) = e^s/(-2\pi\, dU/d\beta)^{\frac{1}{2}}$$

where s is the entropy, equal to the argument of the exponential in (5), evaluated at the saddle point.

In the simplest case of independent particle excitations, the evaluation of ϕ can be given explicitly

$$-\beta\phi = \Sigma_s \log\{1+\exp\beta(\mu - a_s)\} + \log\{1+\exp\beta(\nu - b_s)\}. \tag{6}$$

Eq. (4) remains valid for particles in interaction but evaluation of the partition function prevents great difficulties. Here theoretical nuclear physics has gained much from the theory of superconductivity [4].

The nuclear Hamiltonian may be written as

$$H = \Sigma_s(n_s a_s + z_s b_s) + H_{\text{pairing}} + H_{\text{residual}} \tag{7}$$

and then the effect of pairing interactions may be taken into account, in the B.C.S.-Bogoliubov approximation if the particle energies a_s, b_s are replaced by quasiparticle energies

$$a'_s = \{(a_s - \mu)^2 + \Delta^2\}^{\frac{1}{2}}, \qquad b'_s = \{(b_s - \nu)^2 + \Delta^2\}^{\frac{1}{2}} \tag{8}$$

with Δ calculated self consistently. With these modified energies the partition function is calculated from (6). Many calculations on these lines have been made [5], as well as earlier calculations with a rather simpler pairing model [6].

The pairing gap energy D_0 at zero temperature may be estimated from the pairing energy of the nuclear ground state and Δ falls to zero at a temperature $t \approx 0.3\,\Delta_0$ which turns out to correspond to an excitation energy below neutron binding energy [7]. Thus at neutron binding energy or above, we can deal with the ordinary levels a_s, b_s and use Eq. (6), modified perhaps by the effect of residual interactions. The calculation can be done analytically if the levels a_s are uniformly distributed in the region of importance, near the Fermi level, leading to the result [8]

$$P(U,N,Z) = \frac{1}{12}\left(\frac{6}{g}\right)^{\frac{1}{4}}(U+t)^{-\frac{5}{4}}\exp 2\left(\frac{\pi^2}{6}gU\right)^{\frac{1}{2}} \tag{9}$$

where g is the density of individual particle states over a range of about t on either side of the Fermi level. This approximation has proved adequate to collate a lot of experimental data [5, 8, 9] but is not good for nuclei near closed shells where the individual particle levels are widely spaced.

The shell effects have been successfully described by direct numerical summation of [6] over the individual Nilsson levels. Recent work by BABA [10] gives a good account of the level density at neutron binding energy by this method.

I want to point out that the individual Nilsson levels may be somewhat shifted in an excited nucleus and to give a rough estimate of the effect. This is an aspect of the residual interaction.

Consider the effect of surface waves on the nucleus deforming the surface to

$$R(\Theta\phi) = R_0\left(1 + \Sigma\gamma_{\mu\lambda}\, Y^\lambda_\mu(\Theta\phi)\right),$$

with oscillations described approximately by a collective Bohr Hamiltonian

$$H = \frac{1}{2}\,\Sigma\,(B_{\mu\lambda}|\dot\gamma_{\mu\lambda}|^2 + C_{\mu\lambda}|\gamma_{\mu\lambda}|^2)$$

giving frequencies $\omega_{\lambda\mu} = \sqrt{C/B}$ and mean square amplitude given by

$$C|\gamma_\mu|^2 = h\,\omega\,n = h\,\omega/(e^{h\,\omega/t} - 1)$$

where n is the number of quanta in the oscillator. Then the mean square displacement of the nuclear surface is

$$(|\Delta R|^2) = R^2\,\Sigma\gamma_{\mu\lambda}^2 = R^2\,\Sigma\,\frac{h\omega/C}{e^{h\,\omega/t} - 1}$$

where we have evaluated n assuming that the surface waves are in thermal equilibrium with the other nuclear excitations at temperature t. There is little contribution from complicated modes with large ω/t. Such effects were first calculated by BAGGE [11] and lead to a reduction of the nuclear Coulomb potential barrier which is important for the evaporation of charged particles from highly excited nuclei.

Now at neutron binding energy the nuclear temperature t is above the energy of the quadrupole vibrations, which are strongly excited and contribute approximately

$$(|\Delta R|^2) = R^2\,t\,\Sigma\,1/C = t\cdot\frac{5\,\pi\,r_0^2}{a_s}$$

where $a_s \simeq 16$ MeV is a measure of the nuclear surface tension derived from the term $a_s A^{\frac{2}{3}}$ in the empirical binding energy formula. Note that in this approximation the main quadrupole contribution to ΔR^2 does not depend on ω_λ, so the estimate may not be too bad.

At neutron binding energy, from the empirical result

$$U = A\,t^2/8$$

we find

A	t	$\sqrt{(\Delta R)^2}$	R	$\delta\beta$
64	1 MeV	r_0	$4\,r_0$	0.25
144	0.7 MeV	$0.8\,r_0$	$5\cdot 5\,r_0$	0.15

This is enough variation of β to move one appreciably along the Nilsson diagram, and the nucleus presumably averages over this range. This averaging will tend to reduce, but not remove, the differences between closed shell and other nuclei which would otherwise be calculated from an evaluation of the partition function using the Nilsson levels. It also tends to give the effect of a more uniform distribution between shells, where the Nilsson levels crowd together and cross. Extensive computations of these effects have been carried out by MORETTO [12].

Collective Mode Contribution to Level Density

We come now to the question of the contribution of the collective modes to the free energy and level density. HILL and WHEELER [13] were perhaps the first to draw attention to the fact that a slow, irrotational, incompressible collective velocity can be added to the particle motion in a nucleus

$$v = \nabla \phi, \qquad \nabla^2 \phi = 0$$

and would increase the total nuclear energy by the energy of that hydrodynamic motion. Accordingly, for an excited nucleus, the excitation energy U may be divided between particle and collective modes, to yield a resultant level density

$$P(N, Z, U) = \int \varrho_{\text{particle}}(N, Z u')\, du'\, \varrho_{\text{collective}}(N, Z, U - u').$$

Although in principle the collective modes are expressible in terms of particle motions, this may still be a valid approximation if they come at a lower energy than the individual particle model would suggest and it is known that in even nuclei collective modes appear at energies within the gap of the quasiparticle spectrum [14].

The simplest way to make the folding integral is to add the energies and entropies of the particle and collective motions at temperature t

$$\begin{aligned} U &= U_{\text{particle}} + U_{\text{surface}} \\ &= A\, t^2/f - t + \lambda A^{\frac{2}{3}} t^{\frac{7}{3}} \end{aligned}$$

where, in a Fermi gas model, f is related to the smoothed density g of individual particle states at the Fermi level by

$$A/f = \pi^2 g/6$$

and $\lambda A^{\frac{2}{3}} t^{\frac{7}{3}}$ is the contribution of the surface waves, of all polarizations, to the energy. This yields for the entropy [8]

$$S = 2(A\, u/f)^{\frac{1}{2}} + (3/4)\, \lambda\, (U f)^{\frac{3}{4}} - \log \pi g' t.$$

These contributions to the entropy from collective motion are not large; and cannot well be distinguished through their energy dependence. However, if it becomes possible to calculate the particle contribution with sufficient accuracy, comparison with experiment may reveal the presence of another small contribution to the entropy which might amount to 10% of the main term.

There remains the difficult question of the adequacy of the individual particle model. Many authors have developed perturbation theories for expansion of (4) in terms of the residual interaction between particles, with the conclusion that it seems more profitable to improve the common Hartree-Fock field in which the nucleons move than to expand (4) in terms of the interaction [15]. Nevertheless the residual interactions must lead to a spreading of the actual nuclear levels on either side of the positions given by the approximation used. However this effect is probably masked, by use of the method of steepest descent to evaluate the integral (5) for the density of states, so that BETHE's formula implies an averaging of the entropy over a range of order $\sqrt{(d\,U/d\beta)} = \sqrt{(2\,U t)}$ on either side of the excitation energy U.

References

1. Some review articles are
   Huizenga, J. R., Moretto, L. G.: Ann. Rev. Nucl. Sci. (1972).
   Bodansky, D.: Ann. Rev. Nucl. Sci. **12**, 79 (1962).
   Ericson, T.: Advan. Physics **9**, 425 (1960).
   Le Couteur, K. J.: Nuclear Reactions, Vol. 1, Chap. 7. Amsterdam: North-Holland 1959.
   Bloch, C.: Nuclear Physics, p. 305. New York: Gordon and Breach 1969.
2. Bethe, H. A.: Rev. Mod. Phys. **9**, 69 (1937).
   Lang, J. M. B., Le Couteur, K. J.: Proc. Phys. Soc. (London) A **67**, 585 (1954).
   Lang, D. W.: Nucl. Phys. **77**, 545 (1966).
3. Bethe, H. A.: Phys. Rev. **50**, 332 (1936).
4. Bardeen, J., Cooper, L. N., Schrieffer, J. R.: Phys. Rev. **103**, 1175 (1957).
5. Le Couteur, K. J., Lang, D. W.: Nucl. Phys. **13**, 32 (1959).
   Sano, M., Yamasaki, S.: Progr. Theoret. Phys. (Kyoto) **29**, 397 (1963).
   Moretto, L. G.: Nucl. Phys. A **185**, 145 (1972).
6. Lang, D. W., Le Couteur, K. J.: Nucl. Phys. **14**, 21 (1959).
   Rao, Lakshmana, Rao, B. V. Tirumala, Prasad, P. Rama, Rao, J. Rama: Nuovo Cimento **65** B, 100 (1970).
7. Lang, D. W.: Nucl. Phys. **42**, 353 (1963).
8. Lang, J. M. B., Le Couteur, K. J.: Ref. 2.
9. Lang, D. W.: Nucl. Phys. **26**, 434 (1961).
10. Baba, H.: Nucl. Phys. A **159**, 625 (1970).
11. Bagge, E.: Ann. Phys. (Lpz.) **33**, 389 (1938).
12. Moretto, L. G.: Nucl. Phys. A **182**, 641 (1972).
13. Hill, D. L., Wheeler, J. A.: Phys. Rev. **89**, 1102 (1953).
14. Rowe, D. J.: Nuclear Collective Motion. London: Methuen 1970.
15. Balian, R.: Many Body Problem, p. 133. New York: Academic Press 1962.

Classical Aspects of Spontaneous Emission

ALBERT ROSE

RCA Laboratories, Princeton, N. J./U.S.A.

Abstract

Spontaneous emission is frequently ascribed to the action of zero point vibrations on an excited electron. If this were so, there would be no classical parallel since classical media do not have zero point vibrations. In contrast to the model using zero point vibrations, it is shown that spontaneous emission by a moving electron in a solid can be accounted for in terms of a classical interaction between electron and solid. While the mechanism of the interaction is classical, the evaluation of its magnitude is subject to certain quantum constraints.

1. Introduction

An excited electron relaxes towards lower energy states at a rate, according to the quantum mechanical formalism, proportional to $n+1$ where n is the density of quanta per mode in the field and the term unity is ascribed to "spontaneous emission". The action of the n quanta in the field, both in exciting the electron to higher energy states and in stimulating the emission of energy by an electron, has a natural classical parallel in the action of periodic forces on resonant systems. These forces can, depending on their phase, either excite or damp the vibrations of the resonant system. This leaves the question of whether "spontaneous emission" also has a classical equivalent. The answer should be "obviously, yes" as we will point out below. The literature, however, is far from unanimous in supporting this answer.

Several examples of the interpretations of spontaneous emission are worth citing here. At the outset, the term "spontaneous" was BOHR's translation of EINSTEIN's phrase "emission without excitation by external causes" [1]. Either description was certain to endow spontaneous emission with a certain mystery. Nor does the formalism of quantum mechanics offer any relief. Spontaneous emission is simply one of the consequences of this formalism. Quantum mechanics ignores any questions of "what causes it". Indeed, any answer to the question would belie the name "spontaneous".

A number of authors associate spontaneous emission with the presence of zero point vibrations. The following quotation is from SCHIFF [2] "From a formal point of view, we can say that the spontaneous emission probability is equal to the probability of emission that would be induced by the presence of one quantum in each state of the electromagnetic field. Now we have already seen that the smallest possible energy of the field corresponds to the presence of one-half quantum per state. This suggests that we regard the spontaneous emission as being induced by the zero-point oscillations of the electromagnetic field; note, however, that these oscillations are twice as effective in producing emissive transitions as are real photons and are of course incapable of producing absorptive transitions."

PARK and EPSTEIN [3] derive the Planck distribution by treating the energy of the zero point vibrations on a par with real quanta in their effect on stimulating emission. They avoid the factor of two discrepancy by assigning $1/2\, h\nu$ to each of the two "degrees of freedon" of the zero point vibrations. If the degrees of freedom are taken to mean kinetic and potential energy, then each mode of zero point vibrations will have an energy $h\nu$ rather than the $1/2\, h\nu$ that is usually assigned to the ground state. A recent tutorial paper by SCULLY and SARGENT [4] on "The Concept of the Photon" contains the phrase "... that is, the vacuum fluctuations 'stimulate' the atom to emit spontaneously". The quotes on stimulate are theirs.

The concept that spontaneous emission is caused by the action of zero point vibrations on the excited state has a simple logical consequence. Since classical physics does not contain zero point vibrations, classical physics can not account for spontaneous emission. Spontaneous emission must then be one of the novelties introduced by quantum mechanics and must be completely outside the province of classical physics.

In contrast to these conclusions, we believe that classical physics does provide the physical mechanism for spontaneous emission. This is simply the interaction of an electron with its surrounding medium. Moreover, a semi-classical analysis of this interaction yields the correct average rate of radiation of energy by the electron to its surround. The term semi-classical means here that the classical interaction between electron and medium is retained *as the mechanism or physical cause for spontaneous emission* but that the size of the electron is given by its quantum mechanical uncertainty radius, $\hbar/mv$, where m is, in the case of solids, the effective mass of the electron.

While classical physics provides the physical mechanism for spontaneous emission, and a semi-classical analysis gives the average rate of emission, the actual detailed emission in the form of quanta of energy radiated stochastically in time and space is purely a quantum mechanical phenomenon and has no generally agreed classical source. Stated in other terms, causality, in the deterministic sense, is preserved classically in the average and violated (or not understood) in the fine. This is different from stating that the phenomenon of spontaneous emission is completely outside the sphere of classical physics.

2. Spontaneous Emission in Solids

The quotations from the literature, which we cited, all had in mind the effect of zero-point vibrations in vacuum on causing the spontaneous emission of photons by excited electrons. The quantum mechanical formalism for the spontaneous emission of phonons by energetic electrons in solids completely parallels that for

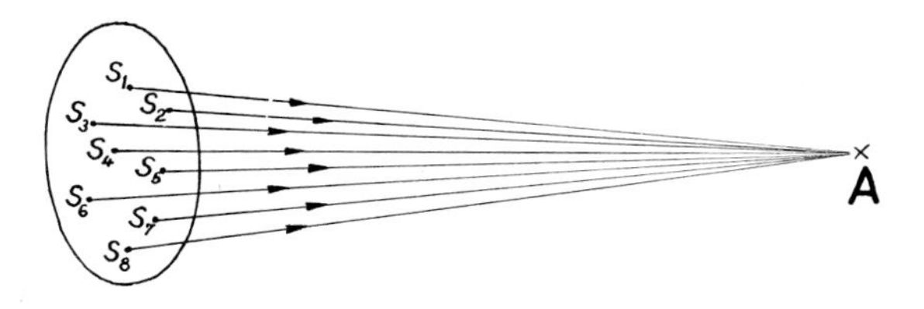

Fig. 6.
Scheme of electric field strength produced at an outside point by a small source of light. S_1, S_2, ... points of light source

Thus in general at a fixed point, because of the random distribution of the phases, one expects an excess

$$N_+ - N_- = \alpha \sqrt{N}.$$

Therefore we expect at a point A taken at a fixed instant of time

$$E_0 \Sigma \cos \varphi_k = \alpha \sqrt{N} E_0$$

$|\alpha|$ order of unity.
The resulting intensity is proportional to the square of the field strength, thus

$$E^2 = \alpha^2 N E_0^2,$$

where E_0^2 is proportional to the intensity of radiation emitted by one atom. The total intensity is proportional to the product of the intensity of one atom and to the number of independent atoms. Since the value of α varies from point to point, we find that the intensity shows a strong spatial fluctuation and the waves also show a strong fluctuation in time. From the law of the conservation of energy, it follows that the average value of α^2 is unity, i.e. the intensity of the beam fluctuates around the average value which would have been obtained if the intensities emitted by individual atoms could be summed without considering interference. The interference, however, causes rapid fluctuations of intensity and these fluctuations cause changes proportional to the intensity itself.

The Fluctuation Experiments

We have repeated the coincidence experiments described above with significant modifications [16].

Recording the coincidences with an arrangement of resolving time τ, we obtain the fluctuations averaged over this period. Thus if τ is much larger than the period of the fluctuations, then the fluctuations are smoothed out. In the original arrangement, when we had $\tau \sim 10^{-6}$ sec, the effect of the fluctuations was not noticeable for this reason. Similarly, if the beam falls on the cathode, the fluctuations on the area of the illuminated spot are not necessarily in phase with each other. If the fluctuations on the various points are incoherent, then the effect of fluctuations is smoothed out, as the average illumination is responsible for the emission of photoelectrons.

In order to obtain observable fluctuations, we used an arrangement with resolving time of the order of 10^{-10} sec and at the same time we used optical apparatus of high quality so as to ensure that the illuminated spots on the PM cathodes were coherent regions.

The detailed theory of the fluctuations of a beam of light states that the period of fluctuations is of the same order as the time the coherence length takes to pass through a point. The latter considerations are quite general, and the result does not depend on how the coherence length of the beam comes about. In particular, they are valid if the coherence length is determined by the mode of emission of the single atoms, but the result is also valid if the coherence length of the beam is determined by the fact that the individual atoms emit different frequencies caused by their heat motion.

For a fixed value of τ it is therefore of advantage to have as long a coherence length as possible.

So as to obtain a beam with a long coherence length, we used special discharge tubes and used the 5570 Å line of crypton.

The coherence length of a beam can be observed with the help of a Michelson interferometer. Going into more detail, one can observe the decoherence curve of the light. Under decoherence curve we understand the plot of relative contrast of the interference pattern as a function of the difference of the two optical paths used in the interferometer. Thus one can measure the contrast k of the pattern as function of the difference Δl of the length of the arms of the interferometer; thus one finds empirically a relation of the form

$$k = k(\Delta l).$$

The mathematical theory of the phenomenon shows that there is a connection between the decoherence curve $k(\Delta l)$ and

$$\Delta \tau = \tau_{\text{eff}} - \tau, \tag{20}$$

the apparent increase of resolving time caused by fluctuations.

The results of a number of experiments are collected in the table below.

Table 1

Number of measuring runs		$\tau_{\text{eff}} - \tau$ measured	$\tau_{\text{eff}} - \tau$ calculated
a)	440	$54 \pm 3 \cdot 10^{-12}$ sec	$55 \pm 2 \cdot 10^{-12}$ sec
b)	634	$53 \pm 4 \cdot 10^{-12}$ sec	$48 \pm 2 \cdot 10^{-12}$ sec
c)	361	$0.4 \pm 3 \cdot 10^{-12}$ sec	0

Results of intensity correlation measurements. The table shows the values of the apparent increase of resolving time of the coincidence arrangement in the case of two series of measurements performed with coherent light beams (a) and (b). When using incoherent beams, no excess coincidences occurred (c).

The experimental results show that there is excellent agreement between the observed excess of coincidences and the estimated excess. This agreement is the more remarkable as the amount of the excess is calculated without making use of adjustable parameters. The excess is determined by measuring the rate of coincidences produced by coherent and incoherent beams. On the other hand, the excess is calculated from the empirically determined decoherence curve. As we see from the table, the two methods lead to substantially the same result. In particular, we note that, in the case where one of the beams is led by a roundabout way to the photocathode, the coherence disappears and the rate of coincidences reduces to that obtained from independent beams.

A Semi-classical Interpretation

The experiments described above can be interpreted in a semi-classical way. We may describe thus the electromagnetic wave in the classical way and suppose that the probability for the emission of a photoelectron from a surface element dS of the cathode in an interval of time dt is equal to

$$dp = \alpha E^2 \, dS \, dt, \tag{21}$$

where $\boldsymbol{E} = \boldsymbol{E}(\boldsymbol{r}, t)$ is the electric field strength in the beam.

The interference experiment described in this manner can be understood clearly. The electromagnetic wave separated by the semi-transparent mirror divides in accord with MAXWELL's equations. If the beams are reunited, then the rate of signals produced on the cathode of the recording PM tube is proportional to the intensity of the part of the pattern by which its cathode is illuminated. The interference pattern is produced by beams reflected on both mirrors and thus there is no difficulty in understanding the phenomenon in a straightforward way.

The first part of the coincidence experiment is accounted for supposing that the rate of signals obtained by the two PM tubes are both proportional to the intensities of the beams falling upon them. The emission of photoelectrons in accord with (21) takes place at *random times* and in particular there is no correlation between the times of emissions of photoelectrons on the two cathodes.

The periodic variation of E in time is so fast that, when recording with any resolving time τ, the probability of receiving a record during this time is

$$p = \int_S \int_0^\tau E^2 \, dS \, dt \tag{22}$$

constant in time in a very good approximation. Therefore the signals can also be regarded as if the cathodes were subjected to a constant excitation to which they yield at random times by emitting a photoelectron. The instants at which one or the other of the cathodes are thus led to make emissions are independent of each other; for this reason the signals show no more coincidences than the rate of expected random coincidences.

The position is more complicated if we take into consideration the time fluctuations of intensity. In the latter case the vector of the electric field can be described as

$$E(t) = A(t) \cos(\omega t + \varphi(t))$$

where $A(t)$ and $\varphi(t)$ are both slowly varying functions of t. Since the electric vectors falling on the two cathodes show the same time dependence, the probability of emitting photoelectrons in a short interval is more than the random

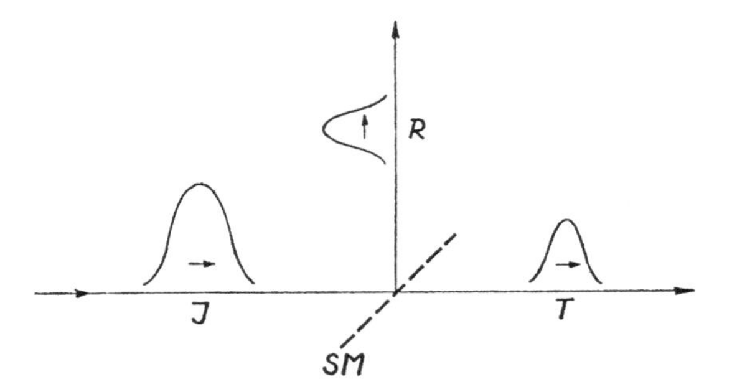

Fig. 7.
Scheme of an electromagnetic pulse being divided by the semi-transparent mirror

probability. Indeed, if $A(t)$ has a particularly high value in a period between t and $t+\tau$, then the probability of emitting photoelectrons during this period in both of the cathodes is more than the average probability. The detailed calculation shows that the fluctuations should in this way lead to a rate of coincidences which can be expressed by a $\tau_{\text{eff}} > \tau$ in accord with Eq. (21).

As we have shown in the previous section, the observed results support this picture.

Summarizing the above considerations, we can roughly say this. If, by the random fluctuation of the intensity of a beam, an intensity exceeding the average is produced somewhere (see Fig. 7), then this maximum is divided by the semi-transparent mirror into two parts which in their turn fall on the two cathodes simultaneously. While these maxima are being absorbed, both cathodes show more than average probability of emitting photoelectrons. In this manner the fluctuating beam produces more coincident signals than a beam of constant intensity would do.

It is important to note that the maxima thus described are not necessarily packets which according to the classical picture contain energy corresponding to several photons. If we calculate the total energy of such a packet, then we find in general that

$$u/h\nu \ll 1.$$

Thus the coincidences are not caused by events where during a short time interval the cathodes are reached by more than average numbers of photons. The excess coincidences are caused at times when, expressed in terms of photons, the probability of a photon reaching the cathode exceeds the average probability.

The Question of Conservation of Energy

The phenomenological description of the results of the experiments gives an accurate description of the observed results. It seems, however, to contradict the law of conservation of energy as it supposes that photoelectrons are emitted at certain intervals although the electromagnetic energy absorbed during the latter may be much less than the energy of the photoelectron.

Whether or not we regard this phenomenon as a real violation of the conservation law is a question of how we define energy. In the quantum-mechanical description of the phenomenon we can define as energy certain quantities which obey conservation. These quantities are certain averages of the quantities used in the semi-classical description.

Indeed, the increase in the rate of coincidences can also be understood by supposing the beam to consist of photons. Consider a short section of the beam of a certain intensity so that the expected number of photons

$$u/h\nu = a$$

where u is the total energy contained in that section. If the beam is split into two components, then the expected numbers of photons in the components are a_1 and a_2 with

$$a_1 + a_2 = a;$$

the probability that one of the beams contains n_1 and the other n_2 photons is thus:

$$p(n_1, n_2) = e^{-a} a_1^{n_1} a_2^{n_2}/(n_1!)(n_2!),$$

and the probability of both beams giving a signal is given by

$$P_K = \sum_{n_1 n_2} p(n_1, n_2)(1-P_1)^{n_1}(1-P_2)^{n_2} = (1-e^{-aP_1})(1-e^{-aP_2}),$$

where P_1 and P_2 are the probabilities that the two PM tubes will respond when absorbing a photon. Thus if $a \ll 1$, then

$$P_K \approx a^2 P_1 P_2.$$

If a changes in time then the rate of coincidences is obtained as

$$K = P_1 P_2 \frac{1}{T} \int_0^T dt \int_{-\tau}^{+\tau} a(t+t')\, a(t)\, dt'. \tag{23}$$

If there is a noticeable correlation between the values of $a(t)$ in an interval of time τ, then the rate of coincidences shows an excess over the rate calculated assuming constant intensities. Indeed, writing

$$N_1 = \frac{P_1}{T} \int_0^T a(t)\, dt \qquad N_2 = \frac{P_2}{T} \int_0^T a(t)\, dt \tag{24}$$

for the average rates of pulses, one finds from (23) and (24) in the case of positive correlation

$$K > 2 N_1 N_2 \tau. \tag{25}$$

Thus the excess of the rate of coincidences observed can be explained in terms of the tendency of the photons to crowd into regions where the amplitude of the wave is large. Alternatively, if we treat the emission of photoelectrons in terms of classical quantum theory, then we can suppose the electrons of the cathodes to be described by a wave function ψ.

The electromagnetic wave falling on the cathode can be taken to be a perturbation, and as the result of this perturbation the wave function ψ shows a perturbation which is caused by the incident waves. The perturbed wave function contains components corresponding to the photocurrent emerging from the cathode. The photocurrent is expected to be modulated in accordance with the slow intensity variations of the incident beam. Thus, taking $|\psi|^2$ as the probability density of the position of the electron, the latter at a point outside the cathode will vary in time in a manner correlated to the intensity fluctuation of the incident beam.

The varying current density emitted by the cathode falling on the further dinods of the multiplier produces the electron avalanche and in the end a macroscopic signal.

Thus, the probability flow of electrons falling on the amplifying system produces macroscopic signals at certain times: the signals arise with higher probability at times when the amplitude of the ψ, wave is larger than at times when it is smaller. The latter process is met with whenever we analyse counting experiments in which elementary particles trigger a macroscopic signal.

In such processes we always meet the same conceptual problem as to how a "probability wave" can produce real macroscopic processes at certain instants of time. One can choose between the various pictures as to which gives the most plausible description of the process.

Remarks on the Dual Nature of Light

The problem of the dual nature of light has had a peculiar history. The problem was several times declared to be cleared up; nevertheless, again and again new investigations were started on the subject. Interesting aspects of the question were discussed by BOHR, KRAMERS and SLATER [17] in their famous article. This article appeared in 1924 before the final formulation of quantum theory and therefore it was generally supposed that the problems raised in that article were all solved by the final formulation of the theory.

Nevertheless, there are authors who are not fully satisfied with the theories of photoeffect and interference phenomena. The questions which arise are twofold in nature.

DONTZOV and BAZ [18] were not satisfied with our own results which showed that interference patterns at very low intensity appear with exactly the same distribution as at higher intensities. Using modern methods, these authors repeated our experiments with intensities much lower than we had used and claimed to have observed a fading-out of interference patterns at the low intensities. These results were, however, apparently erroneous, as was shown by REYNOLD [19] and SCARL [20]; these two authors found, in agreement with our own results, that the interference patterns remain down to the lowest intensities.

Other experiments dealing with a similar subject were those of MANDEL [21]. His experiments showed that two laser beams are capable of producing interference patterns; this experimental finding is in accord with our results concerning the fluctuations of a macroscopic beam. Indeed, both experiments show that the radiations emitted by separate atoms are capable of interfering with each other if the relative phases of the emitters can be kept constant.

Another interesting line of thought, partly stimulated by the above experiments, emphasizes more and more the validity of the semi-classical interpretation of the phenomena thus considered.

The various phenomena can of course be dealt with correctly by quantum electrodynamics; however, the work of e.g. ERNST [22] shows that the semi-classical interpretation of such experiments helps in finding the final formulation of quantum electrodynamics. LAMB [23] has drawn attention to the fact that the photon picture and the second quantization are not necessary parts of the interpretation of the photoeffect. It is possible to treat this and other similar effects by describing the electromagnetic field classically and introducing the quantum effects through the features of matter. Altogether the opinion is growing in broad scientific circles, that the classical features of certain phenomena are of greater importance than was thought to be the case some time ago.

Acknowledgement

I am indebted to GY. FARKAS and M. JÁNOSSY for help in preparing the manuscript.

References

1. TAYLOR, G. I.: Proc. Cambridge Phil. Soc. **15**, 114 (1909).
2. DEMPSTER, A. J., BATHO, H. F.: Phys. Rev. **30**, 644 (1927).
3. FABRIKANT, V., *et al.*: Dokl. Akad. Nauk. SSSR **66**, 185 (1949).
4. VAVILOV, S. I.: Die Mikrostruktur des Lichtes. Untersuchungen und Grundgedanken. Berlin: Akad. Verlag 1954.
5. CSILLAG, L., JÁNOSSY, M., NÁRAY, ZS., SALAMON, T.: Phys. Letters **27** A, 343 (1968).
6. CSILLAG, L., JÁNOSSY, M., NÁRAY, ZS.: Acta Phys. Acad. Sci. Hung. **32**, 275 (1968).
7. JÁNOSSY, M., NÁRAY, ZS.: Phys. Letters **29** A, 479 (1969).
8. ADÁM, A., JÁNOSSY, L., VARGA, P.: Ann. Physik **16**, 408 (1955); —Acta Phys. Acad. Sci. Hung. **4**, 301 (1955) [in Russian].
9. BRANNEN, E., FERGUSSON, H. J.: Nature **177**, 481 (1956).
10. JÁNOSSY, L., NÁRAY, ZS.: Acta Phys. Acad. Sci. Hung. **7**, 403 (1957).
11. JÁNOSSY, L., NÁRAY, ZS.: Nuovo Cimento, Suppl. **9**, 588 (1958).
12. HANBURY-BROWN, R., TWISS, R. Q.: Proc. Roy. Soc. (London), Ser. A **242**, 300 (1957).
13. LORD RAYLEIGH: Theory of Sounds (1894) see e.g. edition New York, 1945.
14. WOLF, E.: Nuovo Cimento **12**, 884 (1954).
    PURCELL, E. M.: Nature **178**, 1449 (1956).
    MANDEL, L.: Proc. Phys. Soc. (London) **72**, 2037 (1958).
15. JÁNOSSY, L.: Nuovo Cimento **6**, 111 (1957); **12**, 369 (1959).
16. FARKAS, GY., JÁNOSSY, L., NÁRAY, ZS., VARGA, P.: Acta Phys. Acad. Sci. Hung. **18**, 199 (1965).
17. BOHR, N., KRAMERS, H. A., SLATER, J. C.: Phil. Mag. S **6**, **47**, 785 (1924).
18. DONTZOV, Y. P., BAZ, A. I.: Zh. Eksperim. i Theor. Fiz. English transl.: Soviet Phys. JETP **25**, 1 (1967).
19. REYNOLDS, G. T., K. SPARTALIAN, D. B. SCARL: Nuovo Cimento **61** B, 355 (1969).
20. SCARL, D. B.: Single Optical Photon Interference. Preprint (1969).
21. MANDEL, L.: Proc. of the Symposium on Modern Optics, p. 143.: Polytechnic Press 1967.
22. ERNST, V.: Z. Physik **229**, 432 (1969).
23. LAMB, W. E., SCULLY, M. O.: Polarization, Matter and Radiation (Jubilee volume in honor of ALFRED KASTLER) p. 363. Paris: Press Univ. de France 1969.

Divergent Integrals in Scattering Theory

R. HUBY

Department of Theoretical Physics, University of Liverpool, U.K.

Introduction

The writer has to thank the catholicity of Herbert FRÖHLICH's interests for his introduction to nuclear theory. While the impetus thus given has carried him to regions of reaction theory apparently remote from most cooperative concepts, it is possible to trace one of those unifying threads dear to physicists from the fact that all kinetic theory rests ultimately on the scattering of many bodies, and the understanding of many-body scattering has to start from the difficulties of three-body scattering, within which lie some of the divergences that form the subject of this article.

Scattering theory is riddled with conceptual problems and "improper" mathematical methods, which, however, have been progressively brought under control. An underlying cause can be seen in the continuous nature of the positive energy spectrum, with its eigenfunctions which do not lie in Hilbert space, but which must be normalised with a delta function. That basic impropriety has been disposed of rigorously by such developments as the theory of distributions; and generally one believes that, since the scattering amplitude is physically finite, there must be discoverable proper ways of calculating it, even though at first a straightforward manipulation of the equations may run into trouble. There are two ways of dealing with an improper mathematical expression when encountered: either to "go round about" by transforming it into a form which has no improper aspect, or to attack it "head on" by finding a valid prescription for handling the expression as it stands—by "regularising" it. The present application of these remarks is to a range of scattering problems recently studied in which divergent integrals have appeared, and it has been found profitable to tackle them head on. A typical divergent integral* is

$$I = \int_0^\infty e^{ikr}\, dr, \tag{1}$$

and the head-on approach is to find a prescription for assigning a finite value to this, analogous to the prescriptions for summing divergent series in the classic book of HARDY [1]—which does also to some extent cover divergent integrals.

* I follow HARDY [1] in describing an integral whose integrand is oscillatory at infinity as "divergent".

The most obvious device, for instance, is to use a "convergence factor", e.g.

$$I = \operatorname*{Lim}_{\alpha \to 0} \int_0^\infty e^{(ik-\alpha)r}\, dr = \frac{i}{k}, \tag{2}$$

which is known as the "Abel" method of regularisation; but such devices must be justified in the physical context*.

It will be shown that these and similar divergent integrals have arisen from what are apparently two different kinds of origin, viz. three-body channels on the one hand and the treatment of resonances on the other, but both of these have coalesced in some calculations of deuteron stripping to unbound levels. I shall describe some of this, without perhaps being able to disclose any completely unifying pattern.

Three-body Final States

Scattering problems involving three-body channels are well known to pose much greater difficulties of convergence than two-body ones, and indeed all attempts at obtaining reliable knowledge about three-body scattering by, say, the Lippmann-Schwinger equation failed, until the seminal work of FADDEEV which was achieved by going "round-about" to a new set of equations. A problem which contains the general features, and which is of particular practical interest, is that of deuteron stripping, for which I shall take an extremely oversimplified model. Let the deuteron consist of two nucleons n and p (spinless), and let the heavy target be represented by a fixed potential field V_n for the neutron only: in studying the (d, p) reaction one may use a model in which the proton has no interaction with the target.

The Hamiltonian is thus

$$H = t_n + t_p + V_{np}(|\boldsymbol{r}_n - \boldsymbol{r}_p|) + V_n(r_n), \tag{3}$$

where V_{np} is the force which binds the deuteron, and the t's are kinetic energies. Suppose the neutron has a bound state ϕ_n in the potential field:

$$(t_n + V_n - \varepsilon_n)\, \phi_n = 0. \tag{4}$$

Then a (d, p) reaction can take place leaving the neutron in this bound state, i.e. a transition occurs to a two-body final channel. The amplitude is given by standard scattering theory as a T-matrix element for which there are two alternative, exact expressions, containing respectively the "prior" interaction V_n and the "post" interaction V_{np}, (with suitable normalisation),

$$\begin{aligned} \langle \phi_n, \boldsymbol{k}_p |\, T\, | \phi_d\, \boldsymbol{k}_d \rangle &= \langle \psi^{(-)}_{n, \boldsymbol{k}_p} |\, V_n\, | \phi_d, \boldsymbol{k}_d \rangle \\ &= \langle \phi_n, \boldsymbol{k}_p |\, V_{np}\, | \psi^{(+)}_{\boldsymbol{k}_d} \rangle, \end{aligned} \tag{5}$$

where $|\boldsymbol{k}_p\rangle$, $|\boldsymbol{k}_d\rangle$ stand for proton and deuteron plane waves with the observed wave-numbers, ϕ_d is the deuteron's internal bound state, $\psi^{(+)}_{\boldsymbol{k}_d}$ is an exact eigen-

* Any such device must of course be such that, when applied to an already convergent integral, it leaves it unaltered.

state of H induced by the deuteron incident wave $|\phi_d, \boldsymbol{k}_d\rangle$, and $\psi^{(-)}_{n,\boldsymbol{k}_p}$ is an exact "anti-scattering" eigenstate of H induced by the plane wave $|\boldsymbol{k}_p\rangle$ incident on the neutron bound-state ϕ_n. Writing out the post expression in full, for example,

$$\langle\phi_n, \boldsymbol{k}_p|\, T\, |\phi_d, \boldsymbol{k}_d\rangle = \int e^{-i\boldsymbol{k}_p\cdot\boldsymbol{r}_p}\, \phi_n^*(\boldsymbol{r}_n)\, V_{np}(|\boldsymbol{r}_p-\boldsymbol{r}_n|)\, \psi^{(+)}_{\boldsymbol{k}_d}(\boldsymbol{r}_n, \boldsymbol{r}_p)\, d\boldsymbol{r}_n\, d\boldsymbol{r}_p.$$

This integral clearly converges rapidly as r_n and r_p become large, and the same is true of the prior integral. In practical applications approximations must be made for the unknown exact wave functions $\psi^{(+)}_{\boldsymbol{k}_d}$ or $\psi^{(-)}_{n,\boldsymbol{k}_p}$ and "distorted wave" refinements may be introduced, etc. However, deuteron break-up can also occur (provided the beam energy is larger than the deuteron binding energy), i.e. there are transitions to three-body channels. The scattering amplitude for such a transition, issuing in wave numbers $\boldsymbol{k}_p$, $\boldsymbol{k}_n$ would be given, in analogy to the prior and post formulae for the bound-state transition, Eq. (5), by

$$\begin{aligned}\langle\boldsymbol{k}_n, \boldsymbol{k}_p|\, T\, |\phi_d, \boldsymbol{k}_d\rangle &= \langle\psi^{(-)}_{\boldsymbol{k}_n, \boldsymbol{k}_p}|\, V_n\, |\phi_d, \boldsymbol{k}_d\rangle \\ &= \langle\phi^{(-)}_{\boldsymbol{k}_n,}\boldsymbol{k}_p|\, V_{np}\, |\psi^{(+)}_{\boldsymbol{k}_d}\rangle,\end{aligned} \tag{6}$$

where $\phi^{(-)}_{\boldsymbol{k}_n}$ is an antiscattering wave function for the neutron in the potential V_n, i.e. a solution of an equation like (4) but with positive energy E_n, and $\psi^{(-)}_{\boldsymbol{k}_n,\boldsymbol{k}_p}$ is an exact eigenfunction of H induced by superimposed plane waves $|\boldsymbol{k}_n\rangle$ and $|\boldsymbol{k}_p\rangle$. Now while the prior integral here converges as before, the post integral does not. This may most clearly be seen in the zero-range approximation, in which $V_{np}\,\phi_d$ is taken proportional to $\delta(\boldsymbol{r}_p-\boldsymbol{r}_n)$. If for illustration we also use a plane wave approximation to $\psi^{(+)}_{\boldsymbol{k}_d}$, the post integral becomes

$$\text{const } x \int e^{-i\boldsymbol{k}_p\cdot\boldsymbol{r}}\, \phi^{(-)*}_{\boldsymbol{k}_n(\boldsymbol{r})}\, e^{i\boldsymbol{k}_d\cdot\boldsymbol{r}}\, d\boldsymbol{r}. \tag{7}$$

Since $\phi^{(-)}_{\boldsymbol{k}_n}$ contains a plane wave and scattered waves, this obviously does not converge. The conclusion is that any formal argument by which the post expression was derived must be defective. We could go round about by eschewing the use of the post integral altogether and working always with the prior one, but this is unsatisfactory because the post form has for various reasons been generally preferred in bound-state stripping calculations. The difficulty has been studied by several authors [2, 3, 4] with the conclusion that the post form is indeed correct if suitably regularised. Always the argument has started from the standpoint that the convergent, prior formula is correct, and this is subjected to transformations which replace it by a regularised form of the post integral.

One procedure used by HUBY and MINES [2] was to expand all the waves (e.g. the plane wave $|\boldsymbol{k}_p\rangle$) in partial waves of angular momentum (e.g. l_p). If we write down T-matrix integrals like those of Eq. (6), but now take for each wave function a single partial wave, then it is found that the partial-wave integrals of post-type are actually convergent, as well as those of prior type. The approximation of Eq. (7) affords an illustration. The proton factor here yields a partial wave which at large distances has the radial factor $r^{-1}\sin(k_p r - \frac{1}{2}\pi l_p)$. The neutron and deuteron waves contribute similar factors, so that the radial partial-wave integral consists a sum of terms with the long-distance form

$$\text{const} \int^{\infty} \frac{e^{\pm ikr}}{r}\, dr, \tag{8}$$

which converges, though only very slowly. We can say that the original, divergent integral has been regularised by expanding the integral in an infinite series (of partial waves) and integrating term by term. We have interchanged the order of summation and integration; and whereas the new procedure of integration followed by summation is convergent, the original, opposite order was not. The partial-wave procedure can be shown to be correct, and indeed it is the natural one for computation, for also in stripping to bound states one has always worked with partial waves. In this sense we can say that the apparent divergence difficulty which originally faced us was illusory. Even so, the very slow convergence of the integrals (8) poses a computational problem, which as we shall see has now been circumvented.

Nevertheless, some work has been done to show that the post integral of Eq. (6) can be regularised more directly. HUBY and MINES [2] introduced an Abel-type convergence factor

$$\lim_{\alpha\to 0} \langle \phi^{(-)}_{\boldsymbol{k}_n}, \boldsymbol{k}_p | \, e^{-\alpha r_n} V_{np} \, | \psi^{(+)}_{\boldsymbol{k}_d} \rangle. \tag{9}$$

However, in their proof (which actually only applied somewhat less generally) this followed at one remove after an initially different form of regularisation. In the latter the integral of Eq. (6) over r_n was first taken out only to a large but finite distance R, but the result was then averaged over R with a weighting function which in the limit only counts infinitely large values of R. This is the Cesàro method of regulalisation which has recently been used again by LIEBER et al. [4]. The Abel and Cesàro methods can be proved equivalent [2, 8]. VINCENT [3] derived Eq. (9) more directly by a transformation of the prior integral in Eq. (6).

In practical Distorted Wave Born Approximation (DWBA) calculations of stripping to unbound states, partial-wave integrals of V_{np} as in Eq. (6) must be computed, but approximate wave functions must be inserted for the deuteron, proton and neutron calculated in their respective potentials, which may contain Coulomb as well as nuclear parts. The very slow convergence of these integrals over r (or r_n) calls for means of expediting the convergence which are similar in spirit to the regularisation of a divergent integral. MINES [5] did this by inserting in the partial-wave integral a convergence factor similar to that in Eq. (9), calculating the result for finite values of α, and extrapolating to $\alpha=0$. VINCENT and FORTUNE [6] gave (with proof) a very powerful method, by deforming the partial-wave integral over r into a contour integral in the complex r-plane. To start with, the integral is carried out over real r from the origin to some point R outside the range of the nuclear potentials. Beyond R each wave function is a linear combination of spherical Hankel functions (or their Coulomb counterparts), and the entire integral is now broken up into a sum of terms in which the overall oscillation becomes asymptotically either purely *outgoing* or purely *incoming*, i.e. the integrals are asymptotically either

$$\text{const} \int^{\infty} \frac{e^{+ikr}}{r}\, dr \qquad \text{or} \qquad \text{const} \int^{\infty} \frac{e^{-ikr}}{r}\, dr, \tag{10}$$

with some positive k (neglecting Coulomb phases). These two groups of terms are now integrated separately. For the former the contour is turned through a right angle into the upper half-plane, and integrated by quadrature over $r=R+iy$

for y from 0 to ∞. For the incoming terms correspondingly the contour is turned into the lower half plane, and integrated over $r=R-iy$ for y from 0 to ∞. In both cases the integrand decays exponentially in y. Loosely we might say that this is regularisation by contour deformation.

So much for divergence problems caused by three-body final channels. However, $a(d, p)$ reaction of this kind is in practice only interesting when the final neutron state $\phi_{k_n}^{(-)}$ represents a sharp resonance or "unbound level", and one wishes to exhibit the close resemblance between this transition and stripping to a bound level. This calls for an appropriate representation of a resonance wave-function $\phi_{k_n}^{(-)}$, and leads on to the study of resonances.

Resonances

While resonances form one of the oldest and most extensive subjects of nuclear theory, I shall review just some aspects which have impinged upon stripping to unbound levels and have involved divergent integrals. One long-standing approach is to associate an unbound level, or resonance, with a Gamow state, that is, an eigenfunction of the Hamiltonian at a discrete, complex energy $W_s=E_s-\frac{1}{2}i\Gamma_s$, obeying the condition that the wave function is finite at all finite points of space, and has purely *outgoing* oscillatory behaviour at infinity. For a particle in a central potential, writing the radial factor in the wave function $r^{-1}u_s(r)$, we have asymptotically

$$u_s(r) \sim \text{const} \exp i(k_1-ik_2)r, \quad (k_1, k_2 \text{ real}), \tag{11}$$

where the wave-number $k=k_1-ik_2$ is given by

$$W_s=E_s-\tfrac{1}{2}i\Gamma_s=\hbar^2(k_1-ik_2)^2/2m. \tag{12}$$

For such solutions k_2 is always positive, i.e. the wave function blows up exponentially at infinity, and it thus appears to be non-normalizable; but Zel'dovich [7] first pointed out that there is a very natural way of normalizing it by the integral of u_s^2, regularized through a suitable convergence factor:

$$\lim_{\alpha\to 0}\int_0^\infty e^{-\alpha r^2}\, u_s^2(r)\, dr=1. \tag{13}$$

The integral of $u_s^2(r)$ is clearly much more divergent than any we have considered before, but nevertheless Berggren [8] showed that the limit in Eq. (13) exists provided $k_2<|k_1|$. Berggren [8] developed this theory and, most importantly, used regularised integrals of this kind to prove that, for a wide class of arbitrary functions $\psi(r)$, the familiar expansion of $\psi(r)$ as a sum over the bound eigenfunctions of the Hamiltonian plus an integral over the continuum eigenfunctions, could be replaced by a formally very similar sum over the bound eigenfunctions plus the Gamow eigenfunctions [normalized according to Eq. (13)], plus an integral of eigenfunctions over a certain contour L^+ in the complex k-plane.

These results rest upon the existence of the limit in Eq. (13), a generalisation of which has been studied by Gyarmati and Vertse [9]. Consider

$$\lim_{\alpha\to 0}\int_0^\infty e^{-\alpha r^n}\, e^{i(k_1-ik_2)r}\, dr, \quad (n>1;\ k_1 \text{ and } k_2 \text{ real}). \tag{14}$$

This is found to exist, and to be independent of n, provided $k_2 < |k_1| \tan \pi/2n$. So, by using a suitable range of n, it is possible to regularise the infinite integral of $\exp ikr$ for any complex value of k_2 other than on the negative imaginary axis. The regularisation at infinity works as well if the integrand contains any power of r attached to the factor $\exp ikr$, and indeed if the integrand is instead a Coulomb wave function. GYARMATI and VERTSE presented an alternative expression for the regularised integral. When k_1 is positive, the convergence factor $\exp(-\alpha r^n)$ may be omitted, and one performs instead a contour integration in the complex r-plane (with the original integrand), taken along a path in the first quadrant which goes to infinity at any angle greater than $\tan^{-1}(k_2/k_1)$ to the real, positive axis. When k_1 is negative, the path must be similar but in the fourth quadrant. Now the contours prescribed by VINCENT and FORTUNE, as described earlier, are just of this kind. This means that the procedure of regularisation by contour deformation due to VINCENT and FORTUNE is meaningful even should the integrand contain asymptotically, besides oscillatory factors and powers of r, an exponentially increasing factor $\exp k_2 r$. However, in the first instance these results serve only to strengthen the existing resonance theory of BERGGREN, and we have now to show that they can be further used constructively.

First, however, note that the writer has developed a theory [10] in which a resonance is associated with a Weinberg function, which is an eigenfunction at real positive energy, in a complex potential, the function having purely outgoing behaviour at infinity, i.e. $u_s(r) \sim \exp ikr$ with k real. The natural normalisation is again formally the same as BERGGREN's Eq. (13), but now the divergence at infinity is much less acute (merely oscillatory).

Stripping to Unbound Resonant Levels

Let us return to the calculation of a partial-wave DWBA matrix element of V_{np} for a (d, p) reaction similar to Eq. (6), but using distorted waves for the deuteron and proton, and a resonant neutron wave-function. If the neutron resonance is approximated using a Weinberg function [10] no new divergence features arise (the integral remains slowly convergent). However, if we use BERGGREN's expansion of $\phi_{k_n}^{(-)}$, one term of this contains a Gamow wave function $u_s^*(r)$*, and near resonance this term acquires a large amplitude. BANG and ZIMANYI [11] did such a DWBA calculation, approximating the neutron wave function simply by a suitable multiple of $u_s^*(r)$. The integral is now strongly divergent, but regularisation as in Eqs. (13) or (14) is justified, because this is applicable to the original convergent integrand. BANG and ZIMANYI computed their integrals by quadrature using the convergence factor $\exp(-\alpha r^2)$ with finite α; but for small α the convergence must be painfully slow because of the exponential growth of the neutron wave function. However, we have seen that in such an integral the use of a convergence factor may be replaced by the complex-r integration method of VINCENT and FORTUNE. The latter could therefore now be applied to the stripping calculation using the Gamow wave function for the neutron resonance, and the ensuing quadratures over complex $r = R \pm iy$ would be just as rapidly, exponentially converging as when using any other approximation for

* The complex conjugate appears because the wave function is an anti-scattering one.

the neutron. However, the use of the Berggren expansion has the disadvantage that very little is known about the effect of the remainder term, expressible as an integral over the contour L^+ in the k-plane. Nevertheless, it is striking how effectively the divergent integrals arising in this complex of problems can be subdued by means of a battery of regularisation devices.

References

1. HARDY, G. H.: Divergent Series. Oxford: Clarendon Press 1949.
2. HUBY, R., MINES, J. R.: Proc. Conference on Direct Reactions and Nuclear Reaction Mechanisms, Ed. by CLEMENTEL, E., and VILLI, C., p. 530. New York: Gordon and Breach 1963; — Rev. Mod. Phys. **37**, 406 (1965).
3. VINCENT, C. M.: Phys. Rev. **175**, 1309 (1968).
4. LIEBER, M., ROSENBERG, L., SPRUCH, L.: Phys. Rev. D **5**, 1330 (1972); D **5**, 1347 (1972).
5. ALTY, J. L., GREEN, L. L., HUBY, R., JONES, G. D., MINES, J. R., SHARPEY-SCHAFER, J. F.: Nucl. Phys. A **97**, 541 (1967).
6. VINCENT, C. M., FORTUNE, H. T.: Phys. Rev. C **2**, 782 (1970).
7. ZEL'DOVICH, YA. B.: JETP **12**, 542 (1961).
8. BERGGREN, T.: Nucl. Phys. A **109**, 265 (1968); — Phys. Letters **37** B, 240 (1971).
9. GYARMATI, B., VERTSE, T.: Nucl. Phys. A **160**, 523 (1971).
10. HUBY, R.: Nucl. Phys. A **138**, 442 (1969); — Phys. Letters **33** B, 323 (1970).
11. BANG, J., ZIMANYI, J.: Nucl. Phys. A **139**, 534 (1969).

Perturbation Methods and Lagrange's Expansion

R. A. Sack

University of Salford, Department of Mathematics, Salford/England

Abstract

Explicit expressions are derived for the higher-order terms in the Rayleigh-Schrödinger perturbation expansions for non-degenerate eigenvalues on the basis of Lagrange's expansion for analytic functions of implicitly defined variables. The series for the perturbed eigenvectors, eigenvalues, analytic functions of the eigenvalues and expectation values are given in a variety of forms depending on the degree of factorization of the perturbed operators, of the projection operators, of the scalar coefficients and of the differentiations involved. Some, but not all, of these expressions are the same as, or equivalent to, series given by other authors. In the limit of complete reduction the sums involve products of matrix elements and derivatives of scalar quotients only. The linked cluster theorem is shown to be a trivial consequence of the additivity of eigenvalues for commuting operators.

1. Introduction

In recent years explicit expressions for the terms in the Rayleigh-Schrödinger expansions for discrete eigenvalues (EVs) and eigenfunctions (EFs) of perturbed linear operators

$$H = h + \varepsilon V \tag{1}$$

with analytically known solutions for the unperturbed operator h have been derived to arbitrary order in ε in a variety of forms. For nondegenerate Hamiltonian operators formulations for the general terms have been given primarily by Kato [1], Bloch [2] and Huby [3] who express their results in terms of projection operators, though many variants and alternative methods have been presented since then. Mostly the derivations are based on contour integration about a pole which is only implicitly defined, though some formulas are deduced as limiting expression of time-dependent perturbations, and a few on *ad hoc* procedures, in particular the establishment of a quadratic equation for the wave operator by Bloch [2] and Löwdin [4], which the former author solves by means of an infinite series and the latter as a continued fraction.

However the introduction of a time variable to derive results which refer to stationary states appears an unnecessary complication (at least from the logical point of view), and integrations in the complex plane can be avoided by means of a theorem, originally due to LAGRANGE (cf. e.g. [5, 6]) which expresses the result of such contour integrals in closed and usually in real form. LAGRANGE's theorem states that a function $F(z)$ of a variable z which is implicitly defined by

$$z = a + \varepsilon\,\phi(z) \tag{2}$$

where $\phi(z)$ and $F(z)$ are analytic in a domain of the complex plane containing $z=a$ and $\phi(a) \neq 0$, can be expanded within a certain circle of convergence for ε in a series

$$F(z) = F(a) + \sum_{1}^{\infty} \frac{\varepsilon^r}{r!} \frac{d^{r-1}}{d a^{r-1}} [\phi^r(a)\, F'(a)]. \tag{3}$$

An alternative expansion, apparently due to Hermite, is

$$\frac{F(z)}{1-\varepsilon\,\phi'(z)} = \sum_{0}^{\infty} \frac{\varepsilon^r}{r!} \frac{d^r}{d a^r} [\phi^r(a)\, F(a)]. \tag{4}$$

In particular, on putting $F(z)=z$, (3) and (4) yield inversion series for the determination of z itself.

The present writer was able, a number of years ago, to derive a number of equivalent perturbation formulas for non-degenerate zeroth order states on the basis of (2)–(4) in which the implicit definition of the perturbed EV corresponding to (2) was provided by the Brillouin-Wigner expansion, though this treatment has not so far appeared in print. Since then similar derivations have been given independently by DES CLOIZEAUX [7] and, on the writer's suggestion, by SILVERSTONE and HOLLOWAY [8]; these authors also consider the case of degenerate unperturbed states. Nevertheless I consider it useful to present this approach afresh, and to show its power, not only as it offers a neat way of systematizing the various formulations in terms of projection operators found in the literature, but also because the formulas can easily be re-expressed in terms of matrix elements and with only minor alterations can be adapted to perturbation problems more general than those given by (1). One of these, the case of multiple perturbations

$$H = h + \sum \varepsilon_l V_l \tag{5}$$

is considered explicitly in Section 6 of this paper; other generalizations will be presented in subsequent publications. Some useful formulas relating to the use of LAGRANGE's formula are summarized in Section 2; the notation used is explained in Section 3. The expansions utilizing projections are derived in Section 4; a number of transformations, including the expansions in terms of matrix elements, are discussed in Section 5. A new proof of the linked cluster expansion [9], [10] is given in the Appendix. The case of degenerate zero-order states is not treated in detail.

The paper is not intended as a review, and though ample comparison is made with the results of other workers in this field, it is not claimed that the bibliography is even approximately complete. The present treatment differs from most previous ones in that Hermiticity conditions are completely irrelevant in the derivations, and the results are valid for both Hermitean and non-Hermitean operators. The convergence of the expansions is only sketchily considered.

2. Formulas Relating to Lagrange's Expansion

As mentioned after (4), putting $F(z) \equiv z$ yields inversion formulas for z itself. Thus on putting D for (d/da) (3) becomes

$$z = a + \sum_1^\infty \varepsilon^r D^{r-1} [\phi^r(a)]/r!, \tag{6}$$

whereas substitution of $F(z) \equiv 1$ in (4) yields

$$B = [1 - \varepsilon \phi'(z)]^{-1} = \sum_0^\infty \varepsilon^r D^r [\phi^r(a)]/r!, \tag{7}$$

which when applied to (4) with $F(z) \equiv z$ and the use of LEIBNIZ' theorem leads to

$$z = a + B^{-1} \sum_1^\infty \varepsilon^r D^{r-1} [\phi^r(a)]/(r-1)!. \tag{8}$$

Note that the individual terms in the sum in (8) are obtained from the corresponding ones in (6) by multiplying by r, the order of the Lagrange inversion, and those in (7) by term-wise differentiation.

The writer has shown [11] that the various forms into which (3), (4) and (6)–(8) can be transformed by factorizing the derivatives of powers of ϕ into products of derivatives are most easily obtained by means of a generating function

$$\Xi(u, v; \varepsilon\phi) = \exp[u \varepsilon \phi(a+v)] = \sum G(r, S; \varepsilon\phi)\, u^r v^S. \tag{9}$$

The coefficients $G(r, S)$ enter into (3) and (4) by means of

$$F(z) = F(a) + \sum_{r=0}^\infty \sum_{S=0}^{r-1} G(r, S; \varepsilon\phi) \frac{(r-1)!}{(r-S-1)!} D^{r-S} F(a), \tag{10}$$

$$F(z) \cdot B = \sum_{r=0}^\infty \sum_{S=0}^{r} G(r, S; \varepsilon\phi) \frac{r!}{(r-S)!} D^{r-S} F(a). \tag{11}$$

The sums over S can be extended to infinity, but all such terms vanish in view of the factorial of negative argument in the denominator. For simplicity's sake we abbreviate

$$r!/(r-S)! = {}^r\gamma_S, \quad {}^{-1}\gamma_0 = 1. \tag{12}$$

With this definition the term $F(a)$ in (10) can be drawn into the sum, so that the individual terms in (10) differ from those in (11) only by the substitution of ${}^{r-1}\gamma_S$ for ${}^r\gamma_S$. Factorization of (9), when applied to (10) leads to *

$$F(z) = \mathsf{S}_{\boldsymbol{t}} \prod_{s=0}^\infty \left[\left(\frac{D^s \phi(a)}{s!} \right)^{t(s)} \frac{1}{t_s!} \right] {}^{r-1}\gamma_S\, \varepsilon^r D^{r-S} F(a) \tag{13a}$$

where the sum is to be taken over all non-negative integer components of the infinite-dimensional vector $\boldsymbol{t} = \{t_0, t_1 \ldots t_s \ldots\}$. The total powers r and S are given by

$$r = \sum_s t_s, \quad S = \sum_s s t_s. \tag{14}$$

---

* As far as possible suffices to subscripts or superscripts will be written as arguments to facilitate type-setting; thus t_s in (13) appears as $t(s)$ in the exponent and t_{ks} in (16) as $t(k, s)$.

Similarly for the more general definition

$$z = a + \sum \varepsilon^k \phi_k(z) = a + \Phi(\varepsilon, z) \tag{15}$$

the generating function analogous to (9) yields

$$F(z) = \mathop{\mathrm{S}}_{\boldsymbol{t}} \prod_k \prod_s \left[\left(\frac{D^s \phi_k(a)}{s!} \right)^{t(k,s)} \frac{1}{t_{ks}!} \right] {}^{r-1}\gamma_S \, \varepsilon^N \, D^{r-S} F(a) \tag{16a}$$

where $\boldsymbol{t}$ now represents the doubly-infinite-dimensional vector with non-negative components t_{ks} ($k=1, 2, 3 \ldots$; $s=0, 1, 2 \ldots$) and the indices N, r and S are given by

$$N = \sum \sum k \, t_{ks}, \qquad r = \sum \sum t_{ks}, \qquad S = \sum \sum s \, t_{ks}. \tag{17}$$

Corresponding factorizations of (11) yield expansions for $F(z) \cdot B$ which differ from (13a) and (16a) only by the substitution of ${}^r\gamma_S$ for ${}^{r-1}\gamma_S$, and which will be denoted by (13b) and (16b). In particular when applied to (6)–(8) one obtains expansions (13) and (16) in which the indices are restricted:

$$\text{For the sums in (6) and (8)}: \; r = S + 1 \tag{18}$$

$$\text{For (7)}: \qquad r = S. \tag{19}$$

3. Definition and Notation

The operators H, h and V are acting in an inner product space of a finite or infinite number of dimensions; as stated at the end of the Introduction, none of these operators are necessarily assumed to be Hermitean, though h should be diagonalizable. Its eigenkets and eigenbras (the latter denoted by the superscript $^+$) are bi-orthogonal and defined by

$$h\,|\psi_i\rangle \equiv h\,|i\rangle = \lambda_i\,|\psi_i\rangle \equiv \lambda_i\,|i\rangle, \tag{20a}$$

$$\langle \psi_i^+|\, h \equiv \langle i^+|\, h = \lambda_i \langle i^+|, \tag{20b}$$

$$\langle i^+ | j \rangle = \delta_{ij}. \tag{21}$$

Only if h is Hermitean are $\langle \psi_i^+|$ and $|\psi_i\rangle$ conjugate to each other. Similarly for the perturbed operator

$$H\,|\Psi_i\rangle = \Lambda_i\,|\Psi_i\rangle, \qquad \langle \Psi_i^+|\, H = \Lambda_i \langle \Psi_i^+|. \tag{22}$$

To avoid an excess of superscripts 0, the analogous quantities relating to the perturbed and unperturbed operators are denoted by the same letters in capitals and lower case respectively; the index 0 refers to that EV of h whose variation with ε is being investigated. By p we denote the projection operator onto $|0\rangle$, i.e. that operator which leaves the inner product of any ket with $\langle 0^+|$ unaltered, but annihilates the inner products with all other eigenbras; the relation

$$\langle i^+|\, p\, |j\rangle = \delta_{i0}\, \delta_{j0}, \qquad p = |0\rangle \langle 0^+| \tag{23}$$

shows that $p = p^+$; by the same reasoning the projection operator P onto the direction of $|\Psi_0\rangle$ also projects on $\langle \Psi_0^+|$; the complementary projection operators are

$$q = 1 - p, \qquad Q = 1 - P. \tag{24}$$

The EF's of H are standardized in the usual way (cf. e.g. [12])

$$\langle \Psi_0^+ | 0 \rangle = \langle 0^+ | \Psi_0 \rangle = 1 \tag{25}$$

and the wave operators Ω and Ω^+ are defined through

$$\Omega\,|i\rangle = |\Psi_0\rangle\,\delta_{i0}, \qquad \langle i^+|\,\Omega^+ = \langle \Psi_0^+|\,\delta_{i0}. \tag{26}$$

The standard expressions for the Brillouin-Wigner expansions

$$\Omega = p + \varepsilon \frac{q}{\Lambda_0 - h - \varepsilon V}\, V p = \sum_0^\infty \left(\frac{\varepsilon q}{\Lambda_0 - h}\, V \right)^k p, \tag{27}$$

$$\Lambda_0 = \lambda_0 + \varepsilon \langle 0^+ |\, V \Omega \,| 0 \rangle = \lambda_0 + \varepsilon \langle V \Omega \rangle \tag{28}$$

(when no confusion can arise the indicators $0^+|$ and $|0$ occurring in expectation values are suppressed from now on) can be deduced in the same way as for Hermitean operators (most easily by partitioning the secular equation [4]); a reversal of the factors in (27) leads to the formula for Ω^+.

Now (28) with (27) is exactly of the form (15) with λ_0 and Λ_0 taking the place of a and z respectively. The purpose of the Rayleigh-Schrödinger expansions is to express quantities defined as functions of the perturbed EV in terms of the unperturbed EV, which is achieved by the various forms of the Lagrange expansion. A concise symbolic notation for the various powers and derivatives of the operators is essential; we put

$$q\,(\lambda_0 - h)^{-n}\, q = :n:, \qquad q\,(\Lambda_0 - h)^{-n}\, q = ;n; \tag{29}$$

$$:0: = ;0; = p = |0\rangle\, \langle 0^+|. \tag{30}$$

In comparison with other notations, such as S^n by Bloch [2] or a^{-n} by Huby [3] this symbolism cuts down the use of higher-order suffices; it corresponds exactly to the graphical method of Salzman [13], the symbol $:n:$ flanked by one or two operators V being equivalent to n bonds attached to one or two circles. Note that in contrast to Kato [1] and Bloch [2] the symbol $:0:$ represents p, and not $-p$. With this notation (27)–(28) can be written

$$\Omega = p + \varepsilon\, ;1;\, V p + \varepsilon^2\, ;1;\, V\, ;1;\, V p + \cdots = \sum \varepsilon^k (;1;V)^k\, p, \tag{31}$$

$$\Lambda_0 = \lambda_0 + \varepsilon \langle V \rangle + \varepsilon^2 \langle V;1;V \rangle + \varepsilon^3 \langle V;1;V;1;V \rangle + \cdots. \tag{32}$$

The rules of differentiation are simply

$$\frac{1}{n!}\left(\frac{d}{d\Lambda_0}\right)^n ;1; = (-)^n\, ;n+1; \qquad \frac{D^n}{n!} :1: \equiv \frac{1}{n!}\left(\frac{d}{d\lambda_0}\right)^n :1: = (-)^n :n+1: \tag{33}$$

and for the powers of the operator the multinomial version of Leibniz' theorem yields

$$\frac{(-)^n}{n!} D^n (:1:V)^k\, p = \mathop{\mathsf{S}}_{\{\nu\}} \{ :\nu_1+1:\, V :\nu_2+1:\, V \ldots :\nu_k+1:\, V p \} \tag{34}$$

$$\frac{(-)^n}{n!} D^n \langle V (:1:V)^n \rangle = \mathop{\mathsf{S}}_{\{\nu\}} \langle V :\nu_1+1:\, V :\nu_2+1:\, V \ldots :\nu_k+1:\, V \rangle \tag{35}$$

where in each case the summation is over all sets of $\{\nu_1, \nu_2, \ldots \nu_k\}$ for which

$$\sum \nu_k = n. \tag{36}$$

It is important to notice that the individual terms on the r.h.s. of (34) and (35), being obtained by differentiations of the l.h.s. occur only as terms in these sums with unit or uniform coefficients. Hence a further shorthand notation for such sums is appropriate; we write Γ_{ku} for the sum of all operators as in (34) with the factor V occurring k times and the sum of all digits between colons equal to u, and similarly $\langle C_{ku}\rangle$ for the sum of expectation values as in (35) under the same conditions, so that

$$\langle C_{ku}\rangle = \langle V\,\Gamma_{k-1,u}\rangle = \langle \Gamma^{+}_{k-1,u}\,V\rangle. \tag{37}$$

This notation simplifies (34) and (35) to

$$(-)^n\,D^n\,\Gamma_{kk}/n! = \Gamma_{k,k+n}; \qquad (-)^n\,D^n\langle C_{k+1,k}\rangle/n! = \langle C_{k+1,k+n}\rangle. \tag{38}$$

If it is necessary to consider the terms in Γ_{ku} and $\langle C_{ku}\rangle$ individually, it is best to add an ordering parameter τ as a third suffix so that

$$\Gamma_{ku} = \sum_{\tau} \Gamma_{ku\tau}, \qquad \langle C_{ku}\rangle = \sum_{\tau}\langle C_{ku\tau}\rangle, \qquad \langle C_{ku\tau}\rangle = \langle V\,\Gamma_{k-1,u\tau}\rangle. \tag{39}$$

If on the left-hand side of (34)–(35) some of the factors $:1:V$ are replaced by $:0:V$, (34) and (35) still hold good with the proviso that the $:0:$ remain in their fixed positions and that any differentiation of such a factor annihilates the term in question. Again terms with a constant number k of V-factors, μ of $:0:$-factors*, and sum u for the digits between colons tend to occur in sums with unit coefficients as in (35), so the number μ is given as a prefix. The analogues of (34)–(35) and (37) become with this notation

$$(-)^n\,D^n\;{}_{\mu}\Gamma_{u+\mu,u}/n! = {}_{\mu}\Gamma_{u+\mu,u+n}, \tag{40}$$

$$(-)^n\,D^n\langle {}_{\mu}C_{u+\mu,u}\rangle/n! = \langle {}_{\mu}C_{u+\mu,u+n}\rangle, \tag{41}$$

$$\langle {}_{\mu+1}C_{ku}\rangle = \langle V\;{}_{\mu}\Gamma_{k-1,u}\rangle = \langle {}_{\mu}\Gamma^{+}_{k-1,u}\,V\rangle. \tag{42}$$

Again, if the individual terms constituting ${}_{\mu}\Gamma_{ku}$ and $\langle {}_{\mu}C_{ku}\rangle$ are required separately, they can be denoted by an ordering parameter τ as final suffix, as in (39). In view of the definition (30) in the non-degenerate case any operator ${}_{\mu}\Gamma_{ku\tau}$ is the product of an irreducible $\Gamma_{k'u'\tau'}$ and $\mu\,\langle C\rangle$-brackets, and a bracket $\langle {}_{\mu}C_{ku\tau}\rangle$ is the product of $\mu\,\langle C\rangle$-brackets; the operators ${}_{\mu}\Gamma_{ku}$ and brackets $\langle {}_{\mu}C_{ku}\rangle$ (i.e. without a final subscript) are sums of such terms.

4. Expansions in Terms of Projection Operators

For the inner product of $\langle \Psi_0^+|$ and $|\Psi_0\rangle$ one obtains from (26), (27)

$$\begin{aligned}\langle \Psi_0^+|\Psi_0\rangle &= \langle \Omega^+\Omega\rangle = \langle p\rangle + \varepsilon^2\langle p\,V\,q\,(\Lambda_0 - h - \varepsilon V)^{-2}\,q\,V\,p\rangle\\ &= 1 - (\partial/\partial\Lambda_0)\,\{\varepsilon^2\langle V\,q\,(\Lambda_0 - h - \varepsilon V)^{-1}\,q\,V\rangle\} = B^{-1}\end{aligned} \tag{43}$$

This is the generalization of the denominator in (4) if (15), represented here by (28), is used instead of (2), and it can be identified with the reciprocal of the appropriate B in (7). Hence on putting

$$|\hat{\Psi}_0\rangle = B\,|\Psi_0\rangle, \qquad \langle \hat{\Psi}_0^+| = B\,\langle \Psi_0^+| \tag{44}$$

---

* For the expectation values $\langle {}_{\mu}C_{ku}\rangle$ (but not for the operators ${}_{\mu}\Gamma_{ku}$) the bra-ket pair $\langle\;\rangle$ is counted as equivalent to one intermediate p-factor.

(21), (25) and (43) yield

$$\langle\hat{\Psi}_0^+|\Psi_0\rangle=\langle\Psi_0^+|\hat{\Psi}_0\rangle=1 \tag{45}$$

$$\langle\hat{\Psi}_0^+|\hat{\Psi}_0\rangle=\langle 0^+|\hat{\Psi}_0\rangle=\langle\hat{\Psi}_0|0\rangle=B \tag{46}$$

so that the circumflexed perturbed EFs are the normalizing partners to $\langle\Psi_0^+|$ and $|\Psi_0\rangle$ and are obtained as projections of the zero order EFs (cf. [2, 7])

$$|\hat{\Psi}_0\rangle=P|0\rangle, \qquad B\Omega=Pp. \tag{47}$$

Hence any expansion formulas for the wave operator Ω derivable by means of (3) leave Ω unaltered, while those deducible by means of (4) lead to expressions for Pp; the former are of the type given by Brueckner [9], Bloch [2] and Huby [3], the latter of the type pioneered by Kato [1]. This relationship has already been pointed out by Des Cloizeaux [7], who, however, takes the perturbed EFs as unique and allows for two sets of zero order EFs; this is a logical approach for Hermitean operators the EVs of which are degenerate in the unperturbed state, though in general non-degenerate in the perturbed state. Des Cloizeaux has also considered expansions which preserve the orthonormality of the EFs, but the resulting expressions are extremely cumbersome; this appears related to the fact that unitary transformations are meaningful only in conjunction with Hermitean operators, whereas the Kato and Brueckner-Bloch formulas can be derived independently of any assumption of Hermiticity. If any initial unitarity is to be preserved under the influence of an Hermitean perturbation it may be preferable to use procedures specially tailored to those conditions (cf. e.g. [14]).

On applying the formulas of Section 2 to the definition (32) a large number of expressions for arbitrary analytic functions of Λ_0, for Λ_0 itself and for the vector-valued function $|\Psi_0\rangle$ can be obtained with various factorizations of the expansion coefficients and of the differentiation operators $D\equiv\partial/\partial\lambda_0$ acting on these. To keep this section as clear and concise as possible, in the following formulas the usual sequential numbering is replaced by a subdivision with alphabetic suffixes where (a)–(h) refer to the type of function to be expanded, (i)–(l) to the definition of parameters and (p)–(r) to ranges of summation (where these are not self-evident from the formulas); the numbers (48)–(54) refer to the possible factorizations of Φ^r. Only the formulas of $F(\Lambda_0)$ and $|\Psi_0\rangle$ based on the expansion (3) are given in detail, the others are found from these by minor modifications.

A simple separation of differentiations of F from those acting on Φ^r yields with (3), (15) and (28)

$$F(\Lambda_0)=\sum_{N=0}^{\infty}\varepsilon^N \mathop{\mathrm{S}}_{\{\nu\}=0}^{1}\sum_S Q'_{rS}(F)\,\frac{D^S}{S!}\,\langle V:\nu_1:V\ldots:\nu_{N-1}:V\rangle \tag{48a}$$

where

$\mathop{\mathrm{S}}\limits_{\{\nu\}=0}^{1}$ indicates the sum over all $\nu_i=0,1 \quad (i=1,2\ldots N-1)$, (48p)

$r-1$ is the number of times $\nu_i=0$ appears as a factor, (48i)

$$Q'_{rS}(F)=D^{r-S}F(\lambda_0)/[r(r-S-1)!], \tag{48j}$$

and S takes the values from 0 to $r-1$ for $r>0$ and $S=0$ for $r=0$.

With the notation for the generalized $\langle C\rangle$-brackets introduced before (40), (48a) can be simplified to

$$F(\Lambda_0) = \sum_N \varepsilon^N \sum_r \sum_S Q'_{rS}(F) \frac{D^S}{S!} \langle {}_r C_{N,N-r} \rangle \tag{49a}$$

which in view of the differentiation rule (41) is equivalent to

$$F(\Lambda_0) = \sum_N \varepsilon^N \sum_r \sum_S Q'_{rS}(F) (-)^S \langle {}_r C_{N,N-r+S} \rangle. \tag{50a}$$

If the individual terms in this sum are required, the $\langle C\rangle$-brackets can be given a third right subscript τ, representing an ordering parameter, and an additional summation over τ carried out, in which the number of terms depends on the other parameters. On reverting to colon notation these terms become

$$F(\Lambda_0) = \sum_N \varepsilon^N \underset{\{\nu\}=0}{\overset{N}{\mathsf{S}}} (-)^S Q'_{rS}(F) \langle V:\nu_1: V \ldots :\nu_{N-1}: V \rangle \tag{51a}$$

where now

$\underset{\{\nu\}=0}{\overset{N}{\mathsf{S}}}$ indicates the unrestricted sum over all $\nu_i = 0, 1, 2 \ldots$ $(i = 1, 2 \ldots N-1)$ (51p)

$$S = \sum_{i=1}^{N-1} [\nu_i - \operatorname{sign}(\nu_i)], \quad r \text{ as in (48i)}. \tag{51i}$$

In the formulas (48)–(51) the individual factors occurring in Φ^r have been left in their original position; if, instead they are factorized, so that only simple (non-prefixed) $\langle C\rangle$-brackets occur, the generating function Ξ of (9)–(11) with the definition (28) yield

$$F(\Lambda_0) = \underset{\{t(k)\}}{\mathsf{S}} \varepsilon^N \sum_{S=0}^{r-1} \frac{D^S}{S!} \prod_{k=1}^{\infty} \frac{\langle C_{k,k-1}\rangle^{t(k)}}{t_k!} \, {}^{r-1}\gamma_S \, D^{r-S} F(\lambda_0) \tag{52a}$$

where (cf. Footnote p. 335 for notation)

$\underset{\{t(k)\}}{\mathsf{S}}$ is taken over (all non-negative integer values of) the components of the infinite vector t_k $(k = 1, 2 \ldots)$, (52p)

$$N = \sum k\, t_k, \quad r = \sum t_k \tag{52i}$$

and ${}^{r-1}\gamma_S$ is defined in (12). The further factorization analogous to (16a) yields in view of (38)

$$F(\Lambda_0) = \underset{\{t(k,s)\}}{\mathsf{S}} \varepsilon^N \prod_{k=1}^{\infty} \prod_{s=0}^{\infty} (-)^S \frac{\langle C_{k,k+s-1}\rangle^{t(k,s)}}{t_{ks}!} \, {}^{r-1}\gamma_S \, D^{r-S} F(\lambda_0) \tag{53a}$$

where

$\underset{\{t(k,s)\}}{\mathsf{S}}$ is taken over the components of the doubly infinite vector t_{ks} and (53p)

$$N = \sum\sum k\, t_{ks}, \quad r = \sum\sum t_{ks}, \quad S = \sum\sum s\, t_{ks}. \tag{53i}$$

This is the formula (16a) as applied to (32). Finally, if the individual terms in (53) are wanted separately, one can again order the $\langle C_{k,k+s-1}\rangle$ by a third parameter τ giving

$$F(\Lambda_0) = \underset{\{t(ks\tau)\}}{\mathsf{S}} \varepsilon^N \prod_{k=1}^{\infty} \prod_{s=0}^{\infty} \prod_{\tau=1}^{T} (-)^S \frac{\langle C_{k,k-1+s,\tau}\rangle^{t(k,s,\tau)}}{t_{ks\tau}!} \, {}^{r-1}\gamma_S \, D^{r-S} F(\lambda_0) \tag{54a}$$

where now

the sum is taken over all components of the triply infinite vector $\boldsymbol{t}$, (54p)

$$N=\sum\sum\sum k\, t_{ks\tau}, \quad r=\sum\sum\sum t_{ks\tau}, \quad S=\sum\sum\sum s\, t_{ks\tau}, \tag{54i}$$

$$T(k, s)={}^{k+s-2}C_s. \tag{54j}$$

The last relation follows from simple combinatorial algebra, being the number of ways in which s differentiations can be distributed over the $k-1$ differentiable factors. The term $\langle C_{10}\rangle\equiv\langle V\rangle$ is a constant; its derivatives vanish in (53) and (54).

The equivalent formulas based on (4) follow from (48a)–(54a) with minor modification as in (13b) and (16b).

$F(\Lambda_0)\,B$ is given by the expansions (48a)–(54a) with $Q'_{rS}(F)$ replaced by $Q_{rS}(F)=D^{r-S}(F)/(r-S)!$ and ${}^{r-1}\gamma_S$ replaced by ${}^{r}\gamma_S$. In addition the summation over S in (48a)–(51a) now extends from 0 to R. (48b)–(54b)

Here the factor B is given, in analogy to (7) and (19), by

$B=$ sum of terms as in (b), restricted to terms with $S=r$, $F(z)\equiv 1$. (48c)–(54c)

For the EV's themselves the analogue of (6) or (8) and (18) becomes

$\Lambda_0-\lambda_0=$ sum of terms as in (a), restricted to $S=r-1$, $F'(z)\equiv 1$. (48d)–(54d)

$(\Lambda_0-\lambda_0)\cdot B=$ terms as in (b), restrictions as in (d). (48e)–(54e)

The Eqs. (48a, b)–(54a, b) require some additional manipulations to be applied to the perturbed EFs, partly because the function F is now vector-valued, but also because even in zero order in Φ, (27) depends explicitly on ε. The procedures follow the same pattern as (48a)–(54a), so only the results are given.

$$|\Psi_0\rangle=\sum_{N=0}^{\infty}\varepsilon^N\sum_{M=0}^{N}\mathop{\mathsf{S}^{(i>M)}}_{\{\nu\}=0}^{1}\sum_{S=0}^{r-1} \times\frac{D^{r-S}}{r(r-S-1)!}\left\{:l_1\colon V:l_2\colon\ldots:l_M\colon V\,|0\rangle\,\frac{D^S}{S!}\,\langle V:\nu_{M+2}\colon\ldots:\nu_N\colon V\rangle\right\} \tag{48f}$$

where r is still defined by (48i), but the summation over $\nu_i=0, 1$ is now restricted to $i>M$, all ν_i being 1 for $i\leq M$. For $r=0$, terms with $S=0$ must be included, the denominator $0\cdot(-1)!$ being again interpreted as unity. If the colon notation is replaced by Γ-operators as defined before (37) and by $\langle C\rangle$-brackets, (48f) becomes

$$|\Psi_0\rangle=\sum_{N=0}^{\infty}\varepsilon^N\sum_{M=0}^{N}\sum_{r=0}^{N}\sum_{s=0}^{r-1}\frac{D^{r-S}\Gamma_{MM}|0\rangle}{r(r-S-1)!}\,\frac{D^S}{S!}\,\langle{}_rC_{N-M,\,N-k-r}\rangle \tag{49f}$$

and with the differentiation rules (38) and (41)

$$|\Psi_0\rangle=\sum_N\varepsilon^N\sum_M\sum_r\sum_s(-)^r\frac{r-S}{r}\,\Gamma_{M,M+r-S}\,|0\rangle\langle{}_rC_{N-M,\,N-M-r+S}\rangle. \tag{50f}$$

The equivalent term-by-term formula becomes

$$|\Psi_0\rangle = \sum_N \varepsilon^N \sum_M \underset{\{\nu\}}{\mathsf{S}} (-)^r \frac{r-S}{r} :\nu_1: V :\nu_2: \ldots :\nu_M: V\,|0\rangle \times \langle V :\nu_{M+2}: V \ldots :\nu_N: V\rangle \tag{51f}$$

with the summation

$$\nu_i = 1, 2 \ldots \quad (i \leqq M), \qquad \nu_i = 0, 1, 2 \ldots \quad (i > M+1) \tag{51q}$$

and the definition (and restriction)

$$S = r - \sum_1^M (\nu_i - 1) = \sum_{M+2}^N [\nu_i - \operatorname{sign}(\nu_i)], \qquad r \text{ as in (48i).} \tag{51k}$$

Factorization of the composite $\langle C\rangle$-brackets leads to

$$|\Psi_0\rangle = \underset{\{t(k)\}}{\mathsf{S}} \sum_M \varepsilon^N \sum_S (-)^r \frac{r-S}{r} \Gamma_{M,M-r-S}\,|0\rangle \frac{D^S}{S!} \prod_k \frac{\langle C_{k,k-1}\rangle^{t(k)}}{t_k!} \tag{52f}$$

with

$$N = M + \sum k\, t_k, \qquad r = \sum t_k. \tag{52k}$$

Again the differentiation rule (38) in conjunction with (16a) yields

$$|\Psi_0\rangle = \underset{\{t(ks)\}}{\mathsf{S}} (-)^r \sum_M \varepsilon^N \frac{r-S}{r} \Gamma_{M,M+r-S}\,|0\rangle \prod_{ks} \frac{\langle C_{k,k+s-1}\rangle^{t(k,s)}}{t_{ks}!} \tag{53f}$$

where

$$N = M + \sum\sum k\, t_{ks}, \qquad r = \sum\sum t_{ks}, \qquad S = \sum\sum s\, t_{ks}. \tag{53k}$$

Finally, when individual terms are wanted, both the $\langle C\rangle$-brackets and the Γ-operators have to be split into their components

$$|\Psi_0\rangle = \mathsf{S} \sum_M \varepsilon^N \sum_\sigma (-)^r \frac{r-S}{r} \Gamma_{M,M+r-S,\sigma}\,|0\rangle \prod_{ks\tau} \frac{\langle C_{k,k+s-1,\tau}\rangle^{t(k,s,\tau)}}{t_{ks\tau}!} \tag{54f}$$

$$N = M + \sum\sum\sum k\, t_{ks\tau}, \qquad r = \sum\sum\sum t_{ks\tau}, \qquad S = \sum\sum\sum s\, t_{ks\tau}. \tag{54k}$$

In view of (44) the expansions for $|\hat{\Psi}_0\rangle$ follow from those for $|\Psi_0\rangle$ in the same way as (48b)–(54b) from (48a)–(54a). The resulting expressions are thus

$|\hat{\Psi}_0\rangle =$ sum of terms as in (f), with $r(r-S-1)!$ replaced by $(r-S)!$ and $(r-S)/r$ by 1. Summation over S extended to $S=r$. (48g)–(54g)

However, a further simplification can be effected in terms of the prefixed Γ-operators introduced before (40) as long as the scalar multiplier is not factorized. By using (4) together with (44) one obtains

$$|\hat{\Psi}_0\rangle = \sum_{N=0}^{\infty} \varepsilon^N \underset{\{\nu\}=0}{\overset{1}{\mathsf{S}}} \frac{D^R}{R!} :\nu_1: V :\nu_2: \ldots :\nu_N: V\,|0\rangle \tag{48h}$$

where the sum is taken as in (48p) $(i = 1 \ldots N)$, but now

R is the number of times that $\nu_i = 0$ occurs as a factor in the operator. (48l)

The equivalent expressions (49)–(51) become

$$|\hat{\Psi}_0\rangle = \sum_N \varepsilon^N \sum_R D^R \{{}_R\Gamma_{N,N-R}|0\rangle\}/R! \tag{49h}$$

$$= \sum_N \varepsilon^N \sum_R (-)^R \, {}_R\Gamma_{N,N}|0\rangle \tag{50h}$$

$$= \sum_N \varepsilon^N \mathop{\mathsf{S}}_{\{\nu\}=0}^{N} (-)^R :\nu_1: V :\nu_2: V \ldots :\nu_N: V|0\rangle \tag{51h}$$

the last again subject to the definition (48l). There are no equivalent formulas (52h)–(54h) because of their inherent factorization.

5. Reductions and Transformations

The main difficulty in applying any of the numerous formulas derived in the preceding section lies in the formation of the reciprocal operators $:\nu:$, and thus it is important to find reductions or transformations which facilitate their evaluation, even at the cost of increasing the number of terms. One method of reduction consists in splitting the projection operator q into a sum of idempotent operators q_j, all of which commute with h

$$q = \sum_1^J q_j, \qquad q_j q_k = q_j \,\delta_{jk}, \qquad q_j h = h q_j \tag{55}$$

where J may be finite or infinite. It is convenient to generalize the colon notation so that

$$q(\lambda_0 - h)^{-\nu} q = \sum_j q_j (\lambda_0 - h)^{-\nu} q_j \tag{56}$$

is written as

$$:\nu: = \sum_{j=1}^{J} \frac{j)\,(j}{:\nu:} \tag{57}$$

and more generally for an arbitrary analytic function G of a set of products of reciprocal operators

$$\mathop{\mathsf{S}}_{\{\nu(i)\}} G[\{:\nu_i:\}] = \mathop{\mathsf{S}}_{\{\nu(i)\}} \mathop{\mathsf{S}}_{\{j(i)\}} G\left[\left\{\frac{j_i)\,(j_i}{:\nu_i:}\right\}\right]. \tag{58}$$

It is also useful to define $q_0 = p$; the appearance of a pair 0)(0 in a numerator implies a :0: in the denominator and conversely. The set of Eqs. (48) and (51) is simply generalized by means of (58). For the other formulas one has to reinterpret the individual products $\langle C_{ku\tau}\rangle$ and $\Gamma_{ku\tau}$ as corresponding not only to a fixed set of $\{\nu_i\}$, but also of $\{j_i\}$, so that the range of summation T in (54j) must be multiplied by $(k-1)^J$, and correspondingly for $\langle {}_rC_{ku\tau}\rangle$ and ${}_r\Gamma_{ku\tau}$. The sums over τ, i.e. $\langle C_{ku}\rangle$, Γ_{ku}, $\langle {}_rC_{ku}\rangle$ and ${}_r\Gamma_{ku}$, remain the same, though they may possibly be evaluated in a different way. The set of formulas (49), (50), (52)–(54) is thus completely unaffected, although the meaning of each $\langle C_{ku\tau}\rangle$ and its associated power must be reinterpreted as explained above.

A special case arises if the reduction (55) is complete, so that each operator corresponds to exactly one EF of h

$$q_j = |j\rangle\langle j|, \qquad h q_j = \lambda_j q_j. \tag{59}$$

Each term on the right hand side of (56) or (57) simply means a division by $(\lambda_0-\lambda_j)^\nu$ and it is preferable to keep the chains of matrix elements separate from the variable scalar factors and let the differentiations act on the product of all scalars simultaneously. The formal analogue to (48a) becomes thus in view of (3)

$$F(\Lambda_0)=\sum_{N=0}^{\infty}\varepsilon^N \mathop{\mathsf{S}}_{\{j(i)\}=0}^{J} \langle 0|\,V\,|j_1\rangle\,\langle j_1|\,V\,|j_2\rangle\ldots\langle j_{N-1}|\,V\,|0\rangle \times\frac{D^{r-1}}{r!}\left[\frac{D\,F(\lambda_0)}{\prod_i'(\lambda_0-\lambda_{j(i)})}\right] \tag{60a}$$

where now

the sum S is to be taken over all $j_i=0,\ldots,J$ $(i=1,\ldots,N-1)$, (60p)

$r-1$ is the number of times $j=0$ appears in the interior of the chain, (60i)

the product $\prod'$ is taken over all i for which $j_i\neq 0$. (60q)

For $r=0$ the operator product $D^{-1}D$ means the identity. The equivalent formula (60b) for $F(\Lambda_0)\cdot B$ is obtained, in analogy to (4), by drawing the last derivative operator D, which in (60a) is inside the bracket, in front of the bracket. The formulas (60c–e) for B, $\Lambda_0-\lambda_0$ and $(\Lambda_0-\lambda_0)\,B$ follow the same recipe as (48c–e). Every term in (60e) is equal to the corresponding term in (60d) multiplied by r, as shown after (8). Like (60a) the equation for $|\Psi_0\rangle$ can be simplified by not explicitly separating the differentiations of the vector from those of the scalar factors, so that

$$|\Psi_0\rangle=\sum_{N=0}^{\infty}\varepsilon^N \mathop{\mathsf{S}}_{\{j(i)\}=0}^{J} |j_0\rangle\prod_{i=1}^{N-1}\langle j_{i-1}|\,V\,|j_i\rangle\,\langle j_{N-1}|\,V\,|0\rangle \times\frac{D^{R-1}}{R!}\left[\prod_{M+1}^{N-1}{}'\frac{1}{\lambda_0-\lambda_{j(i)}}\cdot D\prod_{1}^{M-1}\frac{1}{\lambda_0-\lambda_{j(i)}}\right] \tag{60f}$$

where (60p) applies also to j_0 and

M is the first index i for which $j_i=0$*, (60j)

R is the number of times that $j_i=0$ occurs in the sequence (including j_M). (60k)

For $M=0$ $(j_0=0)$ and $M=1$ $(j_1=0)$, the differentiation of the empty products shows that the only surviving terms have $N=M$. Drawing the last differentiation from inside the last bracket to a position in front of it gives the equivalent formula (60h) for $|\hat{\Psi}_0\rangle$. In this case terms with $M=0$ or $M=1$ show no special behaviour.

If chains of matrix elements in a product are not allowed to contain any $j_i=0$ except at their terminations, but products and powers thereof are permitted, one obtains formulas analogous to (52)–(54). These are not given explicitly here.

Other useful transformations can be obtained if the complementary subspace q is not represented by the bi-orthogonal set (21) which diagonalizes h, but by arbitrary, but complete, basis sets $|j\rangle$ and $\langle j|$ which need not be bi-orthogonal, though $|0\rangle$ and $\langle 0|$ must remain EFs of h. With this basis set the operators h

* This definition differs from the meaning of M in (48f, g)–(54f, g) by one.

and V are represented by matrices $\mathfrak{h}$ and $\mathfrak{V}$ with matrix elements

$$\mathfrak{h}_{ij} = \langle i|\, h\, |j\rangle, \qquad \mathfrak{V}_{ij} = \langle i|\, V\, |j\rangle, \tag{61a}$$

but in addition the overlap matrix $\mathfrak{S}$ with elements

$$\mathfrak{S}_{ij} = \langle i|j\rangle \tag{61b}$$

need not be the unit matrix. With this representation the formulas (48)–(54) can be taken without any alteration, provided we interpret

$$:0: = p, \qquad :\nu: = q\,(\lambda_0\,\mathfrak{S} - \mathfrak{h})^{-1}\,[\mathfrak{S}\,(\lambda_0\,\mathfrak{S} - \mathfrak{h})^{-1}]^{\nu-1}\,q, \tag{62}$$

the matrix inversion being performed in the appropriate number of dimensions. It is clear that h must be representable in this basis set, but that the existence of EFs, other than $|0\rangle$ and $\langle 0|$, is not essential. Similar considerations as for (61)–(62) apply if the reduced subspaces q_j in (55) can represented in a non-biorthogonal basis. Other representations of the operators $:\nu:$ are possible, e.g. by means of GREEN's functions, but the basic Eqs. (48)–(54) are independent of their nature.

6. Multiple Expansions, Expectation Values

If the perturbation depends on several parameters as in (5), the simplest approach is to substitute the sum for εV in any of the formulas of the preceding sections and to sort out the powers of the various ε_l *a posteriori* (cf. e.g. [15]). This implies that each term in the ordered expansions (48)–(51) and (60) with a factor ε^N is now to be replaced by N^L terms where L is the range of l; the denominators $:\nu_i:$ are not affected by this summation over the V_l. The $\langle C\rangle$-brackets and Γ-operators must be redefined so that the first right subscript now represents an l-dimensional vector, though if the same letter also occurs in the following subscript, it remains scalar with a value equal to the sum of all the components of the vector. Except for this minor modification all the formulas (49)–(50) and (52)–(54) remain unchanged; in (52)–(54) the sum rules (p) have to be adapted to the new number of dimensions and the definitions of N in (i) and (k) represent vector equations.

The most important application of Rayleigh-Schrödinger expansions in more than one variable is to the evaluation of expectation values of operators W for the EFs corresponding to a perturbation (1)

$$\langle W(\varepsilon)\rangle = \langle \Psi_0^+|\, W\, |\Psi_0\rangle / \langle \Psi_0^+|\Psi_0\rangle. \tag{63}$$

If (5) is given in the form

$$H = h + \varepsilon V + \eta W \tag{64}$$

and the corresponding EV denoted by $\Lambda_0(\varepsilon, \eta)$, it is well known that [15]

$$\langle W(\varepsilon, \eta)\rangle = \partial \Lambda_0(\varepsilon, \eta)/\partial \eta. \tag{65}$$

If the EV problem on its own is solved for $\eta = 0$, zero must be substituted for η in (65) after the differentiation. Formally this can be written as

$$\langle W(\varepsilon)\rangle = W \frac{\partial}{\partial(\varepsilon V)} \Lambda_0(\varepsilon), \tag{66}$$

implying that expansions for $W(\varepsilon)$ can be obtained from (48d)–(54d) or (60d) by replacing every term by a sum in which each V, one at a time, is replaced by W and at the same time the power of ε is reduced by one. If in an expression powers or products of factors containing V occur, the formal differentiation (66) follows the same rules as normal differentiation

$$W\frac{\partial}{\partial V}\frac{F^t}{t!}=W\frac{\partial F}{\partial V}\frac{F^{t-1}}{(t-1)!};\qquad W\frac{\partial(FG)}{\partial V}=\left(W\frac{\partial F}{\partial V}\right)G+F\left(W\frac{\partial G}{\partial V}\right). \tag{67}$$

The expansions for $\langle W(\varepsilon)\rangle$ derived from (52d)–(54d) are thus sums of terms containing one bracket $W\,\partial\langle C\rangle/\partial(\varepsilon V)$ multiplied by factors $\langle C\rangle^t/t!$; they are thus of the exact form (52f)–(54f) for the EFs with $\Gamma_{M,M+r-S}\,|0\rangle$ replaced by

$$W(\partial/\partial\varepsilon V)\,\langle C_{M+1,M-r+S}\rangle.$$

This result could have been obtained more directly by applying (66) to the implicit definition (32) for Λ_0 and substituting the resulting series as $F(\Lambda_0)$ into (52a)–(54a). The derivations of the expansions for $|\Psi_0\rangle$ and $\langle W(\varepsilon)\rangle$ are then completely analogous. This approach also shows that the same replacement of the Γ operator in (52g)–(54g) yields the expansion for $B\cdot\langle W(\varepsilon)\rangle$, an expression which cannot be derived by means of (65).

An alternative application of the results connected with (5) is the case of non-linear perturbations

$$H=h+\sum\varepsilon^l\,V_l\qquad l\geqq 1 \tag{68}$$

which has been studied in detail by PRIMAS [14] and results from (5) by putting $\varepsilon_l=\varepsilon^l$. The special formulas which arise if the V_l are also equal or proportional to powers of a common V will be discussed in a later publication.

7. Conclusions

The formulas derived in this paper are for the most part not new; some have even been derived before by means of LAGRANGE's expansion. Thus the forms (48c, e, h)–(51c, e, h) in which the ordering of the factors is preserved is originally due to KATO [1]; and (52d, f) where the scalar coefficients, but not the differential operators, are factorized has been given by SILVERSTONE and HOLLOWAY [16]; both sets of authors used contour integrations. In fact, as stated in the Introduction, LAGRANGE's theorem is merely an explicit statement for the results of contour integration; the form (4) and expressions derived therefrom are obtained directly, whereas (3) involves an integration by parts and thus depends on the commutativity of all the variables involved. In consequence, when dealing with degenerate unperturbed states for which the projection p of (23) projects onto several dimensions and for which a product of operators with p as first and last factors is no longer a scalar, generalizations of (48c, e, h)–(51c, e, h) can easily be derived which define the perturbed space and a secular equation therein (KATO [1], DES CLOIZEAUX [7]); by contrast there is no exact equivalent to (48d, f)–(51d, f) in the degenerate case. SILVERSTONE and HOLLOWAY [8, 17] try to overcome this difficulty by diagonalizing h according to the lowest orders in which the degeneracies are successively removed, but they require the use of intermediate operators and do not, in general, obtain explicit expressions. BLOCH

[2] and DES CLOIZEAUX [7] express Ω as $B^{-1} P p$ and obtain formulas in which the products of operators S^{ν} (our $:\nu:$) have to satisfy certain inequalities which reduce the number of terms to a given order in ε compared to those in KATO's expansion. Again this approach can be generalized to degenerate cases. The writer intends to give an alternative treatment of perturbed degeneracies at a later date.

The writer originally arrived at the formulation (60) by means of an expansion theorem by SCHAFROTH [18] who expressed the matrix elements of $F(H)=F(h+\varepsilon V)$ in the representation diagonalizing h in terms of the Newtonian divided differences of F for arguments $\lambda_0, \lambda_{i(1)}, \lambda_{i(2)} \ldots$. Taking the trace of $F(H)$ and retaining only those terms in the divided differences which contain $F(\lambda_0)$ and its derivatives, but no other $F(\lambda_i)$, yields the expansion (60). This retention of a function about one argument only is exactly the effect of contour integration; it can however be justified from arguments of logic and symmetry only (a high power of a small argument may not be swamped by a contribution from a large argument), and by this reasoning the writer deduced (2) and (3), being at that time ignorant of LAGRANGE's work (cf. [6]).

It is interesting to note that none of the formulas (48d, f)–(54d, f) are the exact equivalent of Bloch's expansion [2] (for the non-degenerate case) or the term-by-term corresponding formulation of HUBY [3]. All the terms in the Kato-type expansions (48g, h)–(54g), with the exception of those adding up to $B\,|0\rangle$, occur in (48f)–(54f), only with smaller coefficients. Nevertheless the combinatorial approach through the exponential generating function described in Section 2 considerably reduces the number of terms in the higher orders of the factorized forms (52)–(54). Thus among the 14 fifth-order terms for Λ_0 there occur in Huby's expansion $\langle V a^{-1} \langle V a^{-1} V\rangle a^{-1} \langle V\rangle a^{-1} V\rangle$ plus its mirror image; the equivalent terms in the completely factorized form (54d) are $2\langle V:3:V\rangle\langle V:1:V\rangle\langle V\rangle$; for higher order terms the saving becomes more marked.

More important is the saving effected in the form (53) where all the terms obtained by differentiating $\langle C_{k,k-1}\rangle$ a given number of times remain bracketed together. Thus the fourth-order term for Λ_0 in (53d) contains $-\langle V\rangle(\langle V:2:V:1:V\rangle+\langle V:1:V:2:V\rangle)$; in all higher order terms the two $\langle C\rangle$-brackets in the last parenthesis occur only as powers of the same linear combination. Thus once any such combination has been calculated from the T terms in (54j) in the lowest order in which it occurs, it can be taken as a single unit for all higher orders. The same applies to Γ-operators and derivatives of $W \cdot \partial\langle C\rangle/\partial(\varepsilon V)$. As far as the writer is aware this simplification has not been pointed out before. In SALZMAN's diagrammatic approach [12] this means that all linked graphs containing the same numbers of links and of circles can be grouped together

As the Hermiticity of the operators is nowhere relevant to the present approach, the two terms in the last parenthesis are not necessarily equal. The expansions

(*Note added in Proof:* Professor P. D. ROBINSON has pointed out that the LAGRANGE series has been employed previously [19–22] in the perturbation treatment of the one-dimensional model for H_2^+. In this problem, however, the analytic form of $|\Psi_0\rangle$ is known and only the scale parameter is to be determined; the treatment demonstrates that expansions for Λ_0 may converge for a wider range of ε than those for $|\Psi_0\rangle$.)

cannot be guaranteed to yield real EVs; all that can be said is that if there exists a representation in which all the elements of h and V are real, then within the radius of convergence of the expansion Λ_0 will be real for real ε. The radius of convergence is essentially determined by the smallest modulus $|\varepsilon_c|$ for which $\Lambda_0(\varepsilon_c)$ coalesces with the EV deriving from another λ_i, unless the nature of the interaction curtails the expansion range still further (e.g. for a Coulomb interaction, even in a finite configuration space, the sum of all third order terms diverges). In many cases, including Hermitean ones, the radius of convergence is rather small so that the expansions when ordered in powers of ε are of limited usefulness only. Nevertheless the formulas derived in this paper, especially the formulation (60) in terms of matrix elements (which I also believe to be new), point the way to groupings, other than by powers of ε, which are less prone to divergencies. This will be dealt with in a later paper.

Other possible applications of LAGRANGE's expansion to Rayleigh-Schrödinger perturbation theory is the calculation of the parameter ε in terms of the energy shift $\Lambda_0 - \lambda_0$ and the simultaneous expansion of two or more separated equations with common separation constants. These again will be treated in subsequent papers.

Acknowledgements

I wish to thank a large number of colleagues, in particular Professor H. FRÖHLICH, Professor S. F. EDWARDS, Professor P.-O. LÖWDIN, Dr. R. HUBY and Dr. S. SAMPANTHAR, for useful and stimulating discussions over a period of years.

References

1. KATO, T.: Progr. Theoret. Phys. (Kyoto) **4**, 514 (1949); **5**, 95, 207 (1950).
2. BLOCH, C.: Nucl. Phys. **6**, 329 (1958).
3. HUBY, R.: Proc. Phys. Soc. (London) **78**, 529 (1961).
4. LÖWDIN, P.-O.: J. Math. Phys. **3**, 969 (1962).
5. WHITTAKER, E. T., WATSON, G. N.: A Course of Modern Analysis, 4th Ed., Chap. 7, p. 133. Cambridge: University Press 1927.
6. SACK, R. A.: J. SIAM Appl. Math. **13**, 47 (1965).
7. DES CLOIZEAUX, J.: Nucl. Phys. **20**, 321 (1960).
8. SILVERSTONE, H. J., HOLLOWAY, T. T.: Phys. Rev. A., Ser. 3, **4**, 2191 (1971).
9. BRUECKNER, K. A.: Phys. Rev. **100**, 36 (1955).
10. GOLDSTONE, J.: Proc. Roy. Soc. (London), Ser. A **239**, 267 (1957).
11. SACK, R. A.: J. SIAM Appl. Math. **14**, 1 (1966).
12. MESSIAH, A.: Quantum Mechanics, vol. 2. New York: John Wiley & Sons, Inc. 1965.
13. SALZMAN, W. R.: J. Chem. Phys. **49**, 3035 (1968).
14. PRIMAS, H.: Helv. Phys. Acta **34**, 331 (1961).
15. HIRSCHFELDER, J. O., BYERS BROWN, W., EPSTEIN, S. T.: Advan. Quantum Chem. **1**, 255 (1964).
16. SILVERSTONE, H. J., HOLLOWAY, T. T.: J. Chem. Phys. **52**, 1472 (1970).
17. SILVERSTONE, H. J.: J. Chem. Phys. **54**, 2325 (1971).
18. SCHAFROTH, M. R.: Helv. Phys. Acta **24**, 645 (1952).
19. ROBINSON, P. D.: Proc. Phys. Soc. (London) **78**, 537 (1961).
20. CLAVERIE, P.: Intern. J. Quantum Chem. **3**, 349 (1969).
21. CERTAIN, P. R., BYERS BROWN, W.: Intern. J. Quantum Chem. **6**, 131 (1972).
22. AHLRICHS, R., CLAVERIE, P.: Intern. J. Quantum Chem. **6**, 1001 (1972).

Appendix

Vanishing of Unlinked Clusters

The (admittedly sketchy) treatment of the expansions under the influence of several additive perturbations given in Section 6 provides a means of proving the linked cluster theorem, which is not only trivial but also valid under much more general conditions than is usually assumed [9, 10], in so far as no recourse is made to the concept of Bosons or Fermions. It is only assumed that both the unperturbed operator h and a part of the perturbation V can be split into two or more commuting contributions, i.e.

$$H = H^a + H^b + \varepsilon_{ab} V^{ab} = (h^a + \varepsilon_a V^a) + (h^b + \varepsilon_b V^b) + \varepsilon_{ab} V^{ab} \tag{A 1}$$

with $\varepsilon_a = \varepsilon_b = \varepsilon_{ab} = \varepsilon$ and where the operators with superscript a commute with those marked with b, whereas V^{ab} need commute with none of the other operators. According to Section 6 the terms in the expansions (48)–(54) and (60) which involve powers of ε_a and ε_b only are not affected by the value of ε_{ab}. However for $\varepsilon_{ab} = 0$, H is the sum of the two commuting operators H^a and H^b, and thus its EVs are obtained by addition and its EFs by multiplication of the corresponding characteristic quantities of H^a and H^b.

This leads immediately to the two theorems [the term "chain" is taken in the sense of Eq. (60)]:

Theorem 1. The total contribution to Λ_0 due to chains of matrix elements involving both V^a and V^b, but not V^{ab}, vanishes rigorously.

Theorem 2. Assuming $|\Psi_0^a\rangle$ and $|\Psi_0^b\rangle$ to be the eigenkets for H^a and H^b with standardizations corresponding to (25), the contribution to the EF $|\Psi_0\rangle$ due to chains of matrix elements involving prescribed sequences of elements of V^a and V^b, but none of V^{ab}, is equal to the product of the corresponding contributions to $|\Psi_0^a\rangle$ and $|\Psi_0^b\rangle$.

If a separate proof is required, Theorem 2 can be derived from (60) by induction, and Theorem 1 follows therefrom in view of (28). The usual interpretation of the linked cluster theorem is obtained from Theorem 1 if the momenta (both occupied and unoccupied) involved in a Feynman diagram can be separated into two non-interacting sets. Those terms included in the usual formulation, which do not add up rigorously to zero, but become small with increasing number of particles, are not covered by Theorem 1.

A Simple Lagrangian Formalism for Fermi-Dirac Quantization

G. R. Allcock

University of Liverpool, Department of Theoretical Physics, Liverpool/England

Abstract

A consistent procedure is given, by which the equal-time anticommutators of any Fermi-Dirac system can be calculated directly from its Lagrange function, without the necessity to introduce canonical variables or to solve any equations of motion.

1. Prologue

No tribute to H. Fröhlich would be complete without mention of his important work with Heitler and Kemmer on meson theory and nucleon magnetic moments [1]. In looking back at this early incursion into the relativistic domain one may discern at the same time both the essential unity of theoretical physics and the great seminal value of cooperation between investigators with differing research backgrounds, since the concepts of virtual particle and proper field which were there so basic proved some twelve years later to be extremely fruitful when carried over to various problems of solid-state physics.

The practical utility and validity of quantum field methods in many-body physics is now universally recognised. Yet in field theory itself there remain several unresolved problems. In the present article I propose to dispose of one such problem—a minor one indeed, but one which is nevertheless of some pedagogic and methodological interest.

2. The Problem

The problem on hand concerns the structural or mathematical connection between the Lagrange function of a Fermi-Dirac (F. D.) system and the various quantum equations which collectively define such a system. To demonstrate that there are still non-trivial obscurities here we can hardly do better than sketch out the historical and still popular procedure of quantization for the F.D. oscillator. The Lagrangian normally attributed to this simple system may be abstracted from any text which treats the second quantization of the Schrödinger field by the

Lagrangian method*. All that is necessary is to suppress the space coordinates. The concensus obtained by this recipe is that

$$L = \psi^* (i\dot{\psi} - \omega\psi), \tag{1}$$

where ω is a constant, and ψ and ψ^* are complex dynamical variables, which can be expressed in terms of two real or Hermitian variables by the formulae

$$\psi\sqrt{2} = q^1 + i q^2, \qquad \psi^*\sqrt{2} = q^1 - i q^2. \tag{2}$$

The decomposition (2) is essential if we are to give any proper meaning to the joint variations of ψ and ψ^* in the action integral. Applying (2), and preserving the order of all factors, we find that L decomposes into the sum of two terms:

$$L \equiv L_{\mathrm{B.E.}} + L_{\mathrm{F.D.}} \tag{3}$$ **

with

$$L_{\mathrm{B.E.}} \equiv \tfrac{1}{2} q^2 \dot{q}^1 - \tfrac{1}{2} q^1 \dot{q}^2 - \tfrac{1}{2}\omega (q^1 q^1 + q^2 q^2), \tag{4}$$

$$L_{\mathrm{F.D.}} \equiv \tfrac{1}{2} i (q^1 \dot{q}^1 + q^2 \dot{q}^2) + \tfrac{1}{2} i\omega (q^2 q^1 - q^1 q^2). \tag{5}$$

The accepted procedure at this point is to vary the action integral while treating ψ, ψ^* and their variations as classical (commuting) variables. In this classical context the kinetic part of expression (5) reduces to an irrelevant perfect differential and its potential part vanishes identically, so that, in effect, one is varying an action integral based on expression (4) alone. That is, L and $L_{\mathrm{B.E.}}$ are entirely equivalent. The equations of motion which result from the standard variation procedure are

$$\dot{q}^2 = -\omega q^1, \qquad \dot{q}^1 = \omega q^2, \tag{6}$$

or, equivalently,

$$i\dot{\psi} = \omega\psi. \tag{7}$$

The appropriate classical Hamiltonian is

$$\begin{aligned} H_{\mathrm{B.E.}} &= (\partial L_{\mathrm{B.E.}}/\partial \dot{q}^1)\,\dot{q}^1 + (\partial L_{\mathrm{B.E.}}/\partial \dot{q}^2)\,\dot{q}^2 - L_{\mathrm{B.E.}} \\ &= \tfrac{1}{2}\omega (q^1 q^1 + q^2 q^2) = \omega\psi^*\psi. \end{aligned} \tag{8}$$

Having reached this stage, one exploits the analogy between Poisson brackets and commutator brackets, according to which the quantum B.E. oscillator has as its quantum conditions the equal-time commutation relations

$$[q^1, q^2] = i, \qquad [q^1, q^1] = [q^2, q^2] = 0, \tag{9}$$

or, equivalently,

$$[\psi, \psi^*] = 1, \qquad [\psi, \psi] = [\psi^*, \psi^*] = 0. \tag{10}$$

The next step in this traditional development is to remark that the exclusion principle can be brought in by replacing the commutators of the variables ψ and ψ^* by anticommutators, so that, instead of (10), one writes

$$\{\psi, \psi^*\} = 1, \qquad \{\psi, \psi\} = \{\psi^*, \psi^*\} = 0. \tag{11}$$

---

* An incomplete list of such texts is given under Ref. [2]. WENTZEL's book [3] is the earliest standard work on the subject.

** B.E. means Bose-Einstein.

Finally, one tries for the F.D. Hamiltonian the expression

$$H_{\mathrm{F.D.}} = \omega \psi^* \psi, \tag{12}$$

as suggested by the last term of Eqs. (8), and one is duly gratified to find that this and the rules of anticommutation (11) still generate the equations of motion (6), (7) in the familiar Hamiltonian form

$$i\dot{\psi} = [\psi, H_{\mathrm{F.D.}}], \text{ etc.} \tag{13}$$

Although the final formalism [as expressed by (2), (6), (7), (11), (12) and (13)] is fully consistent, there are a number of strange and arbitrary features in the intervening development. The outcome would have been quite different had we replaced commutators by anticommutators at stage (9) instead of at the equivalent stage (10). For the correct anticommutators (11) can be written as

$$\{q^1, q^2\} = 0, \qquad \{q^1, q^1\} = \{q^2, q^2\} = 1, \tag{14}$$

and these bear no resemblance to (9). Yet, *a priori*, the two procedures would seem to be equally admissible.

Precisely the same sort of difficulty arises with the Hamiltonian, which shows a correspondence when expressed in terms of ψ and ψ^*, but none when expressed in terms of the basic variables q. For (12) reduces to

$$H_{\mathrm{F.D.}} = \tfrac{1}{2} i \omega (q^1 q^2 - q^2 q^1) + \tfrac{1}{2}\omega, \tag{15}$$

which is totally dissimilar to the alternative form for $H_{\mathrm{B.E.}}$ provided by the penultimate member of (8). Thus, as with the anticommutators, the successful development of the formalism depends entirely upon the fortuitous adoption of one of two alternative and different procedures of equivalent *a priori* status.

Again, we may consider the connection between the Lagrangian and the Hamiltonian $H_{\mathrm{F.D.}}$. The great virtue of the Lagrangian method is, or should be, that all the constructs of the theory are summed up in the Lagrange function, from which they can be derived by suitable operations. In this point also the formalism which we have sketched is quite at fault, since it is obvious by inspection that there is no affinity of structure between $H_{\mathrm{F.D.}}$ [Eq. (15)] and $L_{\mathrm{B.E.}}$ [Eq. (4)]. Curiously, there is some suggestion of a connection between $H_{\mathrm{F.D.}}$ and $L_{\mathrm{F.D.}}$. However, it has already been pointed out that $L_{\mathrm{F.D.}}$ could be dropped from L without making the slightest difference. Therefore this suggested connection is purely coincidental at the present stage, and one cannot base anything upon it.

These defects become much more serious in the relativistic domain, because there it is normally the Lagrangian, and not the Hamiltonian, which plays a vital part in the formulation of relativistic quantum conditions, conservation laws, etc. One needs to be able to treat such matters in the Lagrangian manner for Fermions, as for Bosons. The situation becomes particularly bad for the Majorana field with mass [4]. There one has a perfectly good relativistic quantum field theory, yet there is no way whatever to incorporate a mass term in the Lagrangian at the classical level. The only available Lorentz scalars are constructed from skew Dirac matrices, and vanish classically in quite the same way as the potential part of expression (5), so that one cannot even begin to develop the theory along the traditional lines.

Recognising the need for improvement, SCHWINGER [5] introduced a quantum action principle, in which the essentials of F.D. quantization and of the old Dirac-Heisenberg subtraction formalism [3] are both incorporated from the start by using antisymmetric Lagrangian expressions and anticommuting variations. At about the same time PEIERLS [6] independently introduced a new quantization rule which, although very different in appearance from that of SCHWINGER, can also accomodate F.D. systems. The reader may perhaps hardly be surprised that the expression $L_{\text{F.D.}}$ of Eq. (5), which has hitherto played only an incidental and haphazard part in our story, now emerges as the correct Lagrangian for a treatment of the F.D. oscillator by the methods of SCHWINGER or PEIERLS. It is even normalized correctly! It is not quite antisymmetric, but that can be rectified merely by writing

$$q\dot{q} = \tfrac{1}{2}q\dot{q} - \tfrac{1}{2}\dot{q}q + \tfrac{1}{2}d(qq)/dt \tag{16}$$

and discarding the perfect differential.

Although the Schwinger principle provides a link between the Lagrangian and the anticommutators, conservation laws, etc., it is not without problematic features. These were discussed in the early days of the principle [7, 8], but it is doubtful whether the full implications have been widely appreciated. From the discussions it emerged [8] that the principle is sensitive to the addition to the Lagrangian of perfect differentials, and that commutators and anticommutators compatible with the equations of motion and the Hamiltonian will not arise from it except by judicious choice of such additive terms, depending on the type of variation. Some rules for choosing the additive terms have been stated, but nowhere in a manner general enough to cover all of those interesting cases where some of the equations of motion are devoid of time derivatives. From [8] it can besides be shown that the Schwinger principle can never lead to the correct anticommutators for the F.D. oscillator if the variables q^1 and q^2 are varied independently in (5), no matter how (5) be supplemented by additive antisymmetric bilinear perfect differentials! Yet the use of additive perfect differentials can lead to correct anticommutators if ψ and ψ^* are used as variables in the same Lagrangian, provided one disregards Eqs. (2), and formally treats $\delta\psi$ and $\delta\psi^*$ as fully independent variations instead of as conjugates of each other [8]. It is proper to stress that these anomalies are not due to a faulty choice of Lagrangian.

The Peierls method suffers from none of the above drawbacks. It is invariant against perfect differentials, and when it is applied to the same Lagrangian (5) it yields the correct answers (11) while taking full cognisance of relations (2) at all stages [7]. Indeed, it will be shown elsewhere that the Peierls method is always correct [9]. Its only defect seems to be that it is stated in terms which are very difficult to handle and which demand considerable expertise. The present article describes a relatively simple yet quite equivalent approach, with particular emphasis on F.D. systems.

3. The Solution

It is necessary first to settle a little matter of orientation. One would like to have some completely quantal action principle. However, any attempt at such a principle is bound to get into difficulties with the supposed operator properties of the variations. The only practical starting assumption (in the F.D. case) would be that δq^1 and δq^2 each anticommute with both q^1 and q^2. A similar assumption is in fact made [8] in applying the Schwinger principle. Now it is necessary, in exploring the full consequences of any action principle, to use two types of variation, customarily denoted by δq and $\overline{\delta} q$, and related to each other by $\delta q = \overline{\delta} q + \dot{q}\,\delta t$. These cannot both anticommute, once the variables $\dot{q}$ have been quantised. Yet such mutually contradictory properties are needed if one is to obtain both the equations of motion and the formula for the Hamiltonian. The resultant ambiguities are necessarily present both for B.E. and F.D. systems, but are more vexatious for the latter since there is no classical limit in which failure to anticommute can be ignored.

To avoid all such profitless embarrassment we return in this paper to the traditional Dirac approach [10], in which the Lagrangian serves to generate a number of fully consistent relations at a classical level—relations which are then hopefully extended to the quantal level by a process of formal analogy. Our solution must therefore be formulated in terms of an "anticlassical mechanics", in which analogues to anticommutators and commutators will both appear.

To achieve this new mechanics we shall use a real and even Lagrange function $L\left(q(t), \dot{q}(t), t\right)$, in which all the variables q and $\dot{q}$ and the variations to which they are to be subjected are real and anticlassical. By this we mean that the basic variables obey special rules of the form

$$\left.\begin{aligned} A^* &= A, \quad B^* = B, \quad \text{etc.},\\ (AB)^* &= BA = -AB, \quad AA = 0,\\ (ABC)^* &= CBA = -ABC,\\ (ABCD)^* &= DCBA = +ABCD, \quad \text{etc.}, \end{aligned}\right\} \tag{17}$$

and are otherwise subject to the usual rules of algebra. Included in the latter rules we have

$$\{FB = 0, \quad \forall B\} \Rightarrow \{F = 0\}, \tag{18}$$

where F may be even or odd, and this is certainly needed both to exploit the anticlassical action principle, and to ensure that the notion of differential coefficient can be extended uniquely to even and to odd functions.

The author is not aware of any concrete representation of anticlassical variables. Certainly, they cannot be represented by finite matrices (see Appendix). Fortunately, however, the absence of any tangible interpretation in terms of ordinary numbers does not jeopardize our programme, since in setting up the formalism of anticlassical mechanics our concern is only to invent a system of mutually consistent abstract relationships, which are to be taken over by an inexact process of analogy and finally supplanted by the concrete scheme of F.D. quantum mechanics. Precisely the same philosophy could be applied to the problem of quantization of B.E. systems. To the extent that the quantum theory with which one ends is more fundamental than the classical theory from which one starts, it is indeed fully appropriate to think in this abstract way.

The notations of the calculus need some adjustment when dealing with anti-classical variables. The product rule of differentiation becomes

$$\begin{aligned}\delta(A_1 A_2 A_3 \cdots A_n) &= \delta A_1 \cdot A_2 A_3 \cdots A_n + A_1\, \delta A_2 \cdot A_3 \cdots A_n + \cdots \\ &= (-)^{n-1} A_2 A_3 \cdots A_n\, \delta A_1 + (-)^{n-2} A_1 A_3 \cdots A_n\, \delta A_2 + \cdots.\end{aligned} \tag{19}$$

Of the two expansions given above the second is the more easy to compare with others of its type, since all differentials appear in a standard location. Because of this consideration, we introduce an antiderivative ∂_{F} modelled on the last line of (19), in terms of which we can write

$$\delta L = \frac{\partial_{\mathrm{F}} L}{\partial q^\alpha}\, \delta q^\alpha + \frac{\partial_{\mathrm{F}} L}{\partial \dot{q}^\alpha}\, \delta \dot{q}^\alpha + \frac{\partial L}{\partial t}\, \delta t. \tag{20}$$

HAMILTON's principle of stationary action now takes the usual form, but with anticlassical values for the δq^α, and gives the following equations of motion

$$\frac{\partial_{\mathrm{F}} L}{\partial q^\alpha} - \frac{d}{dt} \frac{\partial_{\mathrm{F}} L}{\partial \dot{q}^\alpha} = 0. \tag{21}$$

For the oscillator Lagrangian (5) these equations reduce to (6).

As in ordinary Lagrangian mechanics, it is necessary to consider variations for cases in which the end points t_1 and t_2 of the integration are displaced. Adopting the notation δq to represent the total end-point variation of q, we obtain the familiar result

$$\delta \int_{t_1}^{t_2} L\, dt = (p_\alpha\, \delta q^\alpha - H\, \delta t)\big|_{t_1}^{t_2}, \tag{22}$$

where now

$$p_\alpha = \partial_{\mathrm{F}} L / \partial \dot{q}^\alpha \quad \text{and} \quad H = p_\alpha \dot{q}^\alpha - L. \tag{23}$$

In the usual way we now consider a continuous and closed loop $\mathscr{C}$ in the space of solutions of (21), i.e. a continuous and closed succession of solutions of (21), in which neighbouring members differ one from the next by displacements δ. We specialise to $\delta t = 0$ for the sake of simplicity and we integrate (22) with respect to δ around the loop $\mathscr{C}$. The L.H.S. becomes the integral of a total differential around $\mathscr{C}$, and this vanishes provided L is single valued in the region considered (as we assume). Therefore the terms from the two time limits on the R.H.S. must cancel. We conclude that the *circulation*

$$\oint_{\mathscr{C}} p_\alpha\, \delta q^\alpha \tag{24}$$

is a constant of the motion.

The circulation (24) is the key concept in the theory of contact transformations [11]. If we augment L by a perfect differential $dS(q, t)/dt$, then $p_\alpha \to p_\alpha + \partial_{\mathrm{F}} S / \partial q^\alpha$ and $H \to H - \partial S / \partial t$ but (24) does not change. The same will apply to the concepts which we base on (24).

All finite loops can be built from infinitesimal triangular loops. We take one such, whose vertices, in order of circulation, lie at q, $q + \delta_1 q$ and $q + \delta_1 q + \delta_2 q$, and we evaluate the circulation around it by applying SIMPSON's rule. For the circulation in this special case we introduce the symbol $\frac{1}{2} \delta_2 \times \delta_1$. Thus we find

$$\delta_2 \times \delta_1 = \delta_1 p_\alpha \cdot \delta_2 q^\alpha - \delta_2 p_\alpha \cdot \delta_1 q^\alpha = -\delta_1 \times \delta_2. \tag{25}$$

We remark that the antisymmetry property expressed by (25) has nothing to do with anticommutation.

In classical mechanics the structure $\delta_2 \times \delta_1$ would be called the Lagrange bracket of δ_2 and δ_1. Since L is even, $\delta_2 \times \delta_1$ contains only even products of anticlassical variables and their variations. In general some of these might be velocities. However, it is always possible to introduce sufficient extra variables and Lagrange multipliers to ensure that where velocities appear in L, they appear linearly. This in no way prejudices the magnitude of (25) (see [9]), and for simplicity of notation we shall therefore assume that L does depend upon the velocities in this way. Then neither the p_α nor $\delta_2 \times \delta_1$ will contain any velocities, and we shall have the formula

$$\delta_2 \times \delta_1 = \left(\frac{\partial_F p_\alpha}{\partial q^\beta} + \frac{\partial_F p_\beta}{\partial q^\alpha} \right) \delta_1 q^\beta \cdot \delta_2 q^\alpha. \tag{26}$$

A similar formula holds for classical mechanics, but with the sign inside the vracket negative.

For the F.D. oscillator (5) we find

$$\delta_2 \times \delta_1 = i\, \delta_1 q^1 \cdot \delta_2 q^1 + i\, \delta_1 q^2 \cdot \delta_2 q^2. \tag{27}$$

We now introduce the concept of *contact transformation*. A contact transformation will be defined as a circulation-preserving mapping of the set of all solutions of (21) into themselves. This is a time-independent concept. But here we shall only study it at some fixed time, and at that time it can be regarded as a mapping among the q alone (since the Lagrangian is linear in the velocities, the equations of motion contain no acceleration terms, and therefore every solution is completely specified by a set of values of the q).

Let us consider the change of circulation induced by an infinitesimal displacement field $\delta_1 q^\alpha(q)$. Denoting the change induced in any object by prefixing to the symbol of that object the symbol δ_1, we can write

$$\delta_1 \oint_{\mathscr{C}} p_\alpha\, \delta q^\alpha \equiv \oint_{\mathscr{C}+\delta_1\mathscr{C}} p_\alpha\, \delta q^\alpha - \oint_{\mathscr{C}} p_\alpha\, \delta q^\alpha. \tag{28}$$

Now $\mathscr{C}$ consists of a sequence of displacements δ and $\mathscr{C} + \delta_1 \mathscr{C}$ consists of a similar sequence translocated by the amount δ_1, so that we can fill in the region between $\mathscr{C}$ and $\mathscr{C} + \delta_1 \mathscr{C}$ by small triangles with edges δ and δ_1, after the manner of STOKES' theorem. Doing this, we find at once that (28) can be written as

$$-\oint_{\mathscr{C}} \delta_1 \times \delta, \tag{29}$$

where the act of integration is confined to the differential δ. If δ_1 is to be an infinitesimal contact transformation then the integral (29) must vanish for any closed contour $\mathscr{C}$ whatever. Therefore the corresponding line integral, evaluated from some arbitrary but fixed base point, defines an even infinitesimal function of the q throughout the solution space, and at the time considered:

$$\int^q \delta_1 \times \delta = G_1(q). \tag{30}$$

By very definition of $G_1(q)$ any anticlassical displacement δq in the solution space of the equations of motion will induce in $G_1(q)$ a change $\delta G_1(q)$ equal to

$\delta_1 \times \delta$. Thus

$$\delta G_1 = \delta_1 \times \delta. \tag{31}$$

Eq. (31) typifies a universal relation between infinitesimal generators and the displacements which they induce and can be regarded as a set of equations for the latter. For the linear systems presently under investigation it can be written more explicitly as

$$-\left(\frac{\partial_F p_\alpha}{\partial q^\beta} + \frac{\partial_F p_\beta}{\partial q^\alpha}\right) \delta_1 q^\beta \cdot \delta q^\alpha = \frac{\partial_F G_1}{\partial q^\alpha}\, \delta q^\alpha. \tag{32}$$

In the simplest cases the factors δq^α are arbitrary, and can be cancelled. However, it might be that some of the equations of motion have the form of constraints

$$\phi_a(q) = 0, \qquad (a = 1, 2, \ldots) \tag{33}$$

on the q. In such cases the displacements δq are not entirely arbitrary since, as mentioned, G_1 is defined only in the solution space. The appropriate restrictions are then

$$(\partial_F \phi_a / \partial q^\alpha)\, \delta q^\alpha = 0. \tag{34}$$

Also, by definition, the $\delta_1 q$ are displacements in the solution space, and therefore

$$(\partial_F \phi_a / \partial q^\alpha)\, \delta_1 q^\alpha = 0. \tag{35}$$

It will be shown elsewhere [9] that the combined system (32), (34), (35) can be solved for $\delta_1 q$ with any G_1, and that the solution is unique, provided only that the equations of motion (21) are mutually compatible and fully deterministic.

For the F.D. oscillator (5) there are no constraint equations, and (32) reduces to

$$-i\,\delta_1 q^1 \cdot \delta q^1 - i\,\delta_1 q^2 \cdot \delta q^2 = \frac{\partial_F G_1}{\partial q^1}\, \delta q^1 + \frac{\partial_F G_1}{\partial q^2}\, \delta q^2, \tag{36}$$

which certainly has a unique solution.

The set of linear and coordinate-covariant Eqs. (32), (34), (35) involves a matrix function $\partial_F p_\alpha / \partial q^\beta + \partial_F p_\beta / \partial q^\alpha$ which is symmetric under $\alpha \rightleftarrows \beta$ and even in the anticlassical variables. Because of this, the explicit solution by matrix inversion brings in a second symmetric and even matrix, which we denote by the bracket symbol $\{q^\alpha, q^\beta\}$, writing

$$i\,\delta_1 q^\alpha = -\{q^\alpha, q^\beta\}\, \partial_F G_1 / \partial q^\beta. \tag{37}$$

$\{q^\alpha, q^\beta\}$ is the anticlassical analogue of the anticommutator, and it is uniquely determined provided the Lagrange equations of motion are determinate. Thus the whole problem of assigning anticommutators is reduced to that of finding the solution (37) of the system (32), (34), (35).

In particular, the system in question for the F.D. oscillator is given by (36), and its unique solution is

$$i\,\delta_1 q^1 = -\,\partial_F G_1 / \partial q^1, \qquad i\,\delta_1 q^2 = -\,\partial_F G_1 / \partial q^2. \tag{38}$$

Therefore, comparing with (37), we get

$$\{q^1, q^2\} = 0, \qquad \{q^1, q^1\} = \{q^2, q^2\} = 1, \tag{39}$$

which agree with the known quantum conditions (14).

4. Consistency

One could refer the dynamical system to new anticlassical coordinates q'^{α}, odd functions of the old ones. An equation of the form (37) would still apply, since there is nothing in its derivation to limit the type of coordinates used. The tensorial structure of (37) makes it immediately apparent that we would find

$$\{q'^{\Theta}, q'^{\phi}\} = \frac{\partial_F q'^{\Theta}}{\partial q^{\alpha}} \{q^{\alpha}, q^{\beta}\} \frac{\partial_F q'^{\phi}}{\partial q^{\beta}} . \tag{40}$$

Since any odd functions $X(q)$ and $Y(q)$ might serve here as q'^{Θ} and q'^{ϕ}, it behoves us now to define quite generally

$$\{X, Y\} = \frac{\partial_F X}{\partial q^{\alpha}} \{q^{\alpha}, q^{\beta}\} \frac{\partial_F Y}{\partial q^{\beta}} = \{Y, X\}. \tag{41}$$

The last equality follows from the evenness of the factors $\partial_F X/\partial q$ etc., and the evenness and symmetry of the factor $\{q^{\alpha}, q^{\beta}\}$.

Let us now compare rule (41) with the corresponding product rule for anticommutators of odd products in quantum mechanics, i.e. with the rule

$$\begin{aligned}\{X, q^1 q^2 q^3 \cdots q^{2n+1}\} = \{X, q^1\} q^2 q^3 \cdots q^{2n+1} - q^1 \{X, q^2\} q^3 \cdots q^{2n+1} \\ + q^1 q^2 \{X, q^3\} \cdots q^{2n+1} - + \cdots .\end{aligned} \tag{42}$$

This rule is the same as the anticlassical rule

$$\{X, Y\} = \{X, q^{\alpha}\} \, \partial_F Y/\partial q^{\alpha} \tag{43}$$

in the sense that the various anticlassical terms in the R.H.S. of (43) contain the same factors as do the corresponding quantal terms of (42), and can be made to look exactly like the latter both in order and in sign by suitably changing the order of the anticlassical factors in accord with the basic anticlassical rules of anticommutation. But (43) is a consequence of (41), and by double application it leads back to (41). Therefore the decomposition expressed by (41) is isomorphic to that obtained by the product rule of quantum mechanics, apart from possible anomalies arising from the inequivalence of different signed quantal factor orderings.

Let us now turn our attention to the anticlassical analogues of commutator brackets. These arise when we consider the change $\delta_1 V$ induced in any variable V (even or odd) by the contact transformation (37). We define

$$[V, G_1] \equiv i\,\delta_1 V = -\frac{\partial_F V}{\partial q^{\alpha}} \{q^{\alpha}, q^{\beta}\} \frac{\partial_F G_1}{\partial q^{\beta}} . \tag{44}$$

Two cases now arise, according to whether V is even or odd. When V is even we may write $V = G_2$ (thereby acknowledging that V could serve as a second generator, at least to within an irrelevant infinitesimal constant factor). Then (44) becomes

$$[G_2, G_1] \equiv i\,\delta_1 G_2 = -\frac{\partial_F G_2}{\partial q^{\alpha}} \{q^{\alpha}, q^{\beta}\} \frac{\partial_F G_1}{\partial q^{\beta}} . \tag{45}$$

The outer factors in the last term of (45) are odd, and the central factor is even and symmetric, and therefore this term changes sign when G_2 and G_1 are interchanged. That is

$$[G_2, G_1] = -[G_1, G_2], \tag{46}$$

$$\delta_1 G_2 = -\delta_2 G_1. \tag{47}$$

There is an alternative and more illuminating way to derive these antisymmetry properties, which we can obtain by referring back to the fundamental Eq. (31). In this equation we are free to replace the general variation δ by the special variation δ_2, and if we do this we obtain

$$[G_1, G_2]/i \equiv \delta_2 G_1 = \delta_1 \times \delta_2. \tag{48}$$

PEIERLS' antisymmetry relation (47) follows immediately from this, by virtue of the antisymmetry of Lagrange brackets under $1 \rightleftarrows 2$ [cf. Eq. (25)].

The remarkable identity (48) holds also for ordinary classical mechanics. It equates the Poisson bracket of two generators to the Lagrange bracket of the displacements they respectively induce.

Antisymmetry properties are of less importance when the variable V is odd, since in that case it cannot serve as a generator [we recall that generators are necessarily even, by virtue of their defining Eq. (30)]. Because of this, the anticlassical mechanics contains no context in which $[G_1, V]$ could ever arise, for V odd, and we are free to define it as we please. Therefore, with quantum theory in mind, we define

$$[G_1, V] \equiv -[V, G_1] = + \frac{\partial_F G_1}{\partial q^\alpha} \{q^\alpha, q^\beta\} \frac{\partial_F V}{\partial q^\beta} : V \text{ odd!} \tag{49}$$

The anomalous sign in the last term should be noted [cf. that in (45)].

In quantum mechanics there are besides (42) two product rules for the decomposition of odd-even commutators and one product rule for the decomposition of even-even commutators. These can be analysed in the same way as (42) and with equal facility. They lead to a decomposition isomorphic to that expressed by (44), apart again from problems of ordering of factors. Thus all the rules of composition for the two types of bracket carry over properly from anticlassical mechanics to quantum mechanics.

However, this is not sufficient to ensure the possibility of a quantum analogue to an anticlassical system. The associative property of operator multiplication in quantum mechanics gives rise to two anti Jacobi identities

$$[X, \{Y, Z\}] + [Z, \{X, Y\}] + [Y, \{Z, X\}] = 0, \tag{50}$$

$$\{X, [Y, G]\} + [G, \{X, Y\}] - \{Y, [G, X]\} = 0, \tag{51}$$

and two Jacobi identities

$$[X, [G_1, G_2]] + [G_2, [X, G_1]] + [G_1, [G_2, X]] = 0, \tag{52}$$

$$[G_1, [G_2, G_3]] + [G_3, [G_1, G_2]] + [G_2, [G_3, G_1]] = 0, \tag{53}$$

where the G are even and X, Y and Z are odd. To see whether (51) is obeyed in the anticlassical mechanics we can substitute for G a sum of products of two odd factors, and then apply one of the product rules to decompose any inner brackets containing such products. Having done this we can apply the product rules to the outer brackets. Most terms then cancel, and one finds that (50) ⇒ (51). In the same way (51) ⇒ (52) and (52) ⇒ (53). By repeated decomposition of odd terms into sums of products of three odd factors one can show in much the same way that (50) will always be valid if it is true when $X=q^\alpha$, $Y=q^\beta$, $Z=q^\gamma$. If we apply definition (44) to this special case we find that we have to establish that

$$\{q^\alpha, q^\Theta\}\frac{\partial_F}{\partial q^\Theta}\{q^\beta, q^\gamma\} + \text{cyclic permutations} = 0. \tag{54}$$

This non-linear indentity is however satisfied automatically by virtue of the fact that

$$\frac{\partial_F}{\partial q^\alpha}\left(\frac{\partial_F p_\beta}{\partial q^\gamma} + \frac{\partial_F p_\gamma}{\partial q^\beta}\right) + \text{cyclic permutations} = 0. \tag{55}$$

Eq. (55) is an immediate consequence of the identity

$$\frac{\partial_F^2}{\partial q^\alpha \partial q^\beta} = -\frac{\partial_F^2}{\partial q^\beta \partial q^\alpha}. \tag{56}$$

The derivation of (54) from (55) will be given elsewhere [9]. We may also refer the reader to a work by Pauli [12], which anticipates some of the procedures used in the present article, and provides a proof of the classical version of the anticlassical identity (54) for cases where there are no constraints.

Finally, we shall look at the problem of time dependence, limiting ourselves to systems with time-translational symmetry, i.e. to cases where L does not depend explicitly on t. We shall show that in such cases the natural motion can always be regarded as a contact transformation with generator $H\,\delta t$. Obviously our definition of contact transformation precludes such a conclusion when L does depend explicitly on t. The subtleties of the latter case cannot be treated here.

Since L contains the $\dot{q}$ linearly, it follows from (23) that H does not contain them at all. Therefore

$$\frac{\partial_F L}{\partial q^\alpha} = -\frac{\partial_F p_\beta}{\partial q^\alpha}\dot{q}^\beta - \frac{\partial_F H}{\partial q^\alpha}. \tag{57}$$

Substituting this into (21) the equations of motion appear in the form

$$-\left(\frac{\partial_F p_\alpha}{\partial q^\beta} + \frac{\partial_F p_\beta}{\partial q^\alpha}\right)\dot{q}^\beta = \frac{\partial_F H}{\partial q^\alpha} \tag{58}$$

to which, if there are constraints, we may without prejudice adjoin

$$(\partial_F \phi_a / \partial q^\alpha)\,\dot{q}^\alpha = 0. \tag{59}$$

Clearly these equations contain a subset identical in structure to (32), (34), (35) but with $\delta_1 q$ replaced by $\dot{q}$. Therefore their solution is similar to (37), and is given by

$$i\,\dot{q}^\alpha = -\{q^\alpha, q^\beta\}\frac{\partial_F H}{\partial q^\beta} \equiv [q^\alpha, H]. \tag{60}$$

Although Eqs. (58–60) may fail in the presence of some types of explicit time dependence, they are certainly sufficient to indicate that the present formalism is in agreement with currently accepted quantization rules.

5. Appendix

We shall prove here that anticlassical variables cannot be represented as finite matrices. Assume the contrary, and consider the linear space $\mathcal{S}$ of vectors on which the supposed matrices $A, B, \ldots$, act. If $A\mathcal{S}$ spans $\mathcal{S}$ then so does $AA\mathcal{S}$. But $AA=0$. $\therefore A\mathcal{S}$ does not span $\mathcal{S}$. If $A\mathcal{S}$ is empty, repeat the argument for $B\mathcal{S}$. One cannot have all of $A\mathcal{S}, B\mathcal{S}, \ldots$, zero without contradicting (18). Suppose therefore that $A\mathcal{S}$ is non-empty. Then $BA\mathcal{S}=-AB\mathcal{S}$ and $B\mathcal{S}\subset\mathcal{S}$, $\therefore BA\mathcal{S}\subset A\mathcal{S}$. Thus $A\mathcal{S}$ is a non-trivial subspace of the matrices $A, B, \ldots$, and the representation is reducible. At least one of the reduced sub-matrices must be non-zero by hypothesis, since otherwise (18) is violated. We can use it to effect a further reduction within the reduced sub-matrices, and so forth.

Eventually we reduce to sub-matrices of dimension 1. The equations $AA=0$ then imply that all the one-dimensional reduced sub-matrices must be zero. This contradicts (18), and our assertion is proved.

It is also possible to show from (18) that anticlassical variables cannot be expanded over any finite basis of anticommuting Grassmann symbols. Thus these variables do not constitute a finite Grassmann algebra.

References

1. Fröhlich, H., Heitler, W., Kemmer, N.: Proc. Roy. Soc. (London), Ser. A **166**, 154 (1938).
2. Schiff, L. I.: Quantum Mechanics. New York: McGraw-Hill 1968.
   Roman, P.: Advanced Quantum Theory. New York: Addison-Wesley 1965.
   Bjorken, J. D., Drell, S. D.: Relativistic Quantum Fields. New York: McGraw-Hill 1965.
3. Wentzel, G.: Quantum Theory of Fields. New York: Interscience 1949.
4. Majorana, E.: Nuovo Cimento **14**, 171 (1937).
5. Schwinger, J.: Phys. Rev. **82**, 914 (1951); **91**, 713 (1953).
6. Peierls, R. E.: Proc. Roy. Soc. (London), Ser. A **214**, 143 (1952).
7. Burton, W. K., Touschek, B. F.: Phil. Mag., Ser. 7, **44**, 161, 1180 (1953).
8. Schwinger, J.: Phil. Mag., Ser. 7, **44**, 1171 (1953).
9. Allcock, G. R.: To be published.
10. Dirac, P. A. M.: The Principles of Quantum Mechanics. London: Oxford University Press 1947.
11. Whittaker, E. T.: Analytical Dynamics. London: Cambridge University Press 1937.
    Abraham, R.: Foundations of Mechanics. New York: Benjamin 1967.
12. Pauli, W.: Nuovo Cimento **10**, 648 (1953).

Chapter VII. Synergetic Systems

While most of this book is devoted to cooperative effects in physics, and especially to those in thermal equilibrium, the present chapter contains a number of papers on new developments in other disciplines. Most of the papers are intermittently connected with Fröhlich's ideas. Though at a first sight the various papers deal with quite different phenomena, the reader will observe that those phenomena are in many cases governed by a few very similar principles. The analogies between different systems are so fascinating and important that we see a new field of cooperative phenomena emerging which we may call "Synergetics".

Synergetics—Towards a New Discipline

H. HAKEN

Universität Stuttgart, Institut für Theoretische Physik, Stuttgart/Germany

The interests of H. FRÖHLICH have been extremely widespread, as this book demonstrates once more. One of the areas to which he has devoted his particular interest in recent years is biology: he has clearly pointed out that in biology it is certainly not sufficient merely to understand elementary processes or the structure of molecules but that it is the *cooperation* between individuals which is decisive in bringing about qualitatively novel effects. In our contribution we want to present some thoughts which are, at least in part, closely related to those of FRÖHLICH.

In *physics* scientists have been fascinated in recent years by the phenomenon of phase transitions. Here two features are especially striking:

1. A large system consisting of very many elements may show abrupt changes of certain physical quantities when a certain parameter, e.g. temperature, changes even very slightly. Well-known examples of such transitions are the liquid-gas transition, the ferromagnetic and the superconducting transition.

2. Though these systems are completely different in nature, many of their features show a striking similarity which is reflected in certain "critical exponents". Considerable progress has been achieved by the scaling laws discussed e.g. by KADANOFF [1] and recently by the fundamental papers of WILSON [2]. The scope of the present paper is, as will become clear in a moment, even wider. On the other hand we employ less sophisticated mathematical methods than those used presently in physics, preferring rather to emphasize our basic concepts.

To effect transitions from disorder to order, or transitions between different states of order, is indeed not the privilege of physical systems. It is becoming more and more evident that there are numerous examples in quite different fields like chemistry, biology and sociology, where the disorder-order transition is perhaps even more striking than in physics*: The systems in the ordered state often show actions which one might even call purposeful.

In all the cases we consider here the systems consist of very many subsystems of one or a few types, the subsystems being subject to the random actions of their surroundings. (Examples are listed in Fig. 1.) The ultimate question we seek to answer is then this: Who are the mysterious demons who tell the subsystems how to behave so to create order? Or, in more scientific language: What are the

* For a detailed account we refer the reader to "Synergetics" ed. H. HAKEN, B. G. Teubner, Stuttgart 1973.

Science	System	Subsystem
Physics	ferromagnet	elementary magnets (spins)
	laser	atoms
Chemistry	chemical ensemble	molecules
Biology	biological clocks	molecules
	neural network	neurons
Ecology	groups of animals	individual animals
	forest	individual plants
Sociology	society	human beings

Fig. 1. Examples of multi-component systems

principles by which order is created? To attack this problem, we begin in a modest way by examining the following problems: Which quantities describe multi-component systems in an adequate manner? What equations do these quantities obey? Which properties in these equations describe the order-disorder transitions?

For a physicist it is tempting to approach this problem by way of the concepts so far used in physics. There seems to be, however, at least one fundamental obstacle. Ordered states are obtained in physics by *lowering* the temperature, while in biological systems, for instance, an energy flux into the system is required to establish and maintain "order". Luckily, however, we have found a physical system which is far from thermal equilibrium, whose ordered state can only be maintained by an energy flux through it, and which shows striking similarities to a second-order phase transition [3]. This system is the laser. We will see, however, that there are also cooperative effects in which temperature or energy have at most a formal meaning, or do not appear at all. By means of explicit examples, we outline the basic concepts to deal with multicomponent systems.

The Order Parameter

Consider the ferromagnet. It loses its magnetic field when the temperature T is increased above a certain critical temperature T_c. In an atomic picture the ferromagnet consists of many spins, which are in the disordered state for $T > T_c$ and in an ordered state for $T < T_c$. At $T = T_c$ the phase transition [4] occurs. As is well known, the ordering is explained as follows. There is a certain force (the Coulomb exchange interaction) which tries to bring two spins into parallel alignment, so what one ultimately has to deal with is the interaction of each spin with a number of other spins. The interaction is counteracted by the random temperature motion of the single elementary magnets. If we imagine that there are about 10^{22} spins per cm^3, it at once becomes evident that this many-body problem is extremely difficult and it is clear that it is absolutely hopeless to investigate the motion of each individual spin. Even if we could do so, it would not help us one bit in understanding the behaviour of the whole system. We will do better to discard all unnecessary information about the individual spins* and replace

* This thought has, by the way, very serious consequences in the sociological regime: For a company manager it is absolutely unnecessary to have a detailed knowledge of the internal functioning of all the departments in the company; indeed this would be more likely to confuse him. What he needs to know are just the *relevant* input-output features of each department.

it by a new quantity which relates to the total system. This problem, to which a great deal of effort has been devoted in theoretical physics, has been treated by using the concept of the *order parameter*. Though the modern theory of phase transitions goes beyond this concept, the concept of the order parameter seems well suited to the field of synergetics. It enables the direct interaction between the elementary magnets to be replaced by a two-step procedure. First a fictitious mean field is assumed to be generated by all the individual spins; in the next step this mean field is assumed to react on the individual spins. The order parameter has at least two functions. It describes order, because it is zero in the disordered state and assumes its maximum value in the completely ordered state. On the other hand, it gives orders to the individual elementary magnets by exerting a certain force on them. In more modern language one would say that the order parameter gives *instructions* to the subsystems. Order parameters need not necessarily be fictitious quantities; for instance, in the laser the order parameter is the light field, which is generated by all the atoms. The concept of the order parameter is not limited to systems like ferromagnets which are in thermal equilibrium. It applies equally well to physical systems far from thermal equilibrium. We just mentioned the laser [5], but the order parameter also applies to chemical ensembles and other systems. Examples are listed in Fig. 2.

The order parameter and the subsystems (in our example the spins) form the first step in a hierarchical system. In hierarchical systems there exists also a *hierarchy of time constants* (relaxation times, response times). The subsystems usually have time constants which are much smaller than those of the higher system, i.e. the subsystems immediately follow the changes of the order parameter. This has an important consequence: If the behaviour of the subsystems can be described by differential equations, the coordinates of the subsystems may be eliminated adiabatically. Finding the appropriate order parameters (or macroscopic variables) is sometimes trivial, but in other cases it is the very crux of the problem. Remember that, e.g. in superconductivity, it is not at all obvious that the pair wave function is the order parameter. This may be even more true in the biological domain.

To demonstrate the usefulness of the order parameter concept, we now discuss a few examples of equations for these quantities. Without claiming to present details, we first list some possibilities for deriving such equations:

System	Order parameter
Ferromagnet	mean field
Superconductor	pair wavefunction
Laser	lightfield or photon number
Chemical ensemble	numbers of molecules n_j
Biological clocks	numbers of molecules n_j
Neural network	pulserate
Animals	numbers of animals
Forest	density of plants
Society	number of people of given opinion

Fig. 2. Examples of order parameters

a) by derivation from a microscopic theory (example: ferromagnet, laser, nonlinear optics), by looking for "soft modes" or constructing suitable macroscopic variables or collective coordinates;

b) by the application of general principles, e.g. symmetry principles; this is the basis of the Landau theory of phase transitions;

c) by plausibility arguments (e.g. population dynamics, Volterra-Lotka problem).

Equations for Order Parameters

We present here some typical examples of such equations taken also from other fields besides physics.

a) Population Dynamics (Involving One Species)

We consider an equation which describes the increase of a population, which may be a human or an animal population. This equation, however, holds equally well for a single-mode laser or for an autocatalytic chemical reaction. The dynamics is described by the equation

$$\frac{dn}{dt} = (\alpha - \beta n) n \qquad (+F(t)) \tag{1}$$

where n is the number of individuals (human beings, photons, molecules). The growth rate $\frac{dn}{dt}$ is proportional to the number of individuals n with the rate factor α. If $\alpha < 0$, the "death" rate is greater than the "birth" rate and the population dies out. The solution for the stationary state reads $n_0 = 0$. If $\alpha > 0$, an instability occurs. First the number of individuals increases exponentially. If the resources, which are always limited (food, excited laser atoms, other molecules), become depleted, growth saturation occurs, as described by the factor $-\beta n$ in Eq. (1). The term in brackets, $F(t)$, describes the creation of photons by spontaneous emission, or the random generation of molecules by reactions other than autocatalytic.

b) Population Dynamics (Involving Two Species)

An important example is the famous Lotka-Volterra problem [6]. We consider a biological system with two sorts of fishes. One sort of fish, the predator, lives only on small fishes, the prey. The small fishes are assumed to have an unlimited food supply. Because in the absence of the prey the predators would die out, the rate of change of the predator population is given by $-\gamma_1 n_1$. Conversely, the number of predators will increase if there is food supply due to the presence of the prey. The growth rate is proportional to the number of predator fishes present and to the number of the prey. Thus we obtain for the predators the equation

$$\frac{dn_1}{dt} = -\gamma_1 n_1 + \beta_{12} n_1 n_2 \tag{2}$$

and for the prey by similar considerations

$$\frac{dn_2}{dt} = \gamma_2 n_2 - \beta_{21} n_1 n_2. \tag{3}$$

The above-mentioned authors found that Eqs. (2) and (3) have an oscillatory solution which can be understood as follows. First only a few predators are present, so the number of prey fishes increases. Then the number of predators increases by the fact of eating more and more of the prey fishes. The number of prey fishes then declines and, due to lack of food, finally the number of predator fishes decreases. Then the cycle can start again. Equations of the type (2), (3) also hold for certain autocatalytic chemical reactions. This leads to the conclusion that chemical clocks must exist, as has indeed been demonstrated in beautiful experiments. A somewhat more complicated system found by ZHABOTINSKY [7] shows an oscillatory change of colour from red to blue. Chemical reactions being the foundation of biological events, this model also offers an understanding of the action of biological clocks (and perhaps eventually of the control mechanism of the heart).

c) Equations for Order Parameters in More Complex Systems

The Eqs. (2) and (3) can be generalized in a straight-forward manner for systems with several constituents. They read

$$\frac{dn_j}{dt} = \varepsilon_j n_j - \sum_i \beta_{ij} n_i n_j \vdots + F_j(t)\,. \qquad (4)^*$$

For special choices of the coefficients β_{ij} they have already been treated by VOLTERRA. They describe population dynamics, chemical reactions, and biological cells. They apply to nonlinear transport and to the multimode laser [8]. In these examples the n's are used as adequate numbers or densities. If the β's are slightly generalized, and if the n's are interpreted as pulse rates, Eq. (4) have been used by COWAN [9] to describe neuronal networks.

Let us discuss the multimode laser in slightly more detail. We consider excited atoms in a cavity which start to emit photons in random directions (and also randomly distributed over a certain frequency range). A discussion of Eq. (4) for the laser in which all β's have positive sign then shows the following. First the number of photons of different direction and energy increases, as shown in Fig. 3, but then the nonlinear term becomes dominant and this leads to the selection of a single mode, the one which has the highest gain and the lowest loss [8]. All other modes disappear.

Our laser equations thus describe selection [10]. They apply equally well to autocatalytic reactions. Indeed these equations were more recently used by EIGEN [11] to explain the operation of selection within the process of evolution, which finally led to the genetic code. Contact may be also made with some extremely interesting ideas put out by FRÖHLICH. He suggested at the first Versailles conference ("From Theoretical Physics to Biology") that there may exist in biological systems highly excited longitudinal electric oscillations which play a decisive role in the organization of living entities.

The above examples, to which many others could be added, show convincingly that utterly different phenomena are governed by the same equations or, more generally speaking, by the same basic principles. This discovery has,

* In Eq. (4) the random "forces" $F(t)$ again describe random generation of individuals, e.g. in the laser process or in chemical reactions.

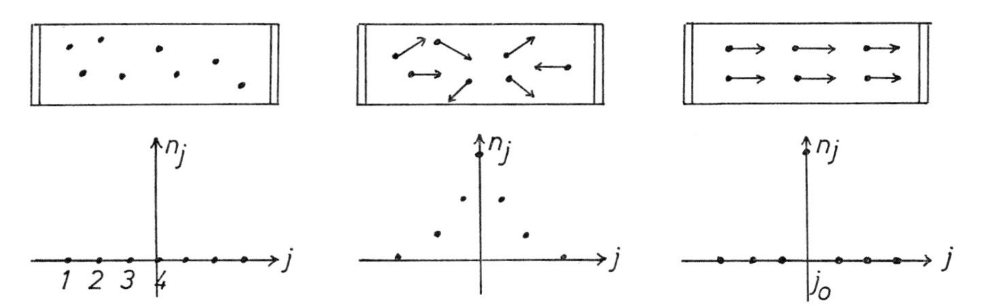

Fig. 3. Photon selection in a laser. The top row sketches a laser with two end mirrors. The dots symbolize excited atoms, the arrows photons emitted in the corresponding directions. The bottom row represents the photon number n_j of the photon kind j. From left to right: Before laser action starts the atoms are excited, no photons are present. Laser action starts, photons are emitted in random positions etc. and are selected according to the different gain. The dots in the middle lower part of Fig. 3 are thus representative of the gain curve. When a stationary state is reached, all kinds of photons but one have died out (right-hand side of Fig. 3)

of course, important consequences. Thus, if one has understood the "mechanism" of a certain phenomenon in one field, one may easily transfer the results to another field. In some cases it may not be possible to solve the nonlinear stochastic equations exactly, yet a wealth of empirical statistical data may be available about one particular field. In this case, the known field can serve as an analogue for another field (e.g. lasers for evolutionary steps).

Systems far from Thermal Equilibrium. Stability, Instability, Broken Symmetry, Fluctuations, Critical Slowing Down, Disorder Transition, Coherence

We now want to show what typical conclusions can be drawn from order-parameter equations. For this purpose we choose the single-mode laser as a simple but nontrivial example. In Fig. 4 the emitted light intensity of a laser ("power output") is plotted against the pump rate ("power input"). For small pump rate the light intensity increases very slowly (small slope of the curve) · in other words, the efficiency of the light source, which is a just perceptible slope, is very small. This is the region of completely incoherent radiation (disordered state).

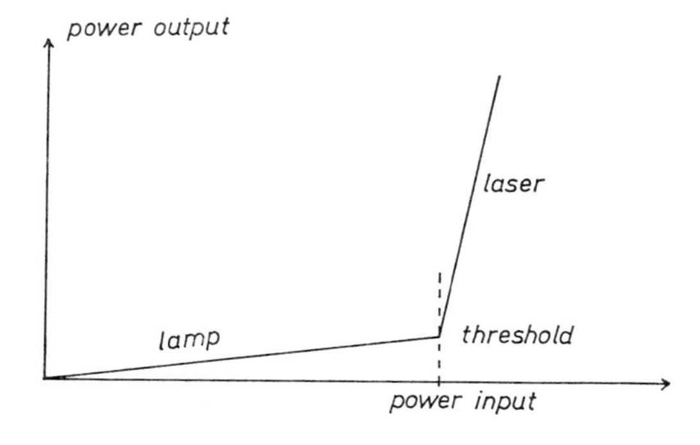

Fig. 4.
Power output versus power input of a laser. Schematic diagram

When a certain threshold of the pump rate is reached, the light intensity increases very rapidly. The efficiency is very high. This is brought about by coherent emission (the ordered state). The analogy to phase transition in thermal equilibrium is striking if we equate the specific heat to the efficiency.

To follow up this analogy more closely, we discuss the order parameter and the equation it obeys. Instead of the photon number n, which occurred e.g. in Eq. (1), we now choose the electric field strength E of the laser light*. E is connected with the photon number n by

$$E = \sqrt{n}\, e^{i\phi} \quad \text{or} \quad n = |E|^2 \tag{5}$$

(besides a constant factor).

Because E depends on the phase ϕ, it contains more information than n. In the microscopic theory one starts from equations for the electromagnetic field, which is coupled to all the laser-active atoms. These equations contain obviously an enormous number of degrees of freedom, which are practically all connected with the rapid decay and pumping processes of the atoms. In distinct contrast to this, there is a macroscopic, slowly varying quantity, the electric field mode. Due to the difference in the time constants, the atomic variables can be eliminated adiabatically and we end up with an extremely simple equation for the mode amplitude E [12]:

$$\dot{E} = \alpha' E - \beta' |E|^2 E \quad (+F(t)) \qquad (6)^{**}$$

for a single mode laser. α' is the so-called unsaturated gain, β' the saturation constant. The meaning of α' and β' becomes obvious when we use Eq. (5). Then Eq. (1) follows with $\alpha = 2\alpha'$, $\beta = 2\beta'$. $F(t)$ is again a fluctuating force [different from that in Eq. (1)] describing spontaneous emission. Eq. (6) is one of the simplest examples, but very instructive for the study of the interplay between fluctuating forces [here $F(t)$] and systematic forces (here $\alpha' E - \beta' |E|^2 E$) or, in MONOD's words [13], of the interplay between "chance and necessity". After we have solved Eq. (6), we can insert E into the equations of motion of each individual atom, where E now acts as a slowly varying parameter by means of which each atom is "slaved" (a technical expression known to electrical-systems engineers).

If we interpret Eq. (6) as the equation of heavily overdamped motion*** of a particle with coordinate E, we can interpret the first two terms on the right-hand side of Eq. (6) as a force with a potential

$$V(E) = -\frac{\alpha'}{2} |E|^2 + \frac{\beta'}{4} |E|^4 \tag{7}$$

* More exactly, if E_{tot} is the electric field strength of the laser light, we put $E_{\text{tot}}(x, t) = (E(t)\, e^{i\omega_0 t} + E^* e^{-i\omega_0 t}) \sin k_0 x$, where ω_0 is the cavity frequency, k_0 the wave number and $E(t)$ the slowly varying amplitude.

** In the case of an infinitely extended laser with a continuous number of modes, the expression ΔE must be added to the right-hand side of Eq. (6). In this case E is a function of time *and space*.

*** To see this explicitly, just add an acceleration term $m\ddot{E}$ with a very small "mass" to the left-hand side of Eq. (6).

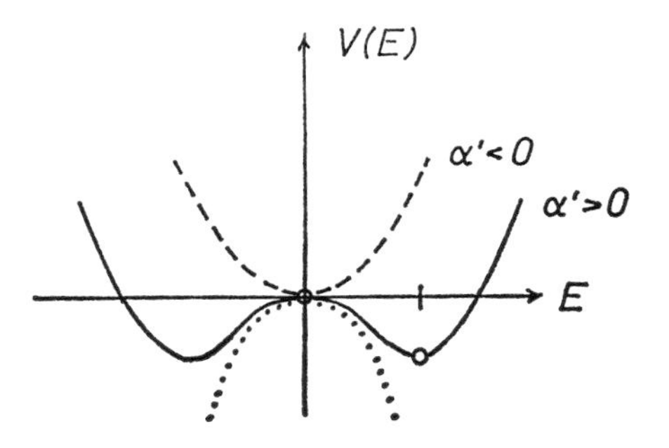

Fig. 5.
The potential $V(E)$, two-dimensional plot. Dashed curve: disordered state, incoherent emission; dotted curve: instability for $\alpha' > 0$; solid line: situation with broken symmetry

which is plotted in Fig. 5. For $\alpha' < 0$, the point $E = 0$ is stable. When the "particle" is pushed away from $E = 0$ by the random force $F(t)$, it falls down the potential wall and returns to $E = 0$. This is the completely incoherent case: the "order parameter" E cannot give any orders ("instructions") to the atoms.

When we let the parameter α' approach zero, the restoring force decreases and $|E|$ comes down more and more slowly ("*critical slowing down*" near an instability). At the same time the fluctuations of E caused by $F(t)$ are growing bigger and bigger ("critical fluctuations").

For $\alpha' \gtrsim 0$, the position $E = 0$ becomes unstable. This *instability is connected with a broken symmetry*. The system has now the choice between all states for which $|E| = E_0$, and an arbitrary optical phase $\phi = \phi_0$. At $\alpha' = 0$, where E changes from $E = 0$ to $E = |E_0| e^{i\phi_0}$, a "phase transition" occurs which is very similar to a second-order phase transition of systems in thermal equilibrium (see below)*.

Eq. (6) was discussed by us some years ago [10] as a model to understand how order is built up from disorder, in particular in biological systems. Consider Fig. 5 and let $E = 0$ (disordered state) at $t = 0$. The system would rest there for ever if there were no fluctuations $F(t)$. But these fluctuations rise because of the "systematic" forces in Eq. (6) to a coherent, macroscopic state.

Because "symmetry-breaking instabilities" play an important role in synergetics, we mention two further examples: molecules which differ in only one property, e.g. optical dichroism, but have identical α, β, and F; DNA or RNA which differ by the arrangement of their constituents but again have the same production factors α. Another example is provided by molecules whose concentration is space- or time-dependent. In the broken-symmetry case, for example, *spatially or temporally periodic* molecule concentrations may occur. Such systems and related ones have been studied in detail by PRIGOGINE [14] and coworkers and were called "dissipative structures". The concept of broken symmetry, which in modern physics plays a fundamental role ranging from elementary-particle physics to solid-state physics, acquires a completely new aspect when applied for example to chemistry or biology. Broken symmetry then means that

* The question as to whether the analogy with the usual second-order phase transition is complete, in particular when the transition is sharp, has been the subject of heated discussions. The answer is this: If we admit a continuum of laser modes, the transition is completely analogous to the smeared-out transition of the one-dimensional superconductor. On the other hand, the *single-mode* laser shows a sharp transition with critical exponents in a well-defined (generalized) thermodynamic limit (DOHM, private communication).

the structure of the system becomes richer on a macroscopic level. In the case of the laser one can show that the structure becomes richer and richer (occurrence of multiple pulse) with increasing energy input. On the other hand, there are certainly systems where increasing energy input simply destroys the structure. Thus we have the important but still unsolved problem: Under what conditions is the ever-increasing differentiation of structure achieved?.

Analogies with respect to instabilities include a large variety of systems, e.g. plasmas [15], or current instabilities [16]. A further example is provided by different states of interstellar matter [17]. Instabilities (or multi-stability) of exactly the type treated here are also found in visual perception and in the sociological model of WEIDLICH [18].

We finally come back to the question how close an analogy can exist between a system in equilibrium and one far from thermal equilibrium. LANDAUER [19] has argued that there is such an analogy when the principle of detailed balance is obeyed. Indeed, it has been shown [20] that in this case the distribution function f can be written by means of generalized thermodynamic potential G in the form

$$f = \mathfrak{N} e^{-G}. \tag{8}$$

A consequence of this is demonstrated in Fig. 6 where the distribution function of laser light with a continuum of modes is compared with that of the pair wave function of the Ginzburg-Landau theory. The analogy is obviously perfect. If the detailed balance principle holds, the fluctuation and dissipation theorem of KUBO [21] can be extended to systems far from thermal equilibrium, as has been demonstrated in different ways by AGARWAL [22] and WEIDLICH [23].

$$f = N \exp\left(-\frac{B}{Q}\right)$$

$$B = \int \left\{\tilde{\alpha}|E(x)|^2 + \tilde{\beta}|E(x)|^4 + \tilde{\gamma}\left|\left(\frac{d}{dx} - i\frac{\omega_0}{c}\right)E(x)\right|^2\right\} dx$$

E electric field strength; $\tilde{\alpha} = \tilde{a}(d_c - d)$; $\tilde{a}, \tilde{\beta}, \tilde{\gamma} > 0$ are laser constants; c velocity of light; ω_0 atomic line frequency; N normalization factor; d atomic inversion; d_c critical atomic inversion; B is a generalized thermodynamic potential; Q is the thrength of the fluctuations.

The distribution function of the laser

$$f = N \exp\left(-\frac{F}{kT}\right)$$

$$F = \int \left\{\alpha|\psi(\underset{\sim}{x})|^2 + \beta|\psi(\underset{\sim}{x})|^4 + \frac{1}{2m}\left|\left(\underset{\sim}{\nabla} - \frac{2ei}{c}\underset{\sim}{A}\right)\psi(\underset{\sim}{x})\right|^2\right\} d^3x$$

ψ pair wave function; $\alpha = a(T - T_c)$ a, β superconductor constants; k Boltzmann's constant; T absolute Temperature; T_c critical Temperature; F the free energy; m, e mass and charge of electrons respectively.

The distribution function of the Ginzburg-Landau theory of superconductivity.

Fig. 6. Comparison of the statistical distribution function of the laser light amplitude with that of the Ginzburg-Landau theory of superconductivity

Conclusion

In our article we have attempted to outline a few aspects of cooperative effects in multicomponent systems. We have tried, quite in HERBERT FRÖHLICH's spirit, to represent these concepts by means of simple equations. We are sure that it will not be difficult to work out more sophisticated mathematical theories.

References

1. See e.g. KADANOFF, L. P., GÖTZE, W., HAMBLEN, D., HECHT, R., LEWIS, E. A. S' PALCIANSKAS, V. V., RAYL, M., SWIFT, J., ASPNES, D., KANE, J.: Rev. Mod. Phys. **39**' 395 (1967).
2. WILSON, K. G.: Phys. Rev. B4, 3174 (1971).
   WILSON, K. G., FISHER, M. E.: Phys. Rev. Letters **28**, 4 (1972).
3. GRAHAM, R., HAKEN, H.: Z. Physik **213**, 420 (1968).
4. For a review on phase transitions see e.g. BROUT, E.: Phase Transitions. New York: Benjamin Inc. 1965.
5. See Ref. [3] and the article by GRAHAM, R., in: Synergetics. Stuttgart: Teubner 1972.
6. LOTKA, A. J.: Elements of Math. Biology, 1924 (Dover Reprint, 1956).
   VOLTERRA, V.: Lécon sur la théorie mathematique de la lutte pour la vie. Paris: Gauthier-Villars 1931.
7. ZHABOTINSKY, A. M.: Biofizika **2**, 306 (1964), Russ. J. Phys. Chem. **42**, 1649 (1968).
8. HAKEN, H., SAUERMANN, H.: Z. Physik **176**, 47 (1963).
9. WILSON, H. R., in: Synergetics, ed. H. HAKEN. Stuttgart: Teubner 1972.
10. The application of laser-type equations to ordering processes in biology has been proposed by HAKEN, H., in: Theoretical Physics and Biology, ed. M. MAROIS. Amsterdam: North-Holland Publ. Comp. 1969.
11. EIGEN, M.: Naturwissenschaften **58**, 10, 465 (1971).
12. HAKEN, H.: Z. Physik **181**, 96 (1964);—Handbuch der Physik, Bd. XXV/2c. Berlin-Heidelberg-New York: Springer 1970 and, in connection with phase transitions, in Festkörperprobleme, Bd. X (1970), Ed. O. MADELUNG, Marburg (Pergamon, Vieweg).
13. MONOD, J.: Le hasard et la necessité, Ed. DU SEUIL. Paris 1970.
14. GLANSDORFF, P., PRIGOGINE, I.: Thermodynamic Theory of Structure, Stability and Fluctuations. New York: Wiley 1971.
15. KLIMONTOVICH, YU. L.: The Statistical Theory of Non-equilibrium Processes in a Plasma. Oxford, London, Edinburg, New York, Toronto, Sydney, Paris, Braunschweig: Pergamon 1967.
16. THOMAS, H., in: Synergetics, Ed. H. HAKEN. Stuttgart: Teubner 1973.
17. GRAHAM, R., Langer, W.: Astrophys. J. **179**, 469 (1973).
18. WEIDLICH, W.: l.c. [16].
19. LANDAUER, R.: IBM Res. RC 2960 (1970).
20. GRAHAM, R., HAKEN, H.: Z. Physik **243**, 289 (1971).
21. KUBO, R.: J. Phys. Soc. Japan **12**, 570 (1957);—Rept. Progr. Phys. **29**, 255 (1966).
22. AGARWAL, G. S.: Z. Physik **252**, 25 (1972).
23. WEIDLICH, W.: Z. Physik **248**, 234 (1971).

Irreversibility and Dissipativity of Quantum Systems

I. PRIGOGINE* and A. P. GRECOS

Université Libre de Bruxelles, Faculté des Sciences, Brussels/Belgium

Abstract

We review some basic properties of the general collision operator as defined in our theory and its relation to the invariants of the motion of a large quantum system. The necessity of the dissipativity condition is shown for the possibility of kinetic description where the $L \cdot t$-invariance of the von Neumann equation is broken. We illustrate this condition by a number of simple models where the collision operator can be calculated in a rigorous (non-perturbative) way.

1. Introduction

It is a pleasure to dedicate this paper to Professor H. FRÖHLICH. The new possibilities which exist today to describe the dynamics of large systems and to investigate the emergence of new properties such as irreversibility in a well-defined asymptotic limit are due to the remarkable progress in many-body theory. This is a field to which Prof. FRÖHLICH has made outstanding contributions. His pioneering work will have a lasting influence on all fields of physics connected with many body problems.

It is a well-established experimental fact that the behaviour of macroscopic systems is "irreversible". The second law of thermodynamics expresses this irreversibility in a phenomenological way through the corresponding positive entropy production. Therefore one of the main problems of non-equilibrium-statistical mechanics is to show that such behaviour is compatible with the reversible nature of the underlying microscopic mechanical laws. In particular, it is important to investigate the appearance of irreversibility when the limit of a "large" system is taken. That such a limit is necessary follows immediately from POINCARÉ's theorem which states that recurrence cycles are inevitable if the system is confined in a finite region of the phase space.

Consider a system with a Hamiltonian H. Its state, given by a density matrix ϱ, evolves according to the von Neumann equation

$$i\, \partial_t \varrho = L \varrho = [H, \varrho]. \tag{1.1}$$

This equation is manifestly reversible. More precisely, it is invariant under the transformation $\{L \to -L,\ t \to -t\}$. However, this "$L \cdot t$-invariance" of the

* Also Center for Statistical Mechanics and Thermodynamics, The University of Texas at Austin, Austin, Texas.

dynamic equation does not mean that its solutions share this property. Indeed, in the integral representation of the solution we have *in addition* to take into account supplementary conditions such as causality. To clarify this important point, let us write the formal solution of Eq. (1.1) in terms of the resolvent of the von Neumann operator L

$$\varrho(t) = \frac{1}{2\pi i} \int_{\vec{C}} dz\, e^{-izt} \frac{1}{L-z}\, \varrho(t=0). \tag{1.2}$$

Here $\vec{C}$ is a contour parallel to the real axis. It lies above $(\vec{C}_+)$ or below $(\vec{C}_-)$ the real axis according to whether we consider $\varrho(t)$ for positive or negative values of the time variable. These two cases lead in general to qualitatively different functional relations between $\varrho(t)$ and its value $\varrho(0)$ at $t=0$. However, if the spectrum of L is discrete, no real distinction can be made between a "retarded" and an "advanced" solution. We may conclude therefore that L must have a continuous spectrum in order to show the irreversible behaviour which would violate the $L\cdot t$-invariance of Eq. (1.1). It is well known that, at least for quantum systems, a continuous spectrum for L is possible only in the limit of an infinite volume and eventually of infinite number of particles (the so-called thermodynamic limit), which of course is in accordance with the content of POINCARÉ's theorem.

For systems with continuous spectra, $[L-z]^{-1}$ as a function of z has a cut along the real axis $(L=L^+)$. In order to study the properties of Eq. (1.2) for $t\gtrless 0$ we need to study the analytic behaviour of the branches of $[L-z]^{-1}$ defined for $Im\, z\gtrless 0$. This behaviour, and in particular the possibility of analytic continuation, depends on the representation with which we are working. The choice of representation is, of course, related to the "language" we use to describe the physical phenomena taking place in the system and the conditions of observation.

Here we assume that the Hamiltonian is the sum of an "unperturbed" one H_0 and a "perturbation" λV

$$H = H_0 + \lambda V \tag{1.3}$$

and that we are working in the representation where H_0 is diagonal. It is obvious that corresponding to Eq. (1.3) we may decompose the von Neumann operator as

$$L = L_0 + \lambda L_1. \tag{1.4}$$

Although this decomposition can be avoided as long as we are dealing with formal aspects only, it is useful in the interpretation of certain results. Furthermore, it appears in a natural manner in several important physical problems.

Up to now we have stressed the importance of the thermodynamic limit in the problem of irreversibility. However, we must also insist that this limit is not by any means a sufficient condition, at least if we want to deduce from our model finite transport properties or even more a microscopic interpretation for the entropy concept of non-equilibrium thermodynamics. In fact, the systematic approach to the problem of a kinetic description for systems with a large number of particles has made it clear that such a description is possible only if the collision operator satisfies in addition the so-called dissipativity condition [1].

In the present article we review some of the properties of the resolvent of L which depend on the asymptotic collision operator Ψ (see § 2) and which are directly related to the stationary solutions of Eq. (1.1), i.e. the invariants of motion.

Although the collision operator does not in general determine the asymptotic evolution of the density matrix (except in the weak coupling limit or the limit of a dilute gas), it yields a relatively simple criterion for the *dissipativity* of the physical system. For many-body systems we must construct Ψ by a perturbation method. Thus the conclusions depend on the convergence of the expansions involved. Nevertheless, in some simple cases where only the infinite volume limit is taken, a rigorous non-perturbative expression can be given for Ψ. In the third section of our paper, we present simple examples. The final section is devoted to a brief discussion of the concept of dissipativity and its relation to the ergodic behaviour of the system.

To end this introductory section let us point out that our considerations apply to classical systems as well. In this case it is possible to construct a non-trivial collision operator even for finite systems [2] by means of a perturbation expansion. This is due to the fact that L, the Liouville operator defined by the Poisson bracket $i\{H, \ \}$, may have a continuous spectrum even if the motion of the system is confined to a finite part of the phase space.

2. The Asymptotic Collision Operator

We consider a projection operator P such that $P\varrho$ is the diagonal part of the density matrix in a given representation. According to the choice we made in the previous section, cf. Eqs. (1.3) and (1.4), P projects into the null space of L_0. We express next the resolvent of L in terms of the resolvent of QLQ [3], where $Q=1-P$,

$$\frac{1}{L-z}=\{P+\mathfrak{C}(z)\}\frac{1}{PLP+\Psi(z)-z}\{P+\mathfrak{D}(z)\}+\frac{1}{QLQ-z}Q\,. \tag{2.1}$$

In writing this decomposition we have defined the collision operator

$$\Psi(z)=-PLQ\frac{1}{QLQ-z}QLP \tag{2.2}$$

the destruction operator

$$\mathfrak{D}(z)=-PLQ\frac{1}{QLQ-z} \tag{2.3}$$

and the creation operator

$$\mathfrak{C}(z)=-\frac{1}{QLQ-z}QLP \tag{2.4}$$

which play an important role in the kinetic description. Note that, because of the definition of P, $PLP=0$. The terminology stems from the fact that the decay of the initial correlations [off-diagonal elements of $\varrho(0)$] depends on the properties of $\mathfrak{D}(z)$, and the new correlations which appear because of the interactions are characterized by $\mathfrak{C}(z)$ (for more details see Ref. [1]).

By introducing Eq. (2.1) into the right-hand side of Eq. (1.2) we obtain the generalized master equation [1]

$$i\,\partial_t\,\varrho_0(t) = PLP\,\varrho_0(t) + \int\limits_0^t d\tau\,G_{00}(t-\tau)\,\varrho_0(\tau) + \frac{1}{2\pi i}\int\limits_{\overrightarrow{C}} dz\,e^{-izt}\,\mathfrak{D}(z)\,\varrho_c(0) \qquad (2.5)$$

which describes the evolution of $P\varrho(t) \equiv \varrho_0(t)$ $(Q\varrho \equiv \varrho_c)$. The operator $G_{00}(t)$ is the inverse Laplace transform of $\Psi(z)$. We may note here that both Eq. (2.5) and Eq. (2.1) are formally valid for any choice of P. However, the possibility of an *asymptotic* description of the physical system in terms of a Markovian equation depends on P. Such an equation can be obtained only if the operators $\Psi(z)$, $\mathfrak{D}(z)$ and $\mathfrak{C}(z)$ have well-defined limits as $z \to +i0$.

The "local" form of the master equation is obtained from Eq. (2.5) by assuming that the term depending on the correlations can be neglected and that the evolution term can be approximated as follows

$$\begin{aligned}\int\limits_0^t d\tau\,G_{00}(\tau)\,\varrho_0(t-\tau) &\approx \left[\int\limits_0^t d\tau\,G_{00}(\tau)\right]\varrho_0(t)\\ &\approx \left[\int\limits_0^\infty d\tau\,G_{00}(\tau)\right]\varrho_0(t) = \Psi(+i0)\,\varrho_0(t).\end{aligned} \qquad (2.6)$$

Thus, to a first approximation, the asymptotic description of the system, i.e. in a time scale characterized by a relaxation time, is governed by the equation

$$i\,\partial_t\,\varrho_0(t) = PLP\,\varrho_0(t) + \Psi(+i0)\,\varrho_0(t). \qquad (2.7)$$

Eq. (2.7) is the simplest example of an equation providing a kinetic description which is not invariant under time inversion and a change of L to $-L$. The $L\cdot t$-invariance of the initial Eq. (1.1) has been broken. To see this in a formal manner let us take the limit $z \to +i0$ in Eq. (2.2), we obtain [4]

$$\Psi(+i0) = -PLQ\frac{\mathscr{P}}{QLQ}QLP - i\pi\,PLQ\,\delta(QLQ)\,QLP. \qquad (2.8)$$

It is clear that the first term in the above expression, which is the Hermitian part of $\Psi(+i0)$, changes sign when $L \to -L$. On the other hand the second term, the anti-Hermitian part, remains invariant under such a substitution. We may call these two parts "odd" and "even" respectively and write

$$\Psi(+i0) = \overset{(0)}{\Psi}(+i0) + \overset{(e)}{\Psi}(+i0). \qquad (2.9)$$

Furthermore it is easy to see that the operator $-i\,\Psi(+i0)$ is a *dissipative operator* because its Hermitian part is non-positive. We formulate then the *dissipativity condition* as

$$\overset{(e)}{\Psi}(+i0) \neq 0. \qquad (2.10)$$

This is a necessary condition for the approach to a stationary state of the solution of Eq. (2.7).

The way we introduced the asymptotic master equation in this section does not permit us to answer in a consistent manner the questions related to the approach to an equilibrium distribution. These questions bring us to the problem of constructing *independent subdynamics* [5], or more generally *causal dynamics* [6], which expresses the symmetry-breaking, i.e. the loss of the $L \cdot t$-invariance of Eq. (1.1), through a non-unitary transformation. An extensive review of the recent work of the Brussels group on these problems may be found in Ref. [7].

3. Properties of the Invariants of Motion

The importance of the properties of the asymptotic operators $\Psi(+i0)$, $\mathfrak{D}(+i0)$ and $\mathfrak{C}(+i0)$ follows also from the fact that these operators, under certain conditions, characterize the null space of L. Consequently they determine the class of possible equilibrium states or more generally the class of stationary states of the system.

The decomposition of the resolvent of the von Neumann operator shows that its behaviour for z near the origin is determined by that of $\Psi(z)$, provided that QLQ has no zero eigenvalue. Suppose that ϕ is an invariant of motion for the system, i.e.

$$L\phi = 0. \tag{3.1}$$

Then it can be shown [8] that its components $P\phi = \phi_0$ and $Q\phi = \phi_c$ satisfy the relations

$$\{PLP + \Psi(z)\}\phi_0 + z\,\mathfrak{D}(z)\,\phi_c = 0 \tag{3.2a}$$

and

$$\phi_c = \mathfrak{C}(z)\,\phi_0 - z\frac{1}{QLQ - z}\,\phi_c. \tag{3.2b}$$

The converse statement is also true, i.e. Eqs. (3.2) imply Eq. (3.1). Furthermore, if $\mathfrak{D}(z)\,\phi_c$ and $[QLQ - z]^{-1}\phi_c$ have finite limiting values as $z \to +i0$, the above equations become

$$\{PLP + \Psi(+i0)\}\phi_0 = 0 \tag{3.3a}$$

and

$$\phi_c = \mathfrak{C}(+i0)\,\phi_0. \tag{3.3b}$$

We may classify the invariants as *regular* or *singular* (with respect to the collision operator) according to whether or not they satisfy Eqs. (3.3). Clearly a non-vanishing collision operator represents a restriction to the class of possible constants of the motion for the perturbed system. We may note also that for finite classical systems Eq. (3.3b) does not admit in general an expansion in the coupling constant λ. This property is related to a well known theorem of Poincaré [9] proving the non-existence of analytic invariants (except trivial ones) for a large class of Hamiltonians.

For a system with discrete spectrum, if P projects *onto* the null space of L_0 and if QLQ is invertible, Eq. (3.3a) measures in a sense the effect of the perturbation L_1 on the invariants of L_0. In effect, because of Eq. (3.3b), the invariants of L are in one-to-one correspondence with the vectors of the null space of $PLP + \Psi(+i0)$. Therefore the dimension of the subspace of the vectors $P\phi$ orthogonal to the solutions of Eq. (3.3a) is equal to the number (finite or even infinite) of the invariants of L_0 which "disappear" because of the perturbation.

Here we may note that, if H_0 and H have non-degenerate discrete spectra, the above argument implies that $\Psi(0)$ vanishes identically. This can be shown also [10] using the properties of the unitary transformation which diagonalizes H. This remark implies that for finite quantum systems there exists no non-trivial collision operator. In certain cases one may construct a non-vanishing Ψ. For instance, if H_0 has a degenerate spectrum and P projects *onto* the null space of L_0, $\Psi(0)$ may be finite. This is due to the fact that, because of the perturbation, the degeneracy of the spectrum may be, at least partially, removed. However, it is evident that the dissipativity condition, Eq. (2.10), is not satisfied because $\Psi(0)$ is simply a Hermitian operator.

4. Elementary Examples

In this section we will illustrate the concept of dissipativity by discussing briefly certain elementary examples. The simplicity of these examples stems from the fact that we may obtain explicit expressions for $(L-z)^{-1}$ and then calculate $\Psi(z)$ using the relation

$$P \frac{1}{L-z} P = \frac{1}{PLP + \Psi(z) - z} \tag{4.1}$$

which follows directly from Eq. (2.1).

We consider first the so-called Friedrichs model [11]. The unperturbed Hamiltonian H_0 has a continuous spectrum and an embedded discrete eigenvalue ω_0, coupled through the perturbation λV. It is evident that this model is related to the Lee model or the Wigner-Weisskopf model when the respective Hamiltonians are restricted to the "two-particle" sector. The model is soluble in the sense that we can calculate explicitly the eigendistributions and the eigenfunctions, if any, of the Hamiltonian and thus obtain the transformation which diagonalizes it. We choose [12] P so that $P\varrho(t)$ is the probability of finding the unstable (discrete) state excited at time t. The resolvent of the von Neumann operator can be obtained by a direct calculation. Using Eq. (4.1) we get $\Psi(z)$ (PLP in this problem as well as in the following ones vanishes identically), and its limiting value is given by

$$\Psi(+i0) = a^{-1} \equiv \left\{ -\frac{1}{2\pi i} \int_{-\infty}^{\infty} dx\, [\eta^+(x)\, \eta^-(x)]^{-1} \right\}^{-1}. \tag{4.2}$$

Here $\eta^{\pm}(x)$ are the limiting values on the real axis of the analytic function

$$\eta(z) = \omega_0 - z - \lambda^2 \int_0^{\infty} d\omega\, \frac{|v(\omega)|^2}{\omega - z} \tag{4.3}$$

(we assume that the continuous spectrum of H_0 and thus of H extends over the positive real axis).

Because P is a one-dimensional projection, $\Psi(+i0)$ as given by Eq. (4.2) is simply an imaginary number. The dissipativity condition is satisfied provided a^{-1} is different from zero. We have shown [12] that this is the case if the coupling parameter is smaller than a critical value, a condition which implies that H has no point spectrum. It is obvious that in the dissipative case there is no regular invariant, which is of course related to the irreversible decay of the unstable state.

It is interesting to consider the continuous spectrum of the Friedrichs model as the limit of a point spectrum. We assume therefore that ω_0 is coupled to a set of discrete eigenvalues ω_m (a row-column bordering model [13]) which are distributed continuously in the limit of an infinite volume, $\Omega \to \infty$, say. With P projecting the diagonal elements, an asymptotic expression for the collision operator [14] may be obtained which is written in a matrix form as

$$\Psi(+i0) = \begin{bmatrix} a^{-1} & -\frac{2\pi}{\Omega} a^{-1} \frac{\lambda^2 |v(\omega')|^2}{|\eta(\omega')|^2} \\ -\frac{2\pi}{\Omega} a^{-1} \frac{\lambda^2 |v(\omega)|^2}{|\eta(\omega)|^2} & \left(\frac{2\pi}{\Omega}\right)^2 a^{-1} \frac{\lambda^2 |v(\omega)|^2}{|\eta(\omega)|^2} \frac{\lambda^2 |v(\omega')|^2}{|\eta(\omega')|^2} \end{bmatrix} \tag{4.4}$$

with a^{-1} as in Eq. (4.2). Again the operator $\Psi(+i0)$ is non-positive. It can be shown that the only regular invariants are functions of the Hamiltonian. This property is responsible for the fact that a certain class of observables tends to an equilibrium value determined by the microcanonical ensemble [14]. In contrast with the previous case, cf. Eq. (4.2), the collision operator describes now not only the decay phenomenon but also the scattering processes taking place in this system.

Another simple example is that of a "codiagonal bordering model" [13, 15]. The Hamiltonian is of the form

$$H = \sum_n a_n^+ a_n + \lambda \sum_n \{a_n^+ a_{n+1} + a_{n+1}^+ a_n\}. \tag{4.5}$$

In the one-particle sector the matrix elements of $\Psi(+i0)$, in the representation where H_0 is diagonal, are given by

$$\langle m|\Psi(+i0)|n\rangle = -i \frac{2|\lambda|}{\pi} \int_{-\pi}^{\pi} d\varphi\, e^{i(m-n)\varphi} \left|\sin\frac{\varphi}{2}\right|. \tag{4.6}$$

It is easy to verify that the general properties we have stated in Section 2 are satisfied here. Furthermore the odd part of $\Psi(+i0)$ vanishes identically. A similar situation appears when we consider an N-body problem. In general, of course, exact calculations are not possible. However, in many cases the collision operator admits of a natural expansion (possibly asymptotic) in terms of a "small" parameter such as the coupling constant or the density. For instance, in the case of a weakly coupled quantum gas [1] with a Hamiltonian

$$H = \sum_p \varepsilon_p a_p^+ a_p + \frac{1}{2\Omega} \sum_{pqkl} v(pqkl)\, \delta_{p+q-k-l}\, a_p^+ a_q^+ a_k a_l \tag{4.7}$$

the asymptotic collision operator to order λ^2 is given by

$$\begin{aligned} \Psi_{(2)}(+i0) = & -i \frac{2\pi\lambda^2}{\Omega^2} \sum_{pqkl} |v(pqkl)|^2\, \delta_{p+q-k-l}\, \delta(\varepsilon_p + \varepsilon_q - \varepsilon_k - \varepsilon_l) \\ & \times \left[N_p N_q N_k N_l - N_p N_q (N_k+1)(N_l+1) \exp\left(-\frac{\partial}{\partial N_p} - \frac{\partial}{\partial N_q} + \frac{\partial}{\partial N_k} + \frac{\partial}{\partial N_l}\right)\right]. \end{aligned} \tag{4.8}$$

One may verify here that only the invariants which are functions of the energy are conserved (to order λ^2) by the collision operator, i.e. they satisfy Eq. (3.3a). However, every invariant satisfies Eq. (3.2a) (even in the limit $z \to +i0$).

5. Concluding Remarks

We have stressed in this article the necessity of the continuous spectrum in order to obtain irreversible behaviour in a quantum system. In general, the continuous spectrum appears in the limit of a large system. Nevertheless, in order to obtain an irreversible kinetic equation, we have to demand a non-vanishing dissipative collision operator which in turn is responsible for finite transport coefficients [16]. Clearly this requirement is stronger than simple ergodic behaviour. Although dissipativity implies ergodicity, the converse statement is not true.

Another aspect of dissipativity is the symmetry breaking of the $L \cdot t$-invariance of the von Neumann equation. The dissipativity condition, if true, implies that the collision operator contains a part which is invariant under the change $L \to -L$. This is an essential property in order to obtain a manifestly irreversible description of a physical system [6, 7] in agreement with the non-equilibrium behaviour described in a phenomenological way by the second law of thermodynamics. In fact, this approach leads to a new and general interpretation of non-equilibrium entropy. Indeed, it is easy to show [6, 7] that in the representation of dynamics which explicitly exhibits the symmetry-breaking terms there exists a Liapunov function which can only decrease in time. This Liapunov function may therefore be used to define the non-equilibrium entropy. The important advance over the classical considerations due to BOLTZMANN is that this definition is independent of any assumption of weak coupling. In this way a new unified formulation of dynamics and thermodynamics has been achieved.

References

1. PRIGOGINE, I.: Non-Equilibrium Statistical Mechanics. New York: Interscience 1962.
2. RAE, J., DAVIDSON, R.: J. Math. Phys. (to appear).
3. BAUS, M.: Bull. Classe Sci., Acad. Roy. Belg. **53**, 1291, 1332, 1352 (1967).
4. BAUS, M.: Thesis, Free University of Brussels (1968).
5. Prigogine, I., GEORGE, CL., HENIN, F.: Physica **45**, 418 (1969).
6. PRIGOGINE, I., GEORGE, CL., HENIN, F., ROSENFELD, L.: To appear Proc. Roy. Swedish Ac. Sc. (1972).
7. PRIGOGINE, I.: The Statistical Interpretation of Non-equilibrium Entropy. Report presented in the conference "100 years Boltzmann Equation", to appear in Austr. Phys. Acta (1972).
8. PRIGOGINE, I., GEORGE, CL., HENIN, F.: Proc. Nat. Acad. Sci. **65**, 789 (1970).
   BALESCU, R., CLAVIN, P., MANDEL, P., TURNER, J.: Bull. Classe Sci. Acad. Roy. Belg. **55**, 50 (1970).
   GRECOS, A.: Physica **51**, 50 (1971).
9. POINCARÉ, H.: Les méthodes nouvelles de la mécanique céleste. Reprinted by Dover Publ., New York 1957.
10. SHIMIZU, T.: Progr. Theoret. Phys. (Kyoto) **47**, 1181 (1972).
11. FRIEDRICHS, K.: Comm. Pure Appl. Math. **1**, 361 (1948).
12. GRECOS, A., PRIGOGINE, I.: Physica **59**, 77 (1972).
13. STEY, G., GIBBERD, R.: Physica **60**, 1 (1972).
14. GRECOS, A., PRIGOGINE, I.: Proc. Nat. Acad. Sci. **69**, 1629 (1972).
15. STEY, G.: Physica, to appear (1973).
16. BALESCU, R.: Physica **27**, 693 (1961).

On the Possibility of Transport-Phase-Transitions in Phonon Systems

MAX WAGNER

University of Stuttgart, Institut für Theoretische Physik
Stuttgart/Germany

Abstract

The Peierls-Boltzmann transport formalism for phonons is re-formulated in such a way that the collision term is represented by the kernel of the master equation. Sequences of transport and statistical postulates lead to higher "golden rules" with non-energy-conserving intermediate states. This reveals the possibility of *cumulative excitations* by transport, in particular if a high-energy local motion is coupled to the transport system. A practical illustration is given in the example of the monoatomic harmonic chain.

1. Introduction

In many regions of the natural sciences the need for applicable knowledge about non-equilibrium statistical physics has grown enormously during recent years. It is therefore of pressing importance at the present time to find a variety of simple systems for which reliable solutions can be obtained.

Yet the present state of non-equilibrium statistical mechanics is not yet sufficiently developed to permit us to investigate problems where very many degrees of freedom can be expected to leave the neighbourhood of thermal equilibrium. Therefore, the systems of particular interest are those where only one or a few degrees of freedom deviate strongly from thermal equilibrium. One of the simplest classes of such problems is that of translationally invariant transport systems which are coupled to one (or a few) additional, singular, local degree of freedom which does not itself take part in the transport.

In this investigation we will specify a translationally invariant phonon system (energy transport) which is coupled to a local excitation. The question then arises, whether it is possible to excite the local degree far beyond thermal equilibrium by *mere energy transport* through the system? Or, to be more precise, the question is whether there is a critical energy flux for which something like a "*transport-phase transition*" exists.

There are already examples of phenomena which may be regarded as "transport-phase transitions". The best-known of these in the microscopic regime is the laser. In this particular case very elegant methods have been developed, and these are exhaustively described in the excellent review article by HAKEN [1]. Many other examples are known in chemical non-linear reaction kinetics and in macroscopic control devices. An illustrative example from the technological field is the control system which stabilizes room temperature. There the "kinetic variables" are the room and heating temperatures, respectively, and the "flux" is given by the size of the heater. If the latter is enlarged continuously, the room temperature rises up to a fixed value at which the thermostat switches off the heating. If the size of the heater is enlarged still further, a *new kind* of behaviour is displayed by the room temperature, which now exhibits a stable fluctuation around the critical value of the thermostat. This illustrates that there is a critical size of heater (a critical "flux") at which a "transport-phase transition" occurs.

Depending on the structure of the kinetic equations, a large variety of such "phase transitions" is conceivable. Starting with an infinitesimally small flux, there is always a stable stationary solution which continuously deviates from thermal equilibrium. This solution is called the "*thermodynamic branch*". At some *critical flux* there may be a stability inversion, i.e. a transition from the thermodynamic branch to either another stable stationary branch or to a stable branch of fluctuating nature (limit cycle). In phenomenological thermodynamics there is already considerable progress in the study of the behaviour of such "open" systems; for further details see GLANSDORFF and PRIGOGINE [2].

It is the aim of the present article to undertake some first steps in the investigation of a non-linear transport theory of phonons on the basis of microscopic first principles. We shall make a sequence of transport postulates, a sequence of statistical postulates and finally a sequence of model assumptions. Some of these presuppositions will be of a more general nature, some of a more specific one. The final result will be to demonstrate the possibility of more than one single stationary excitation of the local, non-transporting degree of freedom for certain flux regions.

2. Phonon Transport Theory

After the fundamental paper of Peierls [3] there was a long period of stagnation in the theory of phonon transport; but in the last decade some very elegant and sophisticated formalisms have been developed [4–6]. These are very suitable to perform realistic calculations for the transport properties of crystals which pertain to the neighbourhood of thermal equilibrium ("thermodynamic" branch of solutions). However, for a first study of effects far from equilibrium, it is absolutely necessary to rely on a formalism of the utmost simplicity. Therefore, we shall utilize a modified version of the Peierls-Boltzmann equation by relating the collision term to the master equation of non-equilibrium statistical mechanics. To this end we introduce the following four "transport" postulates.

A. We assume that a *bilinear* "particle"-density operator can be defined in the form

$$\hat{n}(\boldsymbol{k}\lambda, \mathfrak{r}, t) = \sum_{\boldsymbol{k}'\lambda'} [a^{+}_{\boldsymbol{k}\lambda} a_{\boldsymbol{k}'\lambda'} \chi(\boldsymbol{k}\lambda, \boldsymbol{k}'\lambda') e^{-i(\boldsymbol{k}-\boldsymbol{k}')\cdot\mathfrak{r}} + \text{c.c.}] \tag{1}$$

such that the energy density of the mode $\boldsymbol{k}\lambda$ is given by ($\hbar = 1$)

$$\hat{\varepsilon}(\boldsymbol{k}\lambda, \mathfrak{r}, t) = \omega(\boldsymbol{k}\lambda)\, \hat{n}(\boldsymbol{k}\lambda, \mathfrak{r}, t). \tag{2}$$

Here λ denotes the phonon branches. The function $\chi(\boldsymbol{k}\lambda, \boldsymbol{k}'\lambda')$ restricts the deviation of $\boldsymbol{k}'$ from $\boldsymbol{k}$ in the wave packet and may be chosen as a Wigner function, as was done by KLEIN and WEHNER [6] and others, or as a Gauss function

$$\chi(\boldsymbol{k}\lambda, \boldsymbol{k}'\lambda') = \frac{1}{2V}\,\delta_{\lambda\lambda'}\,e^{-\alpha|\boldsymbol{k}-\boldsymbol{k}'|^2} \tag{3}$$

as was done by HARDY [7], and in many other ways. The choice (3) has the advantage of a simple flexibility in the choice of the space-extension parameter α. The requirement of bilinearity of $\hat{n}(\boldsymbol{k}\lambda)$ is satisfied also in the case of an anharmonic crystal if there is the possibility of a suitable canonical transformation to quasi-harmonicity.

B. The (translationally invariant) phonon Hamiltonian H_0 is such that we have

$$\Phi(\boldsymbol{k}\lambda, \mathfrak{r}, t) = i\langle[\hat{n}(\boldsymbol{k}\lambda, \mathfrak{r}, t), H_0]\rangle = \boldsymbol{v}(\boldsymbol{k}\lambda)\,\nabla\langle\hat{n}(\boldsymbol{k}\lambda, \mathfrak{r}, t)\rangle \tag{4}$$

where $\boldsymbol{v}(\boldsymbol{k}\lambda)$ is the phonon group velocity. This postulate is satisfied in a good approximation if either there is little dispersion (Debye approximation), or if the wave packet (1) has a broad spatial extension. In the HARDY [7] description (3) this would mean that we can neglect terms of the order $O\big((\alpha V)^{-1}\,\partial_{\boldsymbol{k}}^2\omega_{\boldsymbol{k}})\big)$.

C. Defining $\varepsilon\hat{U}$ as the interaction Hamiltonian between the phonon and the local system, where ε is an order parameter, we require the commutator $[\hat{n}(\boldsymbol{k}\lambda, \mathfrak{r}, t), \hat{U}]$ to be replaceable by its spatial average

$$[\hat{n}(\boldsymbol{k}\lambda, \mathfrak{r}, t), \hat{U}] \approx \frac{1}{V}\,[a^+_{\boldsymbol{k}\lambda}\,a_{\boldsymbol{k}\lambda}, \hat{U}]. \tag{5}$$

This seems to be a reasonable assumption, if either $\hat{U}$ is not localized strongly or, again, we choose a broad spatial packet. It should be noted that the spatial average of the "*transport term*" (4) is necessarily zero. The reason for this peculiar fact is that the Born-v. Karman periodic boundary conditions have to be incorporated into any phonon-transport theory starting from translational invariance. To install a space-independent flux it is therefore necessary to think of an energy "sink-source dipole" at the boundary. This will also be assumed in the following, although more careful investigations into this problem would be highly desirable.

D. As the final transport assumption, we postulate the existence of a "master equation" which yields the connection to the results which have been established in non-equilibrium statistical mechanics over the last several years. To be more specific, we will adopt the following form for the master equation:

$$\partial_t\,\varrho_\mu(t) = \varepsilon^2 \sum_\nu \int_0^t d\tau\,[\varrho_\nu(t-\tau)\,W_{\nu\mu}(\tau) - \varrho_\mu(t-\tau)\,W_{\mu\nu}(\tau)] \tag{6}$$

where ϱ_μ are the diagonal terms of the density matrix in the eigenbasis of the uncoupled Hamiltonian. This form (6) is that of ZWANZIG [8] and can be shown to be equivalent to the master equation of PRIGOGINE and RÉSIBOIS [9] and that of MONTROLL [10]. We shall define the kernel $W_{\mu\nu}(\tau)$ in the following section.

The three presuppositions A, B, C are conventional in the theory of phonon transport [3, 11], although partially disputable. The last assumption D is well-adopted in statistical physics. With the aid of these four postulates one arrives at the following system of kinetic equations,

$$\partial_t \langle a^+_{\boldsymbol{k}\lambda} a_{\boldsymbol{k}\lambda}\rangle + \Phi(\boldsymbol{k}\lambda, t) = \frac{\varepsilon^2}{V}\sum_{m=0}^{\infty} m \int_0^t d\tau \sum_\mu \varrho_\mu(t-\tau)\left\{\sum_{\nu}^{N^\nu_{\boldsymbol{k}\lambda}=N^\mu_{\boldsymbol{k}\lambda}+m} W_{\mu\nu}(\tau) - \sum_{\nu}^{N^\nu_{\boldsymbol{k}\lambda}=N^\mu_{\boldsymbol{k}\lambda}-m} W_{\mu\nu}(\tau)\right\} \tag{7a}$$

$$\partial_t \langle a^+_s a_s\rangle = \frac{\varepsilon^2}{V}\sum_{m=0}^{\infty} m \int_0^t d\tau \sum_\mu \varrho_\mu(t-\tau)\left\{\sum_{\nu}^{N^\nu_s=N^\mu_s+m} W_{\mu\nu}(\tau) - \sum_{\nu}^{N^\nu_s=N^\mu_s-m} W_{\mu\nu}(\tau)\right\} \tag{7b}$$

where the latter equation pertains to the local excitation, which is coupled to the phonon system via $\varepsilon\hat{U}$. The detailed derivation of the system of Eqs. (7a, b) will be given elsewhere [12]. It would be emphasized that the use of the kinetic equations (7a, b) has the advantage that it is not necessary to specify the coupling Hamiltonian to the external world. This would be necessary if we had chosen a direct-response formalism [13]. In our case the transport $\Phi(\boldsymbol{k}\lambda, t)$, may be considered to be the external stimulus, which naturally has to be consistent with the definition (4) after the solution has been found.

3. Higher Golden Rules

Since in the collision term of the kinetic Eqs. (7a, b) the connection to the kernel of the master equation has been made, we can utilize the latter via another sequence of postulates. We choose the ZWANZIG form [8] of the kernel of the master equation,

$$\begin{aligned} W_{\mu\nu}(\tau) &= -[\hat{L}_s\, e^{-i\tau(\hat{L}_0+\varepsilon(1-\hat{D})\hat{L}_s)}\,\hat{L}_s]_{\mu\mu\nu\nu} \\ &= -\sum_{ij}\{U_{\mu i}[G_{i\mu j\nu}\,U_{j\nu} - G_{i\mu\nu j}\,U_{\nu j}] + \text{c.c.}\} \end{aligned} \tag{9}$$

where

$$\hat{G}(\tau) = \exp[-i\tau(\hat{L}_0 + \varepsilon(1-\hat{D})\hat{L}_s)] \tag{10}$$

and $\hat{L}_0$, $\hat{L}_s$ and $\hat{L} = \hat{L}_0 + \varepsilon\hat{L}_s$ are the Liouville "tetradic" operators corresponding to H_0, U and $H = H_0 + \varepsilon U$, respectively, e.g.

$$(\hat{L}_s)_{i\alpha j\beta} = U_{ij}\,\delta_{\alpha\beta} - U_{\beta\alpha}\,\delta_{ij} \tag{11}$$

whereas $\hat{D}$ is the diagonalization projector, i.e.

$$(\hat{D}\hat{A})_{i\alpha j\beta} = \delta_{i\alpha} A_{i\alpha j\beta}. \tag{12}$$

If we do not hesitate to apply the theorem of GOLDBERGER and ADAMS [14] which is valid for functional operators, matrix operators and functional matrix operators [15] and which presumably can be easily shown to be valid also for tetradic operators, we have

$$G_{i\alpha j\beta}(\tau) = \sum_{n=0}^{\infty} G^{(n)}_{i\alpha j\beta}(\tau) \tag{13}$$

where

$$G^{(n)}_{i\alpha j\beta}(\tau) = (-i\varepsilon)^n e^{-i(E_i-E_\alpha)} \int_0^\tau d\tau_1 \ldots \int_0^{\tau_{n-1}} d\tau_n \cdot \{[(1-\hat{D})\hat{L}_s(\tau_1)] \ldots [(1-\hat{D})\hat{L}_s(\tau_n)]\}_{i\alpha j\beta} \tag{14}$$

and

$$(\hat{A}(\tau))_{i\alpha j\beta} = (e^{i\hat{L}_0\tau} A\, e^{-i\hat{L}_0\tau})_{i\alpha j\beta}. \tag{15}$$

Our statistical postulates will be the usual ones, as discussed somewhat carefully by JANCEL [16].

I. As an initial condition for the master equation we assume that the nondiagonal part of the density matrix, $\hat{\varrho}_{nd}(0)$, in the chosen eigenbasis of $\hat{H}_0$ disappears at time $t=0$,

$$\hat{\varrho}_{nd}(0) = 0. \tag{16}$$

This condition is already incorporated in the chosen form of the master equation as used for the collision terms in (7a, b). Nevertheless, it is a highly disputable assumption and is still very much under discussion, although for mathematical convenience it is used in nearly all practicall calculations.

II. Also for mathematical convenience, we introduce a kind of Markofficity assumption in the form

$$\varrho_\mu(t-\tau)\, W_{\mu\nu}(\tau) \approx \varrho_\mu(t)\, W_{\mu\nu}(\tau). \tag{17}$$

It leads to considerable simplification, although no principal complications are to be foreseen, if a weaker (higher-order) form of Markofficity is used.

III. Finally we introduce the least problematic assumption, which is used almost uniquely in modern statistical mechanics. It is the assumption that there is a high density of states, or equivalently, a great number of degrees of freedom, such that the Poincaré periodicity t_P is very large as against "physical" times t,

$$t_P \gg t. \tag{18}$$

This postulate leads to *irreversibility*, represented mathematically in the equation

$$\int_0^t d\tau \cos(E_i - E_j)\,\tau \approx \pi\,\delta(E_i - E_j). \tag{19}$$

With the background of these "statistical postulates" we are able to write down higher-order golden rules. For instance, we immediately find the well-known "Fermi golden rule"

$$\int_0^t W^{(0)}_{\mu\nu}(\tau)\, d\tau = 2\pi\, U_{\mu\nu}\, U_{\nu\mu}\, \delta(E_\mu - E_\nu) \tag{20}$$

by using $\hat{G}^{(0)}$ of Eq. (14). This corresponds to the Pauli master equation and there is a direct energy conservation for the transition $\mu \to \nu$, which likewise may be denoted as a *coherent* energy conservation.

To keep the frame of this study within a reasonable limit, we shall not give the derivation of the higher-order golden rules. The details of this derivation will be given in a forthcoming paper by the author [17]. To save space, we furthermore write down only simplified versions of these rules. To be more specific, we think the coherent energy-conserving processes to be predominantly represented by the Fermi golden rule. Whence we assume

$$U_{ij}\,\delta(E_i - E_j) \approx 0 \tag{21}$$

for the higher-order rules. In addition, we assume that the matrix elements U_{ij} are either real or purely imaginary; this latter restriction has simplified the present stage of the derivation, but presumably it is not necessary. We then end up with the results

$$\int_0^t W_{\mu\nu}^{(2)}(\tau)\,d\tau = 2\pi\,\varepsilon^2 \sum_{\alpha\beta} U_{\mu\alpha}\,U_{\alpha\nu}\,U_{\nu\beta}\,U_{\beta\mu}\,\frac{\delta(E_\mu - E_\nu)}{(E_\beta - E_\mu)\,(E_\alpha - E_\mu)} \tag{22}$$

$$\int_0^t W_{\mu\nu}^{(4)}(\tau)\,d\tau = 2\pi\,\varepsilon^4 \sum_{\substack{\alpha\beta\\ \alpha_1\beta_1}} U_{\mu\alpha}\,U_{\alpha\alpha_1}\,U_{\alpha_1\nu}\,U_{\nu\beta_1}\,U_{\beta_1\beta}\,U_{\beta\mu} \cdot \frac{\delta(E_\mu - E_\nu)}{(E_\mu - E_\alpha)\,(E_\mu - E_{\alpha_1})\,(E_\nu - E_{\beta_1})\,(E_\nu - E_\beta)}\,. \tag{23}$$

All rules of odd order disappear. From Expr. (22) and (23) we note that no direct energy conservation is required, but there are one or two intermediate states in each which do not conserve energy. These processes may be looked upon as *incoherent* ones.

4. A Simple Model: The Linear Chain

To fulfill the purpose of this study and to demonstrate the possibility of *cumulative excitations* by transport, we choose a model of the utmost simplicity. The transport system is the monatomic, harmonic linear chain, and there is a local degree with elementary excitation energy ω_s. The following "model" specifications are made.

a) The interaction between the phonon system and the local degree is bilinear,

$$\hat{U} = (a_s + a_s^+) \sum_{\boldsymbol{k}} \lambda_{\boldsymbol{k}}\,(a_{\boldsymbol{k}} + a_{-\boldsymbol{k}}^+), \qquad \lambda_{-\boldsymbol{k}} = -\lambda_{\boldsymbol{k}}. \tag{24}$$

b) For a local excitation ω_s, a three-phonon process is necessary to yield energy conservation,

$$\omega_s = \omega_{\boldsymbol{k}} + \omega_{\boldsymbol{k}'} + \omega_{\boldsymbol{k}''}, \qquad (\omega_s > \omega_{\boldsymbol{k}} + \omega_{\boldsymbol{k}'}). \tag{25}$$

From this it is immediately clear that in the kinetic system (7a, b) only the golden rule (23) (and higher ones) can be utilized, because lower ones do not conserve energy. For the local kinetic Eq. (7b) we then have

$$\partial_t \langle \hat{N}_s \rangle = \sum_{\mu,\boldsymbol{k}\boldsymbol{k}'\boldsymbol{k}''} A(\boldsymbol{k},\boldsymbol{k}',\boldsymbol{k}'')\,\varrho_\mu(t)\,\delta\big(\omega_s - (\omega_{\boldsymbol{k}} + \omega_{\boldsymbol{k}'} + \omega_{\boldsymbol{k}''})\big) \cdot \big[(N_s^\mu + 1)^3\,N_{\boldsymbol{k}}^\mu\,N_{\boldsymbol{k}'}^\mu\,N_{\boldsymbol{k}''}^\mu - (N_s^\mu)^3\,(N_{\boldsymbol{k}}^\mu + 1)\,(N_{\boldsymbol{k}''}^\mu + 1)\big] \tag{26}$$

where

$$A(\boldsymbol{k}, \boldsymbol{k}', \boldsymbol{k}'') = \lambda_{\boldsymbol{k}}^2 \lambda_{\boldsymbol{k}'}^2 \lambda_{\boldsymbol{k}''}^2 \left[\frac{1}{\omega_s - \omega_{\boldsymbol{k}}} \left(\frac{1}{\omega_s - \omega_{\boldsymbol{k}'}} + \frac{1}{\omega_s - \omega_{\boldsymbol{k}''}}\right) + \text{cycl.}\right] \cdot \left[\frac{1}{\omega_s + \omega_{\boldsymbol{k}}} \left(\frac{1}{\omega_s - \omega_{\boldsymbol{k}'}} + \frac{1}{\omega_s - \omega_{\boldsymbol{k}''}}\right) + \text{cycl.}\right]. \tag{27}$$

From Eq. (26) it is immediately evident that the thermal equilibrium

$$\begin{aligned} \overline{N_s} &= [\exp(\beta\omega_s) - 1]^{-1}, \qquad \beta = (k_B T)^{-1} \\ \overline{N_{\boldsymbol{k}}} &= [\exp(\beta\omega_{\boldsymbol{k}}) - 1]^{-1}, \end{aligned} \tag{28}$$

is one stationary solution. To find other possible stationary solutions, we have to choose a special factorization procedure.

c) The factorization procedure has to be chosen in such a way that the thermal equilibrium *remains* one of the stationary solutions (necessary condition). The most straightforward possibility which meets this requirement, is

$$\begin{aligned} &\sum_\mu \varrho_\mu [(N_s^\mu + 1)^3 N_{\boldsymbol{k}}^\mu N_{\boldsymbol{k}'}^\mu N_{\boldsymbol{k}''}^\mu - (N_s^\mu)^3 (N_{\boldsymbol{k}}^\mu + 1)(N_{\boldsymbol{k}'}^\mu + 1)(N_{\boldsymbol{k}''}^\mu + 1)] \\ &= [\overline{(N_s + 1)^3}\, \overline{N_{\boldsymbol{k}}}\, \overline{N_{\boldsymbol{k}'}}\, \overline{N_{\boldsymbol{k}''}} - \overline{(N_s)^3} (\overline{N_{\boldsymbol{k}}} + 1)(\overline{N_{\boldsymbol{k}'}} + 1)(\overline{N_{\boldsymbol{k}''}} + 1)]. \end{aligned} \tag{29}$$

d) As a final model assumption we introduce an "eigen-temperature" T_s for the local excitation,

$$\overline{N_s} = [e^{\beta_s \omega_s} - 1]^{-1}, \qquad \beta_s = (k_B T_s)^{-1} \tag{30}$$

which yields

$$\begin{aligned} \overline{N_s^2} &= \overline{N_s} + 2(\overline{N_s})^2 \\ \overline{N_s^3} &= \overline{N_s} + 6(\overline{N_s})^2 + 6(\overline{N_s})^3 \end{aligned} \tag{31}$$

and from (26) via Eqs. (29) to (31) for the stationary solutions we end up with the local cubic equation

$$(\overline{N_s})^3 + (1 - \sigma)(\overline{N_s})^2 + (\tfrac{1}{6} - \sigma)\overline{N_s} - \tfrac{1}{6}\sigma = 0 \tag{32}$$

with the abbreviation

$$\sigma = b/a = \sigma(\overline{N_{\boldsymbol{k}}}, \overline{N_{\boldsymbol{k}'}}, \overline{N_{\boldsymbol{k}''}}) \tag{33}$$

where

$$b = \sum_{\boldsymbol{k}\boldsymbol{k}'\boldsymbol{k}''} A(\boldsymbol{k}\boldsymbol{k}'\boldsymbol{k}'')\, \delta(\omega_s - (\omega_{\boldsymbol{k}} + \omega_{\boldsymbol{k}'} + \omega_{\boldsymbol{k}''}))\, \overline{N_{\boldsymbol{k}}}\, \overline{N_{\boldsymbol{k}'}}\, \overline{N_{\boldsymbol{k}''}} \tag{34a}$$

$$\begin{aligned} a = &\sum_{\boldsymbol{k}\boldsymbol{k}'\boldsymbol{k}''} A(\boldsymbol{k}\boldsymbol{k}'\boldsymbol{k}'')\, \delta(\omega_s - (\omega_{\boldsymbol{k}} + \omega_{\boldsymbol{k}'} + \omega_{\boldsymbol{k}''})) \\ &\times [1 + \overline{N_{\boldsymbol{k}}} + \overline{N_{\boldsymbol{k}'}} + \overline{N_{\boldsymbol{k}''}} + \overline{N_{\boldsymbol{k}}}\, \overline{N_{\boldsymbol{k}'}} + \overline{N_{\boldsymbol{k}'}}\, \overline{N_{\boldsymbol{k}''}} + \overline{N_{\boldsymbol{k}''}}\, \overline{N_{\boldsymbol{k}}}]. \end{aligned} \tag{34b}$$

In view of Expr. (27), i.e. $A(\boldsymbol{k}\boldsymbol{k}'\boldsymbol{k}'') > 0$, we note that $b > 0$, $a > 0$ and hence $\sigma > 0$. For the discussion of the cubic Eq. (32) we follow the way of Cardan, which is found in any standard textbook. There is the critical quantity

$$\begin{aligned} Q &= (\tfrac{1}{3})^6 [-(\sigma^2 + \sigma + \tfrac{1}{2})^3 + \tfrac{1}{4}(2\sigma^3 + 3\sigma^2 - \tfrac{1}{2})^2] \\ &= -(\tfrac{1}{3})^6 [\tfrac{9}{4}\sigma^4 + \tfrac{9}{2}\sigma^3 + 3\sigma^2 + \tfrac{3}{4}\sigma + \tfrac{1}{16}] \end{aligned} \tag{35}$$

and there is one real and two c.c. roots if $Q > 0$, and three real roots for $Q \leqq 0$, of which at least two are equal for $Q = 0$ and all three different for $Q < 0$.

Considering the quantity σ, given by Def. (33) via Expr. (34a, b), we note that always $\sigma > 0$, from which we have [vid. Eq. (35)] $Q < 0$. This means that we always have three real solutions. However, looking back to the cubic equation (32) and applying Descartes' rule of signs, we note that in the whole variability region $0 \leqq \sigma < +\infty$ there never can be more than a single positive real root.

Summary and Outlook

The purpose of the preceding investigation has been to show the possibility of cumulative excitation of a local degree of freedom by phonon transport. To achieve this, the phonon transport equations have been combined with the master equation through a sequence of "transport postulates". In this way, the flux components $\Phi(\boldsymbol{k}\lambda)$ may be considered as the external stimulus, whereby the definition of a coupling Hamiltonian to the external world can be avoided. Furthermore, the collision terms take the form of the kernels of the master equation. Applicability of the general formalism is reached by another sequence of assumptions ("statistical") by means of which higher "golden rules" can be found. These directly exhibit the feasibility of cumulative (incoherent) excitations via non-energy-conserving intermediate states.

The general formalism is applied to a model system in which a bilinear coupling between the local and the phonon system is assumed. If we further conjecture that a three-phonon process is necessary to make a local excitation, the two lowest "golden rules" are excluded. Finally, a suitable factorization procedure converts the kinetic system into the form of coupled nonlinear differential equations which are of first order in time and contain terms up to the third power in the local occupation number N_s.

The first step in the discussion of such equations is to search for stationary solutions. For these, a cubic equation for N_s with a single parameter σ has to be solved, where σ implicitly contains the still unknown phonon occupation numbers $N_{\boldsymbol{k}}$. Hence, σ depends both on the average temperature of the phonon system and on the energy fluxes $\Phi_{\boldsymbol{k}}$ from outside, but its variability region is restricted to $0 \leqq \sigma < +\infty$. It is found that for the chosen model there exists only one single positive real solution. However, it would be emphasized that this is due to the unrealistic oversimplification of the model. It seems clear that for any high local excitation one would also consider other de-excitation processes besides the one resulting from the bilinear coupling. To discuss the effect of these processes without specifying them in detail, we may describe them by a single relaxation time τ_s, i.e. by an additional term of the form $-(\tau_s)^{-1} N_s$ in the local kinetic Eq. (26). By this means the factor $(\frac{1}{6} - \sigma)$ in the cubic Eq. (32) would be replaced by $(\sigma_0 - \sigma)$, where

$$\sigma_0 = \frac{1}{6}\left(1 + \frac{1}{a\tau_s}\right).$$

Now, if $\sigma_0 > 1$, Descartes' rule of signs would allow three real positive solutions in the region $1 < \sigma < \sigma_0$. Whether all three exist and, if they do, whether one of the two "non-thermodynamic" ones becomes physically effective, requires a lengthy mathematical discussion. After their existence has been established, it has to be ascertained that the non-thermodynamic branches can be *stable*. In this case there would indeed be one or two "transport-phase-transitions" from the thermodynamic to other stationary branches.

It is thus clear that our present investigation has to be supplemented by a detailed mathematical discussion of stability. In general a discussion of this kind is cumbersome, although there are well-developed methods, in particular for non-linear systems of differential equations of the class appearing in kinetic problems. Especially the fundamental work of LIAPUNOV [18] and of POINCARÉ [19] has stimulated great progress, and for an excellent review of this field we refer to the book by ANDRONOV, VITT and KHAIKIN [20]. If it turns out that the non-thermodynamic stationary branches are not stable, it may well be that there are non-stationary stable solutions of a limit-cycle nature for certain energy fluxes, which would indicate another kind of transport-phase transition. An example of this sort is the room temperature control device mentioned earlier.

Both kinds of transport-phase transitions in molecular systems are of high interest for molecular biology, since transport, i.e. *metabolism*, is one of the fundamental characteristics of living matter [21]. A bridge to biology along similar lines has been proposed also by HAKEN [22, 23] and by FRÖHLICH [24, 25]. A specific kind of "transport-phasetransition" has been proposed recently by FRÖHLICH [26] in a model theory of smell.

Returning to physics, one would think that on the one hand, the proposed formalism is simple enough to master the final calculational stages for a given fixed system, and, on the other hand, the sequence of assumptions (except those concerning the specific model) are not so serious as to make the final results unrealistic. To conclude, one naturally would wish to supplement the formalism by experimental results. There is hope that in the near future an experimental investigation will be possible, since very recently the production of almost monochromatic phonons has become feasible [27–29], which allows the variation of the energy flux $\Phi(\boldsymbol{k}\lambda)$ in narrow frequency regions.

References

1. HAKEN, H.: Laser Theory. Im: Handbuch der Physik (Ed. S. FLÜGGE), Vol. XXV/2c. Berlin-Heidelberg-New York: Springer 1970.
2. GLANSDORFF, P., PRIGOGINE, I.: Thermodynamic Theory of Structure, Stability and Fluctuations. London: Wiley-Interscience 1971.
3. PEIERLS, R.: Ann. Physik [5] 3, 1055 (1929).
4. HORIE, C., KRUMHANSL, J. A.: Phys. Rev. **136**, A 1397 (1964).
5. KWOK, P. C., MARTIN, P. C.: Phys. Rev. **142**, 495 (1966).
6. KLEIN, R., WEHNER, R. K.: Phys. Kondens. Materie **10**, 1 (1969).
7. HARDY, R. J.: Phys. Rev. **132**, 168 (1963).
8. ZWANZIG, R.: Physica **30**, 1109 (1964).
9. PRIGOGINE, I., RÉSIBOIS, P.: Physica **27**, 541 (1961).
10. MONTROLL, E. W.: Lectures in Theoretical Physics (Boulder) **3**, 221 (1960).
11. CARRUTHERS, P.: Rev. Mod. Phys. **33**, 92 (1961).
12. WAGNER, M.: Z. Physik (to be published).
13. KUBO, R.: J. Phys. Soc. Japan **12**, 570 (1957).
14. GOLDBERGER, M. L., ADAMS, E. N.: J. Chem. Phys. **20**, 240 (1952).
15. WAGNER, M.: Z. Physik **244**, 275 (1971).
16. JANCEL, R.: Foundations of Classical and Quantum Statistical Mechanics. Oxford etc.: Pergamon 1969.
17. WAGNER, M.: Golden Rules in Transport Theory. Z. Physik (to be published).

18. LIAPUNOV, A. M.: The General Problem of the Stability of Motion. Khar'kov 1892.
19. POINCARÉ, H.: Œuvres. Paris: Gauthier-Villars 1928.
20. ANDRONOV, A. A., VITT, A. A., KHAIKIN, S. E.: Theory of Oscillators. Oxford etc.: Pergamon 1966.
21. WAGNER, M.: Nonlinear Transport as a Possible Key to Physical Understanding in Biology. In ,,Synergetics", ed. H. HAKEN. Stuttgart: Teubner 1973.
22. HAKEN, H.: Proc. 1st Int. Conf. on Theoretical Physics and Biology, Versailles 1967, Ed. M. MAROIS. Amsterdam: North Holland 1969.
23. HAKEN, H.: Introduction to Synergetics, in ,,Synergetics", op. cit. (Ref. 21).
24. FRÖHLICH, H.: Organization and Long-Range Selective Interaction in Biological and other Pumped Systems. In ,,Synergetics", op. cit. (vid. Ref. 21).
25. FRÖHLICH, H.: Proc. 3rd Int. Conf. on Theoretical Physics and Biology, Versailles 1971, Ed. M. MAROIS (to be published).
26. FRÖHLICH, H.: Lecture Notes, University of Stuttgart, 1971 (unpublished).
27. EISENMENGER, W., DAYEM, A. H.: Phys. Rev. Letters **18**, 125 (1967).
28. KINDER, H., LASSMANN, K., EISENMENGER, W.: Phys. Letters **31** A, 475 (1970).
29. DYNES, R. C., NARAYANAMURTI, V.: High-Frequency Monochromatic Phonon Spectroscopy in Ge: Sb. Int. Conf. on Phonon Scattering in Solids, Paris 1972 (unpublished).

Search for Cooperative Phenomena in Hydrogen-Bonded Amide Structures

G. CARERI

Università di Roma, Istituto di Fisica, Roma/Italia

1. One may wonder if the H atoms involved in an extended hydrogen-bonded amide system can become dynamically coupled at low temperature and at higher temperatures can display the phenomenology typical of the order-disorder transitions. From what we already know about H-bonding [1], if this effect can occur at all, it should be observed in highly polarizable H-bonded groups; Amide group networks are therefore good candidates [2] and are particularly interesting in view of developments of biological significance.

With this hope in mind, we have carried out a variety of experiments on a model system, namely crystalline acetanilide CH_3 $CONCH_6$ H_5 (ACN). In ACN, chains of nearly planar H-bonded amide groups run through the crystal, providing a suitable model for an array of H-bonds in one dimension. Infrared and Raman spectra can easily be recorded to investigate the force field acting on the amide group; they also provide a simple and powerful method for investigating slight changes in the electronic ground state induced by temperature variations. If an ordered structure can be built, it should compete with the temperature and must display the usual order-disorder transition phenomenology, provided the "extra" coupling energy of two neighbouring amide groups is of the same order as kT. This "extra" coupling energy of the hydrogen bond can well fall in a suitable range of temperatures, and therefore such a search has been made. Essentially, we have been searching for the spontaneous spectroscopic manifestation of two slightly different amide structures, the ratio of which was expected to be strongly temperature dependent around a critical temperature.

2. In the following we will report only some details of the IR and Raman data of ACN. In both types of spectra, all the amide bands reveal interesting changes on cooling. For brevity, and to be more definite, we will limit ourselves to the amide I band only, which is essentially the in-phase CO stretching mode [3]. It is well known that this mode occurs above 1700 cm^{-1} for a free carbonyl, and that it is shifted to the red by H-bonding; actually this red shift is strictly proportional to the chemical shift of the proton measured by NMR, and can be used as a good measure of the H-bonding.

We do in fact see a drastic change of the amide I band as the temperature is lowered. With decreasing temperature, a new band appears at 1650 cm^{-1} on the red side of the amide I band centered at 1665. Experiments on a single crystal show that these two bands have the same dicroism; in non-crystalline powders the new band is never observed. Quite similar behaviour is seen in the Raman spectra. It is very interesting that on cooling there is not a continuous shift of the amide I band, as one would have expected from the continuous contraction of the crystal, but that a new band grows at a well-defined frequency. Moreover, the total intensity of the two amide I bands remains constant at different temperatures, while the intensity of the band at 1650 cm^{-1} displays a sigmoid curve on cooling. Trivial explanations like Fermi resonance and Davidov splitting can be ruled out, while a parallel investigation of the specific heat, dielectric constant and volume expansion by X-rays excludes the occurrence of rotational isomerism or of a polymorphic transition. Deuterium substitution gives rise to a drastic change in the spectra; similar but less marked phenomena are detected, and the temperature range of interest is raised from that of liquid nitrogen to near the melting temperature at 114 °C.

3. In order to offer a possible explanation for the above facts, we want to recall that the effect of the H bond on the electronic structure of the bonded molecules is very important when these molecules are resonating between close electronic structures. This is precisely the case with the amide groups, where the planarity of the group and the shortening of the C-N bond to 1.32 Å are attributed to resonance between two structures, which contribute about 60% and 40% respectively [2]:

$$\mathrm{H{-}N{-}C{=}O} \qquad \mathrm{H{-}\overset{\ominus}{N}{=}C{-}\overset{\oplus}{O}}$$

It is evident that the electronic structure of an amide group is affected by the positions of the H atoms of the neighbouring groups which are H-bonded to it. Moreover, the effect of H-bonding in inducing changes in the electronic structure of the amide groups will be most intense if these H-bonded groups are arranged in a linear chain, because the cooperative effect between the alternating H-donor and H-acceptor terminals will cause the amide groups to be more "polar" in the field of the two neighbouring H-bond partners. Therefore in these chains one can have a collective electronic state of the chain, when there is a more than pairwise cooperative charge resonance between bonded groups, induced by the H-bonding. The binding energy involved in the formation of this collective electronic state is of the order of magnitude of a few Kcal/mole [1, 4].

$$\cdots \mathrm{O{=}C(CH_3){-}N(C_6H_5){-}H} \cdots \mathrm{O{=}C(CH_3){-}N(C_6H_5){-}H} \cdots$$

It is clear that in this electronic ground state of the chain the H atoms of neighbouring amide groups will show a tendency to move in phase because this correlation will further increase in the binding energy of the system. Then the zero-point motion of the H atoms (essentially the NH stretching mode) must also be affected by this correlation, and a collective zero-point state may result at sufficiently low temperature. This "extra" binding energy ε can be only a fraction of the total energy of the H-bonding energy considered above, an may well be in the range 10 to 100 cal/mole; thus one can have $\varepsilon \simeq k T_c$, where T_c is close to 100 °K, as expected.

We now propose that in ACN a continuous transition occurs between two structures that we will call I and II, with somewhat different strength in H-bonding. Structure I is stable at high temperature and is the only one existing in non-crystalline material; in good crystals, at low temperatures, structure II becomes stable. The two structures differ in their H-atoms correlation in the ground state of the chain; these atoms are statistically independent in structure I, while in structure II they are coupled in a collective, coherent state of motion. A kind of ordering then occurs on cooling, which is reminiscent of an order-disorder transition; however, in ACN the transition cannot be sharp because the geometrical arrangement of the amide groups is nearly linear. To offer a pictorial representation, we may say that the potential energy curve of the H atom has an asymmetrical double-minimum well, the distance between the two minima being about 0.01 Å; then in structure I the H atom will be found closer to the N atom, while in structure II it will be closer to the O atom. Incidentally, the relevance of the zero-point motion for the creation of these double-minimum potential curves has been recently pointed out in simpler systems, like OH ... O [5]. Then the theory should follow the well-known treatment of H-bonded ferroelectrics [6, 7], where the proton lattice interactions are taken into account. Notice, however, that ACN cannot become polarized because the two chains run in opposite directions in the unit cell.

According to the above proposal, the transition between the two structures should not give rise to an observable sharp peak in the specific heat, and the deviation from the Debye profile should be small (because the lattice structure is not drastically affected, and should be distributed over a wide range of temperature. The same holds for the thermal expansion coefficient. As a matter of fact, both these expectations have been fulfilled by the experiments. At the molecular level, a careful determination of the atomic coordinates at different temperatures should reveal a nontrivial change in the bond lengths of the amide group, especially in the H atom position which should be about 0.01 Å closer to the oxygen atom at low temperature. An experimental study of the electronic structure of the crystal by UV, electrical conductivity and ESCA might also reveal that slight changes are taking place in the electronic states.

4. On the basis of the above discussion it seems quite likely that a collective coupling between the H atoms of the amide groups is the physical mechanism which gives rise to the observed features in ACN. The next step should be the detection of these features in more complex amide networks, and particularly in the α and β structures of polypeptides, provided the approach used for ACN can be used at all here. In fact, the transition temperature may well fall at such a high temperature as to escape observation because the material melts or decomposes. This is likely to be the case in the polypeptide structures because in ACN the N ... O distance is 2.94 Å, while in the polypeptide structures this distance is about 2.8 Å and the stronger hydrogen bonding can induce a stronger "extra"

coupling, yielding to a much higher transition temperature. If this statement is correct, it implies that in some biopolymers the H atoms are coupled in their zero-point motion, and that a coherent state must exist. As a matter of fact, the observed splitting of the amide I band of polypeptide chains is still of obscure origin [9], and the effect of deuteration is quite evident for the geometry of the α-helix [10].

A second and experimentally more feasible line of approach can be the detection of changes in the electronic ground state of the amide group, induced by a modification of the hydrogen-bounded system to which this group is connected. For instance, we found [4] that the hydration water of globular proteins markedly affects the IR properties of their peptide groups, and that the state of aggregation of this bound water is in turn affected by the protein, but no detailed investigation of the variations induced in the electronic structure has yet been attempted. This is, however, the line of work which is to be started if we want to understand protein structure and function in modern physical terms.

The experimental work on ACN, briefly outlined above, has been supported by C.N.R. grants, and has been performed in part at the Laboratori Ricerche di Base, Monterotondo, Rome; this work will be published elsewhere. The author acknowledges the contribution of his coworkers, particularly U. Buontempo and E. Gratton for the detection of the IR spectra, and F. Galluzzi and M. Garozzo for the Raman spectra, G. Grillo and E. Scafè for the heat capacity, and M. Cesari for the X-ray measurements.

Finally, the author express his gratitude to this friend Herbert Fröhlich, because he has been a permanent source of inspiration towards an understanding of biological matter from the viewpoint of cooperative phenomena.

References

1. Pimentel, G. C., McClellan, A. L.: The Hydrogen Bond. San Francisco Cal.: W. H. Freeman and Company 1960.
2. Pauling, L.: The Nature of the Chemical Bond, Chap. 8. Ithaca: Cornell Univ. Press 1960.
3. Abbott, N. B., Elliot, A.: Proc. Roy. Soc. (London), Ser. A **234**, 247 (1955).
4. Momany, F. A., McGuire, R. F., Yan, J. F., Scheraga, H. A.: J. Phys. Chem. **74**, 2424 (1970).
5. Anderson, G. R., Lippincot, E. R.: J. Chem. Phys. **55**, 4077 (1971).
6. Blinc, R., Riboric, M.: Phys. Rev. **130**, 1816 (1963).
7. Kobayashi, K.: J. Phys. Soc. Japan **24**, 497 (1968).
8. Konarki, J.: Chem. Phys. Letters **9**, 54 (1971).
9. Tomita, K., Rich, A., Lozé, C. de, Blout, E. R.: J. Mol. Biol. **4**, 83 (1962).
10. Buontempo, U., Careri, G., Fasella, P.: Biopolymers **11**, 520 (1972).

Behaviour of Interacting Protons: The Average-Mass Approach to its Study and its Possible Biological Relevance

G. Aiello, M. S. Micciancio-Giammarinaro, M. B. Palma-Vittorelli and M. U. Palma

Istituto di Fisica dell' Università di Palermo and Gruppi Nazionali Struttura della Materia, Consiglio Nazionale delle Ricerche

1. Introduction

Among all other kinds of ions occurring in the solid, liquid and living states of matter, positive hydrogen ions (which we shall henceforth call protons for brevity) are singled out by their exceptionally small mass. This is the reason for their specific property of tunnelling through potential barriers, which in turn is responsible for such properties as hydrogen bonding. This is also responsible for the unique properties exhibited in many cases by (sub)systems of interacting protons. Examples are the intriguing properties of liquid water and ice [1]; the para-ferroelectric transition in hydrogen-bonded ferroelectrics [2]; the cooperative transitions in ammonium salts [3] and in hexamine halides [4]; a variety of "transitions" observed in organic compounds [5] and many others. In all these cases the interaction among protons prevails over that existing between protons and heavier atoms. This makes the protons behave as a (sub)system having its own properties and characteristics, determining the macroscopic properties of the system as a whole. In a way, this (sub)system is similar to that of electrons in a conductor, even if the energy-band approach is not applicable.

A review of the properties of all the above-mentioned systems is beyond the scope of this article. Rather, we shall briefly recall the average-mass experimental procedure developed and used in our laboratory for obtaining information on the cooperative character of interactions among protons. We shall also see how this approach can provide information on the geometric and motional properties of a proton (sub)system, and how it allows predictions which have been strikingly, if not conclusively, verified. Finally, we shall see how this links up with the suggestion [6] that a considerable advance may be hoped for in our understanding of the living state, by considering all properties of biological molecules and systems in the phase space, rather than in the geometrical space only.

Such an advance might be comparable in importance to the one which marked the passage from crystallography to solid-state physics, when the properties of crystals in the space of momenta were taken jointly into account with those in the geometrical space, thus originating a new body of concepts and a deeper understanding of nature. Since, as we shall see, geometric-motional interdependence can be most marked in a proton (sub)system, we shall confine out attention to these systems.

2. Experimental Procedure

The underlying idea is very simple. Suppose we have a (sub)system of non-interacting protons and we are measuring a physical property of this system which depends upon the (thermally excited) motional state of protons. We shall expect an "isotopic effect" on the measured physical quantity upon substitution of the hydrogen with deuterium ions. In other words, this substitution will affect the temperature dependence of our measured quantity. If we effect a partial substitution (say, 50%) of protons with deuterons, the lack of interaction will make the particles behave individually, so that the result of our measurement will be the same as if we were experimenting on a composite sample consisting of two halves: one containing a (sub)system of protons and the other containing a (sub)system of deuterons.

Suppose now that there exists a strong interaction among protons. In this case, a partial substitution of protons with deuterons will affect the measured physical quantity as if we were now experimenting with a system of particles having a mass value intermediate between that of protons and that of deuterons. The actual "average mass" value will, of course, be determined by the H/D ratio.

Intermediate results between the two extreme cases considered above will be obtained in the case of intermediate interaction strengths or ranges. It is thus clear that a *partial* (and random) deuteration of a proton (sub)system can be a very useful diagnostic method for studying the behaviour of the proton (sub)-system.

A variety of different techniques can be used to measure physical properties sensitive to the geometrical and motional state of the (sub)system of protons. These include ESR and NMR, infrared and photon-scattering spectroscopy, neutron diffraction, specific heat, thermal conductivity, and biological functions. In all cases, one has to look for a stepwise change of the observed property with temperature: a pure shift in the "transition" temperature upon partial or total deuteration will support a strictly collective phenomenon, while a loss of the stepwise character will support, at least partially, non-collective phenomena. In all the cases investigated by the authors the "partial deuteration procedure" has proved to be a powerful tool.

3. Rigid Environment (Fixed Template)

One of the cases extensively studied in our laboratory by the average-mass technique [4f, g, j] is a (sub)system of interacting protons embedded in a crystal matrix (which for our present purposes can be considered as rigid).

The class of crystals studied is $MeX_2 \cdot 6NH_3$, where Me stands for a bivalent iron-group metal ion (Ni^{2+}, Mn^{2+}, etc.) and X for a halide ion. The proton subsystem is that of the $-H_3$ groups, and it was found that these groups undergo

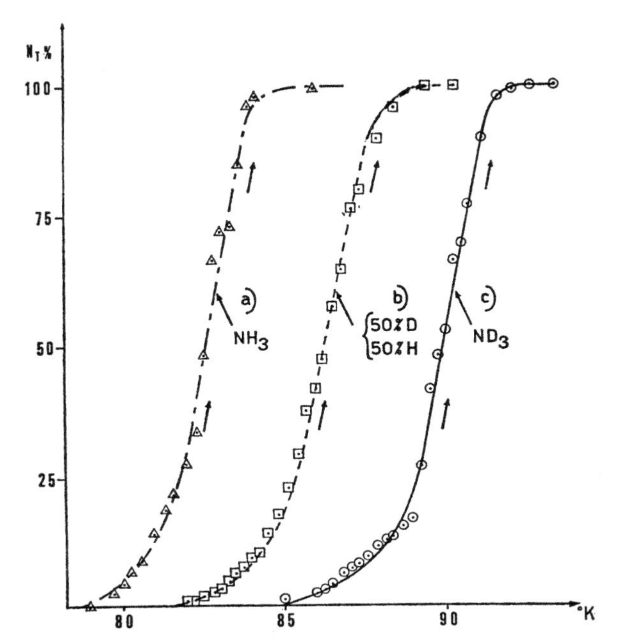

Fig. 1 a–c. The geometric-motional transition of $-H_3$ groups in $NiCl_2 \cdot 6NH_3$ as measured by ESR on the Ni^{2+} ions. The (percent) number of $-H_3$ groups in random motion is plotted for increasing temperatures. Hysteretic loops are omitted for clarity. a) $-H_3$ groups; b) $-H_3$, $-H_2D$, $-HD_2$ and $-D_3$ groups statistically occurring with an overall percentage of 50% H and 50% D; c) $-D_3$ groups. Note that the transition shape in case b) is identical with that of cases a) and c), i.e. that the transition is affected by the *average mass only*. (Reprinted from Phys. Letters **11**, 117 (1964) by permission)

a cooperative motional geometric phase transition and that they behave as an "independent" subsystem within the crystal. At the transition, they switch from a situation of slow and largely correlated non-rotational modes to one of rapid and non-correlated rotational modes, the motional change having essentially an on-off character. The case is particularly clear, and long-range correlations are evidenced by the use of the average-mass technique, as Fig. 1 shows.

The model accounting for this simple case uses the idea (which recalls the Weiss molecular field model for ferromagnetism) of a positive feedback linking the geometric and motional situations within the proton subsystem. In other words, it was found [4 j] that not only does the proton motion depend upon the pattern of the potential barriers and the proton mass, but, in turn, this pattern (as seen by one proton) depends upon the motional state of the remaining protons. This is due to the fact that in the static situation the potential barriers acting on a particular proton are largely due to the system of the remaining protons. In the situation of non-correlated rotations of the $-H_3$ groups (high-temperature phase) this contribution is washed out by motion, leaving only the barriers due to the heavy atoms. These turn out to be more than an order of magnitude lower. This fact is responsible for the on-off character of the switching between the two

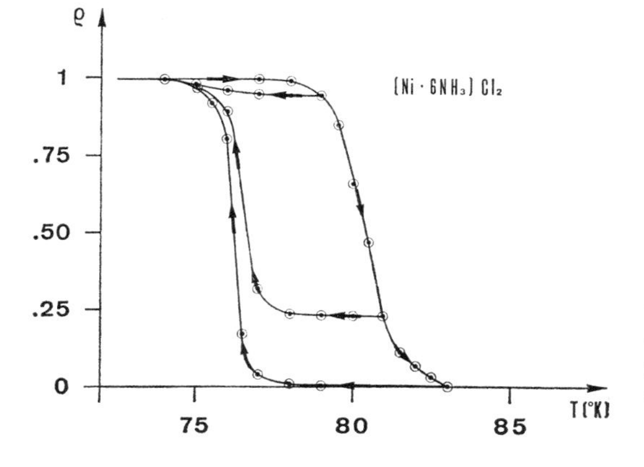

Fig. 2.
The complete hysteretic loop for Fig. 1, case a). Note also the partial loops, indicative of the existence of large domains. (Reprinted by permission of "Collective Phenomena" in press)

configurations. This on-off character makes the transition independent of the "observational frequency" (as experimentally observed) and it is responsible for the hysteretic behaviour.

This model is, of course, only a way to simplify the self-consistent description of the transition in the phase space by decoupling the geometric from the motional picture, and then re-linking them by way of the feedback, which makes the two mutually consistent. It accounts more than satisfactorily [4 j] for the average-mass and hysteretic effects (see Figs. 1 and 2). Under appropriate conditions, we would expect a similar geometric-motional feedback to be effective in enhancing and stabilizing specific geometric-motional structures, imprinted in a proton system by a periodic structure external to it or constraining it [4 j]. In the case of the crystal under consideration, the imprinting structure, or template matrix, will be that of heavier ions in the crystal, which lodges and constrains the $-H_3$ subsystem.

This model should therefore be suitable to describe equally well the properties of crystals such as ice, hydrogen-bonded ferroelectrics, ammonium salts, etc.

4. Mobile Environment (Adjustable Template)

The above-discussed model allows some inferences and speculations. We may, in fact, expect [7, 8] a similar imprinting of a specific structure to be provided by a template matrix which need not necessarily be rigid or fixed like that occurring in crystals. In fact, the enhancing geometrical-motional feedback should be present even in such cases, and it may be expected to be effective, under appropriate circumstances, in stabilizing the imprinted structure. The resulting phenomena may be expected to occur, e.g. also in the case of macromolecules in water, at least for that class of macromolecules having repeat distances which are multiples of a significant distance in liquid water (see Refs. [7] and [8] for a discussion and for a list of relevant literature).

This case will differ from the one that we have already considered. In fact, the matrix of heavier ions in the crystal constitutes a practically rigid structure both above and below the protons ordering temperature. In the case of macromolecules in water, on the contrary, even if only one molecular species is present,

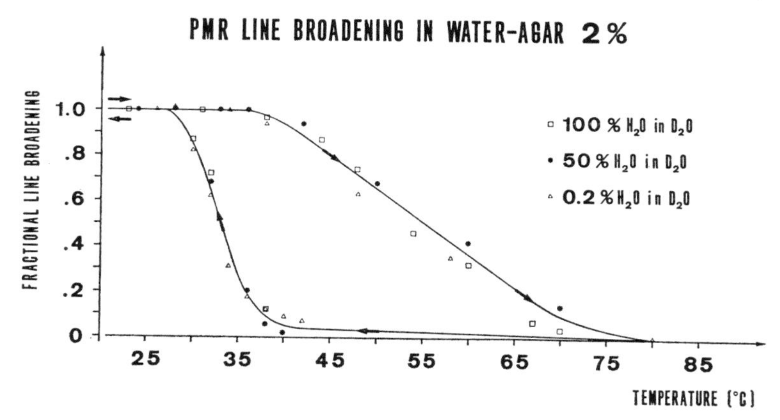

Fig. 3. Hysteretic transition in the geometric-motional state of water protons in a macromolecule-water system as measured by NMR on water protons. (Agar is a polysaccharide compound occurring as cell-wall constituent in some algae). Note the similarity with Fig. 2 (partial loops, not shown here, are also obtained, as in Fig. 2). Note also the absence of isotopic effects, at variance with the case of Fig. 1, also predicted in the model (see text). (Reprinted from Ref. 7 by permission of the Annals N.Y. Acad. Sci. **204** (1973)

macromolecules can undergo random motion, which in general will prevent them from building up a well-defined and concerted three-dimensional potential pattern for the protons subsystem, even if the macromolecular concentration and structure are, in principle, appropriate.

Convincing evidence for the formation and operation of a template matrix for the water protons in the case of a polysaccharide molecular species in water has already been obtained in our laboratory [7, 8]. This is concisely shown in Fig. 3, where the presence of the large hysteretic cycle provides evidence for the operation of the feedback mechanism. No isotopic effects, as measured by NMR are present in this case. This absence is not surprising on the basis of the following considerations:

i) A template matrix (similar to the matrix of heavier ions in the crystals previously discussed) will be obtained in this case only if macromolecules are able (and allowed) to arrange themselves in the appropriate (and "concerted" or self-consistent) disposition in space and if their motion (including internal motion) is slow or concerted [7, 8]. This can only occur below a definite temperature T_M.

ii) This template matrix will in turn be capable of imprinting the corresponding (geometric-dynamical) structure in water, only if the H_2O molecules are in a motional and structural state such that they can accept it and propagate it by the discussed geometric-motional mechanism [7]. This will occur only below a definite temperature T_{H_2O}.

iii) The transition will occur only if both the above conditions are satisfied. Therefore it will be controlled by macromolecules if $T_M < T_{H_2O}$, or by water in the opposite case.

The absence of isotopic effect provides evidence for the transition to be controlled by macromolecules, and therefore is $T_M < T_{H_2O}$ (see Ref. [7]).

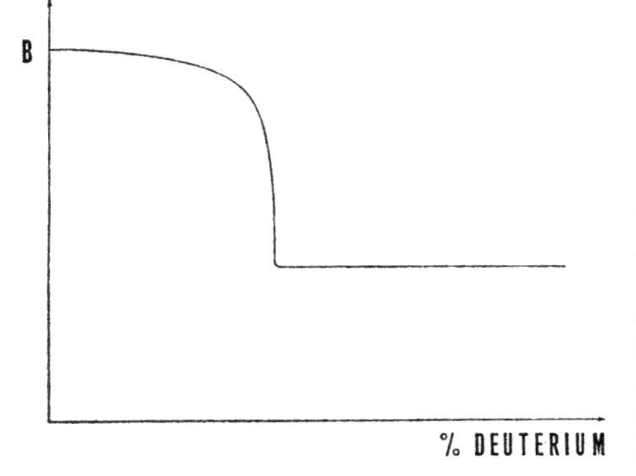

Fig. 4.
One possible hypothetical dependence of a biological property or function (B) of a molecule, upon partial deuteration of the molecule (exchangeable hydrogen ions) and of its aqueous medium. (Reprinted by permission from "From Theoretical Physics to Biology", M. MAROIS, Ed., Proc. 2nd Int'l Conf., Versailles, June, 1969. Editions du C.N.R.S., Paris, 1971)

Another possibility which, as mentioned in the Introduction, can be inferred on the basis of the model discussed in Section 3, is that the dynamical properties and structure of the proton subsystem play a role not only in the geometrical structure of the water-biomolecule system but also (see Ref. [6], p. 29) in a biological property or function (B) of the solute molecule. If this is the case, (B) would be found to depend upon deuteration of water in a spectacularly stepwise manner, as shown in Fig. 4. This would occur if the biomolecule acts as an "adjustable template" and if water does not act trivially as a solvent but rather as a "structural signal transmitter and stabilizer". The above conditions i) and ii) must be satisfied, and, according to condition iii), isotopic effects will be present if $T_{H_2O} < T_M$, or absent in the opposite case.

A very interesting case of such a nature, also investigated in our laboratory by A. CUPANE and E. VITRANO [9], is that of hemoglobin. This protein molecule is often thought of as a "simple" approximation to a regulatory enzyme. The unique oxygen-uptake properties of this protein depend very much [10] upon its ability to convert its own configuration from the "closed" to the "open" form and vice versa (upon oxygen uptake and release). CUPANE and VITRANO thought it worthwhile to study the possible occurrence of the above-discussed imprinting of a geometric-dynamical structure in water by this protein molecule in its different configurations, at concentrations low enough to allow intermolecular interactions to be disregarded [11]. They measured the n_{Hill} parameter, which is a measure of the amount of "cooperativity" in the oxygen uptake. Their reasoning runs as follows: if the imprinting of a geometric motional structure in the surrounding water by hemoglobin plays a role in the biological functions of this protein, this is likely to show up in its n_{Hill} parameter. In this case, a plot of this parameter *versus* the percentage deuteration of the environmental water will have the predicted stepwise dependence.

This indeed what CUPANE and VITRANO found. Some of their results are presented in Fig. 5, which undoubtedly follows very closely the predictions already illustrated in Fig. 4. They further found the "critical average mass" to be temperature-dependent. In fact, by plotting Δn_{Hill} *versus* temperature for different

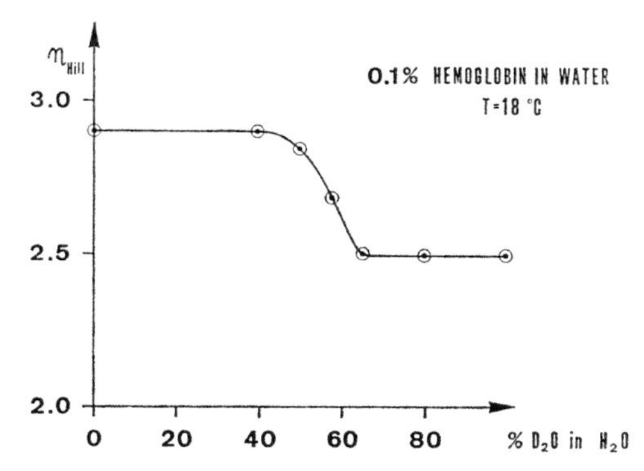

Fig. 5. Stepwise dependence of oxygen-uptake properties of hemoglobin upon deuterium content of solvent water. Hill's constant, n_{Hill}, measures the amount of "cooperativity" within the four subunits of the hemoglobin molecule. It is here plotted versus the percental substitution of water protons by deuterium. Temperature: 18 °C. Concentration ~0.1% pure Hemoglobin (no buffer, no DPG). Note similarity with Fig. 4. Different "critical" deuterium concentrations correspond to different temperatures, so that curves of the type of Fig. 1 are also obtained (Courtesy of Drs. A. CUPANE and E. VITRANO, Refs. 9)

values of the average mass (i.e. of the percentage of D_2O in H_2O, used now as a parameter) they obtained a curve very similar to that of Fig. 1.

It must be emphasized that this similarity, even if very appealing, provides convincing but not yet conclusive evidence.

5. Conclusions

We have reviewed the evidence for the ability of a proton subsystem to undergo a stepwise motional-geometric phase transition under the influence of an appropriate "template matrix". This ability originates as an effect of motional smearing-out or reinforcement of the potential barrier pattern seen by individual protons. The actual value of the "observational frequency" has proved to be irrelevant, thus indicating the on-off character of the transition. The average-mass technique has proved very useful in this type of experiments.

The system of water protons can undergo a similar type of imprinting by appppropriate biomolecules. The interplay between geometry and dynamics in a biomolecule-water system may thus have a role in the biological properties and functions of specific molecules. Experiments performed so far show indeed the interest of proceeding in this direction, although they do not provide, as yet, conclusive evidence.

This is a very appropriate place to remember that H. FRÖHLICH was the first to consider the possible biological significance of the properties of biomolecules in the space of momenta, with his theory of giant oscillations of macromolecules, capable of originating dynamical structure stabilization and pattern recognition [12]. We also benefited from countless discussions with him, through several years, on the line of work summarized in the present paper. It is for us a particular pleasure to point it out here.

References and Notes

1. See e.g.:
   a) DORSEY, N. E. (Editor): Properties of Ordinary Water-Substance. New York: Hafner 1968 (reprinted).
   b) EISENBERG, D., KAUZMANN, W.: The Structure and Properties of Water. Oxford: Univ. Press 1969.
   c) NÉMETHY, G.: Ann. Ist. Sup. Sanità (Rome) **6** (1970). Special Issue [in English].
   d) RIEHL, N., BULLEMER, B., ENGELHARDT, H. (Editors): Physics of Ice. New York: Plenum Press 1969.
2. See e.g.:
   a) FATUZZO, E., MERZ, W. J.: Ferroelectricity. Amsterdam: North-Holland Publishing Co. 1967.
   b) KÄNZIG, W.: Ferroelectrics and Antiferroelectrics. London: Academic Press 1967.
   c) BLINC, R.: Phys. Rev. **147**, 430 (1966).
   d) CUMMINS, Z. M., BRODY, E. M.: Phys. Rev. Letters **21**, 1263 (1968).
   e) FUJIWARA, T.: J. Phys. Soc. Japan **29**, 1282 (1970).
3. a) GUTOWSKY, M. S., PAKE, G. E., BERSHON, R.: J. Chem. Phys. **22**, 643, 651 (1954).
   b) FREUND, I., KOPF, L.: Phys. Rev. Letters **24**, 1017 (1970).
   c) SCHUMAKER, N. E., GARLAND, C. W.: J. Chem. Phys. **53**, 392 (1970).
4. a) PALMA-VITTORELLI, M. B., PALMA, M. U., DREWES, G. W. J., KOERTS, W.: Physica **26**, 922 (1960).
   b) KIM, P. H.: J. Phys. Soc. Japan **15**, 445 (1960).
   c) WATANABE, T.: J. Phys. Soc. Japan **16**, 1131 (1961).
   d) PALMA-VITTORELLI, M. B., PALMA, M. U., PERSICO, F.: J. Phys. Soc. Japan **17**, Suppl. B1, 475 (1962).
   e) GAROFANO, T., PALMA-VITTORELLI, M. B., PALMA, M. U., PERSICO, F.: In: Paramagnetic Resonance, Vol. II (W. LOW, Editor), p. 582. New York: Academic Press 1963.
   f) AIELLO, G., PALMA, M. U., PERSICO, F.: Phys. Letters **11**, 117 (1964).
   g) AIELLO, G., PALMA-VITTORELLI, M. B.: Phys. Rev. Letters **21**, 137 (1968).
   h) BATES, A. R., STEVENS, K. W. H.: J. Phys. (C: Solid State Phys.) **2**, 1573 (1969) (London).
   i) BATES, A. R.: J. Phys. (C: Solid State Phys.) **3**, 1825 (1970).
   j) AIELLO, G., PALMA-VITTORELLI, M. B.: Collective Phenom. (London) in the press.
5. See e.g.:
   a) ANDREW, E. R.: J. Chem. Phys. Solids **18**, 9 (1961) and J. Chem. Phys. (France) **63**, 85 (1966) and literature quoted therein.
   b) BLOEMBERGEN, N., PURCELL, E. M., POUND, R. V.: Phys. Rev. **73**, 679 (1948).
   c) GUTOWSKY, H. S., PAKE, G. E.: J. Chem. Phys. **18**, 162 (1950).
6. MAROIS, M. (Editor): From Theoretical Physics to Biology. Proc. 2nd Int'l Conf., Versailles, June 1969. Paris: CNRS 1971 p. 27ff.
7. MICCIANCIO-GIAMMARINARO, M. S., MICCIANCIO, S., PALMA-VITTORELLI, M. B., PALMA, M. U., MARINO, M. L.: Ann. N.Y. Acad. Sci. **204** (1973).
8. PALMA, M. U., in: From Theoretical Physics to Biology (M. MAROIS, Editor). Proc. 3rd Int'l Conf., Versailles, June 1971 (in the press).
9. a) CUPANE, A.: Thesis, Palermo, June 1972.
   b) VITRANO, E.: Thesis, Palermo, June 1972.
   c) CUPANE, A., PALMA, M. U., VITRANO, E.: To appear on J. Mol. Biol.
   We are indebted to Drs. CUPANE and VITRANO for permission to quote their as yet unpublished results.
10. It is impossible to quote here the vast amount of literature concerning hemoglobin. For the present specific purpose, see: PERUTZ, M. F.: Nature **228**, 734 (1970).
11. It might be objected that different hydration shells correspond to the different quaternary structures, so that what one measures is nothing more than the influence of these hydration shells on the oxygen uptake. The geometric-motional model that we have discussed provides in fact the basis for the microscopic interpretation of the thermo-

dynamical description of "hydrophobic" and "hydrophilic" interactions and for concepts such as "structure-making" or "structure-breaking" (see e.g. Ref. [1, c]). In addition, however, it makes provision for the dynamical part of the picture to enter the microscopic interpretation.

12. FRÖHLICH, H.:
    a) In Ref. 6, p. 13.
    b) Phys. Letters **26** A, 402 (1968).
    c) Int. J. Quantum Chem. **2**, 641 (1968).
    d) In: Proc. 3rd Int'l Conf. on Theor. Phys. and Biology (M. MAROIS, Editor, in the press).
    e) Phys. Letters **39** A, 153 (1972).

Ion Transport Across Cell Membranes

B. W. HOLLAND

University of Warwick, Department of Physics, Coventry/England

1. Introduction

FRÖHLICH [1—4] has recently put forward some exciting new ideas, concerning processes that may arise from cooperative effects between the dipolar units from which many biologically significant macromolecules are constructed. The general hypothesis that living organisms use metabolic energy for the excitation of coherent dipolar oscillations, promises to be a fruitful source of novel proposals concerning the specific mechanisms of many biological processes. Tentative applications have already been made to photosynthesis [5], enzyme action [6] and to the pairing of homologous chromosomes in meiosis [7]. We here advance the hypothesis that the repeating dipole structure of helical proteins determines the vitally important and so far unexplained characteristics of the transport of inorganic ions across the membranes of living cells. Some suggestions in this direction have already been made by ONSAGER [8, 9].

In spite of a wealth of experimental data, and though there exists a very satisfactory phenomenological theory of nerve processes due to HODGKIN and HUXLEY [10], the molecular mechanisms underlying ion transport phenomena in membranes are so far unknown. We are of course hampered by the lack of detailed knowledge of the composition and structure of membranes, and it is arguable that one should wait until such information is available, rather than make speculations on an insecure basis. However, the problem of characteristing in detail the structure of a living membrane is enormous, and in the face of such complexity a frontal assault is likely to be very slow in producing results. One needs to know what to look for, and it is therefore desirable to advance plausible working hypotheses that stimulate the search for specific structural components.

The mechanisms that have been proposed in order to explain the transport properties of cell membranes for inorganic ions fall into two broad classes. On the one hand we have the mechanisms involving specific carrier molecules, which though readily soluble in nonpolar media, have sites for the attachment of ions. Such carriers, in particular certain cyclic polypeptides, have been shown to give rise in artificial membranes to behaviour with some of the characteristics of nerve impulse phenomena [11], and undoubtedly this class of mechanisms is to be taken seriously. As yet however, no such molecules have been positively identified in living membranes. The hypothesis we shall put forward belongs to the other class of pore or channel explanations, that is that ion transport takes place through small isolated regions of the membrane under the control of specific mechanisms.

We first briefly summarise the salient known facts to be explained before proposing a structure and mechanisms which allow us to obtain a qualitative understanding of the various membrane phenomena.

2. Summary of Relevant Membrane Phenomena

The basic structural entity of cell membranes is a lipid bilayer, the lipid molecules in the two layers being oppositely orientated so that the polar groups lie on the two sides of a hydrocarbon chain region typically about 40 Å thick. The static dielectric constant of this region is probably less than 3, while that of water is about 80; hence the energy required to take an inorganic ion from the water on either side of the membrane, and place it in the hydrocarbon environment is very much greater (of order 1 eV per ion) than typical thermal energies. Thus a single lipid bilayer would be virtually impenetrable to inorganic ions. This is a useful property since the concentrations of different ions inside and outside the cell can differ by an order of magnitude or more, so that if the cell membrane were readily permeable, the cell would have to expend metabolic energy at an impossibly high rate, in order to maintain the essential concentration differences. In fact the permeability of cell membranes to inorganic ions is found experimentally to be much higher than expected on simple relative solubility arguments, and to differ markedly for different ions. Evidently specific mechanisms are available for the transport of ions through the membrane, enabling the cell to exert fine control over its internal composition.

The most important inorganic ions for membrane phenomena are those of potassium and sodium. The potassium concentration inside a cell is typically much higher than outside; for sodium the opposite is true. When the membrane is in its resting state the permeability to potassium is roughly one hundred times that for sodium, so that potassium diffuses out more rapidly than sodium diffuses in. Since the membrane is impermeable to the organic anions within the cell, this process gives rise to a charge seperation and a membrane potential which tends to oppose the outward flow of potassium ions. If no other ions were involved an equilibrium would be reached for which the membrane potential and potassium concentrations inside and outside the cell would be related by the Nernst formula. It is in fact found that the potential calculated in this way from the measured potassium concentrations, is quite close to the observed resting potential (of order 100 mV for squid axon under normal conditions).

A fact of the greatest importance, underlying the phenomenon of nerve impulse propagation, is that the permeability to sodium and potassium ions, depends on the membrane potential and on its variation with time (see for example, the book by KATZ [12]). If the potential difference across the membrane is suddenly decreased slightly, the sodium conductance increases rapidly and then decreases practically to zero (within a few m. secs. for squid axon). The potassium conductance increases more sluggishly, remaining almost unchanged until the sodium conductance has passed its maximum, and then increasing and continuing to rise after the sodium conductance has returned to zero, becoming constant if the membrane potential remains constant. When the stimulus is removed, the outward flow of potassium ions restores the resting potential. The changes in conductance for small membrane depolarisations depend on the value of the depolarisation step. But if the decrease in membrane potential exceeds a certain threshold value (about 40 mV for squid axon) and the stimulus is then removed, the depolarisation increases to a value independent of the initial stimulus. For a few m. secs. the permeability of the membrane to sodium ions becomes much greater than that for potassium ions, and the membrane potential is reversed in sign, reaching a value close to that predicted from the Nernst relation, assuming that the membrane is permeable only to sodium. The sodium conductance then rapidly falls to zero ("the sodium gate shuts"), and the slower rise in potassium current restores the resting potential locally (assuming a localised stimulus). While the sodium gate is open at the site of the initial stimulus, the local decrease in sodium concentration outside the membrane, causes diffusion of sodium from the neighbouring region, with a consequent decrease of sodium concentration and depolarisation. This depolarisation of the neighbouring region results in an increase in sodium permeability, culminating in a reversal of the membrane potential. Thus the depolarisation pulse propagates and is self-sustaining.

3. Hypothetical Channel Structure and Transport Mechanisms

3.1. Helical proteins and Reversible Hydrogen Bonds

We postulate that a channel for inorganic ion transport consists of a bundle of α-helical protein strands, placed so that the axes of the helices are perpendicular to the membrane surfaces [9]. The helices have non polar side chains that make the whole structure preferentially soluble in the lipid medium, and stabilise its configuration. The oxygen and nitrogen atoms of the helix then forms chains of electronegative centres traversing the membrane. A positive ion can then pass through the membrane by hopping from one such centre to the next, along the outside of the helix. The energy required for transfer of the ion from the water to such a polar chain, will be much less than that needed for it to pass from water into the lipid region, and may be reduced by cooperative motion of the protons in the hydrogen bonds linking the nitrogen and oxygen atoms. The potential energy of the proton as a function of position between the nitrogen and oxygen atoms will have two strong minima, one near each atom, and though normally the protons sit closest to the nitrogen atoms, they might easily shunt between the two sites to facilitate the motion of a positive ion along the chain. Similar processes, but involving the motion of a bonding defect, have been suggested by ONSAGER [8, 9] and are known to be important in the phenomenon of dielectric loss in hydrogen bonded solids [13].

While ordinarily the stable configuration of the helix is such that the protons of the hydrogen bonds lie near the nitrogen atoms (normal polarisation), alternatively the hydrogen bonds could be oppositely polarised, with protons near the oxygen atoms (anomalous polarisation). We expect this anomalous configuration to be metastable, since configurations between this and the normal one are likely to have higher energies than either, due to proton-proton interactions. Since the anomalously polarised hydrogen bonds will be weakened we expect a lengthening of the helix in the metastable state. If the helix is so orientated that the N-H vector is directed towards the inner surface of the membrane, the transmembrane electric field might well cause the anomalous configuration to become stable, since its dipole moment will be larger than that of the normal configuration (such a possibility has been recently suggested by Fröhlich [14]). A necessary condition for this to occur is that the dipole-field energy change must be large compared to thermal energies. Assuming proton displacements of order 1 Å in the configuration change, the dipole-field interaction energy decrease is of order 0.1 eV per helix, so that the condition is indeed satisfied.

We assume therefore, that in the resting state, the membrane has running between its two surfaces, bundles of α-helices with anomalous hydrogen bond polarisation. Let us see whether such a structure can account at least qualitatively for the permeability properties of membranes to inorganic ions.

3.2. The Hopping Mechanism

We envisage two types of cooperative motion by the protons of the hydrogen bonds during transport of positive ions. When an ion is on an electronegative site (an oxygen or nitrogen atom), the neighbouring hydrogen bonds will be polarised in directions away from this site, thus reducing the electrostatic energy. Hopping of the ion to the next site will also involve a cooperative switching of the polarisation of the intermediate hydrogen band. The activation energy for this cooperative hopping process will determine the rate of diffusion along the chain. The permeability will also depend on the rate of entry to the chain, which will be determined by the activation energy for entry. We expect for alkali metal ions that both these activation energies will be larger the smaller the ionic radius, just as the solvation energy is larger for ions of small radius. Thus the permeabiliy should decrease with decreasing ionic radius as is observed [12].

Notice further, that the activation energy for entry into the channel will depend strongly on the end charge of the dipole [8]. This gives rise to a directional dependence of the permeability, in addition to that arising from the lack of inversion symmetry of the helix. For our postulated structure the inner end charge is positive, and cations will enter the inner end of the channel less easily than they can enter from the outside.

3.3. The Sodium Conductance Changes

In a voltage clamp experiment in which the membrane potential is suddenly decreased, and then held at a constant value, the sodium conductance shows a sharp transient increase lasting typically 2 or 3 m. secs. for squid axon. This can be understood in terms of energy release and dissipation.

Each helix, having a large dipole moment will be subject to stress in the transmembrane field, tending to extend the helix and thus increasing its internal potential energy. When the field is suddenly reduced therefore, energy is released into the longitudinal vibrational modes of the helix, particularly into the polar modes which interact most strongly with the electric field, namely the oscillations involving the protons in the hydrogen bonds. During oscillation the helix will pass through configurations in which the activation energy for hopping is lowered.

Further, the polar modes will tend to couple most strongly to inorganic ions present, and therefore lose energy preferentially to them rather than to the non polar environment. Thus the transport of ions will be temporarily facilitated, by an effective lowering of the activation energy coupled with a raising of the temperature of the ions in the vicinity. It is of interest in this connection that the smallest depolarisation that gives a measurable change in sodium conductance is of order 10 mV, and the energy released in this case for a displacement of somewhat less than 1 Å per proton, is of order kT per helix. Since it is estimated that about 2000 sodium ions per channel cross the membrane in a single impulse [15], this energy release is only sufficient to initiate the onset of conduction. But a single ion passing through the membrane initially aquires an energy of order 0.1 eV. Since the ions interact very strongly with the polar modes, much of this energy will be fed into these modes, tending to maintain the excited state and facilitate the transport of more ions. If the rate at which energy is lost from the helix-ion system exceeds the rate at which it is pumped in, then the conductance falls back to zero without reaching very high values. This is the sub threshold response. But if the initial depolarisation is large enough, the anomalous configuration becomes unstable, and collapses, releasing the energy of the configuration change into the oscillatory modes. Then the initial rate at which sodium ions crossing the membrane pump energy into the polar modes, exceeds the rate of energy loss from the system, and the process becomes self sustaining.

As the flow continues the energy gained from an ion crossing the membrane decreases, hence the rate of pumping energy into the polar modes decreases, and ultimately becomes subcritical, so that the conductance falls to zero again. Thus the time course of the sodium conductance is determined by the time required for sufficient sodium ions to cross the membrane to establish temporary equality of the chemical potentials. The asymmetry of behaviour between sodium and potassium arises from the difference in end changes, which makes it easier for cations to enter from the outside, and also from the weaker coupling of potassium ions to the polar modes, due essentially to the larger ionic radius of this ion.

It is of some interest that this mechanism allows a natural interpretation of the phenomenon of accommodation, i.e. the tendency of slow changes in potential to be relatively less effective, (see for example, Lakshiminarayanaiah [16]). It is found for example, that if a depolarisation of 8 mV (so small that no measurable sodium current is stimulated) is maintained for a few m. secs., and then the depolarisation increased to 44 mV, the sodium current is reduced by about 40% compared with that for a simple 44 mV depolarisation. On the other hand, if the initial step is a hyperpolarisation of 31 mV the sodium current is increased by about 70%. This is just what our mechanism would predict. In the first case some of the energy released in a 44 mV depolarisation has already been allowed to dissipate during the 8 mV depolarisation, so that the energy available to initiate sodium ion conduction is reduced. In the second case more energy is available for the initiation step as a result of hyperpolarisation.

3.4. *Potassium Conductance*

In a voltage clamp depolarisation experiment the potassium conductance rises and remains constant at a higher level as long as the clamp is maintained. We can understand this by noting that the activation energy for hopping between the nitrogen and oxygen atoms at each end of a hydrogen bond, with a polarisation reversal, is likely to decrease strongly with decreasing hydrogen bond length. Because of the dipole-field interaction, the hydrogen bond length will decrease as the polarisation decreases. Hence the activation energy decreases and the conductance increases with a decrease in membrane potential.

The reason for the delay in the increase of potassium conductance as compared to sodium arises from competition between these two ions for the electronegative sites of the conducting channel. Because of the negative outside end charge, and the strong coupling of the helix polar modes to the sodium ions, the first effect of depolarisation is a flood of sodium ions through the channel from the outside to the inside. Only as the sodium current starts to fall off can be the probability for potassium ions to occupy the electronegative sites, increase. On switching off the depolarisation the helix rapidly attains its equilibrium length, and the potassium conductance should start to decrease immediately, as is found.

4. Active Transport

Since the rate of diffusion of ions through the membrane is not actually zero, the inner and outer concentrations will tend to equalise. In order to maintain its internal composition therefore, the cell must be capable of transporting ions against a concentration gradient, a process that obviously involves work by the cell, and is known as active transport. The ability of cells to pump out sodium against a concentration gradient has been studied intensively, but the detailed mechanism for the actual movement of ions remains a mystery.

The most pertinent facts discovered so far are as follows [17]. Under physiological conditions, cell membranes hydrolyse ATP and pump sodium ions out and potassium ions in; roughly three sodium ions are extruded and two potassium ions drawn in for each molecule of A.T.P. hydrolysed. It is widely held as a working hypothesis, that the enzyme involved spans the membrane and undergoes a series of conformational changes that result in translocation of the ions. The pump can only utilise A.T.P. on the inside of the membrane, and requires magnesium ions and sodium ions on the inside for activation. It is therefore supposed that the substrate for the enzyme is a magnesium-ATP complex, and that sodium ions induce a conformational change in the enzyme that allows the formation of a substrate-enzyme complex. However, no hydrolysis takes place without potassium ions in the external medium. It is suggested therefore that adsorption of potassium ions from the outside, induces a further conformational change in the enzyme, that lowers the activation energy for hydrolysis. Hydrolysis itself is accompanied by a conformational change which results in translocation of the ions. A further point of interest is that potassium inside inhibits sodium activation, and sodium outside inhibits potassium activation; evidently the sodium sites inside and the potassium sites outside are not completely selective.

We suggest here that the reversible hydrogen bond polarisation of α-helical proteins is also involved in the mechanism of active transport. Suppose that the α-helices are again gathered into a bundle spanning the membrane, but this time with the N-H vector directed outwards, so that the normal configuration of the hydrogen bonds is stabilized even further by the trans-membrane field. The inner end of the bundle is intimately associated with the enzyme which brings about the hydrolysis of A.T.P. The adsorption of several sodium ions at the inner end of the pump might cause local reversal of the hydrogen bond polarisation, inducing local extension of the helix, which might itself be the conformational change required for the formation of the substrate enzyme complex, or might induce changes in the relative positions of relevant groups linked to the helix. When potassium ions are adsorbed in certain sites at the outer end of the pump allowing hydrolysis to proceed, we assume that the energy is released in such a way as to produce a wave of reverse polarisation of the hydrogen bonds to travel along the α-helix to the outside, driving sodium ions before it. On reversing in direction at the outside, the wave returns driving potassium ions inwards and restoring the normal polarisation. Such a wave might travel up and down the helix many times for a single hydrolysis.

5. Discussion

We have proposed that the inorganic ion permeability properties of cell membranes are determined by channels or pores in the membrane, consisting of bundles of α-helical proteins running from one surface of the membrane to the other. In one type of bundle, associated with active transport, the N-H vectors of the hydrogen bonds along the backbone of the helix are directed outwards, and waves of polarisation reversal are responsible for the transport of ions against a concentration gradient. In the other type of bundle the α-helices are oriented in the opposite sense, and in the presence of the resting potential have an anomalous conformation in which the hydrogen bonds are polarised in a sense opposite to the usual one, with a consequent lengthening of the helix, and an increase in the dipole moment. Return to the normal conformation can be triggered (as it is in a nerve impulse), by a sufficient decrease in membrane potential, and the sodium and potassium conductivities are determined by the resulting energy release and length change, which themselves depend on how the potential changes with time.

We have put forward these hypotheses in order to account qualitatively for the salient features of membrane behaviour, but there is a wealth of ancillary data that is also consistent with them. It is known for example that a large fraction of membrane protein is in the α-helical form [18]. Light scattering and fluorescence studies [19] indicate that nerve activity is associated with conformational changes in the membrane proteins. More important, measurements show that there is a small birefringence decrease following the time course of the impulse, and with optic axis radially directed relative to the axon [20]. This is indeed consistent with the presence of α-helices spanning the membrane, and undergoing polarisation changes during the passage of an impulse. Further, in voltage clamp experiments the birefringence changes are directly dependent on the membrane potential, as we expect. In connection with out mechanism for the initiation of the impulse, it is of great interest that light absorption can stimulate nerve impulses [21]. The initiation of the impulse starts by the injection of energy into the longitudinal polar modes of the helices, normally by changes in the membrane potential. But light energy fed into these modes either directly or indirectly after prior absorption in a pigment in association with the helices could clearly produce the same effect. Yet another piece of evidence is the observation by CALVIN [22] that a transient change in membrane dielectric constant occurs during the onset of the impulse which would be accounted for by the change in dipole moment on switching from the anomalous to the normal hydrogen bond polarisation.

In connection with active transport, a most interesting result was obtained by BARGENON et al. [23] who found in the light beating spectrum of erythrocyte ghosts, a sharp line at about 170 Hz, which disappeared on inhibition of active transport*. We may interpret this result as meaning that the sodium pump operates with a definite frequency, but it is difficult to reconcile this idea with the usual view that the pump translocates the ions by a series of chemical reactions occuring in sequence at a series of locations across the membrane, or with any mechanism involving carrier molecules. On the other hand the frequency is many orders of magnitude lower than any associated with molecular vibrations about a stable configuration. However, it is conceivable that the hydrogen bond polarisation waves might have frequencies in this range. The elementary process here is the movement of a proton from one potential minimum in the hydrogen bond to the other. The frequency with which such a process occurs depends on the activation energy and on the energy of the proton, for example in thermal equilibrium it would be of order $\nu_0 e^{-\varepsilon/kT}$, where ν_0 is the usual vibrational frequency of the proton about its stable position, and ε is the activation energy. In the sodium pump however, the system is far from equilibrium, and it is impossible with present information to give any reasonable estimate of the frequency. We have to deal here with an oscillation between a stable and metastable state, i.e. one that is essentially non linear in the displacements and does not exist at all in the quasi-linear regime. Further, there is actually a system of such units interacting strongly with each other, with an energy and initial state that are not known. We merely emphasize that the question of the frequency of oscillation for such a system, is an open one.

Finally it should be remarked, that while the hydrogen bond polarisation waves are not covered by the model analyses given by FRÖHLICH [1–3, 14], they are indeed consistent with his original and intuitive suggestion that an essential characteristic of living systems, is that they contain entities that support coherent dipolar oscillations, whose excitation involves the dissipation of metabolic energy.

References

1. FRÖHLICH, H.: Theoretical Physics and Biology (MAROIS, M.), p. 1. Amsterdam: North Holland 1969.
2. FRÖHLICH, H.: Int. J. Quantum Chem. **2**, 641 (1968).
3. FRÖHLICH, H.: Phys. Letters **26** A, 402 (1968).
4. FRÖHLICH, H.: Phys. Letters **39** A, 153 (1972).
5. FRÖHLICH, H.: Nature **219**, 743 (1968).
6. FRÖHLICH, H.: Nature **228**, 1093 (1970).
7. HOLLAND, B. W.: J. Theoret. Biol. **35**, 395 (1972).
8. ONSAGER, L.: Science **156**, 541 (1967).

* *Note added in proof:* Unfortunately these authors have been unable to reproduce this result (P. E. R. TATHAM: Private communication).

9. ONSAGER, L.: The Neurosciences (QUARTON, G. C., MELNECHUK, T., SCHMITT, F. O.), p. 75. New York: Rockefeller Univ. Press 1967.
10. HODGKIN, A. L., HUXLEY, A. F.: J. Physiol. (London) **117**, 500 (1952).
11. MUELLER, P. W., RUDIN, D. O.: Nature **217**, 713 (1968).
12. KATZ, B.: Nerve, Muscle and Synapse: New York: McGraw-Hill 1966.
13. DANIEL, VERA V.: Dielectric Relaxation. London: Academic Press 1967.
14. FRÖHLICH, H.: Collect. Phenomena (to be published).
15. ADRIAN, R. H.: The Molecular Basis of Membrane Function, p. 249. TOSTESON, D. C.: Prentice-Hall 1969.
16. LAKSHIMINARAYANAIAH, N.: Transport Phenomena in Membranes. New York: Academic Press 1969.
17. WHITTAM, R.: The Neurosciences (QUARTON, G. C., MELNECHUK, T., SCHMITT, F. O.), p. 313. New York: Rockefeller Univ. Press 1967.
18. SINGER, S. J.: Structure and Function of Biological Membranes (ROTHFIELD, L. I.), p. 145. New York: Academic Press 1971.
19. TASAKI, I., BARRY, W., CARNAY, L.: Physical Principles of Biological Membranes (SNELL, F., WOLKEN, J., IVERSON, G., LAM, J.), p. 17. New York: Gordon and Breach 1970.
20. COHEN, L. B., KEYNES, R. D., HILLE, B.: Nature **218**, 438 (1968).
21. SINGER, I., TASAKI, I.: Biological Membranes (CHAPMAN, D.), p. 347. New York: Academic Press 1968.
22. CALVIN, M.: The Neurosciences (QUARTON, G. C., MELNECHUCK, T., SCHMITT, F. O.), p. 780. New York: Rockefeller Univ. Press 1967.
23. BARGENON, C. B., MCCALLY, R. L., TATHAM, P. E. R., CANNON, S. M., HART, R. W.: Phys. Rev. Letters **28** , 1105 (1972).

Physical Relationships in Architecture

C. B. Wilson

The University of Edinburgh, Department of Architecture,
School of the Built Environment, Edinburgh/Scotland

1. Introduction

Buildings are said to be many things: works of art, historical documents, cultural symbols, social experiments, tools for survival, machines for living in. They become them as interpretations; as objects to which a particular set of values or interests has been applied, and one building has many meanings. The object, however, is physical and a component of a physical system. The system carries information which is capable of interpretation by people in different ways, but the information is physical in origin and physical in transmission. Moreover, the architect specifies only physical characteristics. However he may think about his building, his design is a set of instructions for a particular arrangement of pieces of material.

People respond not only to what they see, but also to whether or not they can see, or to whether they are warm enough or not, and one of the purposes of a building is to provide a place in which human life is comfortably supported and where satisfactory conditions are maintained for various activities. A building is both a means and an end—a thing to see in, and a thing to see—and this ambiguity of purpose is at the root of architecture.

Our terrestrial home is a gas-filled space, and we and our buildings are immersed in, and linked by, the physical fields which exist there: electromagnetic radiation, mechanical vibration, air temperature, air velocity and so on. The presence of a building affects these fields in two ways: it alters their average values (particularly, but not only, inside the building) and it imprints information on them about itself. By expressing the relationship between people and buildings through the fields and the information they carry, it appears to be possible to construct a formal model embodying many of the characteristics of architecture. The exploratory notes which follow suggest a structure for the model and discuss its development and application.

The output of architectural activity is built form. Although a consequence of building may be human satisfaction or pleasure, it is a prediction of that satisfaction, not the satisfaction itself, which influences the form. The form has to exist before the satisfaction and this temporal arrangement of architecture is fundamental and must be reflected in the model.

2. The Structure of the Model

2.1. The Real World

Architecture is complex but it is also familiar, and there is no need to give a detailed account here of the phenomena with which we are dealing. A summary outline of the three main components of the real world with which the model is to be concerned will provide sufficient orientation.

a) a world consisting of a layer of 3-dimensional Euclidean space above a ground surface defined as horizontal by a constant vertical gravitational field. The space is filled with air, and in it exist the familiar physical fields which we shall refer to as environmental or *e*-fields. It also houses human beings and a stock of (solid) building materials which affect the fields according to established laws.

b) human beings, who are sentient creatures with certain requirements and abilities regarding the *e*-fields which range from simple demands of temperature to complex interpretations of spatial and temporal variations within certain frequency ranges of electromagnetic and mechanical spectra. They have needs and desires and a limited ability to foresee the consequences of actions on the world. They have mobility and the power to act on the world by the creation of buildings.

c) buildings, which are spatial arrangements of pieces of building material.

The manipulation of the *e*-fields within and in the neighbourhood of a building is to be taken to mean not only modifications to the average levels of fields existing before or without the building, but also the structuring of the fields with information about the building. J. J. GIBSON [1] has given a detailed treatment of the human senses and environmental information.

2.2. The Human in an Environmental Field

A human requires certain ranges of level, and sometimes frequency, of the *e*-fields in order to function. Outside these ranges, he cannot pick up information from his environment and he may experience discomfort or pain. He responds either to the average levels of the fields—in that he is, perhaps, uncomfortable or unable to do something he wants to do—or to the information he picks up from a field whose level is appropriate to his sensory mechanisms. There are several varieties of models of humans in relation to their environments. Their basis extends from analyses of, for instance, thermal balance, through theories of perception and information, to catalogues of needs, likes and dislikes. Our interest lies in the use that can be made of them in establishing predictors of response to the *e*-fields. For the present discussion there is no need for detail and the interest can be expressed as a requirement of the fields in a simple, "black-box" manner which allows the subsequent introduction of particular models.

At a point x and time t, a human H is immersed in a set of *e*-fields E. We will suppose that H may be in one of two states: H_+ in which he is contented with his world and does not wish to change it, and H_- in which he is discontented and wishes to change it by the creation of a building. We may express the response of H to E by saying that E acts on H in such a way that H goes into H_+ or H_-, or, symbolically

$$E \cdot (H): H \to H_+ \text{ or } H_- \tag{1}$$

It then becomes a requirement of the fields achieved by building that

$$E \cdot (H): H \to H_+$$

2.3. The Building's Effect on the Fields

Suppose an initial ($t < t_1$) world described by $E_0 \equiv (e_j)_0$, where e_j are the individual e-fields, and a building B created at $(x, t) = (x_1, t_1)$. For $t \geqq t_1$, the set of fields is E_1 and we can treat the building as acting upon E_0 to produce E_1, so that

$$E_1 = B \cdot (E_0). \tag{2}$$

This is an act in time, but if the effect of B is only local, covering a spatial region $R: x_1 \in R$, we can find a point $x_2 \notin R$ such that $E(x_2, t \geqq t_1) = E(x_2, t < t_1)$ and transform B from a temporal to a spatial act, so that

$$E(x \in R, t \geqq t_1) = B \cdot \big(E(x_2 \notin R, t \geqq t_1)\big) \tag{3}$$

$$E(x_2 \notin R, t \geqq t_1) = E(x_2 \notin R, t < t_1) \tag{4}$$

If Eqs. (3) and (4) are averaged, in such a way that the informational structuring of the fields is removed, they encompass a familiar formulation of the building envelope's performance as a mediator between the external and internal environments of the building.

2.4. The Effect of the Fields on the Building

Any artefact has not only to perform a function, it has also to survive. The conflict of its survival needs with its functional requirements is often what dictates its form.

Buildings are acted upon by their environments and will be unstable if not designed to withstand that action. We may express this by saying that a requirement of a building is that it must be unchanged by its environment. This may be written

$$E \cdot (B): B \to B \tag{5}$$

(If we wished, it would be possible to introduce an acceptable deviation from an initial state of B and a finite lifetime for B.)

The environment which acts on B has much in common with that which acts on H. For simplicity, we may assume that the symbol $\cdot\,(\;)$ both indicates the action and selects the appropriate aspects of E.

2.5. The Model

We now have three relations which determine the structure of the model:

$$E_1 = B \cdot (E_0) \tag{6}$$

$$E_1 \cdot (H): H \to H_+ \tag{7}$$

$$E_1 \cdot (B): B \to B \tag{8}$$

The model is to be understood as cybernetic in the sense that, H being in state H_- for $E_0 \cdot (H)$, a building is created. If this still leaves H in H_- a further attempt is made, and so on.

Treated as Eqs. (6), (7) and (8) are, of course, to be solved for B. That is the analogue, in the model, of the architect's task. The kind of solution, and the means of approaching it, will depend largely on assumptions which are made about H and $E \cdot (H)$. For instance, if any E other than E_0 puts H into H_+, any B which satisfies (8) is a solution. If no E puts H into H_+, there is no solution. If $E \cdot (H)$ depends on previous actions of E on H, no solution may be possible, though iterative attempts may display a pattern. If, however, the form of $E \cdot (H)$ is independent of E, and a solution exists, it will eventually be achieved even for random choices of B. There are many such examples, but more complex ones demand a knowledge of the large number of inter-relationships implied by the three structural relations.

In terms of established disciplines, (6) corresponds to the physics of buildings and environments, (7) was discussed in Section 2.2, and (8) is a statement of the constructional and structural requirements of buildings.

The formalisation of the theory of building construction and its division into those aspects which contribute to environment and those which relate to stresses on the building's fabric is a task which is, so far, incomplete. The philosophy of structural design, however, fits more readily into the present formulation.

Perhaps the most interesting investigations would derive from (7) seen through the structure of the model, but a discussion of them is beyond the scope of this paper. Instead, in order to develop (6), we will assume that E_1 is specified. As an analogue of certain architectural problems, this is equivalent to assuming an environmental specification, in defined spaces, of temperature range, daylight level, background noise level and so on. It ignores the information transfer possibilities of the model, but that E_1 should meet certain minimum standards is a necessary, if not sufficient, condition for B.

3. Development

3.1. The Description of the Building

The expression $B \cdot (E)$ includes both a description of the building and the laws which govern its effect on the e-fields. Buildings are usually described by their geometry, by components such as doors, windows and walls, and by the materials of which they are constructed. Although some components have a specific environmental function, it is generally true that each piece of material in a building affects all the e-fields—materials which are transparent to any of the fields are rare. For each material there is a constant, or a sub-set of constants, which relates to each e-field and which, through the appropriate laws, defines the change in the field due to the insertion in it of the material.

Suppose, then, that B is made of a (finite) number of pieces of material and that the i-th piece is of material M_i having material constants m_{ij}, where the j-th sub-set of constants relates to the e-field e_j. Suppose also that the size, shape and position of the i-th piece is described by a geometrical factor G_i. We can now write $B = \bigcup_i G_i M_i$.

Alternatively, we can introduce $b_j = \bigcup_i G_i m_{ij}$, so that b_j is the spatial distribution of those material properties which act on e_j, and $B \equiv (b_j)$.

3.2. The Physical Laws

The physical laws which govern the effect of B on E—the laws of thermal conduction, radiative emission, noise transmission, air penetration and so on; there are a great many of them—are the laws which relate field variables, material constants and geometrical parameters. They embrace a wide range of phenomena inside air and solids, and at surfaces, but an important characteristic of them all is that they "pick" certain constants and variables. We may express (6) more explicitly by writing $E_1 = BL(E_0)$ where L is the set of physical laws. Since any one law, or sub-set of laws l_j, relates only m_j and e_j we may write the set of simulatneous relations

$$(e_j)_1 = b_j l_j (e_j)_0 \tag{9}$$

Treated as a set of equations, (9), in combination with (8), is to be solved for the b_j.

The b_j are not independent. Each includes all the geometrical descriptions G_i; and the sub-sets of material constants m_{ij} are inter-related by each choice of M_i. This, essentially, is why walls have windows in them.

3.3. Further Considerations

Expressed in the simple form of the relations (6), (7) and (8), the model has a number of shortcomings.

A basic determinant of built form is the relationship between the mobility and size of people and the sizes and shapes of space they require for their activities. The way in which the building organises space is implied in E_1. It is also expressed indirectly in the G_i, but the relationship between the description in the G_i and the information in E_1 is deeply hidden.

However, there is no objection to redundancy in the formulation and a simple requirement that a set of spatial characteristics S shall be met directly by the set of geometrical descriptions, $(G_i) \Longrightarrow S$, can be added. Any other directly describable physical requirements of the building as an object, which emerge from (7), can be expressed explicitly as additional (and redundant) relations if this adds to clarity in using the model.

Cost is an important aspect of architecture which has not been mentioned. It might be introduced as a conservation requirement applied either locally to H or more generally by closing the resources system. Neither have the mechanical and electrical services of buildings been discussed. As energy sources their representation should not present problems, but as objects their introduction into the model might prove as awkward as does their introduction, in reality, into buildings.

4. Application to Form Generation

An interesting question is whether a specification of the form and materials of a building can be approached by direct solution of Eqs. (9) and (8) rather than by the usual trial and error, synthesis and evaluation process of architectural design. Can a prescription for the necessary attributes of built form be generated directly from a knowledge of $(e_j)_0$, a set of specifications for $(e_j)_1$, and the stability requirement (8)?

There is evidence that it can in at least one class of problems: that of the micro-climatic design of outdoor spaces. These are comparatively simple problems because most building materials are opaque to sun and wind. This means that the distribution of material properties reduces to the distribution of material, and the form/sun and form/wind relationships depend only on the external geometry of the buildings or other enclosing elements. Furthermore, the stability requirement can be solved independently.

In general it appears that the procedure for "solving" (9) would run along the following lines. What might be called the *e*-form/field-effect relationship* must first be derived for each b_j. This is equivalent to transposing each of the relationships of (9). Together with spatial, stability and other requirements, these can then be used to generate a set of *e*-forms for the building. Each *e*-form represents combinations of material constants and geometry and the *e*-forms have then to be synthesized, by choice of materials and geometry, into B.

In practical terms, this may not be quite as formidable as it sounds. Often a small number of environmental performance requirements of the building are dominant and the others may be treated as perturbations. Additionally, E_1 would normally be specified as a set of acceptable *ranges* of the $(e_j)_1$. However, much more theoretical work is needed on this application of the model before it is clear how useful a contribution it can make to practical problems in architectural design. In particular the complexity of the relationships which are involved demands a study of the most appropriate and economical descriptors of both form and field.

5. Postscript

It is worth considering the possibility of extending the model to deal with physical interactions between buildings so that co-ordinated aspects of urban structure could be studied. There are immediate applications to the climatic generation of urban form, of a similar nature to that described in the last section, and a study of the relationship of a development of the model with existing spatial models of urban structure would be interesting.

A further stage would be to introduce many people as well as many buildings. There are already attempts to develop models of society from person-person interactions** and it is tempting to think that some macro-properties of urban areas might be derivable from a model consisting of two interacting and mutually interacting sets of "particles"—people and buildings. An individual experiences both macro- and micro-scale properties of urban areas and their associated societies, and the interaction between the two scales is of particular significance.

References

1. Gibson, James J.: The senses considered as perceptual systems. London: George Allen and Unwin Ltd. 1968.
2. Friedman, Yona: Pour l'architecture scientifique. Paris: P. Belfond 1971.

* Compare the expression "thermal-form" applied to the spatial distribution of thermal constants of a building.

** For an example in an appropriate context, see Friedman [2].

Chapter VIII. Biographical and Scientific Reminiscences

Biographical Notes

By Fanchon Fröhlich

A voracious reader in his youth, once while browsing in a bookshop, Fröhlich happened upon a German translation of the Tao Te Ching by the sixth century B. C. Chinese philosopher, Lao Tzu. The Chinese sign for the somewhat obscure concept of "Tao" consists of a foot stepping tentatively followed by the sign for a long-haired (hence, wise) head.

He was deeply struck by it, feeling not so much that he had come upon an external source of understanding as that he had found in the Tao Te Ching a resonance with his own thought and an intensified expression of it—like having a renewed conversation with an old friend.

In the most general way it characterizes his manner of thought in physics—merging himself with the most non-conscious nature and feeling intuitively the internal tensions and tendencies of the physical system—as well as giving a certain unity to his other diversified fields of interest. The Tao Te Ching perhaps also prefigured his recent interest in biology—being concerned as it is with the organic processes of growth and life, as well as with the more abstract physical order immanent in the world. Philosophically he believes that there is an impersonal, non-individualistic path or Tao embedded both in the world and in the mind, and that at some deep level of insight they coalesce. Thus with respect to modern science, he regards the coalescence of the abstract mathematics done in the mind or on a piece of paper with the elaborate experiments done in a laboratory as a source of wonder and mystery (in contrast to the reductionist who thinks this tautologically trivial). He has frequently said that in the creative process of thinking his mind goes out from his human frame and *becomes* the physical particle and field situation, feeling directly how they tend to behave, but using the techniques of mathematics both to capture this unknown physical situation and as an anchor so the mind can return to his own brain or everyday personality. There after he solidifies what he has found during these mental voyages in calculations. As the Tao Te Ching says "To understand emptiness, first you must become emptiness". His deeper intention might be expressed as the wish to make matter conscious. This is based on a metaphysical belief that once matter has been understood, penetrated by mind, the matter is itself transformed. Such a conviction exhibits parallels with Jung's ideas on Alchemy. He considers that the hard-edged external irony of logical positivism confuses the method of discovery in science with the subsequent method of exposition. This methodologically very important, is the distinction between physics as written for publication, as if it were a logical deduction from certain premises and physics in the process of creation or discovery. One should distinguish between the beauty of the process—a mathematically guided Einfühlung, and the cool elegance of the subsequent mathematical formulation. It is this complete immersion in the trip of discovery, Einfühlung, that constitutes his special unending happiness.

Such tendency also manifests itself in his interest outside physics. He shared with his friend, PAULI, an interest in the Jungian concept of the collective unconscious.

He also delights in observing the development of abstract art. Besides his wife, he has many other friends among the international group of abstract painters centred in Paris at S. W. HAYTER's experimental studio and extended throughout the world. He thinks they too explore a part of the same realm which physics with all the equipment of past theories and mathematical methods investigates, but without themselves having any armour of preforged techniques.

His impersonal attitude toward science, even to art might be summarized as "It is not that 'I think' or 'he thinks', but that 'It thinks in me, or in someone' ". The thought comes and lingers if the mind is empty enough to receive it, and all he does as an individual is to catch it and write it down in the appropriate physical formulaes. So personal ambition among scientists is the most lamentable of calamities, even a metaphysical evil. This impersonal detachment extends also to himself so he regards the particular events of his own life as a fairly amusing adventure film. Appropriately although this is a form of biography, he wanted nothing said about his personal adventures except that he was born in Rexingen, grew up very happily in Munich in which he retains a nostalgic delight, was a student of SOMMERFELD, was an enthusiastic skier and continues to love mountain climbing and forests, and most recently put out a forest fire on a mountain in Italy. The most important things that happen, happen to Mind and only incidentally to one person rather than to another.

Erinnerungen an die gemeinsame Arbeit mit Herbert Fröhlich

W. Heitler

Universität Zürich, Institut für Theoretische Physik, Zürich/Switzerland

Fröhlich kam Mitte der 30er Jahre nach Bristol, wo ich seit 1933 eine Forschungsstelle innehatte. Er war zum zweiten Mal unterwegs. 1933 von Hitler vertrieben, nachdem er sich kurz vorher in Freiburg i. Br. habilitiert hatte, fand er ein Unterkommen in Leningrad, wo Prof. Frenkel für einige Physiker, die Deutschland verlassen mußten, sorgen konnte. Nach einigen Jahren änderte sich die Atmosphäre in Rußland, angeblich weil Stalin glaubte, es könnten Spione unter den Flüchtlingen sein. Fröhlich witterte die Lage rechtzeitig und entkam. Andere, die optimistischer waren, blieben und büßten ihren Optimismus mit dem Leben. Soviel ich weiß, überlebte keiner.

Bald fingen wir an, über Physik zu sprechen. Fröhlichs Gebiet war die Festkörperphysik, für die mein Interesse beschränkt war. Zu jener Zeit bahnten sich aber sehr wichtige Entwicklungen in der Elementarteilchen- und Feldphysik an. Es gelang mir, Fröhlich — oder Herbert, wie ich ihn nun nennen will — zu überreden hier mitzumachen. Die Schauer der Höhenstrahlen waren als rein elektromagnetische Effekte erklärt, die nichts mit Kernphysik zu tun hatten. Unerklärt war die "durchdringende" Höhenstrahlung. Anfangs wollte man sie als Protonen erklären (ein Fehler, in den ich zuerst auch verfiel), aber Blackett konnte zeigen, daß dies nicht möglich war, weil die Ionisationsdichte eine viel kleinere Masse bewies. Blackett wollte Elektronen auch in diesen Teilchen sehen, was der Theorie kraß widersprochen hätte. In dem Dilemma beidseitig fehlerhafter Interpretationen erinnerte sich jemand — ich weiß nicht mehr, wer es war — an die 1935 an versteckter Stelle erschienene kurze Arbeit von Yukawa. Die Kernkräfte sollten durch geladene Teilchen einer Masse von ungefähr 200—300 Elektronenmassen erklärt werden. Nun war die Sache klar, die durchdringenden Teilchen mußten diese Yukawa-Teilchen sein. Wir machten uns an die Arbeit, die Yukawa-Theorie weiter zu entwickeln. — In Wirklichkeit sind die durchdringendenTeilchen das Zerfallsprodukt der Yukawa-Mesonen.

Herberts Stärke bestand in seinem Reichtum an anschaulichen Ideen, durch die er Physik begriff, ohne viel auf tiefgründige Mathematik zurückzugreifen. Der Formalismus war nicht seine Stärke. Darum enthalten manche seiner wichtigsten Arbeiten einen physikalisch wertvollen, richtigen Kern, sind aber in der Ausführung mangelhaft. So z.B. in seinem späteren Ansatz zur Theorie der Supraleitung. Die endgültige Theorie enthält ganz wesentliche Elemente aus Herberts Arbeit. — Eine solche Begabung ist heute leider selten. Formalisten gibt es genug.

Das erste was uns gelang, war eine Theorie der magnetischen Momente von Neutron und Proton, die bekanntlich von den Werten, die die Diracgleichung vorhersagen würde, stark abweichen.

Um diese Zeit gesellte sich auch N. KEMMER zu uns. Er war in London tätig, aber wir trafen uns oft. Das war eine sehr glückliche Ergänzung zu unsern Bemühungen, denn KEMMER beherrschte damals den Formalismus der Quantenfeldtheorie wesentlich besser als wir. Zu dritt konnten wir dann eine Mesontheorie (nach heutiger Bezeichnung) aufstellen. Es war eine Vektormesontheorie, in Anlehnung an die Maxwellsche Theorie. Erst viel später erkannte man, daß das Mesonfeld ein Pseudoskalarfeld war. Wir konnten auch die Existenz des neutralen pi-Mesons vorhersagen (aus den Proton-Proton-Kräften), das später entdeckt wurde. Daraus entwickelte KEMMER dann die ladungssymetrische Theorie, die die Basis für den Begriff Isospin war. Die Vektormesontheorie wurde an 3 verschiedenen Stellen gleichzeitig und in fast identischer Weise entwickelt, nämlich außerdem noch: von YUKAWA, SAKATA und TAKETANI und von BHABHA. Bald darauf konnten wir noch aus unserer Theorie die Tatsache ableiten, daß Proton und Neutron eine Ladungsverteilung mit endlicher Ausdehnung haben mußten. Qualitativ war die Arbeit richtig, quantitativ nicht. Wir glaubten fälschlicherweise durch diese Ladungsverteilung die Linienverschiebung im Wasserstoff erklären zu können, die damals schon durch spektroskopische Messungen ungefähr bekannt war, und die heute als Lambshift bezeichnet wird.

Eine Vohersage, die wir hätten machen können, entging uns. Der erhaltene Ausdruck für die Kernkräfte zeigte eine starke Spin-Bahn-Kopplung. Daraus folgt, daß der Grundzustand des Deuterons ein S-D-Gemisch sein mußte, woraus das Quadrupolmoment des Deuterons folgt. Es wurde bald nachher entdeckt, aber wir hatten es nicht gesehen.

Ich möchte noch eines andern jungen Mitarbeiters gedenken, der sich kurz vor dem Krieg zu uns gesellte: des Holländers B. KAHN. Er war an der Arbeit über die Ladungsverteilung des Nukleons beteiligt. Bei Kriegsausbruch (oder unmittelbar vorher) kehrte er nach Holland zurück, was ihm das Leben kostete.

Natürlich sprachen wir fast täglich über die Politik und schimpften gemeinsam (die Gruppe der deutschen Flüchtlinge war inzwischen in Bristol gewachsen) auf die miserable und dumme Politik der Großmächte. Großbritannien war noch eine Großmacht. Es war die Zeit der Regenschirmpolitik Chamberlains, die wegen seines Versprechens "peace in our time" Unterstützung in der Bevölkerung fand. Ein Jahr später geschah, was vorauszusehen war: der Krieg brach aus.

Als der Krieg ernst wurde — was erst ein halbes Jahr später der Fall war — entstand in manchen enttäuschten Kreisen eine verständliche Hysterie. Plötzlich wurde bekannt (was man vorher verheimlicht hatte), daß es überall eine "fünfte Kolonne" gab. Um diese mit Sicherheit auszuschalten, wurden die Flüchtlinge interniert.

Wir waren in 3 verschiedenen Lagern. Zuerst in einem Hüttenlager, in dem es wenig zu essen gab, das aber in einer herrlichen Gegend in South-Devon lag, mit Blick aufs Meer. Dann in einem reichlich primitiven Zeltlager, irgendwo in den Midlands, in dem es viel zu essen gab. Die Wochen in diesem Lager sollten sich auch wissenschaftlich als besonders fruchtbar erweisen, weshalb ich auch darüber berichte. Zuletzt, als die Zelte dem Herbstwind nicht mehr standhielten, kamen wir auf die Isle of Man, wo wir in Häusern untergebracht waren, was langweilig war.

Herbert war ein Mann mit großer Voraussicht. Er sah nicht nur die Internierung voraus, sondern wußte auch, daß man in einem Lager etwas zum Zeitvertreib und zur Unterhaltung haben mußte. Lektüre, nicht schwer, aber doch gut. So kaufte er sich bei Kriegsbeginn 3 dicke Bände einer ungekürzten Ausgabe der Arabian Nights (deutsch: Tausend und Eine Nacht) in ausgezeichneter englischer Übersetzung und nahm sie ins Lager mit. Die Bände wurden für unser Zelt (ca. 6 Bewohner) eine unschätzbare Unterhaltung. Es gibt schon sehr triftige Gründe, weshalb man für Kinder nur stark gekürzte Ausgaben verwenden kann. Ich vermute, daß die ungekürzte die Schweizer Zensur kaum passieren dürfte. (In England muß es ein Gesetz geben, nach dem alte Werke nicht zensiert werden.) Für das Lager hatte der weise Herbert das richtige getroffen.

Im Zeltlager herrschte ein Militärregime. Es waren kaum die für den Krieg brauchbarsten Offiziere, die man dorthin schickte. Als einmal ein neuer großer Schub von Österreichern ins Lager kam, hörten wir den Oberst, der seine Lorbeeren ohne Zweifel in den Kolonien verdient hatte, zu einem Unteroffizier sagen: "Have you separated the tribes?" Die Heiterkeit im Lager war allgemein.

Nun zurück zur Wissenschaft. Im Zeltlager fand sich eine Gruppe von sehr guten Wissenschaftlern vieler Gebiete zusammen. Wir beschlossen, keine Forschungsarbeit auf unseren Fachgebieten zu betreiben. Um so fruchtbarer waren gegenseitige Information und Kurse für Studenten. Wir gaben Kurse für Physikstudenten über Quantenmechanik, die später von den Universitäten als absolvierte Kurse anerkannt wurden.

Besonders wichtig wurden für uns 2 Vorlesungszyklen, die der Mathematiker-Philosoph WAISMANN aus Oxford gab. Der eine war eine Einführung in die Mengenlehre, der andere über mathematische Logik, besonders über die Sätze von GÖDEL, die damals 10 Jahre alt waren. Wir haben sie gründlich diskutiert und sie haben mein wissenschaftliches Denken bis heute tief beeinflußt. Umgekehrt hielten auch wir Vorträge über verschiedene Gebiete der Physik.

Eine weitere wissenschaftliche Begebenheit möchte ich noch erwähnen, obwohl ich glaube (wenn meine Erinnerung nicht trügt), daß Herbert nicht dabei war. Auf der Isle of Man machten wir Bekanntschaft mit einem Vertreter des Positivismus aus dem sog. "Wiener Kreis". Er erklärte uns, was Physik sei: ein aufgeschlagenes Bilanzbuch, auf dessen linker Seite Meßdaten stehen. Auf der rechten steht oben eine vom Physiker erfundene Formel und darunter die daraus berechneten Zahlwerte. Beide Seiten müssen übereinstimmen, sonst muß eine andere Formel erfunden werden. Ich erinnere mich, daß ich ob solcher Geistlosigkeit fast wütend wurde. Jedenfalls bin ich seither gegen jede Art von Positivismus immun. Die heute in Wien vorherrschende Weltanschauung scheint eher das Gegenteil von Positivismus zu sein.

Im Herbst 1941 waren Herbert und ich wieder in Bristol. Es kamen während des Winters zunächst 12 schwere Bombenangriffe auf Bristol, bei denen das Stadtzentrum und andere Stadtteile weitgehend niederbrannten. Im Frühjahr erhielt ich einen Ruf an das neugegründete Dublin Institute for Advanced Studies. Die Übersiedlungsformalitäten nahmen viel Zeit. So war nicht mehr viel Gelegenheit zu gemeinsamer Arbeit. Mit der Übersiedlung endete — nicht unsere Freundschaft — aber unsere gemeinsame Arbeit — und damit dieser kurze Bericht.

Superconductivity and Superfluidity

K. Mendelssohn

Clarendon Lab., Dept. of Physics, Oxford/England

At the Thirteenth International Low Temperature Conference, held at Boulder, Colorado, in August, 1972, an evening meeting was scheduled at the request of Professor G. E. Uhlenbeck on the origin of the concept of "Quantum Fluids." I admit that initially I was not greatly in favour of this exercise, firstly because I dislike sessions after dinner, and secondly, because much of the material had been made available in popular form in my book *The Quest for Absolute Zero* in twelve languages. My first objection was overruled by my host, John Daunt, who had advanced the time of our dinner by half an hour so that we both could appear at the function as "ancient exhibits." Secondly, I was assured that nothing more would be required except personal recollections.

The session, which had been planned as a small round table discussion, took the organizers somewhat by surprise since more than 500 young physicists attended, filling every available seat and step of a large lecture hall. The speakers turned out to be Jack Allen, John Daunt, John Pellam, Ladislas Tisza and myself, of whom, gratifyingly, only Tisza had provided himself with a sheaf of notes. Other old timers, who did not contribute included Felix Bloch, Cornelius Gorter, Nicholas Kurti, Krien Taconis and Mark Zemansky. Herbert Fröhlich did not attend the Conference. However, in view of his later important contribution to the theory of superconductivity and an early paper on the lambda-point of liquid helium, a personal account of these interesting days may not be out of place in the present volume. Moreover, during the relevant years Fröhlich was at Bristol and we discussed the subject frequently, usually at the Kardomah Cafe in Piccadilly. I recall with particular gratitude his sagacious advice never to be influenced by the opinions of theoreticians, unless they happend to agree with my own views.

It seems to have been tacitly agreed at the evening session that a quantum fluid is an assembly of particles which, owing to the high zero point energy of the system, are prevented from forming a crystal. They are the metal electrons (at any temperature), the two liquid stable helium isotopes and—to some extent—hydrogen. Curiously enough, no distinction was made between this property and the anomalous transport properties shown by some of them.

The first indication of quantum effects was, in fact, provided by liquid hydrogen which Dewar in 1898 found to evaporate a good deal faster than the excellence of the newly invented vacuum flask, which bears his name, seemed to justify.

It became clear that TROUTON's rule, according to which the latent heat of evaporation should be proportional to the absolute boiling temperature, did not hold for hydrogen. This failure became even more apparent when ten years later KAMERLINGH ONNES liquefied helium, who also was surprised by the astonishingly low density of the liquid. This, and particularly the density maximum discovered by him in 1911 [1], led him, when receiving the Nobel Prize in 1913, to a vague suggestion that it might be connected with the quantum principle.

However, it was NERNST who first recognized the real root of the trouble. Due to the work which led him in 1906 to enunciate his third law of thermodynamics, he was probably better qualified than anyone else to recognize the significance of the quantum prinicple at low temperatures. In 1916 he drew attention to the deviations from TROUTON's rule for hydrogen and helium. This was in the middle of the first world war and after its end he suggested to his co-workers, KURT BENNEWITZ and FRANZ SIMON [2], to investigate the question more thoroughly. In a paper, published in 1923, they pointed out that the data could be explained if the latent heat was reduced by a parameter, connected with a high "internal energy". In 1934 SIMON [3] returned to the particular problem of liquid helium, pointing out that this parameter, representing the high zero point energy of helium was responsible for keeping the substance, under saturation pressure, in the liquid phase down to absolute zero. This same conclusion had already been arrived at by the Leiden workers on purely experimental grounds.

This led to the first crop of helium theories. Like SIMON, the brothers FRITZ and HEINZ LONDON had by then moved to Oxford and, in addition to their phenomenological theory of superconductivity, FRITZ LONDON had been urged by SIMON to try his hand at the helium problem. His aim was to find a highly mobile atomic assembly of low entropy, and he hit on an open type of diamond lattice [4] which, through its crystalline nature, would provide him with the necessary high degree of orderliness. He suggested that it might be verified by *X*-ray analysis. The theory had no place for the lambda point, which by then (1936) had been well established by the huge specific heat anomaly discovered by KEESOM and CLUSIUS [5]. In the following year FRÖHLICH [6], fully aware of this shortcoming, tried to remedy the situation by introducing into the London theory an order-disorder transformation which, as he hinted at, might not show up in the *X*-ray work. Such investigation was, in fact, at that time in progress at Leiden, and it yielded a pure liquid structure [7]. Nevertheless, the authors, KEESOM and TACONIS, were so much under the spell of the theoreticians that they, somewhat reluctantly, conceded that their result might not be incompatible with a loose diamond structure. Actually, they preferred a slightly different structure of their own in which helium atoms were able to pass through the lattice by tunnel-like apertures. A model, brought to the famous discussion meeting at the Royal Society in 1938, attracted little attention, and not only because it had become hopelessly tangled up during the Channel crossing. The whole aspect of the helium problem had suddenly been changed by the experiments carried out at Oxford, Cambridge and Moscow during the preceding months. Nearly all of them were reported in that memorable volume of Nature 141, 1938.

In retrospect it seems quite inconceivable that it took 25 years, from the observation of the density maximum in liquid helium to the discovery of superfluidity. Moreover, signs and portents were not lacking but they had been studiously overlooked by experimentalists and theoreticians alike. In those years a thousand or more experiments with liquid helium had been performed in a dozen different laboratories and duly published. However, unexplained experimental difficulties tend to be left unmentioned and there are before 1938 no, or almost no, records of superfluid (lambda) leaks spoiling some experiment.

After I finished my thesis work with him in Berlin, SIMON had been very generous in letting me follow my own research. In its course an American guest worker, JOHN CLOSS, and I ran into vacuum trouble which was so unusual that we decided to publish an account of it together with our results [8]. It only occurred when the copper calorimeter had been cooled to below 2 °K indicating either a sudden leak or desorption of a helium film, or both. Since we then had no reason to suspect the former, we decided on the second alternative which clearly required further investigation. However, I was just then moving to Breslau where a suitable apparatus was constructed but the work had to be discontinued and shelved indefinitely when I went to England early in 1933. When visiting Leiden on my way, KEESOM, whom I asked about a similar occurrence at a Leiden calorimetric measurement, felt sure that everything could be explained to satisfaction.

The first determination of the viscosity of liquid helium was carried out in 1935 in Toronto [9] and an appreciable drop was found at the lambda-point but it was not more spectacular than the large anomaly in the specific heat. A similar result was obtained in 1938 by MACWOOD in Leiden [10]. In both investigations oscillating cylinders or discs were used which precluded the observation of superfluidity.

The sudden disappearance of boiling in helium which was cooled below the lambda-point, and which must have been noticed by many observes, was strangely enough not connected with an increase in thermal conductivity. Indeed, in 1932 MCLENNAN, SMITH and WILHELM [11] drew particular attention to this strange effect without, however, drawing the obvious conclusion. Finally, during a repetition of the specific heat measurement, KEESOM and his daughter [12] were driven to the assumption that at the lambda-point the heat conductivity must rise strongly because the observed temperature drifts changed suddenly. They therefore tried in 1936 to measure the heat conductivity by a conventional method [13] and found that below the lambda-point it "surpassed a few hundred times that of copper." An elegant and more accurate method was adopted in 1937 by ALLEN, PEIERLS and ZAKI UDDIN [14] in Cambridge who simply used a glass bulb, containing a heater, which was connected to the helium bath by a capillary. They thus could determine the temperature difference along the tube by the vapour pressure in the bulb read as the depression of the liquid level inside the bulb against that of the bath. The authors, just as KEESOM and MISS KEESOM, found that the heat conductivity of liquid helium reached enormous values below the lambda-point which decreased as the temperature was lowered. However, even more astonishing was their observation that the thermal conductivity, if it could be called such, increased at all temperatures with diminishing heat current. They ended their paper by saying that they could offer no explanation for these results "for which there is no analogy in the behaviour of other substances."

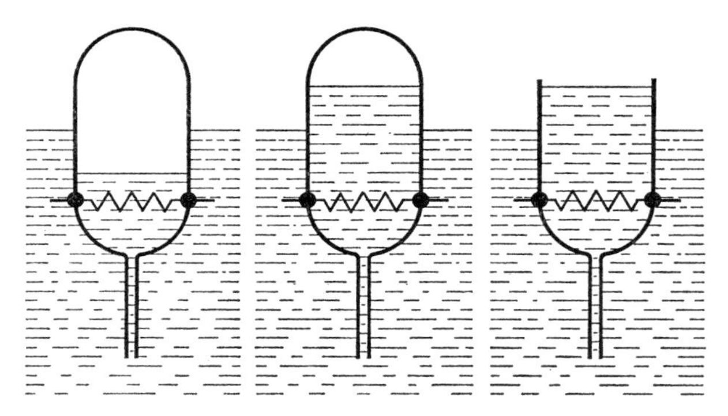

Fig. 1. The stages which led from the heat conductivity measurement of ALLEN, PEIERLS and UDDIN to the discovery of the thermo-mechanical effect in He II

It was PEIERLS' only incursion into the helium research and neither did UDDIN, whom ALLEN remembers as severe introvert who spent hours squatting on the floor of the laboratory in contemplation, take any further part in the work. However, this letter to Nature, published in July 1937, set off that remarkable series of research communications to the same journal in which nearly all the transport phenomena in liquid helium were discovered and largely investigated. ALLEN, using the same apparatus, noted the almost unbelievable effect that at very low heat flows the inner level, far from being depressed, seemed in fact to rise above that of the bath. He therefore opened the top of the bulb to the vapour above the bath and found that it was perfectly true. The thermo-mechanical effect, a mass flow of helium opposing the heat current, was discovered. Improving on the arrangement by using a powder-filled bulb, open at the bottom and with a narrow orifice at the top, they were able to produce a fountain several centimetres high when the powder was heated by shining a light on it. When ALLEN and JONES published these results [15] in February 1938, they tried to connect mass flow and heat conduction, but were looking in vain for a convincing explanation.

Meanwhile something even more staggering had happened. In the January issue there had appeared two communications, printed back to back. One was by KAPITZA [16] in Moscow and the other by ALLEN and MISENER [17] in Cambridge. They both reported on viscosity measurements of He II by flow through a narrow slit or capillaries. These both yielded the same result; the viscosity below the lambda-point fell to values of 10^{-9} c.g.s. units and, moreover, it was not proportional to the pressure gradient. The flow velocity increased, becoming less pressure-dependent for the narrower flow channels and KAPITZA, for the first time, used the term "superfluidity." DAUNT's and my own entry into the helium field was quite accidental. We were busy with work on superconductors, particularly their entropy. In its course we tried (unsuccessfully) to see whether there was any energy absorption in r.f. fields. However, in order to make the observations, we had to attach to the bottom of our metal cryostat a copper-glass seal holding a glass ampoule of a few cc. I was delighted to see liquid collect in it since, in ten years of helium work with metal equipment, I had never set eyes on it. I then recalled a curious observation mentioned by KAMERLINGH ONNES in 1922 when he saw a rapid "distillation" of helium between concentric dewar vessels. This observation was never satisfactorily explained nor had it been repeated, and DAUNT and I decided to have a shot at it. We replaced our ampoule by two of them, connected by a narrow tube. Nothing spectacular happened but, after waiting for more than half an hour, a rise in the lower level of 0.2 cm was recorded.

Since there was no rapid distillation, I returned to the idea of surface effects which had troubled CLOSS and me six years earlier and which had remained unexplained. Moreover, in 1936 ROLLIN [18] had suggested that his failure to reach very low temperature might be due to the high thermal conduction of a helium film covering the inside wall of his pumping tube. This, as it soon turned out erroneous, explanation was restated by KIKOIN and LASAREW [19]. DAUNT and I, therefore, cut the connecting tube between our ampoules, inserted a wick of copper wires, re-sealed it and repeated the experiment two days later. The success was immediately apparent; the levels now changed with much greater speed. It thus became clear that a film did, in fact, cover all solid surfaces in contact with He II; however, the flow was not one of heat but of mass. The evaporation in ROLLIN's cryostat had come from the top of his tube and not from the liquid at its bottom.

The next step was to replace our ampoules by a little helium bath and a beaker which could be lowered into it. Crouching on the floor—the only position from which the arrangement could be viewed—DAUNT and I spent hours registering the changing height of the helium levels by the light of a low power neon lamp. We saw the liquid flowing in and out and when we raised the beaker above the bath level if formed with clockwork regularity drops at its bottom, falling back into the bath [20]. By then it was between 2 and 4 a.m. and we though that a second opinion on this bizarre effect was desirable. We got from the basement a

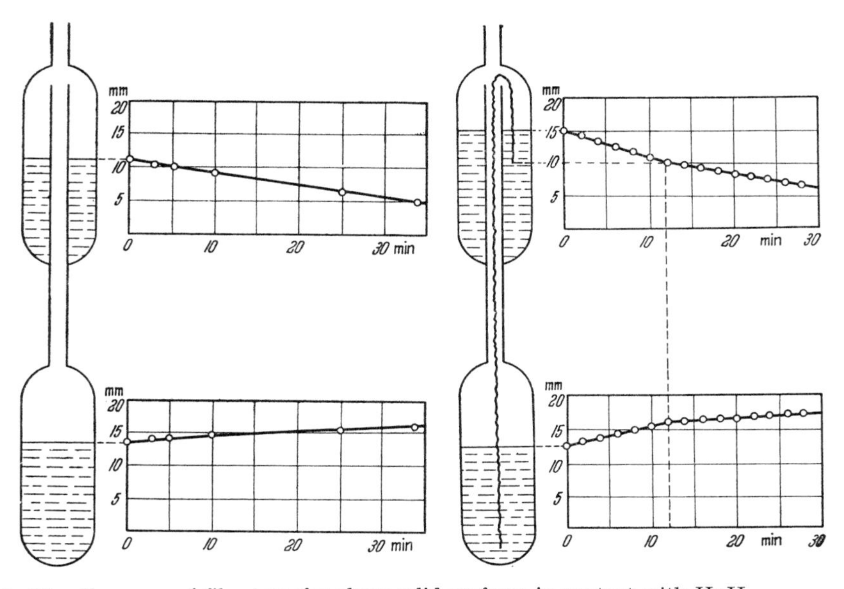

Fig. 2. The discovery of film transfer along solid surfaces in contact with He II

somnolent nuclear physicist, engaged in some experiment of long duration. Unfortunately, at this stage of the account, DAUNT's and my recollections differ. While I seem to remember that the witness agreed to have seen the drops, John maintains that the man's eyes were far too bleary to see anything.

The rest of the story is too well known to require detailed repetition. We had seen straight away that the film transfer was independent of the level difference, the height of the rim or the length of the path traversed. It also varied with temperature and, since its own heat conductivity was found to be negligibly small, we explained its heat transfer as the product of transfer rate and heat content. Its thickness was found to be about 5×10^{-6} cm. The phenomenology of the transfer was studied on a wide variety of beaker geometry. The most important feature of our experiments was that the helium film, being so much thinner than KAPITZA's or ALLEN's capillaries, provided the pattern of superfluidity [21], *par excellence*, including the ideal explanation of the thermo-mechanical effect [22].

The whole investigation had taken us less than six months and by the early summer of 1938 a bewildering variety of the most bizarre observations had been obtained in Moscow, Oxford and Cambridge. A discussion meeting was arranged in the summer of 1938 at the Royal Society at which its President, Sir WILLIAM BRAGG, referred to an Alice in Wonderland world. Many people were frankly incredulous, and I remember Cockcroft greeting me on arrival with the question: "Well, does it still run over the rim?" Among those present were all the low temperature physicists from Oxford, Cambridge and Bristol. KEESOM and TACONIS had come from Leiden and brought with them the ill-fated crystal model. Theoreticians included F. and H. LONDON, TISZA and many others.

Only a year separated this memorable meeting from the outbreak of war, and none of us could then know that in the preceding six months most of the basic phenomena had been discovered. That year, in fact, turned out to be mainly a sorting-out and mopping up operation. ALLEN and his associates gathered a large amount of quantitative data both on flow through capillaries [23] and on the thermo-mechanical effect [24]. The one remaining phenomenon which everybody knew must still exist was its reversal, the mechano-caloric effect, but its discovery was made difficult by the high heat conductivity of He II. ALLEN, using various capillary sizes, had been unable to detect any temperature difference when the liquid was forced through them. At the Boulder meeting he recalled that sometime in the spring of 1939 we met at a Chinese restaurant when I told him that DAUNT and I had observed the effect in flow through fine powder [22].

The first six months of 1938 had brought a revolution in liquid helium theory which was as thorough as that in the experiments. The spell of the quasi-crystalline model had finally been broken by FRITZ LONDON himself who had gradually realized that it would be unstable. Instead, he turned to an old paper of EINSTEIN's [25] in 1925 in which the latter had postulated a strange condensation phenomenon to be expected for an ideal Bose gas at low enough temperatures. The condensate would not be a crystal but manifest itself in momentum space instead. Two years later UHLENBECK [26], in his doctoral thesis, had called this result in doubt and there it stayed until LONDON [27] rescued it in a letter to Nature in April 1938. He not only changed UHLENBECK's [28] mind about the validity of EINSTEIN's result but he also, very cautiously, suggested that the condensation phenomenon might be connected with the lambda-point. Unfortunately, the calculated specific heat did not much resemble the latter but, as LONDON pointed out [29], there is a vast difference between an ideal gas and a liquid.

LONDON, who by then had moved to Paris, discussed his work with a young Hungarian theoretician, LADISLAS TISZA, working at the College de France. TISZA, being young and enthusiastic, suddenly saw a wonderful opportunity of applying LONDON's tentative ideas to the newly discovered transport phenomena—and he seized it. In a letter printed in the May issue of Nature [30], he proposed what has now become known as the "two-fluid model" and applied it, with impressive success, to the observations on helium. It explained the reason for the different results in the two types of viscosity measurements, the thermo-mechanical effect and the heat conduction as a counter-current. TISZA's ideas were taken up immediately by HEINZ LONDON [31], then at Bristol, who had been playing unsuccessfully with a KNUDSEN gas model. Heinz now tackled the thermo-mechanical effect thermodynamically, relating the fountain pressure directly to the entropy. Although TISZA had made fullest acknowledgement to FRITZ LONDON's inspiration and did probably more than anyone else to popularize the Bose-Einstein condensation, FRITZ was livid. He somehow felt that TISZA had jumped the gun—moreover, that he had done it mathematically not very soundly but most successfully as far as phenomenology went. During the next years Fritz published two more detailed papers on Bose-Einstein condensation [29, 32] with, to some extent, identical wording in which he dealt more rigorously with the basic idea, but made hardly any reference to the observed phenomena. Both papers contain the same brief, patronizing and not very complimentary reference to TISZA's work. Heinz came off a little, but not very much, better.

It would be wrong to assume that the experimental work of 1938/9 was much influenced by the new theoretical ideas. These generally tended to follow the observations rather than vice versa. It is significant that JONES [33] still tried to explain the fountain effect with the lattice model and that none of the experimental notes mention TISZA until DAUNT and I made reference to his work in our note on the mechano-caloric effect [22], published in April, 1939. Tisza's prediction of thermal waves [34] was published too late to be tested before war broke out.

No Russian low temperature papers reached us after the KAPITZA and LASAREW letters to Nature in 1938. Scientific contact, which had been uncertain since 1936, was completely lost and not restored until several years after the war. They might have lived on another planet, except for one instance. The issue of the Physical Review of 15th August, 1941, carried two articles on liquid helium, one by LANDAU [35] and the other by KAPITZA [36]. The papers had been received on 23rd June, which agrees well with the submission date (May 21, 1941) of two similar, if longer, articles [37, 38] to the Soviet Academy Journal of Physics. It has to be remembered that Germany invaded Russia on June 22nd, 1941. KAPITZA's paper gave an account of measurements of the heat of transport while that of LANDAU gave the first account of his theory of liquid helium. In it, as is well-known, he proposed a kind of two-fluid model in which the excited atoms were replaced by thermal excitations. In it he dismissed "TISZA's well-known attempt to consider Helium II as a degenerate Bose gas" because "it cannot be accepted as satisfactory—even putting aside the fact that liquid helium is not an ideal gas".

Landau showed that his theory led to another wave propagation besides oridnary sound waves which he called "second sound". Tisza's prediction of thermal waves evidently had not reached the Russians because they later tried to detect second sound unsuccessfully with a microphone until E. M. Lifshits pointed out that these were temperature waves. At very low temperatures the propagation velocity of these waves should rise steeply according to the Landau model whereas Tisza, who had assumed that the superfluid would retain a phonon entropy, predicted a drop to zero. The critical experiment made by Pellam and Scott [39] in 1949 showed that Landau's prediction was correct. This, incidentally, fitted in well with our own postulate that the entropy of the superfluid should be strictly zero.

While Landau quoted Tisza, he never referred to Fritz London's original work. It, moreover, remained an open question why ^{3}He did not become superfluid. There was no reference in Landau's paper to this dilemma and, as mentioned earlier, postal communication was non-existent. I was therefore delighted to be able to put the question to Landau personally at our first meeting in Moscow in 1957. His answer was that, while he was convinced that Bose statistics were necessary for superfluidity, he was equally convinced that the lambda-point had no connection with the Bose-Einstein condensation phenomenon. He did not elucidate this oracular statement further but it seems justified by the fact that while in superconductivity the Cooper electron pairs have statistically to be considered as bosons, they "condense" out of a Fermi gas.

When war broke out all experimental low temperature work in England came to a stop. Daunt and I, although engaged on different war projects, both stayed in Oxford, which allowed us to continue the assessment of our results. Shortly before we took up the film research in 1938 we had measured the entropy carried by a persistent supercurrent [40] and had found that it was zero, even at finite temperatures. This had a direct bearing on our decision to observe the mechanocaloric effect and we were delighted when we were able to show experimentally that the entropy of superflow was appreciably lower than that of the liquid as a whole. Our measurements, which had to be discontinued, were not accurate enough to demonstrate that it was zero, but this was subsequently found to be true by Kapitza [36]. Comparing our data on the superfluid helium film with those on superconductivity we became increasingly convinced of a fundamental analogy between the two phenomena of superconductivity and superfluidity and in a note to Nature in November, 1942, we enumerated the striking similarities of the basic phenomena [41]. As reason for this similarity we postulated a condensation in momentum space into a state of zero entropy, saying: "This new form of aggregation of matter evidently follows quite general rules so that the fact that the particles are electrons in one case and atoms in the other is only of secondary importance."

There still remained the need for thinking up an experiment with superfluidity that would be strictly analogous to the basic current potential measurement on a superconductor. I recall long solitary walks in the blackout, and the moment when I found the solution. Unfortunately, years of war still had to pass

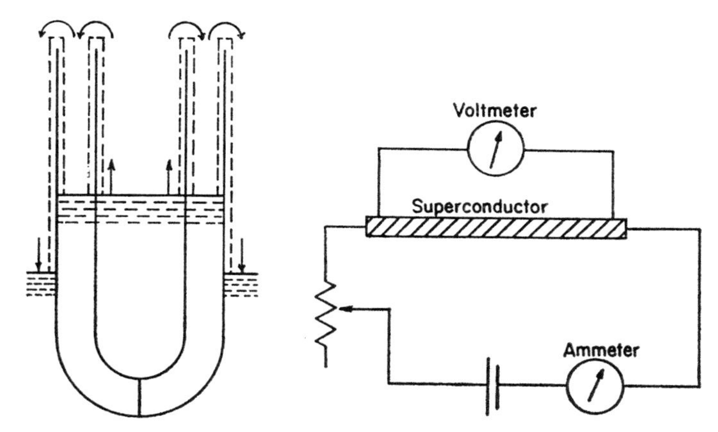

Fig. 3. The double beaker experiment in film transfer which is the analogue to the current potential measurement with a superconductor

before it could be tried. Even then we could not do it in Oxford because our cryogenic plant had been completely dismantled. However, JACK ALLEN in Cambridge offered to DAUNT and me hospitality at the Mond Laboratory for the crucial experiment for which we had so long been waiting. The apparatus, which was extremely simple, had been prepared in Oxford and all that was now needed was a few hours of liquid helium. The arrangement consisted of two concentric glass beakers and helium was to be transferred by film flow from the inner into the outer beaker and from there into the helium bath. The level difference between inner beaker and bath corresponds to the battery potential and the current strength is given by the flow rate. For reasons of geometry the outer beaker serves as the resistor, limiting the current to below the critical value, while the inner beaker is the superconductor with the level difference between inner and outer beaker representing the potential difference. The experiment turned out to be completely successful; the levels between the two beakers stayed together. Helium was being transferred over the rim of the inner beaker under zero potential [42].

The only person who referred to our 1942 note was FRITZ LONDON. In 1945 he wrote [43] that we had "assembled some evidence for an interesting analogy between the surface flow of liquid helium II and the electric surface currents in a superconductor." Then, with the mixture of erudition and condescension which he used to adopt towards experimentalists, FRITZ goes on: "However, they gave no hint why such a connection should exist; nor did they give any support to their view by presenting empirical evidence concerning the number of superfluid or superconductive particles.

While this so far quite qualitative analogy may appear rather accidental, it might be of interest to draw attention to the fact that the parallelism goes possibly further than at first suspected" ...

This was probably as much praise as anyone could ever expect from FRITZ LONDON. However, he did not like it when he learned that the flow formula based on the quantum constant which he suggested as the real reason for the analogy had been given by me at about the same time [44]. He certainly made no reference to our "interesting analogy" when a few years later he published "Superfluids", Volumes I and II, but I received complimentary copies of both. I was even more pleased by his reference in Volume II to superfluid helium as "a particular fourth state of aggregation, besides the solid, liquid and gaseous state" since it seemed to have a familiar ring. Curiously enough ,London seemed to restrict this state to liquid helium and I would have liked to know what was in his mind. Unfortunately, when the book appeared, FRITZ LONDON was dead.

Finally, one may ask why the analogy between superfluidity and superconductivity as a new form of aggregation had not been raised earlier. However, it has to be recalled that the number of people working on these phenomenon at the time was very small and, while some worked on helium and others on superconductors, nobody worked on both. DAUNT and I certainly happened to be on the right road when we investigated the entropy of a supercurrent but, except for the accidental pleasure of seeing liquid helium for the first time, we would not have gone into the film experiments.

A description of those early years would be incomplete without mentioning the part played by F. A. LINDEMANN (later Viscount Cherwell). LINDEMANN, NERNST's favourite pupil, had become Professor of Experimental Philosophy in Oxford, and it was his great ambition to establish there a Low Temperature laboratory. He invited me to set up a helium liquefaction plant at the Clarendon Laboratory early in 1933 and I began work in the spring. When later LINDEMANN brought SIMON, KURTI, the brothers LONDON, and eventually, SCHRÖDINGER, to Oxford, the Clarendon became one of the foremost centres of low temperature research, a position which was maintained until after LINDEMANN's death in 1957. LINDEMANN was in every respect a stout conservative and tended to admit, grudgingly, quantum mechanics and relativity merely as corrections but never as substitutes for classical physics. He was convinced that superconductivity was the condensed phase of the electron gas but refused to accept the notion of momentum space. On the other hand, he always supported and never interfered with my work. He was an acute, and sometimes formidable, critic, but always a reliable friend to any member of his low temperature school.

What I did not know when I accepted his invitation to initiate low temperature work at the Clarendon was that one of LINDEMANN's aims had been to steal a march on Cambridge, KAPITZA and RUTHERFORD, who had just obtained a lot of money for the Mond Laboratory. However, KAPITZA's initial reserve changed into a good and lasting friendship when he learned that I had been quite ignorant of LINDEMANN's little scheme.

References

1. ONNES, H. K.: Commun. Leiden No. 119 (1911).
2. BENNEWITZ, K., SIMON, F.: Z. Physik **16**, 183 (1923).
3. SIMON, F.: Nature **133**, 529 (1934).
4. LONDON, F.: Proc. Roy. Soc. (London), Ser. A **153**, 576 (1936).
5. KEESOM, W. H., CLUSIUS, K.: Commun. Leiden No. 219e (1932).
6. FROEHLICH, H.: Physica **4**, 639 (1937).
7. KEESOM, W. H., TACONIS, K. W.: Physica **5**, **161**, 270 (1938).
8. MENDELSSOHN, K., CLOSS, J. O.: Z. Phys. Chem. B **19**, 291 (1932).

9. Wilhelm, J. O., Misener, A. D., Clark, A. R.: Proc. Roy. Soc. (London), Ser. A **151**, 342 (1935).
10. MacWood, G. E.: Physica **5**, 374, 763 (1938).
11. McLennan, J. C., Smith, H. D., Wilhelm, J. O.: Phil. Mag. (7) **14**, 161 (1932).
12. Keesom, W. H., Keesom, A. P.: Physica **2**, 557 (1935).
13. Keesom, W. H., Keesom, A. P.: Physica **3**, 359 (1936).
14. Allen, J. F., Peierls, R., Zaki Uddin, M.: Nature **140**, 62 (1937).
15. Allen, J. F., Jones, H.: Nature **141**, 243 (1938).
16. Kapitza, P. L.: Nature **141**, 74 (1938); — USSR J. Physics **4**, 181 (1941).
17. Allen, J. F., Misener, A. D.: Nature **141**, 75 (1938).
18. Rollin, B. V.: Act. 7 Int. Congr. Refrig. **1**, 187 (1936).
19. Kikoin, A. K., Lasarew, B. G.: Nature **141**, 912 (1938).
20. Daunt, J. G., Mendelssohn, K.: Nature **141**, 911 (1938); — Proc. Roy. Soc. (London), Ser. A **170**, 423, 439 (1939).
21. Daunt, J. G., Mendelssohn, K.: Nature **142**, 475 (1938).
22. Daunt, J. G., Mendelssohn, K.: Nature **143**, 719 (1939).
23. Allen, J. F., Misener, A. D.: Proc. Roy. Soc. (London), Ser. A **172**, 467 (1939).
24. Allen, J. F., Reekie, J.: Proc. Cambridge Phil. Soc. **35**, 114 (1939).
25. Einstein, A.: Ber. Berl. Akad. 261 (1924); 3 (1925).
26. Uhlenbeck, G. E.: Thesis, Leiden 1927.
27. London, F.: Nature **141**, 643 (1938).
28. Uhlenbeck, G. E., Kahn, B.: Physica **5** (1938).
29. London, F.: Phys. Rev. **54**, 947 (1938).
30. Tisza, L.: Nature **141**, 913 (1938).
31. London, H.: Nature **142**, 612 (1938); — Proc. Roy. Soc. (London), Ser. A **171**, 484 (1939).
32. London, F.: J. Phys. Chem. **43**, 49 (1939).
33. Jones, H.: Proc. Cambridge Phil. Soc. **34**, 253 (1938).
34. Tisza, L.: Compt. Rend. **207**, 1035, 1186 (1938).
5. Landau, L. D.: Phys. Rev. **60**, 356 (1941).
36. Kapitza, P. L.: Phys. Rev. **60**, 354 (1941).
37. Kapitza, P. L.: USSR J. Physics **5**, 59 (1941).
38. Landau, L. D.: USSR J. Physics **5**, 71 (1941).
39. Pellam, J. R., Scott, R. B.: Phys. Rev. **76**, 869 (1949).
40. Daunt, J. G., Mendelssohn, K.: Nature **141**, 116 (1938); — Proc. Roy. Soc. (London), Ser. A **185**, 225 (1946).
41. Daunt, J. G., Mendelssohn, K.: Nature **150**, 604 (1942).
Mendelssohn, K.: Proc. Phys. Soc. (London) **57**, 371 (1945).
42. Daunt, J. G., Mendelssohn, K.: Nature **157**, 839 (1946).
43. London, F.: Rev. Mod. Phys. **17**, 310 (1945).
44. Cf. Daunt, J. G., Mendelssohn, K.: Phys. Rev. **69**, 126 (1946).
Mendelssohn, K.: Phys. Soc. Camb. Conf. Rep. 35 (1947).

Some Reminiscences of Research in Liverpool in 1950

J. G. Powles

The University Canterbury, Physics Laboratories, Kent/England

I first met Fröhlich at the Farady Society Discussion meeting on Dielectrics [1] in Bristol in 1946. Perhaps I should apologise right away for using the personal pronoun but I intend in this contribution to try to give a rather personal account of the atmosphere in Fröhlich's Department during my time there, 1950–51, and of the state of our knowledge of dielectric phenomena at that time. I must also apologise, if that is necessary, for saying "Fröhlich" pure and simple because I have personally never heard him referred to in any other way. I believe even his wife, in those days at least, called him Fröhlich. It was therefore a surprise to me, on receiving Professor Wagner's letter asking me if I would contribute to this volume, to discover, after all these years, that the *H* in fact stands for Herbert and is not just a symbol—Hamiltonian for instance would be appropriate. As I was saying, Fröhlich impressed me tremendously in 1946 with his grasp of the fundamental issues in Dielectrics and this was amply confirmed when I saw his book [2] in 1949. Being a callow graduate student and an experimentalist at that, this contact with Fröhlich opened up new horizons for me and I formed the resolve to work with Fröhlich as soon as I could. However at the same meeting I also met Edmond Bauer and Michel Magat and the prospect of dielectrics research in Paris after the war years in Manchester was too tempting.

By the time I managed to get to Fröhlich he had moved from Bristol to Liverpool. This was at first a disappointment to me as Bristol is almost Gloucestershire, my home county, whereas Liverpool is somewhere "up North". However geographical preferences must yield to scientific opportunity and with some misgivings I went to "Abercromby Square". The misgivings were also of a scientific nature and I feel now were well justified as I was not well qualified for theoretical physics having been trained as an Electrical Engineer and "finished" in a Physical Chemistry Laboratory. However I suppose that more important than knowledge is a willingness to learn, an enthusiasm for the subject and good teachers. Anyhow Fröhlich never revealed to me any irritation about my evident ignorance of quantum mechanics and statistical mechanics. He followed what seemed to be his normal method of tossing out a few ideas and leaving me to sort it out for myself, with occasional pertinent but cryptic suggestions as to how to proceed. In fact I was able to make some suggestions for topics myself. One of mine, which Fröhlich thought could not be done, worked and one which Fröhlich thought could be done has since proved to be wrong. Fröhlich will excuse me mentioning this because it impressed me at the time and it is in some ways more interesting to study why a great man is sometimes wrong than why he is right. Twenty years ago the graduate students were, in my experience, much less critical of the "experts" than they are now. On the other hand the experts were, I think, more overtly critical of each other. I recall with fascination the ding-dong battles at the 1946 Faraday Conference, my first, and I regret the more subtle denigration which is more common today. Perhaps if scientists more obviously hated each other, and said so, the profession of physicist would be more interesting to the young people of today. An artificial war of some sort seems necessary for their fulfilment. We, who have experienced a real war, value tranquillity more.

The Theoretical Physics Department in Liverpool in 1950 was physically an old converted house in Abercromby Square which was really very pleasant. No doubt it has now been torn down to make way for some modern "efficient" building. How any building could possibly be more efficient at producing theoretical physics I do not know. I gather they are now at the top of a Tower. The community at Abercromby Square was rather self-contained. I sometimes wondered if the rest of the University knew it existed. I only ever saw one undergraduate there that year, the only one considered good enough to do Theoretical Physics. The main events of the day at Abercromby Square were morning coffee and afternoon tea. I have ever since tried, anywhere I have been, to achieve the same standard of discussion at coffee and tea and have always judged the "Abercrombie coefficient" to be a good guide to the standard of physics being done. The conversation was of course mainly about Physics and a fascinating stream of ideas emerged. Many of these were quickly shown to be false, usually on the blackboard. On the other hand a great deal of time was wasted in arguing about matters which could have been immediately settled by resort to a reference book. However this way of resolving a dispute did not seem popular and I have always noted this trait in theoretical physicists. As an experimentalist I found this irritating but only once did I settle an argument by nipping up to the library and finding the answer. Never again. It broke up that morning's discussion in disarray—both sides were wrong! I suppose an experimentalist is less able to afford a mistake than a theoretician—he may not live through it. Perhaps they were right though in that an argument about some "known" matter brought out all sorts of ideas which were useful. Perhaps indeed we had inadvertently discovered the "think tank". For me at least this was the most important "cooperative phenomenon" developed by FRÖHLICH. One thing is certain, we never did come to any agreement whatever on matters of religion, HUANG [3] being a communist and O'Dwyer a Roman Catholic. I remember one morning we got to arguing about the atomic bomb which was a closely guarded secret at the time. In a very short time we worked out on the board what size it would have to be and how it should be detonated. Subsequent publications showed that we were almost exactly right! I have never since had any respect for secrecy in scientific matters. Once it is known a thing can be done it is simple to find out how.

I have never discovered why FRÖHLICH went to Liverpool since surely he would have had the offer of a Chair in more congenial surroundings. I have heard it said that he went there because, at the time, it had the best Chinese restaurants in Europe. That is true and all of FRÖHLICH's pupils must learn to appreciate chinese food. As it turned out Liverpool was in fact rather a fascinating place. It was no surprise to me that in the middle 50's the Yellow Submarine, i.e. the Liverpool sound and the Beatles, emerged from the river Mersey. Another advantage of Liverpool is the ready access to North Wales. We had many splendid, breathless, discussions of Physics pursuing FRÖHLICH's long legs over the mountains.

The situation in Dielectrics in 1950 was that the field had by then absorbed and accommodated itself to the very significant wartime advances in technique which among other things had extended the accessible frequency range from some 100's of MHz to some 10000's of MHz. This was a substantial advance in experimental technique which, by and large, was not matched on the theoretical side.

I believe it was a major grant from the Electrical Research Association * which enabled FRÖHLICH to have a substantial team on dielectrics and in particular to employ me. They should receive credit for doing this because I have heard it said that FRÖHLICH refused to fill up the usual forms which asked, among other things, what he proposed to do with the money. He said "theoretical research on dielectrics" and refused to say more. How nice it would be to do the same sort of thing to the Science Research Council Physics Committee today. I doubt if even a Nobel Prizeman would get away with it. If this story is not true it ought to be.

The theories in 1950 were still virtually those of DEBYE's classic book of 1929 [4] except for some progress made by the Kirkwood school. This was no doubt appreciated by FRÖHLICH who made a substantial contribution with his book [2] and in particular with his theory of the static dielectric constant (Ref. [2], § 7 and Ref. [5]) which is a much tidier and acceptable analysis, in my opinion, than KIRKWOOD's [6]. However this theory had never been applied in detail to any explicit substance to give a confrontation with experiment. Wartime work in our team suggested to me that ice was a good candidate and I proposed to FRÖHLICH to actually calculate the static dielectric constant of ice microscopically. No such explicit microscopic calculation had previously been attempted. FRÖHLICH was rather dubious about this especially as it clearly involved a lot of computation and we must remember that computers were rudimentary in 1950. Moreover it was very definitely not in the spirit of that Department to actually get numbers. I believe I am right in saying that I was the only person there who possessed a slide rule and who had ever consulted "KAYE and LABY" [7]. The only "computer" available was a desk machine of the type where one turns a handle and even sets the numbers by means of levers. In the event this was so tedious that I had to invent some tricks and in the end I did practically all the calculations on a slide rule. I think in fact there is a good deal of truth in the assertion that if it cannot be worked out on a slide rule then one is not doing it the right way. The advent of the fast digital computer is a mixed blessing. My students spend a lot of computer time fitting straight lines to experimental data. It seems to me that if one can't draw a good straight line through the experimental points by eye one ought to go back to the laboratory and do a better experiment. Be that as it may, I persisted and somewhat to FRÖHLICH's surprise I got quite a good answer [8] in a few months. It would now take about three minutes on an IBM 360!

* The Research Associations then got half their money from the Government and half from the appropriate Industry.

FRÖHLICH had shown [5] that the static dielectric constant, ε_s, is given by,

$$\varepsilon_s - n^2 = \frac{3\varepsilon_s}{(2\varepsilon_s + n^2)} \cdot \frac{4\pi N_0}{3} \left(\frac{n^2+2}{3}\right)^2 \frac{\langle \underset{\sim}{m} \cdot \underset{\sim}{m}^* \rangle}{kT} \tag{1}$$

where the electronic polarisation is represented by n^2.

$\underset{\sim}{m}$ is the dipole moment of any small spherical region which is at first fixed. $\underset{\sim}{m}^*$ is the average dipole moment of a spherical region containing $\underset{\sim}{m}$ which is large enough that the material outside can be treated macroscopically. The average $\langle \underset{\sim}{m} \cdot \underset{\sim}{m}^* \rangle$ is then to be taken, as was the first, according to statistical mechanics in the absence of any external field. It is a fair approximation that $\underset{\sim}{m}$ be taken as one molecule and one can then take the enclosing volume, to obtain $\underset{\sim}{m}^*$, larger and larger until either $\langle \underset{\sim}{m} \cdot \underset{\sim}{m}^* \rangle$ becomes independent of the size of that volume or one's patience is exhausted. The various configurations taken were those which PAULING [9] used in his famous calculation of the residual entropy of ice, except that they were properly weighted according to the energy, which was assumed to be electrostatic. This is quite an undertaking because everything, including reaction fields, has to be accounted for. Incidentally, PAULING's value for the entropy, $k \ln 3/2$, is not much changed. Fortunately $\langle \underset{\sim}{m} \cdot \underset{\sim}{m}^* \rangle$ approaches an asymptotic value reasonably well for about 25 molecules in the sphere and so both the static dielectric constant and its temperature variation were quite well reproduced. Another interesting argument of the theory is that the formula for a free sphere as opposed to the immersed one of Eq. (1) is,

$$\frac{\varepsilon_s - n^2}{\varepsilon_s + 2} = \frac{4\pi N_0}{3} \frac{(n^2+2)}{3} \cdot \frac{\langle \underset{\sim}{m} \cdot \underset{\sim}{m}^*_{\text{vac}} \rangle}{3kT}. \tag{2}$$

For ice $\varepsilon_s \gg n^2$ and $\langle \underset{\sim}{m} \cdot \underset{\sim}{m}^* \rangle$ is very different from $\langle \underset{\sim}{m} \cdot \underset{\sim}{m}^*_{\text{vac}} \rangle$. This is because the reaction field energy of $\underset{\sim}{m}^*$ with the surrounding dielectric in (1) is absent. This difference was shown explicitly in the calculation. It illustrates very vividly the difficulties which arise in dielectric theory due to the relatively long range nature of the dipolar interaction. It would be interesting to repeat this calculation with modern computers.

Another problem which was not well understood at the time was the relation between the actual microscopic relaxation time for reorientation of a dipole and the observed macroscopic relaxation time of the dielectric polarisation. The former is strictly speaking the decay of the time correlation of the dipole vector $\underset{\sim}{\mu}(t)$ i.e. $\langle \underset{\sim}{\mu}(t) \cdot \underset{\sim}{\mu}(t+\tau) \rangle$ which for a stationary process may be written,

$$\langle \underset{\sim}{\mu}(t) \cdot \underset{\sim}{\mu}(t+\tau) \rangle = \langle \underset{\sim}{\mu}^2(t) \rangle \, \Phi(\tau). \tag{3}$$

Usually the correlation time τ_c is defined as,

$$\tau_D \equiv \int_0^\infty \Phi(\tau)\, d\tau. \tag{4}$$

If, as is often a good approximation, $\Phi(\tau)$ is exponential we have

$$\Phi(\tau) = \exp - \tau/\tau_D \tag{5}$$

which satisfies (4). However the dipoles inside a dielectric sample on which we are doing a measurement are not "free". They are subject to an electric field due to the macroscopic polarisation of the dielectric which is itself decaying. This bootstrap effect retards the decay of $\underset{\sim}{\mu}(t)$ and the decay of the macroscopic polarisation, $P(t)$, is surely slower than τ_c and indeed is not necessarily exponential even if $\Phi(\tau)$ is. It is often observed however that the decay of polarisation is exponential to a good approximation,

$$P(t) = P(0) \exp - t/T. \tag{6}$$

Actually the decay of $P(t)$ is itself a function of the external conditions which, if I remember rightly, was well recognised by Fröhlich although I am not able to find it in his book. In any case it is easy to show for instance that if T_E is the (exponential) relaxation time of a dielectric subject to a constant field then it also relaxes exponentially when the charge on the condenser is kept constant but with the different relaxation time T_Q. Moreover,

$$T_Q/T_E = \varepsilon_\infty/\varepsilon_0 \tag{7}$$

where ε_0 and ε_∞ are the low and high frequency limits of the dielectric constant ε. For highly polar dielectrics this ratio is substantial. I have never seen this result confirmed by an actual experiment but surely someone must have done it. Both these matters, the relation of microscopic to macroscopic relaxation and the dependence of macroscopic relaxation on the external constraints, are quite general. The latter is nicely discussed for instance by Davies and Lamb [10]. The problem of the relation of T to τ_D is also quite general in relaxation processes. The difference is usually small but for polar dielectrics there could be a substantial effect. For instance in 1950 we had only Debye's result (Ref. [4], p. 94) which said that,

$$T/\tau_D = (\varepsilon_0 + 2)/(\varepsilon_\infty + 2). \tag{8}$$

For highly polar dielectrics the factor is substantial, for water it is about 18. No-one believed Debye's result and most people took T to be τ_D. In fact it can be shown, by an indirect measurement of τ_D by nuclear magnetic resonance relaxation that for water [11] the factor between T and τ_c is certainly smaller than that given by Debye and must indeed be of order unity. The same is true (having regard to the possible factor of up to 3 difference discussed below) for many substances having very high static dielectric constants. A striking example of this is for the region of temperature above and below the III $\leftrightarrow$ II phase transition in solid hydrogen bromide [12] where the static dielectric constant has a lambda type anomaly and ε_0 attains very large values [13].

It was not difficult to show, by a rather hand-waving argument [14] that if both $\mu(t)$ and $P(t)$ decay exponentially then,

$$T/\tau_D \simeq 3\,\varepsilon_0/(2\,\varepsilon_0 + \varepsilon_\infty). \tag{9}$$

Consequently T and τ_D differ at most by a factor 3/2. More recent developments in dielectric theory using, in effect, the fluctuation-dissipation approach confirm that this relation is in fact correct [15]. This result is of some importance in ascertaining the actual mode of molecular reorientation since one of the ways of doing this is to compare correlation times from different methods. The dielectric correlation time τ_D is that of $\cos\theta(t)$ whereas the correlation time from nuclear magnetic resonance is that of $[3\cos^2\theta(t)-1]$ which we call τ_n. If the molecule reorients by reorientational diffusion $\tau_D=3\tau_n$ whereas if it reorients by "jumps" in angle greater than say 90° then $\tau_D\sim\tau_n$. Thus to study the process of reorientation a correction factor of up to 1.5 on τ_c is crucial.

Perhaps, to be honest, I should also mention some ideas which did not work or which proved to be of little interest since there is no doubt that theoretical physicists spend a lot of time that way. I have always thought that we ought to give the undergraduates a course on things that do not work since this is realistic and one can learn a great deal from mistakes. Such a course would be quite different from one on the history of science which tends to trace the development of ideas which were accepted at the time but which were *subsequently* abandoned. In this respect I return to our coffee and tea sessions which I regard as the best part of my education in theoretical physics since it was there I learned how some 99% of ingenious ideas do not work for one reason or another. I remember Szigeti was particularly good at shooting down apparently good ideas by pure cold logic. An example of an apparently useless exercise is my generalisation of FRÖHLICH's theory of the static dielectric constant to the case of anisotropic materials [16]. To my knowledge this has never been used because most polar crystals which have a substantial static dielectric constant are cubic or else they are extremely difficult to get in single crystal form. An example is one of the lower temperature phases of hydrogen bromide where the dipole reorientation and hence the dielectric constant is surely anisotropic [12]. Another example of failure, also carried out in Liverpool at that time, was the proposed crystal structures of the hydrogen halides [17] which has only recently been shown to be most probably wrong by the determination of the deuteron positions by slow neutron diffraction [18].

Another thing that sometimes happens is that one gets an idea which looks promising but one does not follow it up at the time and one sees it later (sometimes sooner than later!) developed into something quite good. One of the things of this sort which I considered in 1950 but did not follow up was the question of the fluctuations in dipole moment of a macroscopic specimen. This point of view is fundamental to FRÖHLICH's approach to dielectrics [2] and indeed is the precursor to Eqs. (1) and (2). FRÖHLICH showed (Ref. [2], p. 42) that the mean square dipole moment, $\overline{M^2}$, of a macroscopic dielectric sphere in its own medium is given by,

$$\overline{M^2}=\frac{3}{2}k_B T\frac{V}{2\pi}\frac{(2\varepsilon_0+1)(\varepsilon_0-1)}{3\varepsilon_0}, \tag{10}$$

and the corresponding case for a sphere *in vacuo*. If the dielectric forms part of a condenser such fluctuations can be regarded as generating the thermal electrical noise voltage V_n. It says in the textbooks on statistical mechanics that a condenser of any capacitance corresponds to one degree of freedom and so its classical mean thermal energy is $\frac{1}{2}k_B T$. The capacitance referred to is not well defined but is presumably the "static" value, i.e. $\varepsilon_0 C_g$, where C_g is the geometrical capacitance. Hence,

$$\overline{V_n^2}=k_B T/\varepsilon_0 C_g. \tag{11}$$

The measurement of $\overline{V_n^2}$ therefore determines ε_0? However this is a theoreticians' result, if you will excuse my saying so, because this $\overline{V_n^2}$ cannot actually be measured. For one thing ε_0 contains a contribution from electronic polarisation and the fluctuation noise voltage from this is presumably in the ultraviolet. This is not measured in an electrical experiment and in any case is beyond the quantum mechanical cut-off at a frequency of $k_B T/\hbar$. It is tempting therefore to replace ε_0 in (11) by $\varepsilon_0 - \varepsilon_\infty$. However this would be too naive since what we actually have is more nearly two condensers in parallel, a noise-free one $\varepsilon_\infty C_g$ and a noisy one $(\varepsilon_0 - \varepsilon_\infty) C_g$ and it is not immediately obvious how these fit together. It is shown below that the result is

$$\overline{V_n^2} = \frac{k_B T}{C_g}\left(\frac{1}{\varepsilon_\infty} - \frac{1}{\varepsilon_0}\right). \tag{12}$$

If $\varepsilon_0 = \varepsilon_\infty = 1$ we have $\overline{V_n^2} = 0$ for a vacuum filled condenser, which is correct—classically—but is inconsistent with the statement above that the mean thermal electrical energy is $\frac{1}{2} k_B T$. However again this is an idealisation because any actual "vacuum" condenser we could actually make would have finite losses and the $\frac{1}{2} k_B T$ comes back.

I was also naturally led to consider the time autocorrelation function of V_n (or $\underset{\sim}{M}$) and its power spectrum which I felt ought to be related somehow to the frequency dependence of the dielectric constant. Of course we now have a much clearer picture of how the spectrum of the fluctuations are connected to losses or dissipation than we did in 1950. Indeed I like to delude myself that I almost discovered the fluctuation-dissipation theorem [19], which was clearly in the air at the time. For dielectrics the relation is quite simple if one uses the conventional equivalent circuit of a relaxing Debye dielectric (a condenser $\varepsilon_\infty C_g$ in parallel with a series combination of a condenser $(\varepsilon_0 - \varepsilon_\infty) C_g$ and a resistor $T/(\varepsilon_0 - \varepsilon_\infty) C_g$, where T is the macroscopic relaxation time). This and Nyquist's theorem enables one to show that,

$$\overline{V_n^2}(\omega) = \frac{2}{\pi} \frac{k_B T}{C_g} \frac{(\varepsilon_0 - \varepsilon_\infty)}{\varepsilon_0^2} \frac{T}{1 + \left(\omega \frac{\varepsilon_\infty}{\varepsilon_0} T\right)^2}. \tag{13}$$

The integral of (13) yields (12).

I proposed to justify this expression using Fröhlich's theory of fluctuations as applied to dielectrics and possibly to generalise it. It seems that this was only done as late as 1964 [20]. I also thought it would be of great interest from a fundamental, and possibly even a practical, point of view to confirm relations such as (13) experimentally but I suspected that experimental techniques in 1950 were not good enough for this. I never got around to it because after leaving Fröhlich I went to work with Charlie Smyth and got involved once more with plastic crystals, a pastime—twenty years later—now becoming very popular with physicists. In fact it seems this experiment was not done until 1966 [21]*.

Some who are not familiar with dielectrics might be puzzled by the fact that in (13) the apparent relaxation time of the emission spectrum is $\frac{\varepsilon_\infty}{\varepsilon_0} T$ and not T. However this is just one more example of the effect of constraints on relaxation which applies to spontaneous as well as to forced fluctuations. One sees the same

* I am grateful to Dr. Claude Brot for drawing my attention to this paper.

factor in Eq. (7). The result (13) applies to a dielectric connected to an infinite impedance voltage detector and so to relaxation at constant charge, i.e.

$$T_Q = \frac{\varepsilon_\infty}{\varepsilon_0} T. \tag{14}$$

If instead of measuring $\overline{V_n^2}$ we used a current detector of zero impedance then we find instead of (13),

$$\overline{I_n^2}(\omega) = \frac{2}{\pi} k_B T C_g (\varepsilon_0 - \varepsilon_\infty) \frac{\omega^2 T}{(1 + \omega^2 T^2)}. \tag{15}$$

Thus the relaxation time is now T. Since the short circuited dielectric corresponds to constant E then T is in fact T_E. This together with (14) gives the same equation as (7) but for constraints on fluctuations.

Perhaps I should complete my reminiscences of FRÖHLICH's Department in 1950 by a brief account of some of the other Physics which was going on. No doubt this will be amply discussed by other contributors but it may not be without interest to the reader to notice its effect on me. I never did understand what Huang was doing but I now know that his trips to Edinburgh to see MAX BORN gave rise to the classic treatise on vibrations in crystals [3]. While I was in Liverpool FRÖHLICH went off to the States for a few months and returned with a theory of superconductivity enshrined in several hundred closely written sheets of algebra. I was impressed that HUANG read and understood this in about two days and moreover discovered a factor of two error. This was my first introduction to the traditional error by a factor of two which almost inevitably appears in the first version of any theory. This theory is, I believe, now recognised as one of the great near misses of theoretical physics since undoubtedly FRÖHLICH had the mechanism of superconductivity [22] and had laid the foundation for the now accepted BCS theory [23]. I was unwise enough to choose this topic for the "deuxieme thèse" for my Doctorat ès Sciences so I looked at it rather closely. Fortunately the "jury" at my examination did not find out, what I have since realised, that I did not understand it very well—or at least they were nice enough not to allow themselves to reject the three years of experimental work in the main thesis on that account.

While I was in Liverpool we organised a Conference on Dielectrics. Perhaps because I was put in charge of the social arrangements that is about the only part I remember. I organised a coach trip to North Wales to the region of Llangollen and stopped about ten miles away to let off the *strong* walkers to be led over the mountain by FRÖHLICH. I proposed to lead the weaker brethren by an easier and shorter route. It was my first experience of trying to get a large group of physicists to behave rationally. They nearly all went with FRÖHLICH, including Mrs. FREYMANN who is lame and Professor FREYMANN who is not built for mountain climbing. Much to my dismay I saw them disappear into the mist far above with FRÖHLICH well in front as usual. To my relief they all got to Llangollen safely. I remember at the same Conference I was sitting next to EDMOND BAUER and he asked me if I wanted to listen to the lecture, which was unusually boring. I said, "No" and so he said, "Let's go and see the ships!". So off I went down to the docks with BAUER on the back of my motorbike. He must have been at least seventy at the time but still full of boyish enthusiasm. FRÖHLICH was very relieved when we got back safely after a delightful afternoon talking about Physics and drinking tea with the dockers who were very kind to us. We were all poorer and happier in those days.

Just to show that we also did other things as well as Physics I should also mention that four of us got married about that time. I need not comment on my own marriage nor on that of Le Couteur, both of which were entirely conventional. Of more interest perhaps is the fact that during that year HUANG returned to China which, during his stay in England, had been overrun by Chairman Mao. Sometime later our Secretary-Physicist Avril announced at morning coffee that she was off to China on the following day to marry HUANG. This was a complete surprise to me and, I understand, also to her parents! Another piece of excitement that we had was that one day FRÖHLICH did not come in to coffee and this was unusual since it was known that he was not away on a trip. Nor was BELA SZIGETI in for that matter but this excited no comment as he had never been known to be in *before* coffee and frequently arrived in time for lunch. I hasten to add that he was usually there long after I had left. In the afternoon at tea SZIGETI turned up but no FRÖHLICH, although he was known to be working upstairs. In response to our enquiries regarding these strange happenings SZIGETI explained that FRÖHLICH had got married that morning and was now catching up with the working time he had consequently lost. I always warn the fiancées of my graduate students that they must face the prospect of becoming "physics widows". However never have I known a wife become a physics widow so rapidly as Mrs. FRÖHLICH. Thus in this, as in all matters connected with physics, FRÖHLICH has clearly excelled. "Hommages au maître."

References

1. A general discussion on Dielectrics. Trans. Faraday Soc. **42** A (1946).
2. FRÖHLICH, H.: Theory of Dielectrics. Oxford: Clarendon Press 1949.
3. BORN, M., HUANG, K.: Dynamical Theory of Crystal Lattices. Oxford: Clarendon Press 1954.
4. DEBYE, P.: Polar Molecules. Dover reprint (1945).: Chemical Catalog Co. Inc. 1929.
5. FRÖHLICH, H.: Trans. Faraday Soc. **44**, 238 (1948).
6. KIRKWOOD, J. G.: J. Chem. Phys. **7**, 911 (1939).
7. KAYE, G. W. C., LABY, T. H.: Physical and Chemical Constants.: Longmans 1944.
8. POWLES, J. G.: J. Chem. Phys. **20**, 1302 (1952).
9. PAULING, L.: J. Amer. Chem. Soc. **57**, 2680 (1935).
10. DAVIES, R. O., LAMB, J.: Proc. Phys. Soc. (London) **69** B, 293 (1956).
11. SMITH, D. W. G., POWLES, J. G.: Mol. Phys. **10**, 451 (1966).
12. NORRIS, M. O., STRANGE, J. H., POWLES, J. G., RHODES, M., MARSDEN, K., KRYNICKI, K.: J. Phys. C **1**, 422 (1968).
13. SMYTH, C. P., HITCHCOCK, C. S.: J. Amer. Chem. Soc. **55**, 1830 (1933).
14. POWLES, J. G.: J. Chem. Phys. **21**, 633 (1953).
15. GLARUM, S. H.: J. Chem. Phys. **33**, 1371 (1960).
16. POWLES, J. G.: Trans. Faraday Soc. **51**, 377 (1955).
17. POWLES, J. G.: Trans. Faraday Soc. **48**, 430 (1952).
18. SANDOR, E., FARROW, R. F. C.: Nature **213**, 171 (1967).
19. CALLAN, H. B., WELTON, T. A.: Phys. Rev. **83**, 34 (1951).
20. DAVIS, L.: J. Appl. Phys. **35**, 2004 (1964).
21. LE BOT, J., RIAUX, E., GROSVALD, G., OLLIVIER, R.: Compt. Rend. **262**, 822 (1966).
22. FRÖHLICH, H.: Phys. Rev. **79**, 845 (1950).
23. BARDEEN, J., COOPER, L. N., SCHRIEFFER, J. R.: Phys. Rev. **108**, 1175 (1957).

Bibliographical Notes

G. J. HYLAND

University of Warwick, Department of Physics, Coventry England

Whilst upon superficial perusal of this bibliography of H. FRÖHLICH, one is indeed struck by the great diversity of his contributions, it is only upon a deeper awareness of their contents that the fundamental significance and profound influence of his work can be fully appreciated; this publication in his honour provides a natural opportunity for cultivation of such an awareness. This can in no better way be catalysed than by reflection on his own constant awareness of the possible relevance of concepts to branches of physics in which they did not at first arise.

Analysis of his bibliography reveals two particularly important illustrations of this breadth of his outlook—an outlook so refreshing in these days of over-specialization in theoretical physics. The first illustrates the fruits of the fusion of ideas and techniques from two entirely different branches of physics (I refer, of course, to the introduction of field theoretical concepts into solid state physics) whilst the second emphasizes the rôle of a single concept (non-locality) in areas so diverse as particle physics and biology.

In persuance of these considerations it is instructive to trace the genesis of FRÖHLICH's fundamental contribution to the microscopic theory of superconductivity—a contribution which proved to be the key to the deadlock that had prevailed for the preceding 20 years. Already in the course of his early work [18] on dielectric breakdown in 1937—a field in which his interest steadily expanded during the suceeding decade, culminating in the publication of his eminent treatise [2] "Theory of Dielectrics", of which Russian and Japanese translations now exist—the interaction of an electron with an ionic crystal was treated by considering the ions not as individual particles but in terms of the polarization of the material; the following year on the other hand, saw one of the first applications [20] of meson theory (still at that time couched in terms of "heavy electrons") by himself and HEITLER to the anomalous magnetic moments of the proton and neutron. It then occurred to FRÖHLICH that this polarization could be treated as a field just like in relativistic field theories (e.g. the meson theory just referred to). This immediately implied, in addition to emission and absorption of (in this case) vibrational quanta (phonons)—a process already inherent in BLOCH's theory of electrical conductivity—the existence of an electron self-energy (now nondivergent) and in the case of more than one electron an interaction between them. The rest is well-known; the former effect, the first to be investigated, led to the development [62] of polaron theory, whilst the second led to the possibility [67] of a phonon-induced attraction between electrons which proved basic to the development of the microscopic theory of superconductivity. These developments and their ramifications are described in FRÖHLICH's own inimitable way in Refs. [80, 97, 106, 112] and in Ref. [121], especially.

Further details need not concern us here; the point at issue is that this fundamental advance was contingent upon an awareness and familiarity with concepts and techniques peculiar (at that time) to entirely different branches of physics; indeed, only a qualitative familiarity was necessary. Quantitative analyses soon followed, however, in a paper [72] wherein, for the first time, modern field theoretical techniques were employed in solid state physics—a new era had begun.

The potentiality of such techniques was quickly appreciated and methods already well-known in relativistic field theory, quickly adapted. During the ensuing years, however, FRÖHLICH became increasingly convinced of a rampant misuse of such methods and lamented the neglect of general considerations evidenced in many applications of these *tour de force* approaches, to the many-body problem—"Such efforts often overlook that a number of important general relations can be derived from microphysics, and applied to macrophysics, in a manner which does not require detailed solution of the N-body problem" (Ref. [125]).

In no way is this better illustrated than in connection with the derivation of macroscopic equations from microphysics—"Surely it must be possible to achieve this from the quantum-mechanical many-body problem without going into the complications exhibited in so many papers on this subject" (Ref. [115]—adapted). He was quick to point out, however, that progress in this manner would require the introduction of new concepts.

One such concept has proved to be that of non-locality. Arguing [97] that the introduction of certain non-local features into hydrodynamics could well circumvent the necessity of invoking atomistic considerations to remove difficulties in quantum hydrodynamics, we later find the same concept of non-locality arising in a very different setting in his contribution [109] to the Yukawa memorial issue, in a paper entitled "Geometrical Interpretation of Electrodynamics". For this analysis, from which, for the first time, was field quantization seen to be imposed by geometrical considerations, is based on a bilocal field theory which, in its local limit, contains the Maxwell equations. Stimulated by the implications of such non-local considerations he ends the paper with a profound statement which imposes an exacting criterion on any basic theory of particles "... establishment of a metric would require the existence of measuring instruments of length. They would consist of particles, whose existence cannot be postulated in a theory whose aim would be to derive this existence as one of its main consequences".

The necessity of incorporating non-local considerations into hydrodynamics was re-iterated in a more intensified form the following year, and finally received formulation [115] in 1967 in terms of the reduced density matrices—the appropriate macroscopic entities in terms of which the connection between micro and macro physics could be established in a very general way with the use of little more than symmetry considerations—"This does not, of course, make detailed treatments (using various approximate methods) superfluous but puts them into their proper place which is the calculation of magnitudes and other properties (e.g. temperature dependence) of various parameters entering in the formulation of macroscopic equations" (Ref. [125]). His early concern over the mis-use of such detailed treatments was thus completely vindicated.

Progress did not rest there, however, for he was quick to appreciate the implications of "off-diagonal-long-range-order" (O.D.L.R.O.), a new concept introduced earlier by YANG relating the reduced density matrices to the macroscopic wave-functions, characteristic of superfluids and superconductors; there now thus existed a well defined procedure for the derivation of macroscopic wave equations, from which the properties of flow in these materials follow, in a very general way, independent of the details of any particular microscopic model. The existence of O.D.L.R.O.—an extreme asymptotic property of the non-locality of the reduced density matrices—is a feature common to both superfluids and superconductors; that such a common feature should exist had been repeatedly stressed by FRÖHLICH on many occasions since 1961 (vide Refs. [97, 112, 121]). In terms of macroscopic wave-functions the existence of O.D.L.R.O. is equivalent to the existence of long-range-phase correlations and an associated coherency over macroscopic distances.

On first thoughts the relevancy of such coherency to biological systems might seem remote. Stimulated by his ubiquitous outlook, however, and by knowledge of the peculiar dielectric properties of membranes he was led to conjecture the possibility of coherent states becoming stabilized at ordinary temperatures in such biological systems which are "relatively stable yet in some respects far from thermal equilibrium" (Ref. [121]). In this case the implications of such cross-fertilization between seemingly different branches of physics could well be epochal; already experimental investigations to check the predictions of the detailed theory are encouraging whilst at an international level a bienniel conference on Theoretical Physics and Biology has been established (Refs. [127, 135]).

The fruits of FRÖHLICH's broad outlook and constant awareness of the possible relevance of concepts and techniques to branches of physics in which they did not at first arise are then well manifest, standing as "a strong indictment against fragmentation and overspecialization in theoretical physics" (Ref. [112]).

It remains but to highlight a few papers not referred to above but which in their own right have had a great influence on current outlook; since all of FRÖHLICH's papers have something of importance to say, the choice is largely subjective. First and foremost must be mentioned his book "Elektronentheorie der Metalle", which although first published in 1936—when it ranked as one of the first texts on solid state physics—contains succinct yet penetrating accounts of some problems which are still incompletely resolved. The theory of low energy electron diffraction is a case in point; for the years immediately following a great revival of interest in this field some 30 years later were spent on rediscovering the results of 1936!

The book subsequently became unavailable during which time one German physicist told FRÖHLICH proudly—"I wrote it all out with my own hand"—to which FRÖHLICH replied—"I wrote it with my own hand too". Of his prewar publications, two, in particular stand out:

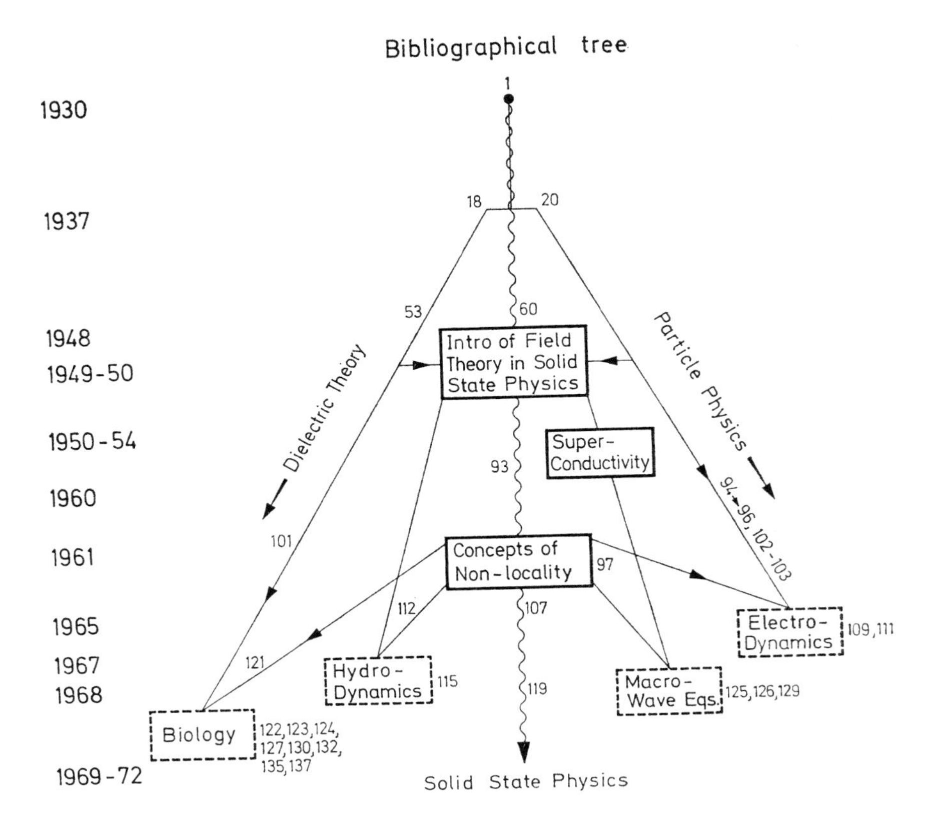

i) "Shot effect in quantum mechanics" [4] and

ii) "Electronic specific heat of small metal particles" [17].

The former is probably the first paper in solid state physics to use creation and annihilation operators, whilst the problem enunciated in the latter is a topic of current interest and one which is still basically unresolved. From the post war period should be mentioned:

i) "Energy distribution and stability of electrons in electric fields" [53]—an offshoot from his work in dielectric theory, in which the idea of a field-dependent electron temperature was first introduced, a concept basic to Hot Electron physics.

ii) "Nuclear ferromagnetism" (with F. R. N. NABARRO), [60]— in which the process now known under the names of RUDERMAN and KITTEL was first formulated.

iii) "Electric conduction in semiconductors" (with G. L. SEWELL) [92]—a short paper in which the breakdown of the band model due to strong interaction of a single electron (in a non-metal) with the ions of the lattice, and the consequent onset of hopping conduction was first formulated. Later developments of this work (now known as small polaron theory) led to a complete reformulation of the electron-phonon interaction in which the coupling of an electron to the displacement of the ion nearest to it is already included in zeroth order (Refs. [104, 105, 112]); the implications for superconductivity in materials with narrow bands (e.g. d-bands in transition metals) is discussed in Ref. [120]. This paper follows one in which a mechanism for superconductivity in materials which do not exhibit an isotope effect (materials with incomplete inner shells), is presented. It is based on a screening of the d-band plasma by the s-elections such that for long waves the frequency of the d-band plasma becomes proportional to the wave number which accordingly then represents a new acoustic branch, which like the ordinary (ion) acoustic branch leads to the possibility of an attractive interaction between the free electrons, and hence to superconductivity.

Space does not, unfortunately, permit further discussion upon these works; suffice it to say that they stand witness to the penetrating physical insight and versatility of a truly "complete" physicist.

Publications of H. Fröhlich from 1930—1972

Books

1. Elektronentheorie der Metalle. Berlin: Springer 1936; reprinted, Ann Arbor 1943, republished, by Springer 1969 [in German].
2. Theory of Dielectrics. Oxford: Clarendon Press 1949; 2nd Ed. 1958. Translations in Russian and Japanese.

Original Papers and Review Articles

3. Photoelectric effect in metals: Ann. Physik **7**, 103–128 (1930) [in German].
4. Shot effect in quantum mechanics: Z. Physik **71**, 715–719 (1931) [in German].
5. Theory of secondary emission from metals: Ann. Physik **13**, 229–248 (1932) [in German].
6. Light absorption and selective photoelectric effect: Z. Physik **75**, 539–543 (1932) [in German].
7. Determination of energy levels of metallic electrons from their optical constants: Naturwissenschaften **20**, 906 (1932) [in German].
8. Position of the absorption spectra of photochemically coloured alkali halide crystals: Z. Physik **80**, 819–821 (1933) [in German].
9. Absorption of metals in the visible and ultraviolet: Z. Physik **81**, 297–312 (1933) [in German].
10. Magnetic interaction of metallic electrons. Criticism of Frenkel's theory of superconductivity (with H. A. Bethe): Z. Physik **85**, 389–397 (1933) [in German].
11. The calculation of the work function in Sommerfeld's model of a metal: Z. Physik Sov. **7**, 509–510 (1935) [in German].
12. The number of free electrons in a metal: Proc. Cambridge Phil. Soc. **31**, 277–280 (1935).
13. Inner photoelectric effect in semiconductors: Z. Physik Sov. **8**, 501–510 (1935) [in German].
14. Time orientation of nuclear spins in a magnetic field (with W. Heitler): Z. Physik Sov. **10**, 847–848 (1936) [in German].
15. Time effects in the magnetic cooling method (with W. Heitler): Proc. Roy. Soc. (London), Ser. A **155**, 640–652 (1936);—Proc. VII Intern. Congr. of Refrigeration (with W. Heitler and E. Teller), p. 1–2 (1936).
16. Quantum mechanical discussion of the cohesive forces and thermal expansion coefficients of the alkali metals: Proc. Roy. Soc. (London), Ser. A **158**, 97–110 (1937).
17. Electronic specific heat of small metal particles at low temperatures: Physica **4**, 406–412 (1937).

18. Theory of electrical breakdown in ionic crystals, I: Proc. Roy. Soc. (London), Ser. A **160**, 230–241 (1937).
19. Theory of the λ-point of helium: Physica **8**, 639–644 (1937).
20. Magnetic moments of the proton and the neutron (with W. HEITLER): Nature **141**, 37 (1938).
21. Nuclear forces and the magnetic moments of the neutron and proton (with W. HEITLER and N. KEMMER): Proc. Roy. Soc. (London), Ser. A **166**, 154–177 (1938).
22. Solution of the Schrödinger equation by perturbation of the boundary conditions: Phys. Rev. **54**, 945–947 (1938).
23. Deviation from the Coulomb law for the proton (with W. HEITLER and B. KAHN): Proc. Roy. Soc. (London), Ser. A **171**, 269–280 (1939).
24. Mean free path of electrons in polar crystals (with N. F. MOTT): Proc. Roy. Soc. (London), Ser. A **171**, 496–504 (1939).
25. Theory of electrical breakdown in ionic crystals, II: Proc. Roy. Soc. (London), Ser. A **172**, 94–106 (1939).
26. Dielectric breakdown in ionic crystals, III: Phys. Rev. **56**, 349–352 (1939).
27. Dielectric breakdown in solids: Rept. Progr. Phys. **6**, 411–430 (1939).
28. Photodisintegration of deuteron in meson theory (with W. HEITLER and B. KAHN): Proc. Roy. Soc. (London), Ser. A **174**, 85–102 (1940).
29. Orientation of nuclear spins in metals (with F. R. N. NABARRO): Proc. Roy. Soc. (London), Ser. A **175**, 382–391 (1940).
30. On the dielectric strength of mixed crystals: Proc. Roy. Soc. (London), Ser. A **178**, 493–505 (1941).
31. Dielectric loss in dipolar solids. Solution of dipolar molecules in solid paraffin: Rep. Brit. Elec. Allied Indust. Res. Assoc.* Ref. L/T 121, 12 pp. (1941).
32. Theory of the dielectric properties of dipolar solids: E.R.A. report Ref. L/T 124, 55 pp. (1941).
33. Electric breakdown in ionic crystals: Phys. Rev. **61**, 200–201 (1942).
34. Meson theory and the magnetic moments of protons and neutrons: Phys. Rev. **62**, 180 (1942).
35. Dielectric loss in paraffin wax solutions: Proc. Phys. Soc. (London) **54**, 422–428 (1942).
36. Theory of the dielectric properties of dipolar solids: calculation of dielectric losses: E.R.A. report, Ref. L/T 132, 8 pp. (1942).
37. Theory of dielectric breakdown: Nature **151**, 339–340 (1943).
38. Theory of the dielectric properties of dipolar solids. Influence of dipolar interaction: E.R.A. report, Ref. L/T 142, 12 pp. (1942).
39. Dielectric properties of dipolar substances (with R. A. SACK): Proc. Roy. Soc. (London), Ser. A **182**, 388–403 (1944).
40. Theory of the dielectric constant and energy loss in solids and liquids: J. Inst. Elec. Eng. (London) **91**, 456–463 (1944).
41. Phase transitions of solid paraffins and the flexibility of hydrocarbon chains: Trans. Faraday Soc. **40**, 498–502 (1944).
42. Theory of dielectric breakdown in amorphous solids: E.R.A. report, Ref. L/T 153, 7 pp. (1945).
43. Dielectric and thermal properties of long-chain substances: E.R.A. report, Ref. L/T 156, 7 pp. (1945).
44. Dielectric properties of solids at very high frequencies: E.R.A. report, Ref. L/T 157, 6 pp. (1945).
45. Dielectric properties of solids at ultra-high frequencies: E.R.A. report, Ref. L/T 163, 5 pp. (1946).
46. Dielectric properties of dipolar solids: Proc. Roy. Soc. (London), Ser. A **185**, 399–414 (1946).
47. Theory of rheological properties of dispersions (with R. A. SACK): Proc. Roy. Soc. (London), Ser. A **185**, 415–430 (1946).
48. Shape of collision-broadened spectral lines: Nature **157**, 478 (1946).

* Hereinafter denoted by E.R.A. report.

49. Theoretical Physics in industry (Royal Institution Lecture): Nature **158**, 332–334 (1946).
50. Dipolar interaction: Trans. Faraday Soc. **42**A, 3–7 disc. 36–39 (1946).
51. Light absorption and selective photo-effect in adsorbed layers (with R. A. SACK): Proc. Phys. Soc. (London) **59**, 30–33 (1947).
52. On the theory of dielectric breakdown in solids: Proc. Roy. Soc. (London), Ser. A **188**, 521–532 (1947).
53. Energy distribution and stability of electrons in electric fields: Proc. Roy. Soc. (London), Ser. A **188**, 532–541 (1947).
54. Decay of negative mesons in matter: Nature **160**, 255 (1947).
55. The binding energies of very light nuclei (with K. HUANG and I. N. SNEDDON): Proc. Roy. Soc. (London), Ser. A **191**, 61–82 (1947).
56. General theory of the static dielectric constant: Trans. Faraday Soc. **44**, 238–243 (1948).
57. A quantitative discussion of the interaction between nuclear particles (with W. H. RAMSAY and I. N. SNEDDON): Rep. Intern. Conf. on "Fundamental Particles" (1946) in Phys. Soc. (London) **1**, 166–175 (1948).
58. Polarization of dielectrics by slow particles (with H. PELZER): E.R.A. report, Ref. L/T 184, 8 pp. (1948).
59. Decay and capture of slow mesons in dielectrics (with R. HUBY, R. KOLODZIEJSKI and R. L. ROSENBERG): Nature **162**, 450–451 (1948).
60. Nuclear ferromagnetism (with F. R. N. NABARRO): Rep. Intern. Conf. on "Low Temperatures", Cambridge 1946, in Phys. Soc. (London) **2**, 130–134 (1948).
61. Time dependence of electronic processes in dielectrics (with J. O'DWYER): E.R.A. report, Ref. L/T 219, 5 pp. (1949); Proc. Phys. Soc. (London) A **63**, 81–85, and p. 299 (1950).
62. Properties of slow electrons in polar materials (with H. PELZER and S. ZIENAU): E.R.A. report, Ref. L/T 221, 12 pp. (1950); Phil. Mag. **41**, 221–242 (1950).
63. Conduction electrons in non-metallic solids: Research (London) **3**, 202–207 (1950).
64. Isotope effect in superconductivity: Proc. Phys. Soc. (London) A **63**, 778 (1950).
65. Notes on the theory of dielectric breakdown in ionic crystals (with F. SEITZ): Phys. Rev. **79**, 526–527 (1950).
66. Intrinsic dielectric breakdown in solids (with J. H. SIMPSON): Advanc. Electronics **2**, 187–217 (1950).
67. Theory of the superconducting state, I. The ground state at the absolute zero of temperature: Phys. Rev. **79**, 845–856 (1950).
68. Theory of the superconducting state, II. Magnetic properties at the absolute zero of temperature: Proc. Phys. Soc. (London) A **64**, 129–134 (1951).
69. Crystal structure and superconductivity: Nature **168**, 280 (1951).
70. Superconductivity and the effective mass of electrons: Nature **168**, 280–281 (1951).
71. On the theory of dielectric breakdown in ionic crystals: E.R.A. report, Ref. L/T 277, 14 pp. (1952).
72. Interaction of electrons with lattice vibrations: Proc. Roy. Soc. (London), Ser. A **215**, 291–298 (1952).
73. Theory of the superconductive state: Proc. Washington Conf. (1952).
74. Superconductivity and lattice vibrations: Proc. Lorentz-Kamerlingh Onnes Conf., in Physica **19**, 755–764 (1953).
75. Energy loss of moving electrons to dipolar relaxation (with R. L. PLATZMAN): Phys. Rev. **92**, 1152–1154 (1953).
76. Rotational transitions in solids: 2e Reunion de Chimie Physique Paris, p. 231–234 (1953) [in French].
77. Remarks on the theory of superconductivity: Proc. Intern. Conf. of Theoretical Physics, Kyoto and Tokyo, 1953, p. 909–915, publ. by Science Council of Japan (1954).
78. Electrons in dielectrics: Proc. Intern. Conf. of Theoretical Physics, Kyoto and Tokyo, 1953, p. 805–809, publ. by Science Council of Japan (1954).
79. On the theory of superconductivity: the one dimensional case: Proc. Roy. Soc. (London), Ser. A **223**, 296–305 (1954).
80. Electrons in lattice fields: Advan. Physics **3**, 325–361 (1954).

81. Dielectric polarization in polar substances. Remark on a paper by F. E. HARRIS and B. J. ALDER: J. Chem. Physics **22**, 1804–1806 (1954).
82. Heat conduction in semiconductors. Remark on a paper by G. BUSCH (with C. KITTEL): Proc. Amsterdam Conf. on Semiconductors, in Physica **20**, 1086 (1954).
83. Plasma oscillations and energy loss of charged particles in solids: E.R.A. report, Ref. L/T322, 14pp. (1955); Proc. Phys. Soc. (London) A**68**, 525–529 (1955).
84. Theory of secondary electron emission from solids: Proc. Phys. Soc. (London) B**68**, 657–660 (1955).
85. Dielectric breakdown in solids (with B. V. PARANJAPE): Proc. Phys. Soc. (London) B**69**, 21–23, 866 (1956).
86. The influence of interelectronic collisions and of surfaces on electronic conductivity (with B. V. PARANJAPE, C. G. KUPER and S. NAKAJIMA): Proc. Phys. Soc. (London) B**69**, 842–845 (1956).
87. Plasma interaction and conduction in semiconductors (with S. DONIACH): Proc. Phys. Soc. (London) B**69**, 961 (1956).
88. Remark on the calculation of the static dielectric constant: Physica **22**, 898–904 (1956).
89. Debye loss in ionic solids: Arch. Sci. (Geneva) **10**, 5–6 (1957) [in French].
90. Speculations on the masses of particles: Nucl. Phys. **7**, 148–149 (1958).
91. A survey of the theory of dielectrics: Proc. 8th Colloque Ampère, in Arch. Sci. (Geneva) **12** (special number) 5–8 (1959).
92. Electric conduction in semiconductors (with G. L. SEWELL): Proc. Phys. Soc. (London) **74**, 643–647 (1959).
93. Phenomenological theory of the energy loss of fast particles in solids: Contribution to "Max Planck Festschrift, 1958", Ed. W. FRANK, p. 277–284. Berlin: VEB Deutscher Verlag der Wissenschaften 1959.
94. Space-time reflexions, isobaric spin and the mass ratio of bosons: Proc. Roy. Soc. (London), Ser. A**257**, 147–164 (1960).
95. Light quanta and heavy bosons: Proc. Roy. Soc. (London), Ser. A**257**, 283–290 (1960).
96. Space-time reflexions, light quanta and heavy bosons: Helv. Phys. Acta **33**, 803–828 (W. PAULI Memorial Issue) (1960).
97. The theory of the superconductive state: Rept. Progr. Phys. **24**, 1–23 (1961).
98. Breakdown in non-polar semiconductors: Soviet Phys. Solid State **3**, 491 (1961).
99. New heavy bosons: Proc. Phys. Soc. (London) **77**, 1223 (1961).
100. The structure of momentum space, the neutrino and the Pauli principle: Nucl. Phys. **26**, 324–337 (1961).
101. Hot electrons: E.R.A. report, Ref. L/T414, 11pp. (1961).
102. Isobaric spin algebra (with C. M. TERREAUX): Nucl. Phys. **42**, 21–26 (1963).
103. On isobaric spin space: Nucl. Phys. **45**, 609–613 (1963).
104. Low mobility materials and Debye dielectric loss due to electrons: E.R.A. report 5003, 8pp. (1963); with S. MACHLUP and T. K. MITRA: Phys. Kondens. Materie **1**, 359–366 (1963).
105. Electron-phonon interaction, superconductivity and the third law of thermodynamics: Phys. Letters **7**, 346–347 (1963).
106. Introduction to the theory of the polaron: Contribution to "Polarons and Excitons", Eds. C. G. KUPER and G. D. WHITFIELD, p. 1–32. Edinburgh and London: Oliver and Boyd 1963 (Scottish Universities' Summer School 1962).
107. The Boltzmann Equations in electron-phonon systems (with A. W. B. TAYLOR): Proc. Phys. Soc. (London) **83**, 739–748 (1964).
108. Quasi-superconductive transitions in strong magnetic fields (with C. M. TERREAUX): Proc. Phys. Soc. (London) **86**, 233–236 (1965).
109. Geometrical interpretation of electrodynamics: Progr. Theoret. Phys. Suppl. dedicated to H. YUKAWA on the 30th Anniversary of meson theory, p. 1–13 (1965).
110. Macroscopic wave-functions in superconductors: Proc. Phys. Soc. (London) **87**, 330–332 (1966).
111. Generation of dual transformations through fields in electrodynamics: Progr. Theoret. Phys. **36**, 636–647 (1966).

112. Superconductivity and the many body problem: Contribution to "Perspectives in Modern Physics", Ed. R. E. MARSHAK, p. 539–552. New York: Interscience 1966 (Essays in honour of HANS A. BETHE).
113. Limits of the band model and transitions to the metallic state: Contribution to "Quantum Theory of Atoms, Molecules and the Solid State", Ed. P.-O. LÖWDIN, p. 465–468. New York: Academic Press 1966 (A Tribute to JOHN C. SLATER).
114. A contradiction between quantum hydrodynamics and the existence of particles: Physica **34**, 47–48 (1967).
115. Microscopic derivation of the equations of hydrodynamics: Physica **37**, 215–226 (1967).
116. Dielectric Instabilities: Contribution to "Ferroelectricity", Ed. E. F. WELLER, p. 9–15. Amsterdam: Elsevier Publ. Co. 1967 (Dedicated to P. J. W. DEBYE).
117. Proposal of crucial experiments for superconductors with incomplete inner bands: Phys. Letters **26**A, 169–170 (1968).
118. Localized versus band model of electrons in solids: Helv. Phys. Acta **41**, 838–839 (G. BUSCH Festschrift).
119. Superconductivity in metals with incomplete inner shells: J. Phys. C (Proc. Phys. Soc.), Ser. 2, **1**, 544–548 (1968).
120. Superconductivity and the magnitude of the electron-phonon interaction (with T. K. MITRA): J. Phys. C **1**, 548–549 (1968).
121. Theoretical problems in superconductivity: Phys. Bull. **19**, 209–212 (1968) (Jubilee Article).
122. Bose condensation of strongly excited longitudinal electric modes: Phys. Letters **26**A, 402–403 (1968).
123. Long-range coherence and energy storage in biological systems: Intern. J. Quantum Chem. **2**, 641–649 (1968).
124. Storage of light energy and photosynthesis: Nature **219**, 743–744 (1968).
125. General remarks on the connection of the laws of micro and macro physics: Proc. Intern. Conf. on Statistical Mechanics, Kyoto 1968, in J. Phys. Soc. Japan **26** (Suppl.), 189–195 (1969).
126. Macroscopic wavefunctions and wave equations: Contribution to "Problems of Theoretical Physics", p. 373–378. Moscow: Nauka 1969 (Essays in honour of N. N. BOGOLIUBOV).
127. Quantum mechanical concepts in biology: Proc. 1st Intern. Conf. on Theoretical Physics and Biology, Versailles, 1967, Ed. M. MAROIS, p. 13–22. Amsterdam: North Holland Publ. Co. 1969.
128. Proposed model experiments on the storage of light energy in photosynthesis: Nature **221**, 976 (1969).
129. The macroscopic wave equations of superfluids: Phys. Kondens. Materie **9**, 350–358 (1969).
130. Theoretische Physik und Biologie: Contribution to "Wohin führt die Biologie", Ed. M. LOHMANN, p. 147–173. München: Verlag Karl Hanser 1970 (Transcript of a German Radio Broadcast) [in German].
131. Possibility of a second acoustic branch in transition metals: Mat. Res. Bull. **5**, 607–609 (1970) (N. F. MOTT Festschrift).
132. Long-range coherence and the action of enzymes: Nature **228**, 1093 (1970).
133. What future for superconductivity: Nature **228**, 1145-1146 (1970).
134. Superconductivity, lattice stability and phonon frequencies: Phys. Letters **35**A, 325–326 (1971).
135. From theoretical physics to biology: Proc. 2nd Intern. Conf. on Theoretical Physics and Biology, Versailles, 1969, Ed. M. MAROIS, p. 17–19. Paris: Edition du C.N.R.S. 1971.
136. Wohin führt die Physik: Bild der Wissenschaften, Vol. 8, Heft 1, p. 64–70 (1971) [in German].
137. Selective long-range dispersive forces between large systems: Phys. Letters **39**A, 153–154 (1972).

Subject Index